Intelligent Systems Reference Library

Volume 283

The aim of this series is to publish a Reference Library, including novel advances and developments in all aspects of Intelligent Systems in an easily accessible and well structured form. The series includes reference works, handbooks, compendia, textbooks, well-structured monographs, dictionaries, and encyclopedias. It contains well integrated knowledge and current information in the field of Intelligent Systems. The series covers the theory, applications, and design methods of Intelligent Systems. Virtually all disciplines such as engineering, computer science, avionics, business, e-commerce, environment, healthcare, physics and life science are included. The list of topics spans all the areas of modern intelligent systems such as: Ambient intelligence, Computational intelligence, Social intelligence, Computational neuroscience, Artificial life, Virtual society, Cognitive systems, DNA and immunity-based systems, e-Learning and teaching, Human-centred computing and Machine ethics, Intelligent control, Intelligent data analysis, Knowledge-based paradigms, Knowledge management, Intelligent agents, Intelligent decision making, Intelligent network security, Interactive entertainment, Learning paradigms, Recommender systems, Robotics and Mechatronics including human-machine teaming, Self-organizing and adaptive systems, Soft computing including Neural systems, Fuzzy systems, Evolutionary computing and the Fusion of these paradigms, Perception and Vision, Web intelligence and Multimedia.

Indexed by SCOPUS, DBLP, zbMATH, SCImago.

All books published in the series are submitted for consideration in Web of Science.

Gilberto Rivera • Witold Pedrycz • Álvaro Labella •
Rogelio Florencia
Editors

Recent Advances in Artificial Neural Networks

Case Studies and Applications

Springer

Editors
Gilberto Rivera
Departamento de Ingeniería Eléctrica y Computación
Universidad Autónoma de Ciudad Juárez
Ciudad Juárez, Mexico

Witold Pedrycz
Department of Electrical and Computer Engineering
University of Alberta
Edmonton, AB, Canada

Álvaro Labella
Departamento de Informática
Universidad De Jaén
Jaén, Spain

Rogelio Florencia
Departamento de Ingeniería Eléctrica y Computación
Universidad Autónoma de Ciudad Juárez
Ciudad Juárez, Mexico

ISSN 1868-4394 ISSN 1868-4408 (electronic)
Intelligent Systems Reference Library
ISBN 978-3-032-13496-7 ISBN 978-3-032-13497-4 (eBook)
https://doi.org/10.1007/978-3-032-13497-4

This Springer imprint is published by the registered company Springer Nature Switzerland AG
The registered company address is: Gewerbestrasse 11, 6330 Cham, Switzerland

Preface

Artificial Neural Networks (ANNs) form the cornerstone of an expanding ecosystem of deep learning models and hybrid architectures that have become essential for addressing the increasingly complex digital age challenges. Their ability to extract meaningful insights from large-scale data, capture nonlinear relationships, and generalize to previously unseen scenarios has profoundly transformed contemporary approaches to forecasting, pattern recognition, and data-driven decision-making across a wide range of domains.

In recent years, integrating ANNs with emerging techniques—such as attention mechanisms, transformer architectures, multimodal learning, and ensemble modeling—has led to substantial advancements. These innovations have moved beyond theoretical exploration, giving rise to practical, high-impact applications in agriculture, healthcare, intelligent transportation, surveillance systems, and human-centered technologies.

This edited volume, *Recent Advances in Artificial Neural Networks—Case Studies and Applications*, brings together 20 rigorously peer-reviewed chapters that collectively showcase the dynamic and evolving role of deep learning methodologies. Each contribution addresses a concrete real-world challenge, emphasizing methodological innovation and practical relevance with clear societal impact. To provide a coherent structure and facilitate navigation through the diverse topics covered, the chapters are organized into three thematic sections:

Part I. *Forecasting and Predictive Modeling*. This section comprises seven chapters that examine the application of neural architectures and machine learning models to a range of predictive tasks, including time series forecasting, disease prediction, agricultural yield estimation, and hydrological modeling. The studies featured in this section employ diverse methodologies, such as long short-term memory (LSTM) networks, transformer-based masked language models, standard neural networks, hybrid CNN–BiLSTM architectures, and ensemble frameworks enhanced with attention mechanisms. The practical applications addressed include sales forecasting, sugarcane production estimation, prediction of atherosclerotic heart disease mortality, health monitoring, detection of schizophrenia and tuberculosis, and streamflow forecasting in river basins.

Part II. *Computer Vision and Multimodal Recognition.* This section comprises six chapters centered on object detection, multimodal learning, image analysis, industrial inspection, vision-language integration, and intelligent transportation systems. The contributions present advanced methodologies, including single-shot multi-box detectors, Siamese networks, efficient channel attention mechanisms, convolutional neural networks, and Swin Transformer–GPT encoder-decoder architectures. The application scenarios addressed range from aircraft and ship detection to image-text similarity, fine-grained component recognition in printed circuit boards, automatic image captioning, and real-time traffic sign recognition.

Part III. *Innovations in Deep Learning Architectures and Applications.* The final section comprises seven chapters that explore integrating neural networks with complementary paradigms in domains such as driver monitoring systems, large language models, aerial surveillance, precision agriculture, image steganography, stress detection, and plant health assessment. The methods employed include deep learning combined with fuzzy inference systems, probabilistic decoding strategies for language models, object detection frameworks, convolutional and attention-based architectures, a DNN–CycleGAN–CNN hybrid model, CNN-BiLSTM networks, and U-Net-based segmentation techniques. These contributions exemplify emerging trends in intelligent driver assistance, narrative generation, abstractive summarization, real-time aerial monitoring for airspace security, automated detection of apple leaf diseases, secure data transmission through image-based steganography, physiological stress analysis, and UAV-assisted crop monitoring.

This volume results from a collaborative effort among researchers affiliated with academic institutions and professional organizations worldwide. Each chapter has undergone a rigorous double-blind peer review process to ensure scientific rigor, originality, and relevance to the field. The book is intended to serve as a valuable reference for researchers, graduate students, and practitioners interested in state-of-the-art deep learning applications. By integrating theoretical foundations with practical case studies, this volume aims to bridge the gap between academic research and real-world implementation, fostering further innovation in developing intelligent systems capable of addressing some of the most pressing challenges of our time.

Ciudad Juárez, Mexico — Gilberto Rivera
Edmonton, AB, Canada — Witold Pedrycz
Jaén, Spain — Álvaro Labella
Ciudad Juárez, Mexico — Rogelio Florencia

Contents

Part I
Forecasting and Predictive Modeling

Chapter 1
A Comparative Analysis of LightGBM and ES-RNN Applied to Time Series Forecasting

Antonio Mata-Alvarado, Mirna P. Ponce-Flores, Salvador Ibarra-Martínez, Jesús David Terán-Villanueva, and Julio Laria-Menchaca

Abstract Advancements in machine learning are reshaping time series forecasting, with the M competitions serving as crucial benchmarks. The M4 and M5 competitions highlighted the strengths of methods like ES-RNN and LightGBM, respectively. This study compares the M4 competition winner (ES-RNN) and a LightGBM forecasting approach inspired by its prevalence among top M5 solutions. We evaluate their performance on a challenging subset of 2724 daily frequency time series from the M4 dataset. Using multiple error metrics (sMAPE, MAE, MAPE, RMSE), our results indicate that LightGBM outperforms ES-RNN in a slight majority (approx. 54–55%) of the series. However, a key finding is that LightGBM's superior performance is *conditional* and strongly linked to a discernible trend in the time series. LightGBM demonstrates a significant advantage and higher forecast correlation for series with clear upward or downward trends. Conversely, the hybrid ES-RNN model exhibits greater robustness for series lacking strong trends or displaying higher variability. We introduce a quantitative trend measure and employ non-parametric statistical tests to statistically validate the significant impact of the trend on the correlation between LightGBM forecasts and actual values. We conclude that while LightGBM is highly effective, particularly for trended data common in domains like retail sales, its broader application requires careful consideration of the underlying data characteristics. Understanding the time series trend is crucial for practitioners selecting between these powerful forecasting methods.

Keywords Machine learning · Forecasting · Time series · Trend · LightGBM

A. Mata-Alvarado · M. P. Ponce-Flores (✉) · S. Ibarra-Martínez · J. D. Terán-Villanueva · J. Laria-Menchaca
Facultad de Ingeniería Tampico, Universidad Autónoma de Tamaulipas, Tampico, Mexico
e-mail: antonio.mata21.95@gmail.com; mirna_poncef@hotmail.com; sibarram@docentes.uat.edu.mx; jdteran@docentes.uat.edu.mx; jlaria@uat.edu.mx

G. Rivera et al. (eds.), *Recent Advances in Artificial Neural Networks*, Intelligent Systems Reference Library 283, https://doi.org/10.1007/978-3-032-13497-4_1

1.1 Introduction

A key challenge in time series forecasting is identifying the most suitable method for a given application. To address this, researchers have held several competitions to compare the performance of different forecasting methods, including classical and machine learning approaches, these competitions, known as M Competitions or Makridakis competitions, began in 1982 and initially included time series from diverse areas and frequencies [1, 2]. Over time, these datasets have become important benchmarks, with the M4 competition in 2018 being one of the most significant [3, 4].

The M5 competition, held in 2020, focused just on sales forecasting, marking the first of its kind. Its primary objectives included advancing prediction theory and promoting its use among businesses and non-profit organizations [5].

Findings from these competitions reveal the growing importance of machine learning techniques in forecasting. These techniques sometimes outperform classical methods, prompting many businesses to adopt them. However, some companies still prefer classical techniques due to their ease of interpretation and lower computational cost [6].

One of the primary advantages of machine learning methods lies in their capacity to handle large datasets. Businesses and institutions struggle to store as much data as possible for optimized decision-making. While statistical methods may have yielded comparable or even superior results with smaller datasets, machine learning techniques have become increasingly dominant with the growth in data volume [7].

The LightGBM method, a gradient boosting-based approach [8, 9], has recently gained popularity among researchers for prediction tasks. While other methods have yielded positive results in temporal sales forecasting [10–13], Its success in the M5 sales forecasting competition on Kaggle has solidified its position [14, 15]. The primary objective of this study is to evaluate whether the LightGBM method, the winner in the M5 competition and other sales forecasting competition can generate competent forecasts across various domains beyond sales. To achieve this, we use the M4 daily dataset, which comprises 4227 time series from diverse fields such as industry, demographics, macroeconomics, microeconomics, and finance, among others. This research, therefore, focuses on comparing LightGBM's performance with time series from other domains and seeks to validate whether it achieves similar results as reported in previous studies [16, 17]. This approach allows us to investigate whether the effectiveness of LightGBM is maintained when applied to contexts with distinct characteristics and inherent sector-specific complexities. This study reveals that LightGBM's competitiveness against the M4 winning method is conditional, mainly depending on the time series trend, performing better with positive trends. Consequently, we investigate the trend's influence on these methods' forecasting capabilities.

The remainder of this paper unfolds as follows: We begin in Sect. 1.2 by laying the theoretical groundwork for time series analysis and forecasting. Sect. 1.3 surveys the relevant literature and establishes the current state of the art. Our specific approach is detailed in Sect. 1.4, covering the M4 and M5 competition datasets, the experimental setup for comparing LightGBM against the ES-RNN benchmark, the evaluation metrics, and our method for analyzing time series trend. In Sect. 1.5, we present and meticulously analyze the results, comparing the methods' performance and exploring the crucial link between trend and forecast accuracy via correlation and statistical tests. Recognizing the study's boundaries, Sect. 1.6 discusses its limitations and charts potential paths for future research. Finally, Sect. 1.7 draws together the main conclusions, emphasizing the conditional nature of LightGBM's effectiveness and the pivotal role of trend in practical forecasting applications.

1.2 Theoretical Foundations

1.2.1 Time Series

A time series is a sequence of data points indexed in chronological order, representing measurements or observations taken at successive time intervals [18]. The primary goal of time series (TS) analysis is to identify underlying patterns and forecast future values based on historical data. Time series data typically consist of several key components:

Trend Component
The trend component captures the long-term movement of the series, which may exhibit increasing, decreasing, or stationary patterns over an extended period. Trends reflect the underlying direction of the data and are crucial for understanding the overall changes in the observed phenomenon.

Seasonality
This component refers to recurring patterns that occur at regular intervals, such as daily, weekly, monthly, or annually. These patterns are driven by systematic influences, including climatic conditions, holidays, or cyclical behaviors in consumer activities.

Cycle Component
It represents longer-term fluctuations in the TS that do not follow a fixed period. These cycles often correspond to economic expansions and contractions, spanning several years and reflecting macroeconomic trends.

Residual/Random Component
The residual or random component accounts for irregular variations in the TS after removing trend, seasonality, and cyclical effects. This component is typically unpredictable and may result from random external influences, or measurement errors.

Understanding these components is essential for effective time series modeling and forecasting. By decomposing a TS into its constituent parts, analysts can develop more accurate predictive models and gain deeper insights into the underlying data dynamics [19].

1.2.2 Types of Time Series

Time series data can be categorized into various types [20], each influencing the appropriate choice of forecasting methodology. Key classifications include:

Stationary Series
A stationary series exhibits statistical properties, such as mean and variance, that remain constant over time. These series do not contain trends or seasonal patterns, making them suitable for simpler forecasting methods.

Non-Stationary Series
Non-stationary series display statistical properties that change over time. These series often include trends, seasonality, or cyclical components, requiring more sophisticated modeling techniques to account for their dynamic nature.

Series with Trend
Series with a trend exhibit a persistent upward or downward movement over time. Proper modeling of these series necessitates incorporating trend-adjustment techniques to accurately capture directional changes.

Series with Seasonality
Series with seasonality exhibit repeating patterns at regular intervals, such as daily, monthly, or annually. Forecasting such series requires methods capable of effectively capturing and modeling seasonal variations.

The selection of an appropriate forecasting method depends significantly on identifying the underlying patterns within the data [21]. Stationary series can often be modeled using simpler approaches, while non-stationary series, particularly those exhibiting trends or seasonality, may necessitate advanced techniques, such as decomposition-based methods, to accurately model their components.

1.2.3 Challenges in Time Series Forecasting

Accurate forecasting of TS data presents several inherent challenges that must be addressed to enhance predictive performance [22].

Non-Linearity
Many time series exhibit complex, non-linear relationships that traditional linear models fail to capture effectively. Machine learning algorithms, particularly deep learning techniques, offer a promising alternative by modeling these intricate dependencies more accurately.

Variability
Time series data often exhibit substantial variability due to external factors, structural changes, or inherent randomness. This variability complicates pattern recognition and requires robust forecasting methods capable of distinguishing meaningful trends from noise.

Presence of Noise
Random fluctuations within time series data can obscure underlying patterns, reducing forecasting accuracy. Advanced models equipped with noise-reduction strategies and adaptive learning mechanisms are crucial for mitigating these effects.

Complexity of Patterns
Time series often consist of multiple interacting components, including trends, seasonal effects, cyclic variations, and stochastic noise. Effective forecasting necessitates sophisticated modeling techniques, such as decomposition methods and hybrid approaches, to capture these complex structures accurately.

Addressing these challenges is essential for developing reliable forecasting models. The selection of an appropriate methodology depends not only on the nature of the time series patterns but also on the ability of the chosen approach to effectively manage the complexities and uncertainties inherent in the data [23].

1.2.4 Time Series Forecasting

Forecasting involves predicting future events or values based on historical data and patterns. By using past observations, models are built to extrapolate future trends. Essentially, forecasting is a proactive approach to planning and decision-making, seeking to understand potential future outcomes by analyzing past data. This predictive capability is essential in various fields, enabling organizations to make informed adjustments to expected future conditions [24].

Forecasting plays a critical role across different sectors, offering significant benefits for both strategic and operational decision-making. The accuracy of forecasts is crucial in numerous contexts, influencing key areas such as resource allocation, risk management, and operational efficiency.

Business Operations
Businesses rely on forecasting to predict demand, manage inventory, optimize supply chains, and allocate resources efficiently. Accurate sales forecasts enable companies to maintain optimal stock levels, meeting customer needs while avoiding both shortages and unnecessary inventory costs. Forecasting also informs production planning, marketing strategies, and pricing decisions [25].

Economic Analysis
Governments and financial institutions use economic forecasting to assess trends in key indicators such as inflation, interest rates, and GDP. These insights guide policymaking, financial risk assessment, and public expenditure planning.

Supply Chain Management
Effective supply chain management depends heavily on accurate demand forecasting, which aids in coordinating logistics, optimizing manufacturing schedules, and reducing lead times. This ensures products are available at the right time and location, while minimizing operational costs [26].

Financial Markets
Investors and financial analysts use forecasting to predict asset performance, including stocks, bonds, and currencies. These forecasts help guide investment strategies and support financial risk management [27, 28].

Resource Allocation
Organizations, such as healthcare providers and government agencies, use forecasting to allocate resources effectively and address emerging needs. For instance, hospitals forecast patient inflows to optimize staffing and resource distribution.

Accurate forecasting is vital to effective decision-making across multiple sectors, ensuring enhanced efficiency, reduced uncertainty, and improved strategic planning.

1.2.5 Types of Forecasting

Forecasting can be classified based on the time horizon it addresses, with each type serving distinct functions in planning and decision-making.

Short-Term Forecasting
Short-term forecasting typically covers a time frame of days to a few months. It is primarily used for operational planning, such as inventory management, scheduling, and daily resource allocation. Due to its shorter time horizon, short-term forecasting tends to yield higher accuracy and plays a critical role in facilitating immediate decision-making [8, 29].

Medium-Term Forecasting
Medium-term forecasting spans a period of several months to a year. It is particularly valuable for mid-range decision-making, such as capacity planning, budgeting, and marketing strategies. This type of forecasting captures seasonal trends and economic cycles, providing a balance between specific predictions and broader trend analysis [30, 31].

Long-Term Forecasting

Long-term forecasting extends beyond a year and is employed for strategic planning, capital investments, and significant organizational changes. Given the extended time horizon, long-term forecasting is subject to greater uncertainty and focuses more on broad trends than on precise short-term fluctuations [9].

Selecting the appropriate forecasting time frame is crucial and should align with specific operational and strategic goals. Short-term forecasts support immediate operational adjustments, medium-term forecasts inform mid-range planning decisions, and long-term forecasts guide broader strategic and policy development [32]. Each type of forecasting employs different methodologies, input variables, and levels of detail, tailored to the specific application at hand.

1.2.6 Forecasting Methods: Classical Statistical Approaches and Machine Learning Techniques

This section explores methodologies for time series forecasting, contrasting classical statistical approaches with modern machine learning techniques.

1.2.6.1 Classical Statistical Methods for Time Series Forecasting

Classical statistical methods have long been employed in time series analysis due to their interpretability and reliance on well-defined statistical properties [33]. These approaches are commonly used for forecasting tasks and provide a solid foundation for understanding time-dependent data.

One widely used method is the ARIMA (Autoregressive Integrated Moving Average) model, which captures temporal dependencies by combining three key components [34]. The autoregressive (AR) term models relationships between past and present values, while the moving average (MA) component accounts for errors from previous observations. Additionally, the integrated (I) term ensures stationarity by applying differencing techniques. ARIMA is particularly effective when a time series exhibits strong autocorrelation patterns.

Exponential Smoothing methods provide an alternative approach by recursively modeling time series components while assigning greater importance to recent observations [35]. Holt's method extends basic exponential smoothing by incorporating a trend component, making it suitable for linear trends. Winters' method further enhances this approach by introducing a seasonal component, allowing it to handle data with both trends and seasonality.

Another fundamental technique is time series decomposition, which separates a TS into its core components: trend, seasonality, cycle, and residuals. By isolating these elements, decomposition methods facilitate independent modeling of each component, ultimately improving forecast accuracy.

Despite their advantages, classical statistical methods present certain limitations [7]. Their interpretability and computational efficiency make them well-suited for large datasets, yet they struggle to capture complex, non-linear relationships. Additionally, their performance diminishes in high-noise environments and when faced with rapidly evolving patterns. These constraints have led to the development of machine learning-based approaches, which offer greater adaptability for forecasting in dynamic and complex datasets.

1.2.6.2 Machine Learning Methods for Time Series Forecasting

Machine Learning (ML) algorithms have significantly advanced time series forecasting by capturing complex and non-linear relationships within data [36]. Unlike classical statistical methods, ML models do not rely on strict assumptions regarding data distribution or underlying patterns. Instead, they leverage historical data to dynamically adjust their parameters, improving forecasting accuracy in scenarios with high variability and structural complexity.

Supervised learning plays a crucial role in time series forecasting by training models on historical data. This method incorporates both input features, such as lagged values and external variables, and observed target values to uncover intricate dependencies within the dataset. By learning these patterns, ML models enhance predictive performance and adaptability in real-world forecasting applications.

1.2.6.3 LightGBM Forecasting Method

LightGBM, short for Light Gradient Boosting Machine, is a machine learning framework developed by Microsoft [37]. Based on decision trees, LightGBM is suitable for classification and regression tasks [38].

Its key advantages include minimal optimization requirements, parallel training capabilities, support for various loss functions, regularization techniques, and bagging. This practicality makes LightGBM efficient in terms of speed and memory utilization. The method utilizes histogram-based algorithms, grouping continuous values into discrete bins, further optimizing memory usage and training speed.

Figure 1.1 illustrates a key difference between LightGBM and other decision tree methods. LightGBM constructs its tree vertically, resembling the growth of a tree's leaves, and selects the leaf with the highest loss for further expansion [39]. This contrasts with most other methods that build horizontally or level-by-level (depth-wise).

This approach, combined with its scalability for large datasets and ability to capture both linear and non-linear patterns, positions LightGBM as a versatile tool for real-world forecasting applications, particularly in scenarios demanding efficiency and accuracy.

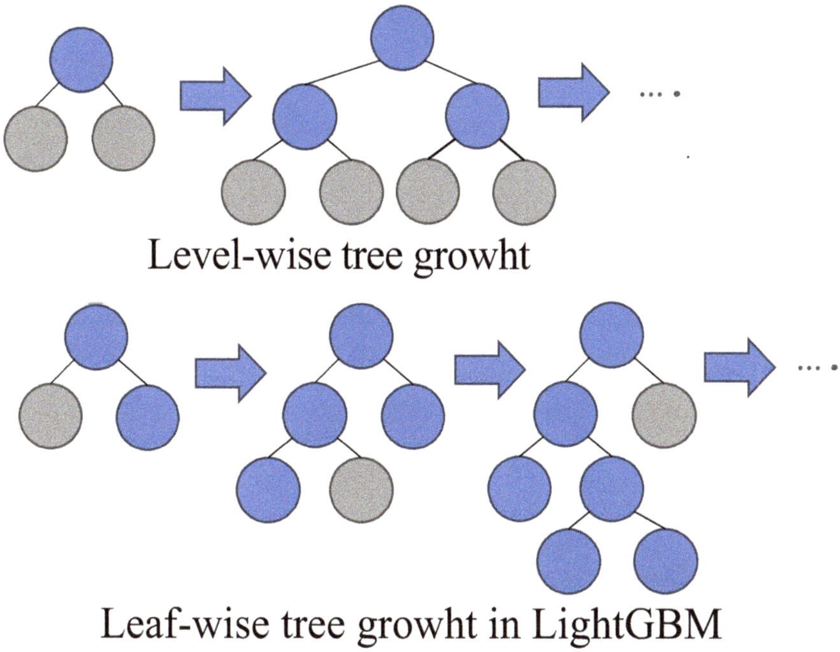

Fig. 1.1 Different growth strategies in decision trees [40]

1.2.6.4 Neural Networks

Neural networks are an artificial intelligence method inspired by the functioning of the human brain. Designed to learn and recognize patterns within data, they are widely applied in various fields, including image processing, natural language processing, speech recognition, computer vision, and forecasting. Their ability to model complex relationships has made them a fundamental tool in modern machine learning applications.

The foundations of artificial neural networks and deep learning were unknowingly established by Warren McCulloch and Walter Pitts [41]. A crucial advancement in this field was the development of the backpropagation algorithm, which effectively addressed the exclusive logic problem and significantly improved the training efficiency of multi-layer neural networks.

Backpropagation operates by adjusting the "weights" of the connections between artificial neurons based on the difference between the obtained and desired outcomes [42]. This process enables neural networks to refine their predictions over multiple iterations, enhancing their learning capabilities and overall performance.

1.2.7 *Recurrent Neural Networks (RNN)*

Recurrent Neural Networks (RNNs) are a type of artificial neural network where connections between nodes form sequential patterns [43]. This structure allows them to exhibit temporal dynamic behavior, making them particularly suited for time-dependent data.

The key characteristic of RNNs is their ability to capture temporal dependencies within data. By modeling and understanding patterns over time, they are highly effective in tasks where the sequence and order of information are crucial, such as natural language processing, speech recognition, and machine translation.

There are three fundamental types of recurrent neural networks: simple recurrent neural networks (RNNs), gated recurrent units (GRUs), and long short-term memory (LSTM) networks. Each of these architectures enhances the ability of RNNs to process sequential data by addressing challenges such as vanishing gradients and long-term dependencies.

1.2.7.1 Long Short-Term Memory (LSTM)

Long Short-Term Memory (LSTM) networks are a variant of Recurrent Neural Networks (RNNs) that have proven highly effective in processing long sequences and capturing long-term dependencies [44]. Designed to address the gradient degradation problem in standard RNNs, LSTMs prevent information loss as it propagates through recurrent layers by incorporating a more complex structure with gated mechanisms.

An LSTM unit consists of three key gates: the input gate, the forget gate, and the output gate. These gates work together to regulate the flow of information, determining what is stored, discarded, and passed forward through the sequence.

The gating mechanism in LSTM networks ensures that recurrent connections retain information over extended time periods, mitigating the gradient degradation issue. This allows LSTMs to capture long-range dependencies and maintain a memory state crucial for effective sequence processing.

Figure 1.2 illustrates the structures of the three types of recurrent neural networks.

1.2.7.2 ES-RNN (Exponential Smoothing Recurrent Neural Network)

The ES-RNN (Exponential Smoothing Recurrent Neural Network) is a hybrid forecasting model that combines the interpretability of exponential smoothing with the predictive power of Recurrent Neural Networks (RNNs). Developed by Slawek Smyl, this method gained significant recognition after winning the M4 competition, showcasing its superior forecasting performance [45].

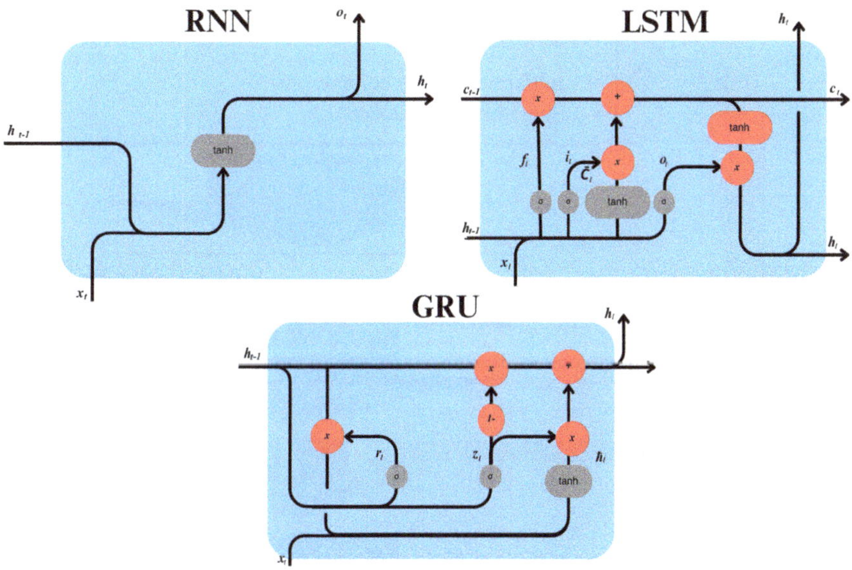

Fig. 1.2 Different structures of recurrent neural networks

Figure 1.3 shows the data flow in the ES-RNN method for generating forecasts. This approach outperforms both classical statistical methods and single-method machine learning approaches. ES-RNN utilizes three key elements: adaptive deseasonalization and normalization, forecast generation, and ensemble averaging.

The model first applies exponential smoothing to decompose the time series and normalize its values, followed by an LSTM-based recurrent neural network to generate forecasts. The final predictions are derived by recombining the modeled components.

The model captures short-term patterns using exponential smoothing, while leveraging the long-term dependencies that RNNs are known for. This hybrid architecture provides a flexible and effective approach for forecasting diverse types of time series data.

When selecting an appropriate forecasting methodology, several factors must be considered:

- **Data Complexity**: More intricate patterns in the data may require advanced deep learning models to capture their nuances effectively.
- **Interpretability**: Classical methods and hybrid approaches like ES-RNN provide greater transparency, which can be crucial for understanding model behavior and results.
- **Computational Resources**: Some machine learning models, particularly deep learning-based ones, require substantial computational power, which can impact their practicality in real-world applications.

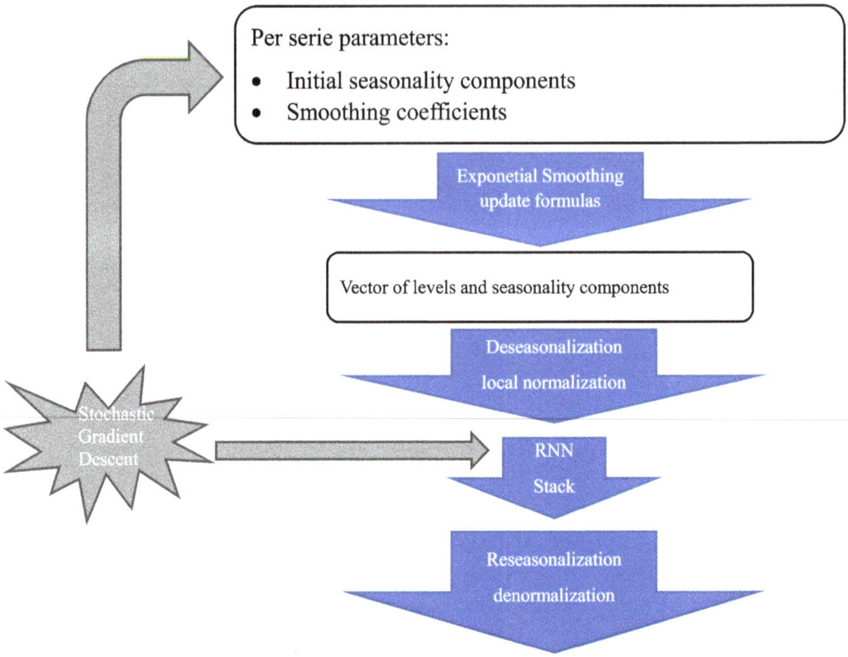

Fig. 1.3 Data flow of the ES-RNN method [45]

By leveraging a combination of machine learning techniques, forecasting strategies can be tailored to a wide range of time series patterns, ensuring enhanced accuracy and reliability across various domains.

1.3 State of the Art

Kilimci et al. [10] developed an intelligent demand forecasting system aimed at optimizing stocks, reducing costs, and improving sales and customer loyalty. The study integrates time series analysis techniques, the support vector regression algorithm, and deep learning models using a novel decision integration strategy. Additionally, it adapts a boosting ensemble strategy to enhance demand forecasting accuracy. The system was tested on real-life data from SOK Market in Turkey, which operates 6700 stores and 23 distribution centers. Experimental results demonstrate that the proposed model outperforms state-of-the-art methods, with significant accuracy improvements due to the inclusion of support vector regression, deep learning, and the novel integration strategy.

Ju et al. [8] proposed a wind power forecasting model integrating a convolutional neural network (CNN) and LightGBM to address the volatility and uncertainty of wind power. The study constructs new feature sets from time series data of wind fields and adjacent wind fields. CNN is used to extract information, with network parameters adjusted based on actual results. To enhance accuracy and robustness, LightGBM is integrated to overcome the limitations of single-convolution models. Experimental results demonstrate that the proposed fusion model outperforms support vector machines, standalone LightGBM, and CNN in both accuracy and efficiency.

Similarly, Weng et al. [9] introduced a supply chain sales forecasting model combining LightGBM and LSTM to enhance accuracy, efficiency, and interpretability. The model leverages LSTM's ability to extract high-level temporal features while incorporating LightGBM's efficiency and strong interpretability. To evaluate performance, experiments were conducted on three representative supply chain sales datasets. Results demonstrate that the proposed model achieves high forecasting accuracy and efficiency, making it suitable for industrial applications. This study highlights the role of big data and AI in long-term sales forecasting, providing key technical support for supply chain management.

Ariannezhad et al. [11] presented a dual-branch neural network architecture for demand forecasting in retail, addressing the challenge of non-plannable features—factors available during training but not at prediction time. The model integrates a privileged information (PI) branch to incorporate non-plannable features during training, while a single-branch network with a simulation component is used at prediction time. Experiments on two real-world datasets demonstrate that the proposed approach outperforms state-of-the-art methods in terms of mean absolute error and symmetric mean absolute percentage error. Additional visualizations and experiments validate the contribution of each model component.

Sun et al. [16] developed a cryptocurrency price trend forecasting model using the Light Gradient Boosting Machine (LightGBM) algorithm. The study integrates daily data from 42 major cryptocurrencies with key economic indicators to enhance prediction accuracy. Experimental results demonstrate that LightGBM outperforms other methods in robustness, and the overall strength of cryptocurrencies influences forecasting performance. The findings provide valuable insights for investors in portfolio construction and risk management.

Lim et al. [46] introduced the Temporal Fusion Transformer (TFT), an attention-based model for multi-horizon forecasting that balances high performance with interpretability. TFT integrates recurrent layers for local processing and self-attention layers for capturing long-term dependencies. It incorporates specialized components for feature selection and gating layers to suppress irrelevant information. Experimental results on multiple real-world datasets demonstrate that TFT outperforms existing benchmarks. Additionally, the study presents three practical interpretability use cases, providing insights into temporal dynamics in forecasting scenarios.

Likewise, Zhang et al. [17] formulated a VMD-TPE-LightGBM model to improve the accuracy and stability of metro passenger volume prediction. The method starts by regrouping the dataset both by station and chronological order, with a 10-min time interval. Time features are extracted by analyzing the variation trends of passenger volume. To enhance precision, the variational mode decomposition algorithm decomposes the original data series, and each mode is input into a LightGBM model optimized using the Tree of Parzen Estimators and K-fold cross-validation. Experimental results show that the proposed model outperforms other methods, achieving exceptional results in short-term metro passenger volume forecasting.

Ma and Fildes [12] introduced a meta-learning framework utilizing deep convolutional neural networks for retail sales forecasting across thousands of products and stores. The framework automatically learns feature representations from raw sales data and links these features to a set of weights that combine multiple base-forecasting methods. Experimental results using IRI weekly data show that the proposed meta-learner outperforms state-of-the-art methods, though the accuracy improvements over more sophisticated meta-ensemble benchmarks are modest, and the learned features lack interpretability. The study recommends building a diverse pool of base forecasters to optimize forecast combinations rather than focusing on the best individual method.

Dairi et al. [47] conducted a comparative study on machine learning methods for forecasting COVID-19 transmission, focusing on deep learning models such as LSTM-CNN, GAN-GRU, GAN, CNN, LSTM, and RBM, alongside baseline models like logistic regression (LR) and support vector regression (SVR). The study highlights the effectiveness of hybrid models (LSTM-CNN and GAN-GRU) in improving forecasting accuracy. Using COVID-19 time-series data from seven countries (Brazil, France, India, Mexico, Russia, Saudi Arabia, and the US), the results demonstrate that deep learning models, particularly LSTM-CNN, outperform baseline models, with LSTM-CNN achieving a mean absolute percentage error of 3.718%. This research underscores the potential of machine learning for optimizing resources and managing hospital overcrowding during pandemics.

Mitra et al. [48] compared the performance of five machine learning regression techniques—random forest (RF), extreme gradient boosting (XGBoost), gradient boosting, adaptive boosting (AdaBoost), and artificial neural networks (ANN)—against a proposed hybrid model (RF-XGBoost-LR) for sales forecasting in a retail chain. The study used weekly sales data from a US-based retail company, considering factors such as regional temperature and store size. The results showed that the hybrid RF-XGBoost-LR model outperformed the other methods in terms of forecasting accuracy. This research provides valuable insights for industry decision-makers to enhance sales forecasting techniques.

Pirani et al. [49] focused on comparing the performance of Long Short-Term Memory (LSTM), Gated Recurrent Unit (GRU), and Bidirectional Long Short-Term Memory (BiLSTM) models for time series prediction, highlighting their ability to outperform traditional regression-based methods like ARIMA. The study explores the unique advantages of LSTM-based models, including their ability to handle longer sequences of data with additional gates. BiLSTM, being a two-way model, provides extra training by processing data in both directions, while GRU, with fewer gates, offers a simpler architecture. The analysis evaluates which model delivers the best predictions with minimal error, alongside a comparison of their behavioral performance and hyperparameter standardization.

Ensafi et al. [35] analyzed the sales history of a retail store to forecast furniture sales using a range of forecasting models. Classical methods like Seasonal Autoregressive Integrated Moving Average (SARIMA) and Triple Exponential Smoothing were first applied, followed by advanced models including Prophet, Long Short-Term Memory (LSTM), and Convolutional Neural Network (CNN). The models were evaluated based on accuracy metrics such as Root Mean Squared Error (RMSE) and Mean Absolute Percentage Error (MAPE). The study found that the Stacked LSTM method outperformed the others, with Prophet and CNN also showing strong performance.

Kurani et al. [50] investigated the role of Artificial Neural Networks (ANN) and Support Vector Machines (SVM) in stock market forecasting, emphasizing their hybrid variants and applications in financial risk management. The study reviewed traditional methods like sentiment analysis, decision trees, moving average algorithms, and data mining, alongside advanced techniques such as ANN-MLP, GARCH-MLP, and SVM integrated with Random Forest or Genetic Algorithms. By synthesizing recent research, the authors demonstrated that hybrid ANN models (e.g., combining Backpropagation with Multilayer Feed-forward networks) achieve superior performance in handling missing data and non-linear patterns. Meanwhile, basic SVM models attained 60%–70% accuracy, with further improvements observed when combined with ensemble methods. The paper also highlighted challenges including time constraints, data scarcity, and cold start problems but argued that integrating ANN/SVM with novel techniques (e.g., hybrid methodologies) could mitigate these limitations. This work underscores the potential of hybrid models to advance stock prediction frameworks, offering researchers and practitioners actionable insights for refining forecasting systems.

Jewel et al. [51] conducted an empirical comparison of LSTM and LightGBM for retail sales forecasting using data from an American multinational corporation. The study employed rigorous optimization techniques, including memory optimization, feature engineering, and parameter tuning for both models. Evaluation using metrics such as RMSE, MAE, WMAPE, and WRMSEE revealed that LightGBM consistently outperformed LSTM in capturing sales patterns and generating accurate forecasts. The authors identified LightGBM as the superior choice for retail demand forecasting, emphasizing its efficiency and adaptability to retail data dynamics. This work underscores the critical role of model selection in practical machine learning applications, positioning LightGBM as a robust framework for retail sales prediction.

Rani et al. [36] conducted a comprehensive comparison of traditional and advanced deep learning models for time series forecasting, focusing on ARIMA, Random Forest, RNNs, LSTMs, and GRUs. The study adheres to open science principles, using publicly available datasets to ensure reproducibility. It emphasizes preprocessing, model architecture, and hyperparameter tuning, with Root Mean Square Error (RMSE) as the primary metric for forecasting accuracy. The research provides insights into the performance of each model under different conditions, highlighting when each excels or falls short based on dataset size and complexity. It also explores model interpretability, offering recommendations for optimal model selection, contributing to both theoretical understanding and practical forecasting applications.

Balamwar et al. [52] forecasted the monthly steel consumption for a prominent metal forming industry in Southeast Asia using an aggregated LightGBM method. The model is trained on daily consumption data of 35 unique steel types over five years (2014–2019), incorporating multiple critical temporal features. Daily predictions are aggregated to a monthly level to provide an in-depth analysis of the consumption behavior, addressing the challenge of optimal inventory management in a market driven by complex supply and demand dynamics.

Malakouti et al. [53] addressed the urgent need for renewable energy by forecasting wind speed and energy output in Turkey using the Light Gradient Boosting Machine (LightGBM). They optimized the model's hyperparameters via grid search and evaluated its performance using root mean square error and R^2 metrics. Notably, the model achieved an R^2 value of 0.98 for wind speed forecasting within 25 s and an R^2 of 1 for predicting wind turbine power output in 90 s.

Bian et al. [54] presented an integrated data-driven model (LSTM-LightGBM) for runoff prediction by combining Long Short-Term Memory and Light Gradient Boosting Machine techniques with a reciprocal error method. The model is applied to forecast annual runoff at the Caiqi hydrological monitoring station in the Shiyang River Basin. Performance evaluation using Error of Peak (EP), Nash-Sutcliffe Efficiency (NSE), Root Mean Squared Error (RMSE), and Mean Absolute Error (MAE) indicated that the integrated model outperformed the individual LSTM and LightGBM models (NSE of 0.92, RMSE of 0.075 million m^3, and MAE of 0.046 million m^3). Interpretability analysis, conducted through permutation importance and SHAP values, identified four key feature variables influencing runoff. Future runoff predictions for 2025–2075 under 12 climate scenarios showed generally steady trends, with the GFDL RCP 6.0 and IPSL RCP 2.6 scenarios yielding mean annual runoffs of 3.12×10^8 m^3 and 3.04×10^8 m^3 respectively—representing decreases of 24.09% and 26.03% compared to the baseline period (1955–2017). These results provided a scientific basis for future water resources planning in the downstream Shiyang River Basin.

Song and Chen [55] introduced an innovative two-stage hybrid forecasting model for financial time series in the international shipping market. The variational mode decomposition method is initially applied to extract multi-scale predictive features from the time series data. Subsequently, LightGBM is employed to capitalize on its efficiency and scalability in handling high-dimensional data. Empirical evaluations on sixteen distinct time series from the global shipping market demonstrated that the proposed hybrid model outperformed state-of-the-art methods in forecasting accuracy and reliability. The research provided valuable insights for stakeholders such as shipowners, investors, brokers, and shipyards, highlighting the potential of hybrid machine learning techniques for financial time series forecasting in complex and dynamic markets.

1.4 Methodology

This section describes the two datasets used in the proposed methodology. It outlines the machine learning methods that were winners in the M4 and M5 competitions, forming the basis for the comparative analysis.

The first dataset we selected is from the M5 competition. This competition marks the first time a competition utilizes time series specifically related to sales provided by Walmart [5]. This dataset contains unit sales data for 3049 distinct products categorized into three groups: hobbies, food, and household.

The sales data originates from ten stores located in Texas, California, and Wisconsin in the United States. The time series span from January 29, 2011, to June 19, 2016, and are classified by the level of aggregation, resulting in a total of 42,840 time series. Table 1.1 presents the distribution and description of these time series.

The second dataset comprises time series from the M4 competition, specifically the daily subset, with 4227 time series. These series have different lengths, from 92 to 9919 observations, and cover six distinct domains: macroeconomics, microe-

Table 1.1 Number of M5 series per aggregation level

Level id	Level description	Aggregation level	Number of series
1	Unit sales of all products, aggregated for all stores/states	Total	1
2	Unit sales of all products, aggregated for each state	State	3
3	Unit sales of all products, aggregated for each store	Store	10
4	Unit sales of all products, aggregated for each category	Category	3
5	Unit sales of all products, aggregated for each department	Department	7
6	Unit sales of all products, aggregated for each state and category	State-category	9
7	Unit sales of all products, aggregated for each state and department	State-department	21
8	Unit sales of all products, aggregated for each store and category	Store-category	30
9	Unit sales of all products, aggregated for each store and department	Store-department	70
10	Unit sales of product *i*, aggregated for all stores/states	Product	3049
11	Unit sales of product *i*, aggregated for each state	Product–state	9147
12	Unit sales of product *i*, aggregated for each store	Product–store	30,490
Total			42,840

conomics, demographics, industry, finance, and others. Given that competition datasets are generally provided in a pre-cleaned state, our preprocessing focused primarily on selecting the appropriate subset: we first filtered the data to retain only daily frequency series to allow for comparison with M5 methods. Then, we selected only time series with at least 750 observations to ensure a minimum of two years of historical data for analysis. This selection resulted in a total of 2724 time series.

Our proposed methodology analyzes the sales time series used in the M5 competition. We replicated three winning methods from this competition and used one to generate forecasts for the M5 time series and the daily subset of various categories from the M4 competition. This approach allows us to explore the feasibility of using this forecasting method in areas beyond sales. We used the symmetric mean absolute percentage error (sMAPE) to evaluate the accuracy of both forecasting methods (M5 winner - LightGBM and M4 winner - ES-RNN). This error measure is commonly used in the literature for evaluating forecasts [56] and is defined in Eq. (1.1):

$$SMAPE = \frac{100\%}{n} \sum_{t=1}^{n} \frac{|F_t - A_t|}{(|A_t| + |F_t|)/2} \tag{1.1}$$

Where A_t represents the actual future value of the time series at point t and F_t is the forecast being evaluated, and n is the size of the training sample.

Figure 1.4 outlines our structured methodology to compare the LightGBM and ES-RNN forecasting methods. This multi-stage process integrates insights from both M5 and M4 competitions.

Initially, we performed parallel preparation steps. We replicated key M5 methods to select an efficient and accurate LightGBM approach and establish performance baselines. Simultaneously, we selected the relevant M4 daily time series subset and identified the benchmark ES-RNN forecasts.

These preparations fed into the Core Comparison Process. We preprocessed the M4 data and generated forecasts using the chosen LightGBM configuration. We then quantitatively compared these LightGBM forecasts against the ES-RNN benchmark forecasts using metrics like sMAPE.

Finally, we conducted an in-depth analysis and validation phase to deepen our understanding beyond simple error comparisons. This involved calculating the correlation between LightGBM forecasts and actual M4 data, computing a proposed trend measure for the series, analyzing the trend's influence on forecast correlation, and executing non-parametric statistical tests to confirm the significance of observed differences.

We present the detailed experimentation based on this methodology, encompassing the results of each stage depicted in Fig. 1.4 in Sect. 1.5.

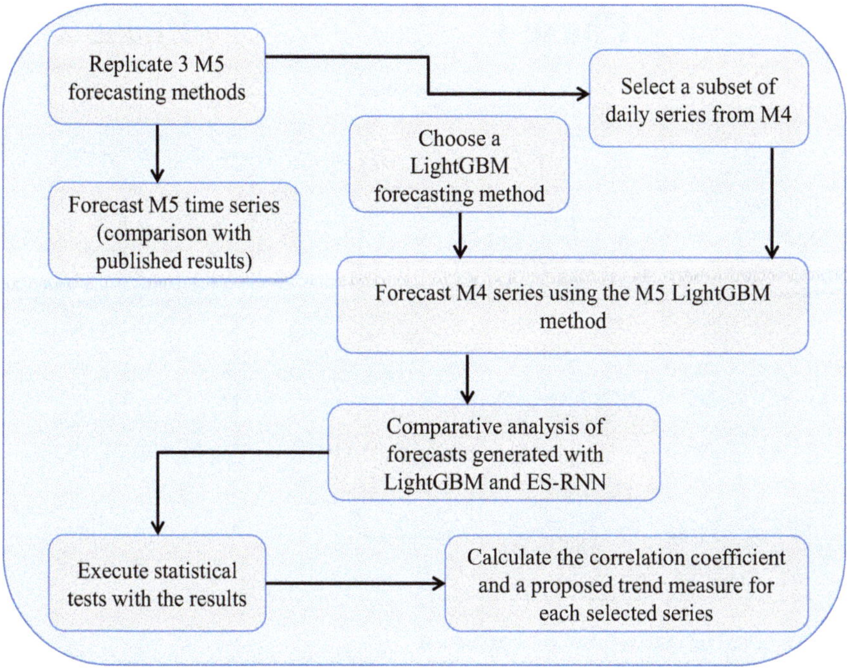

Fig. 1.4 Methodological scheme of the comparative analysis of the LIGHTGBM method for time series of areas different from sales

1.5 Results

Figure 1.5 shows the two phases of the experimentation process. It details the steps taken to perform the comparative analysis between the two forecasting methods and presents a detailed analysis of the forecasts generated using LightGBM.

1.5.1 Phase One: Selection and Replication of Three M5 Competition Methods

Replicating the winning method (A1) from the M5 competition involved a total training time of 16.5 h using 220 different models. These models were distributed as follows: 10 models for the aggregation level, 3 for sales by store, 30 models for the aggregation level, 8 for sales by store and category, 70 models for the aggregation level, and 9 for sales by store and department. This resulted in 110 recursive models and 110 non-recursive models. The forecasts were obtained in approximately 8.5 h. Adjusting some parameters allowed us to reduce the training time while achieving similar predictions to those reported.

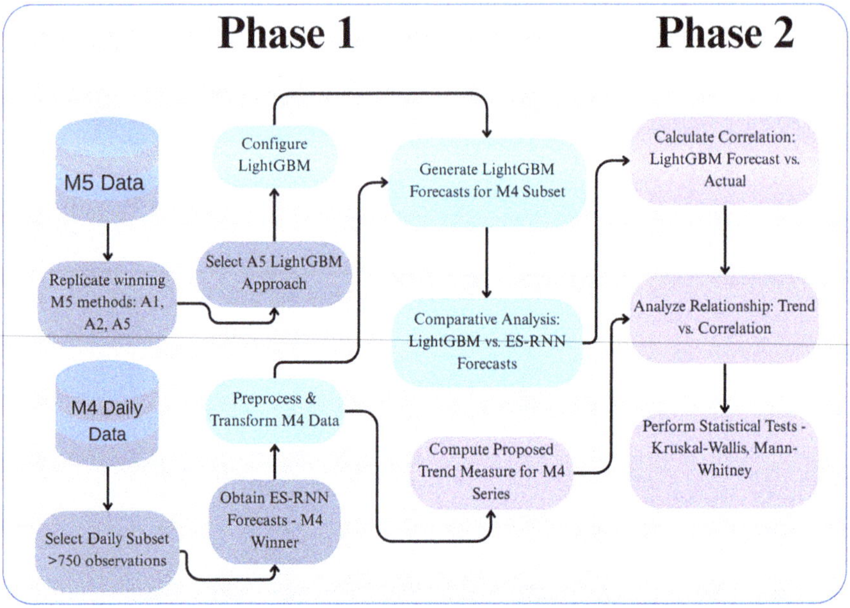

Fig. 1.5 Diagram of the experimentation process

Table 1.2 Performance of the winning methods in terms of time and accuracy, reported in the M5 competition

Method	Time in hours	WRMSSE
A1	≈ 24	0.400
A2	≈ 27	0.416
A3	N/R	0.421
A4	≈ 16	0.418
A5	≈ 3	0.414

The second-place method (A2) from the M5 competition required a total execution time of 27.75 h for training and prediction. We retained the original parameters, resulting in replicated results consistent with the competition.

Finally, replicating the fifth-place method (A5) from the M5 competition, the total execution time was less than 2.5 h for data processing, training, and prediction.

Table 1.2 shows the total time reported in the competition, representing the combined time for training and generating sales forecasts for the top five methods. It highlights that method 5 is the most time-efficient in terms of training and achieved score, which was evaluated using the weighted scaled mean squared error (WRMSSE). The competition used WRMSSE as the standard metric for evaluating forecasts at aggregation level 3, representing total sales for a store.

Figure 1.6 shows the forecasts generated by the top five methods in the M5 competition (A1, A2, A3, A4, A5) and the real values. The y-axis represents the total sales for the first store in California (CA1), while the x-axis represents the 28-

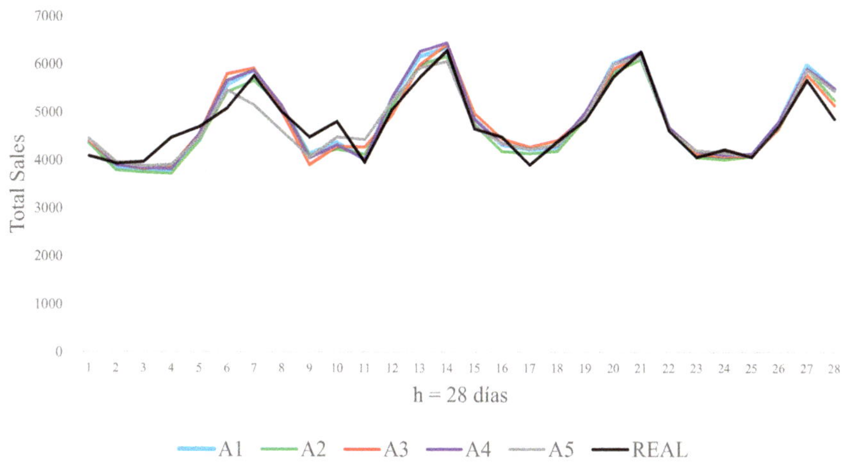

Fig. 1.6 Forecasts of the top five positions and the actual values of the time series in store CA1

Table 1.3 Characteristics of the three replicated M5 forecasting methods

Method	Performance	LightGBM	Mix of methods	Quickness
A1	✓	✓		
A2		✓	✓	
A5		✓		✓

day forecasting horizon used in the competition. The forecasts the top five methods generate exhibit similar patterns and closely align with the actual values.

We chose three of the five winning methods from the M5 competition for our method selection. The first method was selected due to its performance and exclusive reliance on LightGBM to create multiple models with various parameters. The second method, combining LightGBM with two other methods, aimed to achieve a superior forecast. Lastly, we selected the fifth-place method for its simplicity, speed, and impressive results. Table 1.3 summarizes these selections.

1.5.2 Forecasts for M4 Daily Time Series

To facilitate the comparison with the M4 winning method (ES-RNN), we applied a LightGBM forecasting approach, inspired by the successful methods in the M5 competition, to the selected M4 daily subset. The specific process, illustrated in Fig. 1.7, involved several key steps:

We generated LightGBM forecasts on the M4 daily subset using the following process:

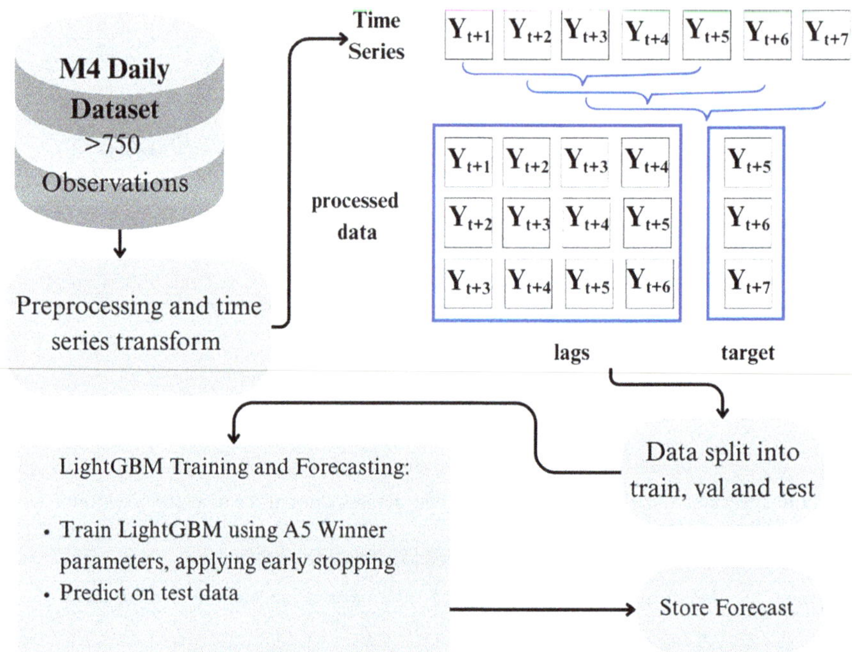

Fig. 1.7 Forecasts with LightGBM process

Dataset Selection and Preprocessing: We selected a time series from the M4 Daily dataset containing more than 750 observations to ensure sufficient historical data.

Feature Engineering via Time Series Transformation: Using a sliding window approach, we transformed each univariate time series into a supervised learning format compatible with LightGBM. We configured the window to use the six preceding observations (lags = 6) as input features for predicting the next immediate time step (t). This process yielded a feature matrix where each row contained six input features (values from t-6 to t-1) and one target output (value at t).

Train/Validation/Test Splitting: To ensure robust evaluation and prevent data leakage while maintaining temporal order, we partitioned the transformed data for each series into three distinct sets:

(a) **Test Set:** We held out the final 14 observations as the final test set, corresponding to the M4 competition's 14-step forecast horizon.
(b) **Validation Set:** We designated the most recent 20% as the validation set from the data preceding the test set.
(c) **Training Set:** The remaining data points constituted the training set.

LightGBM Model Training and Tuning: We employed the LightGBM framework. We configured key hyperparameters based on the setup reported by the A5 Winner of the M5 competition. These LightGBM training and forecasting processes for the M4 subset were executed within a standard Kaggle kernel environment, highlighting the method's feasibility on readily available computational platforms. For the comparative benchmark, ES-RNN forecasts were obtained directly from the official M4 competition repository; consequently, the computational resources required for their generation were not assessed in this study.

Training Process: We trained the LightGBM model exclusively on the training set:

– Tuning via Early Stopping: We monitored the model's performance on the validation set. To select the optimal number of trees and prevent overfitting, we used early stopping: the training process was halted if performance did not improve for 200 consecutive boosting rounds.

Forecasting: After training the model, we used it to generate the 14-step forecast by predicting the features of the held-out test set.

Storing Forecasts: We stored the resulting 14-step forecast for each time series. We subsequently used these forecasts for performance evaluation against the actual test set values and for comparison against the ES-RNN benchmark results.

After generating forecasts using LightGBM for the M4 daily subset, we compared these with the forecasts generated by the winning ES-RNN method from the M4 competition. Table 1.4 summarizes the results for the entire daily subset (2724 time series). LightGBM achieved a lower sMAPE in 55.00% of the time series, while ES-RNN achieved a lower sMAPE in 45.00% of the cases.

While sMAPE was the primary metric following the M4 competition standard, we conducted a broader evaluation using additional common forecasting error metrics as Mean Absolute Error (MAE), Mean Absolute Percentage Error (MAPE), and Root Mean Squared Error (RMSE) to assess the robustness of our comparative findings. This multi-metric analysis helps ensure that the observed performance differences are not merely an artifact of the chosen error measure.

Table 1.5 summarizes the win counts and corresponding percentages for ES-RNN and LightGBM across all four evaluated metrics over the 2724 time series subset.

Figure 1.8 provides a visual comparison of these win percentages.

Beyond error metrics, examining the correlation between forecasts and actual values provides further insight into model behavior. Figure 1.9 presents histograms comparing the distribution of these correlation coefficients for both LightGBM and ES-RNN across all 2724 analyzed series.

The distributions differ markedly. LightGBM's correlations ($n = 1498$ series where it won on sMAPE, absolute mean correlation $= 0.62$) are strongly skewed towards high positive values, indicating its forecasts often capture the direction and movement of the actual data well. In contrast, ES-RNN's correlations ($n = 1226$ series where it won on sMAPE, absolute mean correlation $= 0.33$) are more widely dispersed and centered much closer to zero, suggesting a less consistent alignment with the actual data dynamics on average across this dataset.

Table 1.4 Distribution of TS with the lowest sMAPE errors by forecasting method analyzed

Categories	Daily frequency	TS > 750 obs	LightGBM (total TS)	LightGBM (%)	ES-RNN (total TS)	ES-RNN (%)
Macro	127	62	23	37.09	**39**	**62.90**
Micro	1476	1186	**751**	**63.32**	435	36.60
Demographic	10	10	3	30.00	**7**	**70.00**
Industrial	422	43	20	46.51	**23**	**53.48**
Finance	1559	1388	681	49.06	**707**	**50.93**
Other	633	35	**20**	**57.14**	15	42.85
Total	4227	2724	**1498**	100.00	1226	100.00

Table 1.5 ES-RNN vs. LightGBM win counts and percentages by error metric

Metric	ES-RNN (total TS)	LightGBM (total TS)	ES-RNN %	LightGBM %
sMAPE	1226	**1498**	45.00	**55.00**
MAE	1231	**1493**	45.19	**54.80**
MAPE	1234	**1490**	45.30	**54.69**
RMSE	1262	**1462**	46.32	**53.67**

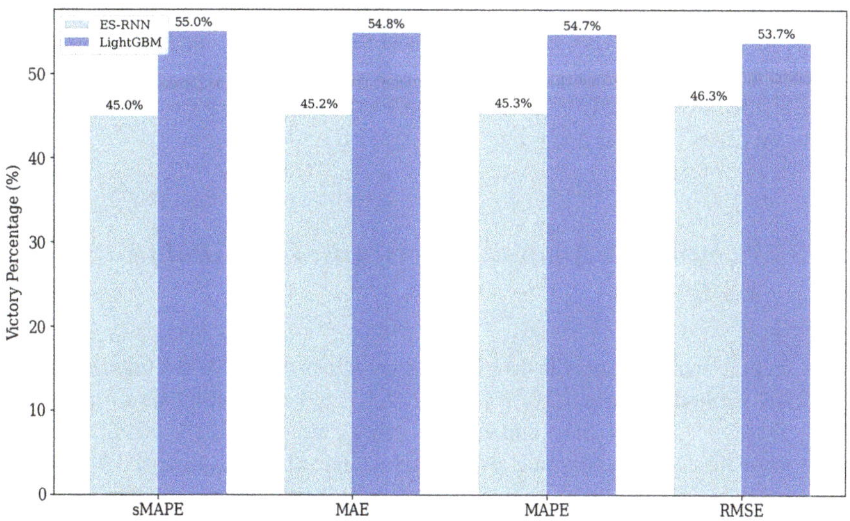

Fig. 1.8 Percentage comparison of wins for ES-RNN and LightGBM across different error metrics

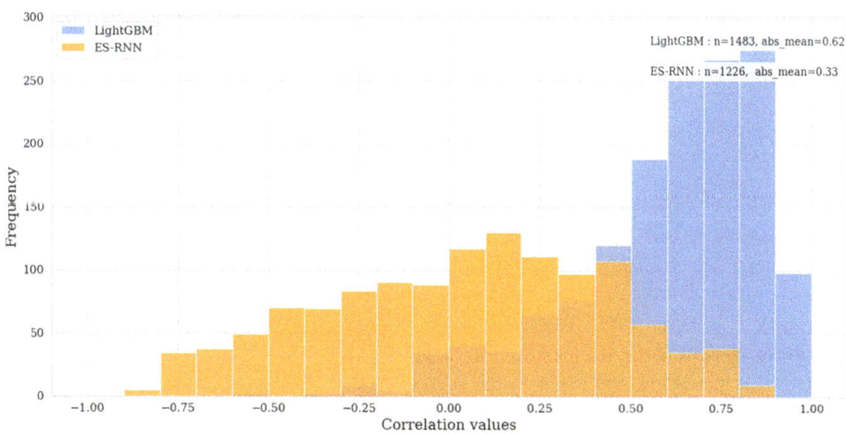

Fig. 1.9 Distribution of correlation coefficient between ES-RNN and LightGBM

The comparison of the M4 daily TS subset with over 750 observations (N = 2724) reveals a competitive landscape between LightGBM and ES-RNN. Across the evaluated error metrics (sMAPE, MAE, MAPE, and RMSE), LightGBM consistently secures wins in a slight majority of the series. Specifically, LightGBM achieves win percentages ranging from 53.67% (RMSE) to 55.00% (sMAPE), corresponding to the 1462 and 1498 series, respectively. Conversely, ES-RNN demonstrates superior performance in the remaining series, with win percentages between 45.00% (sMAPE) and 46.32% (RMSE). While the performance difference is narrow, the overall trend indicating a slight advantage for LightGBM remains robust across metrics. This finding motivates the subsequent analysis to focus on understanding the characteristics of the series where LightGBM excels. Based on these results, we randomly selected and analyzed five time series where LightGBM produced better forecasts according to the sMAPE error measure.

1.5.3 Comparative Analysis of Forecasts: LightGBM vs. ES-RNN

This section analyzes specific M4 time series where LightGBM outperforms the ES-RNN method. Figures 1.10, 1.11, 1.12, 1.13, and 1.14 depict series numbered D36, D175, D1610, D1686, and D2566, representing the daily series from the macroeconomic, microeconomic, demographic, industrial, and financial categories, respectively.

Figure 1.10 shows that the actual data exhibits a consistent linear downward trend. LightGBM accurately tracks this trend throughout the forecast horizon,

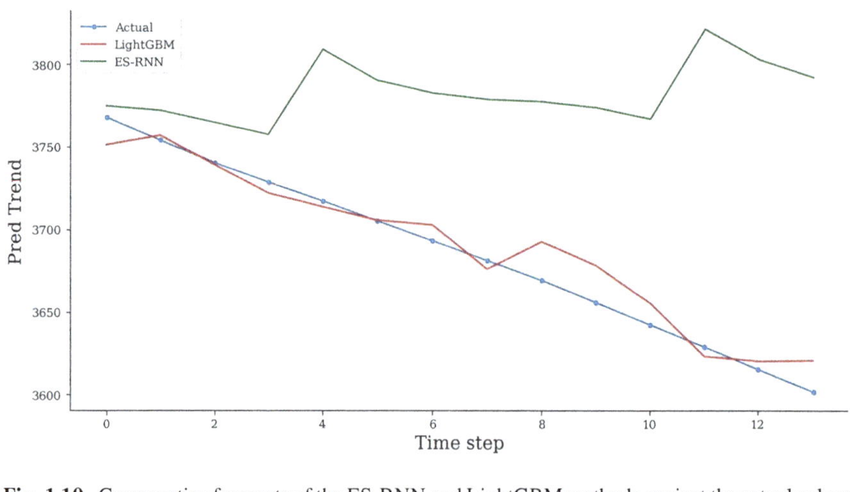

Fig. 1.10 Comparative forecasts of the ES-RNN and LightGBM methods against the actual values in D36

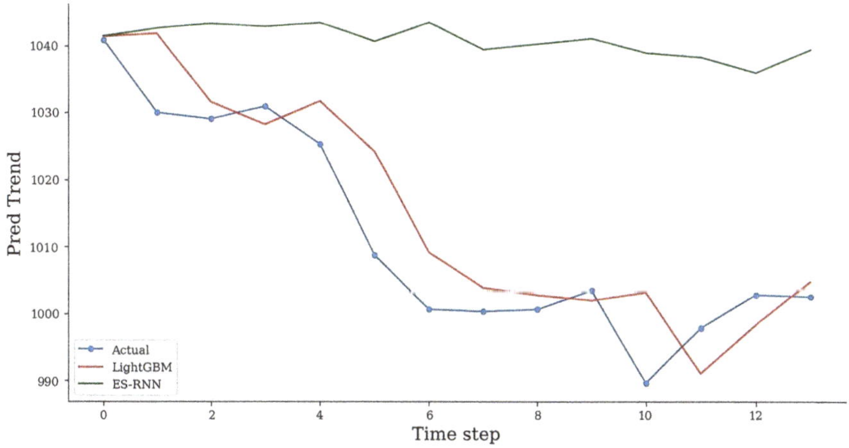

Fig. 1.11 Comparative forecasts of the ES-RNN and LightGBM methods against the actual values in D175

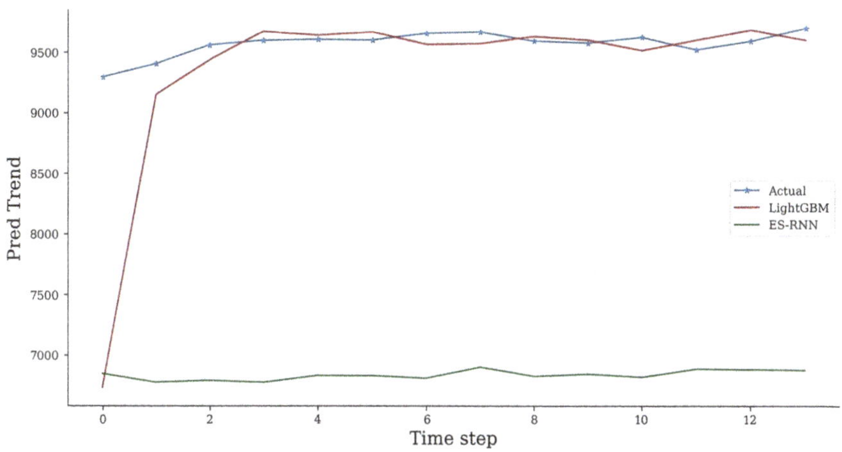

Fig. 1.12 Comparative forecasts of the ES-RNN and LightGBM methods against the actual values in D1610

yielding a close fit. In contrast, ES-RNN deviates significantly after the initial steps, overestimating the values and failing to capture the sustained decline.

Figure 1.11 shows that LightGBM effectively captures the initial steep decline shown in the actual data. In contrast, ES-RNN fails to model this trend, producing a nearly flat forecast significantly detached from the actual values throughout the horizon.

Figure 1.12 shows how the actual series rises sharply and stabilizes significantly. ES-RNN provides a flat forecast well below the actual values. LightGBM, despite an

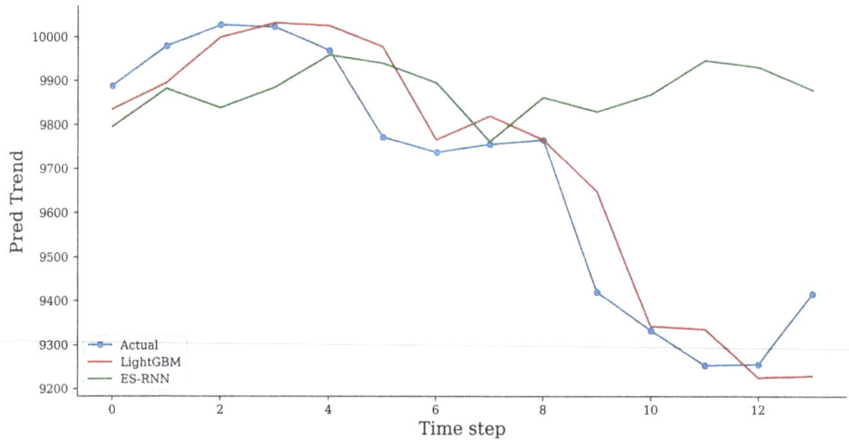

Fig. 1.13 Comparative forecasts of the ES-RNN and LightGBM methods against the actual values in D1686

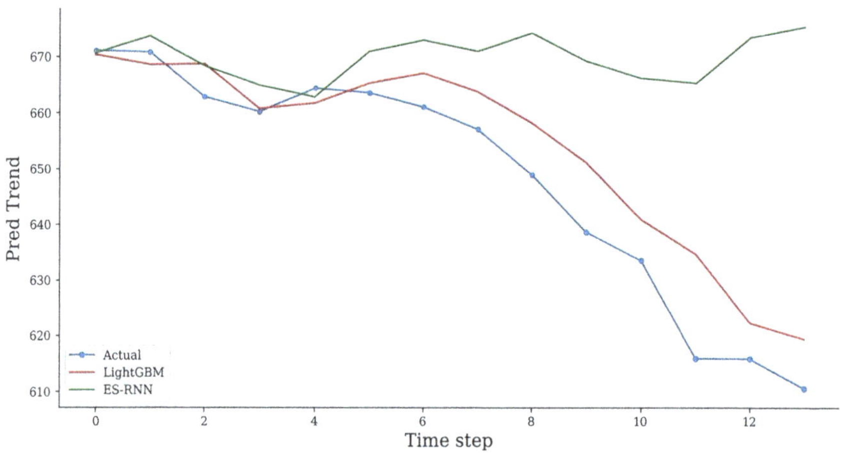

Fig. 1.14 Comparative forecasts of the ES-RNN and LightGBM methods against the actual values in D2566

inaccurate start, quickly adapts to capture both the upward trend and the subsequent plateau accurately.

Figure 1.13 shows that for this volatile series with multiple turning points, both models exhibit notable deviations from the actual data. LightGBM tracks the initial peak and subsequent downturn more closely, while ES-RNN displays smoother predictions that often lag or misjudge the amplitude of the fluctuations. Neither model achieves a consistently accurate fit.

Finally, in Fig. 1.14, The actual data declines sharply after initial stability. LightGBM captures this downward trend, whereas ES-RNN erroneously predicts

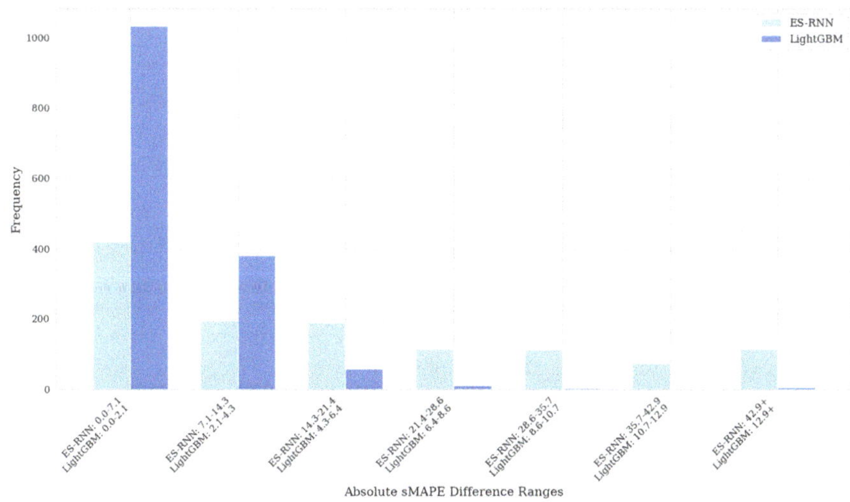

Fig. 1.15 Distribution of Absolute sMAPE difference between ES-RNN and LightGBM forecasts on the M4 daily subset

an upward movement, leading to significant divergence. LightGBM remains closer to the actual values for most of the forecast horizon.

To provide a more granular view of the performance comparison, Fig. 1.15 presents a histogram illustrating the frequency distribution of absolute sMAPE errors for the ES-RNN and LightGBM forecasts across the M4 daily subset. The figure groups the errors into distinct magnitude ranges on the x-axis, and the y-axis displays the corresponding frequency counts.

Notably, LightGBM (darker blue) demonstrates a significantly higher frequency in the lowest absolute error bin (LGBM range: 0.0–2.1) compared to ES-RNN (light blue, ~420 series in its corresponding 0.0–7.1 range). Conversely, in all subsequent bins representing larger absolute error magnitudes, ES-RNN consistently shows higher frequency counts than LightGBM. This indicates that while LightGBM achieves very low errors more often, ES-RNN has more series falling into moderate-to-higher absolute error brackets than LightGBM, whose error frequency drops sharply as error magnitude increases.

Table 1.6 shows the sMAPE error difference between the ES-RNN and LightGBM forecasts for the five selected time series.

These results show close competition between LightGBM and ES-RNN on this diverse M4 daily subset. While LightGBM achieves a slight majority of wins across all evaluated error metrics as shown in Table 1.5 and Fig. 1.8, the sMAPE difference distribution (Fig. 1.15) reveals that many of these victories, for both models, occur within relatively narrow margins, particularly clustered around zero difference. This highlights the competitive nature of forecasting tasks in many series.

Furthermore, performance exhibits context dependency based on the time series category (Table 1.4). LightGBM demonstrates a clear advantage in the Microeco-

Table 1.6 Difference of sMAPE error of the five selected time series

		Winner	
Time series names	SMAPE Difference	ES-RNN	LightGBM
D36	2.366	X	✓
D175	2.212	X	✓
D1610	29.492	X	✓
D1686	1.745	X	✓
D2566	1.826	X	✓

nomic category (63.32% wins), whereas ES-RNN performs better in the Macroeconomic (62.90% wins) and Demographic (70% wins) categories. Performance is closely contested within the Industrial and Finance categories. This variability underscores that neither model universally dominates, and the best choice may depend on the specific data characteristics.

1.5.4 Phase Two: Correlation Between LightGBM Forecasts and Actual Data

This phase focused on analyzing the relationship between the forecasts generated by LightGBM and the actual test data using the Pearson correlation coefficient. We aimed to understand factors influencing how well the forecasts tracked the real movements, particularly investigating the impact of the underlying time series trend.

However, it was not possible to compute a meaningful correlation coefficient for all 2724 series initially analyzed for error metrics. Specifically, 217 series were excluded from this correlation analysis. This exclusion was primarily due to instances where the LightGBM model generated completely flat (constant) forecasts over the 14-step horizon. While such forecasts could sometimes yield favorable error metrics, they prevent meaningful correlation analysis. As a result, this phase focused on the 2507 series suitable for this analysis. To do this, we:

Calculated the Pearson correlation coefficient between the forecasted and actual data.

Calculated the trend of the series using a proposed measure.

This measure evaluates the trend of a time series by analyzing the differences between successive values. First, we compute the changes between consecutive points and determine the net accumulated change (i.e., the sum of the differences) and the total magnitude of changes (i.e., the sum of the absolute differences). Next, we derive a ratio by dividing the net change by the total magnitude, which yields a value between -1 and 1. This ratio reflects the overall directional tendency of the series—positive values indicate an upward trend, while negative values indicate a downward trend. Finally, the ratio is mapped onto a predetermined scale to provide a quantifiable classification of the trend's intensity and direction.

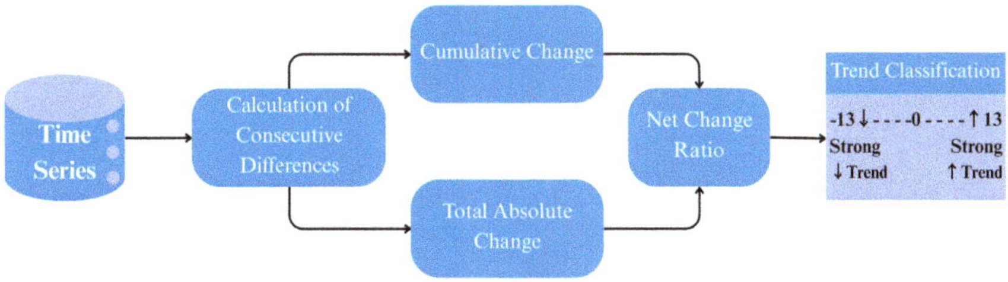

Fig. 1.16 Trend classification process

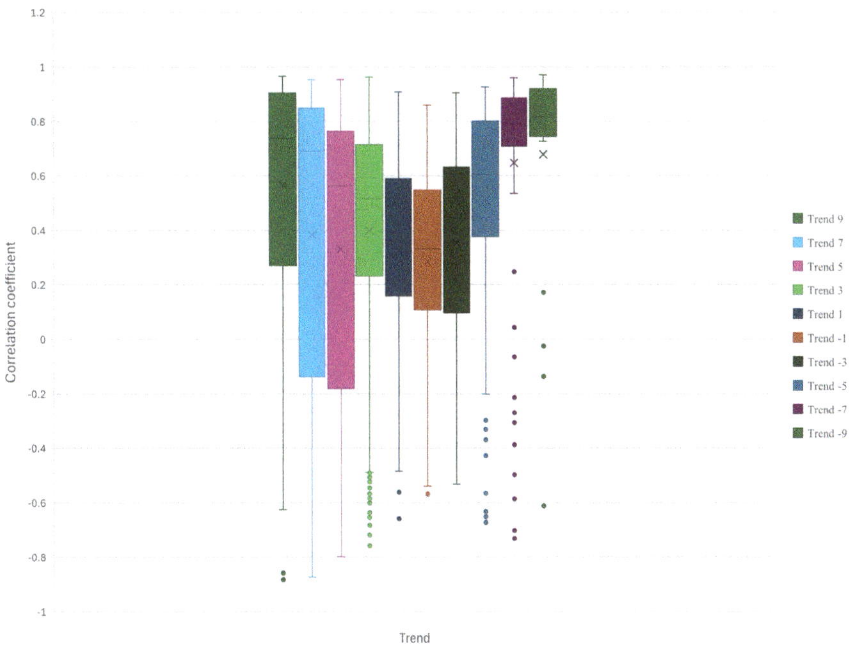

Fig. 1.17 Trend and correlation of the forecasts generated by LightGBM compared to the test data

Figure 1.16 shows the trend classification process.

Figure 1.17 displays boxplots illustrating the distribution of the correlation coefficient between the LightGBM forecasts and the actual test data, categorized by the calculated series trend. This figure visually highlights the impact of the series trend on forecast correlation between +9 and − 9 trend values, omitting higher trend values due to the small number of time series within those groups.

Series exhibiting stronger downward trends (−5 to −9) tend to show higher median correlation coefficients and more compact distributions than series with upward trends (5–9); where the latter in addition to showing higher median correlation it also presents higher variability. This suggests a more consistent and accurate forecast performance with a downward directional trend. Conversely,

Ranks

	N	Mean Rank
CREL -13	10	1177.80
-11	19	999.42
-9	29	1917.45
-7	86	1873.35
-5	141	1492.46
-3	197	1145.71
-1	280	983.11
1	412	1069.84
3	596	1271.00
5	454	1267.35
7	186	1391.35
9	68	1643.62
11	23	1217.35
13	6	2110.50
Total	2507	

Test Statistics

	CREL
Chi-Square	210.72
df	13
Asymp. Sig.	.000

Fig. 1.18 Kruskal–Wallis statistical test

upward directional trends can yield accurate forecasts with a shortcoming of higher variability.

Additionally, time series with weaker trends (−3 to +3) display markedly lower median correlation coefficients and median variability showing lower accuracy and reliability.

We then conducted the Kruskal–Wallis and Mann–Whitney non-parametric statistical tests. The Kruskal–Wallis test determines if groups of data originate from the same population, while the Mann–Whitney test verifies the heterogeneity of two samples.

The Kruskal–Wallis test (Fig. 1.18) yielded a highly significant result, confirming that the distribution of the correlation coefficients differs significantly according to the trend group of the series. This statistical finding is further reinforced by the visual evidence, providing a strong basis for the conclusion that the underlying trend is a key factor influencing the correlation of LightGBM's forecast with the actual values in this dataset.

Figures 1.19 and 1.20 present the results of the Mann–Whitney test. We excluded trends thirteen and eleven from this analysis due to the limited number of time series representing these trends. Figure 1.19a, b show no significant difference between trend −9 and trend −7 and trend −9 and trend 9.

However, Fig. 1.20a, b indicate a significant difference between trend −9 and trend −5 and significant differences persist with other compared trends.

a)

Ranks

		N	Mean Rank	Sum of Ranks
CREL	-9	29	63.21	1833.00
	-7	86	56.24	4837.00
	Total	115		

Test Statistics

	Mann-Whitney U	Wilcoxon W	Z	Asymp. Sig. (2tailed)
CREL	1096.00	4837.00	-.97	.331

b)

Ranks

		N	Mean Rank	Sum of Ranks
CREL	-9	29	53.55	1553.00
	9	68	47.06	3200.00
	Total	97		

Test Statistics

	Mann-Whitney U	Wilcoxon W	Z	Asymp. Sig. (2tailed)
CREL	854.00	3200.00	-1.04	.298

Fig. 1.19 Mann–Whitney statistical test, (**a**) the result of the test between trend −9 and −7; (**b**) result of the test between trend −9 and 9

a)

Ranks

		N	Mean Rank	Sum of Ranks
CREL	-9	29	116.21	3370.00
	-5	141	79.18	11165.00
	Total	170		

Test Statistics

	Mann-Whitney U	Wilcoxon W	Z	Asymp. Sig. (2-tailed)
CREL	1154.00	11165.00	-3.69	.000

b)

Ranks

		N	Mean Rank	Sum of Ranks
CREL	-9	29	141.48	4103.00
	7	186	102.78	19117.00
	Total	215		

Test Statistics

	Mann-Whitney U	Wilcoxon W	Z	Asymp. Sig. (2-tailed)
CREL	1726.00	19117.00	-3.12	.002

Fig. 1.20 Mann–Whitney statistical test, (**a**) the result of the test between trend −9 and −5; (**b**) result of the test between trend −9 and 7

We have made the comparative analysis code available in the following repository: https://github.com/AntonioMata21/Forecasting-M4-Daily-with-LightGBM/tree/main

1.6 Limitations & Directions for Future Research

While this study provides valuable insights into the comparative performance of LightGBM and ES-RNN, we must acknowledge several limitations inherent in its scope and design. These limitations, however, also highlight promising avenues for future investigation.

1.6.1 Limitations

Dataset Scope: We confined our analysis to subsets of the M4 and M5 competition datasets. Although chosen for their benchmark status and domain diversity, the lack of validation on external, non-competition datasets limits the definitive generalizability of our findings. Furthermore, while enabling a fair comparison, our focus on daily frequency data restricts conclusions regarding performance on other time granularities (e.g., hourly, weekly).

Method Selection We replicated three M5 specific methods driven by LightGBM's prevalence and code availability. This decision excluded other potentially competitive methods from M5 and the forecasting literature in general.

1.6.2 Directions for Future Research

The previous limitations naturally suggest several directions for extending this work:

Broader Empirical Validation Future studies should validate these findings on a wider range of external datasets, encompassing diverse domains, characteristics (e.g., intermittency, volatility), and multiple time frequencies (hourly, weekly, monthly) to assess the generalizability of LightGBM's performance profile more robustly.

Expanded Method Comparison Incorporating a broader set of state-of-the-art forecasting methods beyond those prominent in M5, including other gradient boosting variants, deep learning architectures (like Transformers or advanced RNNs), and potentially refined classical methods, would provide a more robust comparative landscape.

Advanced Trend and Seasonality Analysis Developing and employing more sophisticated techniques for trend and seasonality decomposition and measurement could help to more precisely quantify their impact on the relative performance of LightGBM and ES-RNN, potentially revealing more complex interactions. By addressing these areas, future research can build upon our findings to provide a more comprehensive understanding of LightGBM's role in the time series forecasting toolkit and guide practitioners in selecting the most appropriate methods for their needs.

1.7 Conclusions

This study compared the LightGBM forecasting method, a key component in the winning solutions of the M5 competition, against the M4 competition winner, the hybrid ES-RNN model. Utilizing a diverse subset of 2724 daily time series from the M4 dataset, our primary objective was to assess LightGBM's performance generalizability beyond its recognized strength in retail sales and investigate the influence of the time series trend on its relative effectiveness.

Our findings, evaluated across multiple error metrics (sMAPE, MAE, MAPE, RMSE), indicate a nuanced competitive landscape. While LightGBM secured a slight majority of wins, achieving victories in approximately 54–55% of the series (metric-dependent), the performance margins were often narrow, suggesting close competition. Furthermore, performance varied by series category (as shown in Table 1.4), underscoring that neither model universally dominates this heterogeneous dataset.

Crucially, this research highlights a decisive conditional factor influencing LightGBM's performance: its tendency to produce upward-biased and downward-biased forecasts translates into a competitive advantage primarily on time series exhibiting distinct trends. Graphical analysis (Figs. 1.10, 1.11, 1.12, 1.13, and 1.14), which shows LightGBM often tracked directional trends better than ES-RNN, supports this observation, and statistical tests (Kruskal–Wallis, Mann–Whitney) confirm it, showing a significant link between series trend and forecast correlation (Figs. 1.18, 1.19, and 1.20) confirm this link (see Fig. 1.17), showing a significant relationship between series trend and forecast correlation. Specifically, LightGBM's forecasts, when outperforming ES-RNN, displayed a significantly higher positive correlation with the actual values.

Conversely, LightGBM's relative performance decreased for series with high variability or a lack of a clear trend, where the adaptability of the hybrid ES-RNN proved more robust overall. These results underscore that the suitability of the M5-derived LightGBM approach is context-dependent. While powerful and efficient, especially for trended data common in retail, its direct application to broader domains requires careful consideration of the specific time series characteristics, particularly trend behavior.

Despite the limitations outlined in Sect. 1.7, this study provides valuable insights into LightGBM's performance profile. As discussed in Sect. 1.7, future research directions, mainly focusing on more advanced trend analysis and wider empirical validation, will be essential for further delineating its strengths and weaknesses across the diverse landscape of time series forecasting.

We conclude that LightGBM is a highly effective forecasting method for time series with discernible upward and downward trends, but its broader applicability warrants caution. The choice between LightGBM and robust hybrid models like ES-RNN should be informed by understanding the underlying data dynamics, particularly trend behavior, to optimize forecasting performance in specific applications.

AI Use Statement Some portions of this manuscript were enhanced with the assistance of generative AI tools. The authors carefully reviewed, edited and approved all AI-generated content. The authors take full responsibility for the accuracy and integrity of the entire chapter.

References

1. Makridakis, S., Andersen, A., Carbone, R., Fildes, R., Hibon, M., Lewandowski, R., Jewton, J., Parsen, E., Winkler, R.: The accuracy of extrapolation (lime Series) methods: results of a forecasting corn petition. J. Forecast. **1**, 111–153 (1982). https://doi.org/10.1002/for.3980010202
2. Makridakis, S., Chatfield, C., Hibon, M., Lawrence, M., Mills, T., Ord, K., Simmons, L.R.F.: The M2-competition: a real-time judgmentally based forecasting study. Int. J. Forecast. **9**, 5–22 (1993). https://doi.org/10.1016/0169-2070(93)90044-N
3. Makridakis, S., Hibon, M.: The M3-competition: results, conclusions and implications. Int. J. Forecast. **16**, 451–476 (2000). https://doi.org/10.1016/S0169-2070(00)00057-1
4. Makridakis, S., Spiliotis, E., Assimakopoulos, V.: The M4 Competition: 100,000 time series and 61 forecasting methods. Int. J. Forecast. **36**, 54–74 (2020). https://doi.org/10.1016/j.ijforecast.2019.04.014
5. Makridakis, S., Spiliotis, E., Assimakopoulos, V.: M5 accuracy competition: results, findings, and conclusions. Int. J. Forecast. **38**, 1346–1364 (2022). https://doi.org/10.1016/j.ijforecast.2021.11.013
6. Makridakis, S., Spiliotis, E., Assimakopoulos, V.: Statistical and machine learning forecasting methods: concerns and ways forward. PLoS One. **13**, 1–26 (2018). https://doi.org/10.1371/journal.pone.0194889
7. Cerqueira, V., Torgo, L., Soares, C.: Machine learning vs statistical methods for time series forecasting: size matters. arXiv Prepr. arXiv1909.13316. (2019). https://doi.org/10.48550/arXiv.1909.13316
8. Ju, Y., Sun, G., Chen, Q., Zhang, M., Zhu, H., Rehman, M.U.: A model combining convolutional neural network and lightgbm algorithm for ultra-short-term wind power forecasting. IEEE Access. **7**, 28309–28318 (2019). https://doi.org/10.1109/ACCESS.2019.2901920
9. Weng, T., Liu, W., Xiao, J.: Supply chain sales forecasting based on lightGBM and LSTM combination model. Ind. Manag. Data Syst. **120**, 265–279 (2019). https://doi.org/10.1108/IMDS-03-2019-0170
10. Kilimci, Z.H., Akyuz, A.O., Uysal, M., Akyokus, S., Uysal, M.O., Atak Bulbul, B., Ekmis, M.A., Silva, T.C.: An improved demand forecasting model using deep learning approach and proposed decision integration strategy for supply chain. Complexity. **2019** (2019). https://doi.org/10.1155/2019/9067367
11. Ariannezhad, M., Schelter, S., de Rijke, M.: Demand Forecasting in the Presence of Privileged Information. In: Lemaire, V., Malinowski, S., Bagnall, A., Guyet, T., Tavenard, R., Ifrim, G. (eds) Advanced Analytics and Learning on Temporal Data. AALTD 2020. Lecture Notes in Computer Science(), vol 12588, Springer, Cham (2020) https://doi.org/10.1007/978-3-030-65742-0_4
12. Ma, S., Fildes, R.: Retail sales forecasting with meta-learning. Eur. J. Oper. Res. **288**, 111–128 (2021). https://doi.org/10.1016/j.ejor.2020.05.038
13. Jeon, Y., Seong, S.: Robust recurrent network model for intermittent time-series forecasting. Int. J. Forecast. (2021). https://doi.org/10.1016/j.ijforecast.2021.07.004
14. Anderer, M., Li, F.: Hierarchical forecasting with a top-down alignment of independent-level forecasts. Int. J. Forecast. (2022). https://doi.org/10.1016/j.ijforecast.2021.12.015
15. Bojer, C.S., Meldgaard, J.P.: Kaggle forecasting competitions: an overlooked learning opportunity. Int. J. Forecast. **37**, 587–603 (2021). https://doi.org/10.1016/j.ijforecast.2020.07.007
16. Sun, X., Liu, M., Sima, Z.: A novel cryptocurrency price trend forecasting model based on LightGBM. Financ. Res. Lett. **32** (2020). https://doi.org/10.1016/j.frl.2018.12.032
17. Zhang, Y., Zhu, C., Wang, Q.: Lightgbm-based model for metro passenger volume forecasting. IET Intell. Transp. Syst. **14**, 1815–1823 (2020). https://doi.org/10.1049/iet-its.2020.0396
18. Liu, M., Zeng, A., Lai, Q., Xu, Q.: Time series is a special sequence: forecasting with sample convolution and interaction. ArXiv. abs/2106.0. (2021)
19. Valli, L.N., Sujatha, N.: Time series analysis in data science: forecasting trends and anomalies. In: 2024 5th International Conference on Smart Electronics and Communication (ICOSEC), pp. 15–20 (2024)

20. Jones, E., Harden, S., Crawley, M.J.: Time Series. In: The R Book, pp. 715–739. Wiley, Chicheste (2022)
21. Sun, Y., Xie, Z., Eldele, E., Chen, D., Hu, Q., Wu, M.: Learning pattern-specific experts for time series forecasting under patch-level distribution shift. arXiv. (2024)
22. Aria, S.S., Iranmanesh, S.H., Hassani, H.: Optimizing multivariate time series forecasting with data augmentation. J. Risk Financ. Manag. **17** (2024). https://doi.org/10.3390/jrfm17110485
23. Shrestha, M., Bhatta, G.: Selecting appropriate methodological framework for time series data analysis. J. Financ. Data Sci. (2018). https://doi.org/10.1016/J.JFDS.2017.11.001
24. Kumar, S., Nayak, M.A.: Predictive analytics for demand forecasting: a deep learning-based decision support system. Int. J. Eng. Comput. Sci. **13**, 26291–26299 (2024). https://doi.org/10.18535/ijecs/v13i07.4853
25. Lei, C., Zhang, H., Wang, Z., Miao, Q.: Multi-model fusion demand forecasting framework based on attention mechanism. Processes. **12**, 1–20 (2024). https://doi.org/10.3390/pr12112612
26. Spiliotis, E., Makridakis, S., Kaltsounis, A., Assimakopoulos, V.: Product sales probabilistic forecasting: an empirical evaluation using the M5 competition data. Int. J. Prod. Econ. **240**, 108237 (2021). https://doi.org/10.1016/j.ijpe.2021.108237
27. Jung, G., Choi, S.Y.: Forecasting foreign exchange volatility using deep learning autoencoder-LSTM techniques. Complexity. **2021** (2021). https://doi.org/10.1155/2021/6647534
28. Lv, J., Wang, C., Gao, W., Zhao, Q.: An economic forecasting method based on the LightGBM-optimized LSTM and time-series model. Comput. Intell. Neurosci. **2021** (2021). https://doi.org/10.1155/2021/8128879
29. Qian, W., Gu, C., Zhu, C., Jiang, Z., Han, B., Yu, M.: Short-term load forecasting based on multi-model fusion of CNN-LSTM-LGBM. In: 2021 International Conference on Power System Technology (POWERCON), pp. 934–939 (2021)
30. Huang, H., Chen, W., Zhong, Y., Zhang, S., Chen, J., Pan, Y., Xu, R., Fang, H.: Medium-long term industry load forecasting method based on normalized mutual information and BiLSTM. In: 2024 China International Conference on Electricity Distribution (CICED), pp. 616–622 (2024)
31. Dudek, G., Pelka, P., Smyl, S.: A hybrid residual dilated lstm and exponential smoothing model for midterm electric load forecasting. IEEE Trans. Neural Netw. Learn. Syst. **33**, 2879–2891 (2022). https://doi.org/10.1109/TNNLS.2020.3046629
32. Kumar, S.R.: Long-term vs. short-term forecasting data scientist's strategies for different horizons. Int. J. Sci. Res. **10**, 1796–1797 (2021). https://doi.org/10.21275/SR24213014558
33. Kumari, P., Vekariya, P.R., Garde, Y.A.: Computational methods of time series forecasting in data analysis. In: Futuristic Trends in Computing Technologies and Data Sciences, vol. 3, Book 3 edn, pp. 143–158. IIP Series (2024)
34. Athiyarath, S., Paul, M., Krishnaswamy, S.: A comparative study and analysis of time series forecasting techniques. SN Comput. Sci. **1**, 1–7 (2020). https://doi.org/10.1007/s42979-020-00180-5
35. Ensafi, Y., Amin, S.H., Zhang, G., Shah, B.: Time-series forecasting of seasonal items sales using machine learning—A comparative analysis. Int. J. Inf. Manag. Data Insights. **2**, 100058 (2022). https://doi.org/10.1016/j.jjimei.2022.100058
36. Rani, S., Kaur, R., Desai, C.: Enhancing time series forecasting accuracy with deep learning models: a comparative study. Int. J. Adv. Res. **12**, 315–324 (2024). https://doi.org/10.21474/ijar01/19257
37. Machado, M.R., Karray, S., De Sousa, I.T.: LightGBM: An effective decision tree gradient boosting method to predict customer loyalty in the finance industry. In: 14th Int. Conf. Comput. Sci. Educ. ICCSE 2019, pp. 1111–1116 (2019). https://doi.org/10.1109/ICCSE.2019.8845529
38. Wang, D., Zhang, Y., Zhao, Y.: LightGBM: an effective miRNA classification method in breast cancer patients. ACM Int. Conf. Proc. Ser. **7–11** (2017). https://doi.org/10.1145/3155077.3155079
39. Ke, G., Meng, Q., Finley, T., Wang, T., Chen, W., Ma, W., Ye, Q., Liu, T.-Y.: LightGBM: a highly efficient gradient boosting decision tree. Curran Associates, Red Hook (2017)

40. Microsoft Research: LightGBM documentation. https://lightgbm.readthedocs.io/en/latest/Features.html
41. McCulloch, W.S., Pitts, W.: A logical calculus of the ideas immanent in nervous activity. Bull. Math. Biophys. **5**, 115–133 (1943). https://doi.org/10.1007/BF02478259
42. Werbos, P.J.: Beyond Regression: New Tools for Prediction and Analysis in the Behavioral Sciences. Harvard University (1975)
43. Csordás, R., Potts, C., Manning, C.D., Geiger, A.: Recurrent neural networks learn to store and generate sequences using non-linear representations. arXiv2408.10920. (2024). https://doi.org/10.48550/arXiv.2408.10920
44. Hu, Y., Wei, R., Yang, Y., Li, X., Huang, Z., Liu, Y., He, C., Lu, H.: Performance degradation prediction using LSTM with optimized parameters. Sensors. **22**, 1–16 (2022). https://doi.org/10.3390/s22062407
45. Smyl, S.: A hybrid method of exponential smoothing and recurrent neural networks for time series forecasting. Int. J. Forecast. **36**, 75–85 (2020). https://doi.org/10.1016/j.ijforecast.2019.03.017
46. Lim, B., Arık, S., Loeff, N., Pfister, T.: Temporal fusion transformers for interpretable multi-horizon time series forecasting. arXiv. 1–27 (2019)
47. Dairi, A., Harrou, F., Zeroual, A., Hittawe, M.M., Sun, Y.: Comparative study of machine learning methods for COVID-19 transmission forecasting. J. Biomed. Inform. **118**, 103791 (2021). https://doi.org/10.1016/j.jbi.2021.103791
48. Mitra, A., Jain, A., Kishore, A., Kumar, P.: A comparative study of demand forecasting models for a multi-channel retail company: a novel hybrid machine learning approach. Oper. Res. Forum. **3**, 1–22 (2022). https://doi.org/10.1007/s43069-022-00166-4
49. Pirani, M., Thakkar, P., Jivrani, P., Bohara, M.H., Garg, D.: A comparative analysis of ARIMA, GRU, LSTM and BiLSTM on financial time series forecasting. In: 2022 IEEE International Conference on Distributed Computing and Electrical Circuits and Electronics (ICDCECE), pp. 1–6 (2022)
50. Kurani, A., Doshi, P., Vakharia, A., Shah, M.: A comprehensive comparative study of Artificial Neural Network (ANN) and Support Vector Machines (SVM) on stock forecasting. Ann. Data Sci. **10**, 183–208 (2023). https://doi.org/10.1007/s40745-021-00344-x
51. Jewel, R.M., Linkon, A.A., Shaima, M., Badruddowza, Sarker, M.S.U., Shahid, R., Nabi, N., Rana, M.N.U., Shahriyar, M.A., Hasan, M., Hossain, M.J.: Comparative analysis of machine learning models for accurate retail sales demand forecasting. J. Comput. Sci. Technol. Stud. **6**, 204–210 (2024). https://doi.org/10.32996/jcsts.2024.6.1.23
52. Balamwar, A., Mitra, R., Tiwari, M.K., Verma, P.: Prediction and analysis of seasonal dynamic metal consumption using aggregated lightgbm—a case study. IFAC-PapersOnLine. **55**, 725–730 (2022). https://doi.org/10.1016/j.ifacol.2022.09.494
53. Malakouti, S.M.: Prediction of wind speed and power with lightgbm and grid search: case study based on scada system in Turkey. Int. J. Energy Prod. Manag. **8**, 35–40 (2023). https://doi.org/10.18280/ijepm.080105
54. Bian, L., Qin, X., Zhang, C., Guo, P., Wu, H.: Application, interpretability and prediction of machine learning method combined with LSTM and LightGBM-a case study for runoff simulation in an arid area. J. Hydrol. **625**, 130091 (2023). https://doi.org/10.1016/j.jhydrol.2023.130091
55. Song, X., Chen, Z.S.: Enhancing financial time series forecasting in the shipping market: a hybrid approach with Light Gradient Boosting Machine. Eng. Appl. Artif. Intell. **136**, 108942 (2024). https://doi.org/10.1016/j.engappai.2024.108942
56. Schnaars, S.P.: Long-range forecasting: From crystal ball to computer: J. Scott Armstrong, 2nd ed. (Wiley, New York, 1985) £22.95 (paper), pp. 689. Int. J. Forecast. 2, 387–390 (1986). doi: https://doi.org/10.1016/0169-2070(86)90059-2

Chapter 2
Analysis and Forecasts of Sugarcane Production in Tons Using Machine Learning

Adhara A. Cavazos-Matsumoto, Salvador Ibarra-Martinez, José A. Castán-Rocha, Alejandro H. García-Ruiz, and Jesús D. Terán-Villanueva

Abstract Sugarcane is a cornerstone of Mexico's rural economy, supporting over 2 million jobs and contributing 12% of agricultural GDP (Gross Domestic Product). Despite ranking as the world's sixth-largest producer, the industry faces declining yields due to climate variability, structural challenges, and limited technology adoption among smallholders. To address critical forecasting gaps, we developed a predictive model using public mill data from CONADESUCA (47 mills) and compared machine learning approaches: Long Short-Term Memory (LSTM), Random Forest, and Dense Neural Networks. LSTM demonstrated superior performance in capturing production trends, with key mills achieving RMSE values as low as 0.153 and R^2 up to 0.981. While climate variables provided marginal improvements, results suggest operational factors may dominate yield variability. The study highlights both the potential of AI-driven forecasting for mill planning and persistent data limitations—particularly the lack of field-level records from small producers.

Keywords Data science · LSTM · Random forest · Time series · Artificial neural network · Agricultural forecasting

2.1 Introduction

The first sugarcane (*Saccharum officinarum*) plantations in Mexico date back to colonial times, with Hernán Cortés among the first farmers in 1524. Soon, established villages where milled sugarcane stalks and extracted juice then became sugar. These facilities were known as *trapiches*, now referred to as sugar mills [16].

A. A. Cavazos-Matsumoto · S. Ibarra-Martínez · J. A. Castán-Rocha · A. H. García-Ruiz (✉) · J. D. Terán Villanueva
Centro Universitario Tampico Madero, Universidad Autónoma de Tamaulipas (UAT), Tampico Tamaulipas, México
e-mail: almatsumoto@uat.edu.mx; sibarram@uat.edu.mx; jacastan@docentes.uat.edu.mx; ahgarcia@docentes.uat.edu.mx; jdteran@docentes.uat.edu.mx

G. Rivera et al. (eds.), *Recent Advances in Artificial Neural Networks*, Intelligent Systems Reference Library 283, https://doi.org/10.1007/978-3-032-13497-4_2

Today, considered a global leader, Mexican sugar agro-industry ranks sixth in sugarcane production and seventh in sugar production [36]. As of 2025, the Mexican sugar industry operates 47 mills across 15 states [12], generating direct and indirect employment for more than two million people. The industry involves more than 180,000 producers and generates a primary production value of approximately 30 billion pesos [35]. This crop has significant social and economic impacts. Agribusiness accounts for 12% of GDP in the agricultural sector and 7% in the country's food industry. Socially, sugarcane has supported rural communities, providing employment and strengthening local traditions [37]. Approximately Mexico has 783,597 hectares (ha) under cultivation nationwide, involving the participation of more than 165,000 small producers [39].

The sugarcane agroindustry faces significant challenges, including high sugar inventories, declining crop quality, stagnant productivity, low mill efficiency, and limited sugar consumption growth. In recent years, the industry has experienced considerable fluctuations in production driven by climate change and economic conditions [42].

Although Mexico maintains a significant sugarcane industry, its average agricultural productivity –measured as tons of cane produced per hectare (TCH)– remains lower than that of global leaders such as Brazil or Australia. While these countries often achieve average yields above 100 million TCH, average Mexican production over the fifteen years has ranged between 5 and 7 million tons, as shown in Figs. 2.1 and 2.2. A consistent decline below 4 million tons would

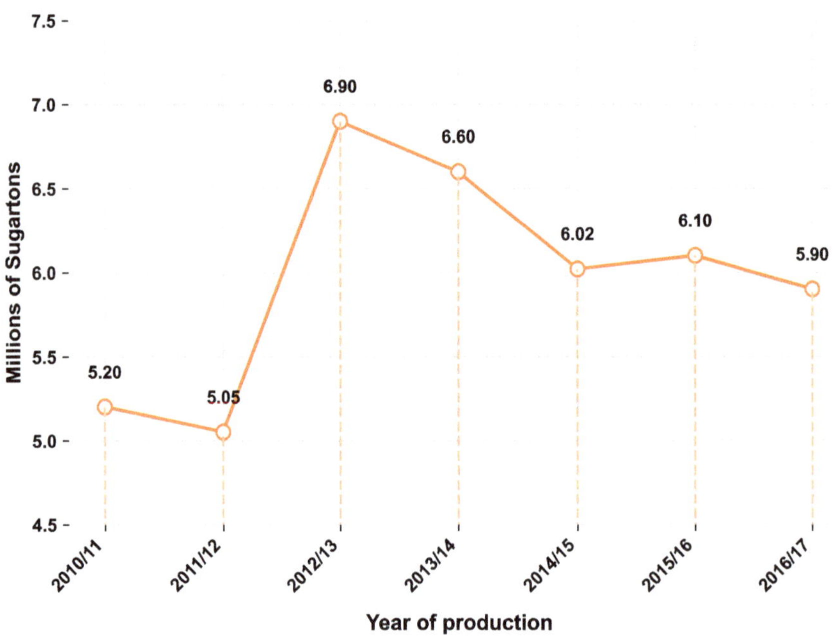

Fig. 2.1 Tons of cane per year of production (1)

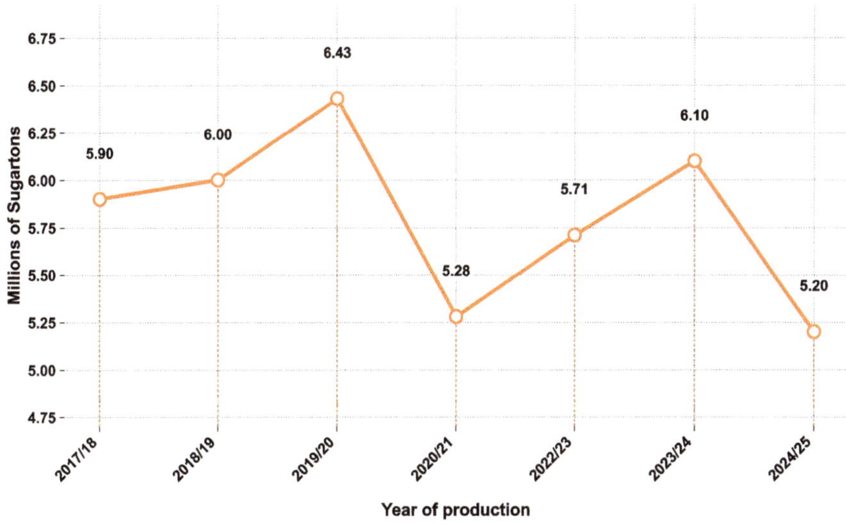

Fig. 2.2 Tons of cane per year of production (2)

indicate more serious structural problems within the sector, such as long-term damage to infrastructure or agricultural practices [11]. According to the Annual Report of the National Committee for the Sustainable Development of Sugarcane (CONADESUCA), Mexican production projections for 2021 were approximately 5.8 million tons. However, it has declined in the subsequent years, 2022/23 and 2023/24 [10].

Current production at the national level has processed around 80% of the expected cane. The initial projection ranged between 5.8 and 6 million tons, now estimated between 5.5 and 5.7 million tons a slight decrease due to uneven yields [38].

Despite the importance of the industry, forecasting sugarcane production remains a significant challenge, especially for small-scale growers. Unlike large producers or mills—who typically have access to advanced farm management systems and predictive analytics—many smallholders in Mexico lack the necessary technological and financial resources to implement complex solutions. Additionally, there is a limited culture of systematic record-keeping for agricultural activities, further hindering the development of accurate and accessible predictive models [42].

Studies have shown that digital technology adoption within Mexico's agrifood sector remains limited, with only 5.57% of producers using computers and 7.88% having internet access [6]. As part of this study's initial exploratory phase, we have an informal discussion with a group of sugarcane producers in Pánuco, a municipality in Veracruz. This meeting revealed a critical limitation among small-scale producers: the absence of systematic documentation of field activities. Many participants cited cultural, economic, and technological barriers to formally recording practices such as planting, fertilization, pest control, and harvesting.

This lack of documentation presents a significant obstacle to developing predictive models tailored to specific plots, as such models require robust historical data on agricultural practices and their evolution. Due to the lack of producer-level historical data, this study relies on publicly available datasets from sugar mills, which offer a broad overview of the behavior of Mexico's sugar agro-industry over time, with a focus on the agricultural component.

Due to data limitations, this study proposes a model to predict the tons of cane produced per hectare and the total tonnage delivered to mills so that the sugar mills can plan their agricultural cycles and have a forecast for next year's harvest. Using the techniques of Long Short-Term Memory (LSTM) and Random Forest (RF). In addition, we will use a Dense Neural Network (DNN) for extra testing and comparison with the other models.

The remainder of this paper is organized as follows: Sect. 2.2 provides background on the key factors affecting sugarcane production and the agents involved in the process. Section 2.3 reviews the related work in this domain. Section 2.4 describes the dataset preparation methodology, while Sect. 2.5 details the model training process. Section 2.6 presents the experimental results and discusses the findings and future work directions in Sect. 2.7. Finally, Sect. 2.8 concludes the paper by summarizing its key contributions.

2.2 Background

Accurately forecasting sugarcane production is a complex task influenced by multiple interacting factors. Before detailing the specific variables and methods considered in this study, it is crucial to understand the broader agronomic, climatic, and operational conditions that shape sugarcane productivity. This section provides an overview of these elements, establishing the foundation for developing predictive models.

2.2.1 Factors Affecting Sugarcane Production

Sugarcane production is an extensive and multifactorial process influenced by agronomic conditions and weather, economic, and agricultural management factors [34]. Among the primary metrics used are agricultural yield—measured as TCH—and productivity, understood as the system's efficiency in converting inputs and resources into usable sugarcane production [27].

Both concepts are distinct in the literature, and the two terms converge in practice since both seek to reflect the success of agricultural management and growing conditions. Therefore, achieving optimal yields depends on multiple factors [40], such as:

- Weather conditions: temperature, precipitation, soil and environmental humidity, solar radiation, evapotranspiration, rainfall frequency, and intensity, among others.
- Human factor: access to agricultural technologies, technical training, availability of credit, public support policies, application of agricultural techniques, availability of labor, and fluctuation of prices for the purchase and sale of products.
- Agricultural management: Among the main ones are soil conditions, fertilization, irrigation, crop rotation, crop control, and pest management.

These factors interact dynamically throughout the agricultural cycle, generating important variations in the observed yield from one cycle to another. Farmers aim to maximize yield, ensuring crop quality and production efficiency [40].

2.2.2 Agents Involved in the Forecasting Process

In Mexico, the principal actors in production forecasting are:

- Producers: responsible for all field activities, from soil preparation to harvesting [46].
- Sugar mills: record data on cane reception (weight, quality, delivery date) and carry out industrial processing activities [43].
- Government institutions: The Secretariat of Agriculture and Rural Development (SADER) and CONADESUCA are responsible for compiling, systematizing, and publishing agroindustrial statistics in compliance with the Sugarcane Sustainable Development Law, which seeks to ensure that good practices and efficient management, ensuring safety and seeking more sustainable practices during the agroindustrial process [18].

Developing predictive models at the plot level is difficult because many small producers do not formally document their agricultural activities. For this reason, this research uses mill records, which make it possible to infer general trends in the sector from a structured and standardized data set [40].

2.3 Related Work

Recent studies have increasingly addressed the challenge of forecasting sugarcane yield and productivity, aligning closely with the objectives of this research. Many of these efforts leverage Machine Learning (ML) techniques, applying them to agronomic variables such as weather conditions, soil characteristics, crop management practices, and industrial parameters like sucrose content and cane maturity. It is important to note that while yield and productivity are occasionally treated differently in academic literature, the variables considered in predictive modeling efforts tend to overlap significantly. This convergence reflects a growing interest in developing comprehensive models integrating data from multiple stages of the sugarcane value chain, aiming to enhance decision-making from field operations to industrial processing.

2.3.1 *Agricultural Yield Prediction Models for Sugarcane*

To model early yield prediction in Mexico, [25] proposed a model integrating time series of vegetation indices extracted from satellite imagery and weather data in conjunction with the Random Forest technique, which showed the metric R2R2 0.84 and RMSE 0.034. Similarly, [41] proposes a model combining ML techniques and remote sensing information in India, using spectral indices such as GNDVI, NDVI, and LSWI, where Random Forest achieved the best performance (R2R2 of 0.94). Expanding on remote sensing, [32] compares weather variables and EVI/NDVI vegetation indices in West Africa, finding that the combined model (Random Forest) achieved the highest R2R2 (0.61) with a 4-month lead time.

In contrast, [17] proposes the combination of ML techniques with electronic devices to increase productivity in Brazil, where SVR outperformed RF and DLN with an RMSE of 0.0018. Meanwhile, [33] uses climate parameters and historical data in India, demonstrating that Random Forest fits the data better than KNN, with a MAPE of 10%. Further exploring climatic influences, [44] evaluates monthly climatic variables in Karnataka, India, with ANN yielding the best accuracy among SVM, RF, and SMLR.

For trend-based approaches, [28] integrates exponential smoothing, MLP, and LSTM in China, achieving superior results (MSE: 0.011, R2R2: 0.979). Similarly, [45] optimizes environmental variable modeling in China using LSSVM-WCA, outperforming PSO (RMSE: 5.385 vs. 5.726). On the other hand, [24] employs ARIMA for weather forecasting and SVM for yield prediction in Sri Lanka, achieving an R2R2 of 87%.

In the context of agricultural management, [29] uses parameters like cultivation area and irrigation source in India, with XGBoost and GBM (66% accuracy) slightly outperforming RF (65%). Likewise, [30] applies satellite imagery in India, finding GBR the best performer (R2R2 = 0.66, RMSE = 7.15 t/ha). Focusing on soil data, [15] develops a ML model for sugarcane in Brazil using soil variables and decision trees, achieving 75

For large-scale statistical modeling, [21] creates a multivariate linear regression model for India, achieving 97.23% accuracy with variables like cane production and factory data. In Argentina, [20] tests RF, DNN, and SVM, with DNN excelling in hectare prediction (MSE: 0.0081) and RF in production tons (MSE: 0.0550). Finally, [8] compares RF and XGBoost for plot-level yield in Thailand, with RF marginally outperforming (71.83% vs. 71.64%).

Table 2.1 Studies on the use of agro-industrial and agronomic variables in sugarcane predictive models

Author(s)	Objective	Technique	Results
[2]	Combining remote sensing data and ML algorithms	Random forest	The R^2 obtained values between 0.92 and 0.99
[7]	Forecast the optimal time to harvest, knowing the Brix degrees and the purity of the sugar cane	ANN, RF, MLR	Brix best fits RF (R^2 0.68) Purity best fits ANN (R^2 0.74)
[4]	Model for predicting agricultural yields	IOF-LSTM	IOF-LSTM obtained significant lower RMSE compared to the other algorithms. Mitigated overfitting
[26]	Model for soil moisture prediction at a depth of 10 cm, 20 cm and 30 cm of the surface layer	LSTM	The best model performance for 10 cm depth was an R^2 of 0.999, a loss of 0.022
[14]	Soil moisture content prediction	Multihead LSTM	Trained with data at different scales. The model forecasts up to one month, with R^2 of 95.04%
[31]	Crop yield prediction	Random Forest	The model obtained the lowest error compared to others
[23]	Soil temperature prediction model	LSTM	The model showed that it can be successfully applied for temperature measurement
[1]	Prediction of soil nutrients (N,P,K) in crops	LSTM,SVM,RNN	The model obtained 97% accuracy

2.3.2 Studies on Agro-Industrial and Agricultural Variables in Sugarcane Production

In Table 2.1, relevant studies on agriculture and agro-industrial variables in the sugarcane sector are summarized, highlighting the objectives, applied techniques, and key findings of each work. This review provides context for the current research and helps identify common approaches and gaps in the existing literature.

2.3.3 Current Challenges and Contribution of This Research

Several researchers study the prediction of sugarcane yield and productivity using ML techniques and agro-industrial variables; the analysis shows that there are still significant areas of opportunity. In particular, there is a need for models that more effectively integrate field data and industrial processes, especially in contexts where access to systematized information is limited. In addition, most current approaches focus on specific regions or use restricted data sets, limiting their general applicability.

Among the techniques employed, the Random Forest algorithm stands out as one of the most widely used methods in this type of application due to its ability to handle data with high variability and its robustness to overfitting. However, given that agricultural production tends to have a temporal behavior, neural network-based models such as Long Short-Term Memory (LSTM) offer an additional advantage by capturing sequential patterns and dependencies over time.

Therefore, with these challenges and opportunities, the present research proposes a complementary approach that takes advantage of open data and advanced prediction techniques to develop models that are more accessible, accurate, and adaptable to the real conditions of sugarcane producers.

2.4 Data Description and Preprocessing

In Mexico, the CONADESUCA developed the Information System for the Integration of the Sugar Balance (SIIBA) [13]. This portal provides timely information about the field and factory processes carried out weekly by the sugar mills in the country. This information is accessible to both field specialists and internet users. Users can consult historical field records (datasets) dating back to 2010 and factory datasets dating back to the 2005/06 sugarcane processing season, grouped under titles that highlight the activities they cover.

2.4.1 Data Selection

The study aims to apply ML techniques to develop a forecasting model based on data collected during the agricultural cycle by Mexican sugar mills. Table 2.2 shows the field's datasets with their respective attributes. The files contain information on basic field tasks such as planting, fertilization, weather data, harvest information, and sugar production reports. We tested the different attributes in the field records and chose those most relevant to basic field tasks.

As mentioned in the previous section, we focused on the agricultural process of sugarcane. Hence, we downloaded the corresponding files for the 49 mills currently operating, with records available on the website up to 2022, when this study began, but today, just 47 sugar mills operate. Table 2.3 shows the mills located in Mexico, divided by state. Those marked are those mills for which no data were found or have already closed. Thus, we obtained 46 mills to process. Figure 2.3 shows the general structure of the files, each of which includes the name of the mill, the harvest or year it operated, the weeks, and their corresponding N attributes related to the activities

Table 2.2 SIIBA's datasets and their attributes

Filename	Attributes	
Basic operations progress	–Planting –Cultivated area –Irrigation –Stubble	–Fertilization –Weed control –Cutting area
Fertilizer receipt and progress	–Fertilizer tonnage received –Fertilizer application per hectare	–Fertilizer application in tons
Harvested and milled cane balance	–Cane from other mills –Cane used for seed –Non-scheduled milled cane –Total cane harvested –Cane for other purposes –Contracted cane manual cutting –Contracted cane mechanized cutting	–Free cane (not contracted) –Cane sent to other mills –Total gross milled cane
Weather	–Average 8-hour temperature –Areas affected by frost –Rainfall (precipitation)	–Areas affected by flooding –Areas affected by drought
Freshness of milled cane	–Green-cut cane tonnage	–Cane freshness measurement
Field-factory relationship parameters	–Milling hours –Rain-related downtime –Harvesting loss –Net cane KARBE/KABE	–Field downtime S- ucrose % Cane) –Mixed juice purity –Impurity sampling
Pests and diseases	–Scald –Sugarcane borer –Sugarcane rat –Brown rust	–Banded sugarcane aphid –Cutworm –Orange rust –Fusarium
Average daily harvesting resources	–Average tons harvested per cutter/day –Average tons harvested per harvester/day –Number of cutters, trucks, loaders, harvesters, carts, and trailers used in fieldwork	–Harvesting front

performed. However, the reports presented some issues; the headers needed to match the information only covered some working weeks completely, and a significant portion of the records contained zeros. Considering that containing zeros means that there is no record in that month.

Having a substantial number of zero values makes it difficult to generate accurate forecasts. Therefore, after conducting various tests to select and adjust the relevant attributes for the model, the team decided to group them by month.

Table 2.3 Sugar mills by state

State	Sugar mills	
Veracruz	Central El Potrero	Central La Providencia
	Central Motzorongo	Central Progreso
	Central San Miguelito	CIASA
	Constancia	El Higo
	El Modelo	La Gloria
	Mahuixtlan	El Carmen*
	Panuco	San Cristobal
	San Jose de Abajo	San Nicolas
	San Pedro	Tres Valles
Jalisco	Bellavista	Jose Maria Morelos
	Melchor Ocampo	San Francisco Ameca
	Tala	Tamazula
San Luis Potosi	Alianza Popular	Plan de Ayala
	Plan de San Luis	San Miguel del Naranjo
Oaxaca	Adolfo Lopez Mateos	El Refugio
	La Margarita	
Michoacan	Lazaro Cardenas	Perdernales
	Santa Clara	
Tabasco	Presidente Benito Juarez	Santa Rosalia
Morelos	Central Casasano	Emiliano Zapata
Chiapas	Cia. La Fe	Huixtla
Puebla	Atencingo	Calipam*
Nayarit	El Molino*	Puga
Tamaulipas	El Mante	
Sinaloa	Serapio Group (Eldorado)	
Campeche	La Joya	
Colima	Queseria	
Quintana Roo	San Rafael de Pucte	

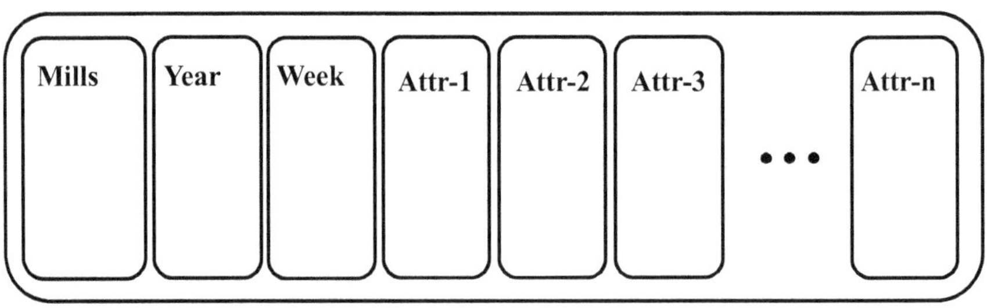

Fig. 2.3 Basic structure of SIIBA reports

2.4.2 Data Preparation and Cleaning

Field activities last for 52 weeks, and tasks range from planting, fertilizer application, pesticides, and irrigation, among others. Therefore, when grouping the information by month, it is necessary to consider the year (whether it is a leap year or not) and the number of Sundays each month. Thus, for each file, regardless of the year, the data can be grouped according to the number of weeks, allowing us to treat each file accordingly. Equation 2.1 proposes the summation of all values of attribute i that belong to the set of weeks W worked by a mill in a given year.

$$a_i(y_j, m_k, mo_t) = \sum_{\forall w_l \in mo_t} a_i(y_j, m_k, w_l) \tag{2.1}$$

Where:

$A = \{a_1, a_2, \ldots, a_r\}$ *the set of attributes.*
$Y = \{y_1, y_2, \ldots, y_n\}$ *the years*
$M = \{m_1, m_2, \ldots, m_p\}$ *the sugar mills*
$W = \{w_1, w_2, \ldots, w_q\}$ *the weeks worked in the year* y_j
$MO = \{mo_1, mo_2, \ldots, mo_t\}$ *a set that contains a set of weeks* w_l, *for a particular month* mo_t

The files Basic Work Progress and Pests and Diseases contain between 20 and 80 columns, with each attribute repeated more than twice due to the division of attributes based on the cane's growth phase (plant cane, ratoon cane, second ratoon). Therefore, in Basic Work Progress, we grouped the information by attribute.

In the Pests and Diseases file, we grouped the information as mentioned [13], identifying diseases that producers can address as soon as possible to avoid losses and those that are uncontrollable; we did the same with pests (Table 2.4).

In the file Harvested and Milled Cane Balance, since the focus is on forecasting the tons of cane entering the mill, we added the column Total Harvested, which sums up the total sugarcane obtained during the agricultural cycle. For the summation, we excluded attributes defining activities unrelated to the mill. Lastly, we added the column Total Milled, representing the mill's tons processed. These two columns

Table 2.4 Division pests and diseases

Pests	Diseases
Easier to control:	Easier to control:
– Scald	–Orange rust
–Banded sugarcane aphid	–Brown rust
–Sugarcane borer	
–Cutworm	
More complex to control:	Easier to control:
–Sugarcane rat	–Fusarium
–Others	–Others

Table 2.5 Division harvested and milled

Total harvested	Total milled
Contracted cane manual cutting	Non-scheduled milled cane
Contracted cane mechanized cutting	Total gross milled cane
Cane from other mills	
Free cane (Not contracted)	
Cane used for seed	
Cane sent to other mills	
Cane intended for other purposes	
Total cane harvested	

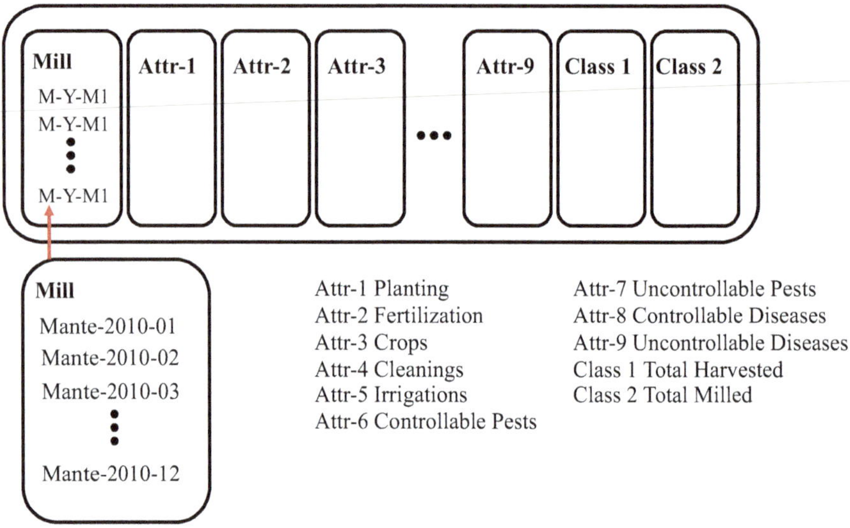

Fig. 2.4 Dataset structure

correspond to the class of the dataset. Table 2.5 shows the variables. In this way, the dataset with the attributes shown in Fig. 2.4.

Moreover, we conducted tests by adding weather information to observe how the data behaves with and without it. However, due to the limited information in the SIIBA's Weather file, we queried the Weather API portal, a website for acquiring meteorological data such as real-time information, forecasts, historical data, and other options. Since the dataset used in this study had global information for each mill, the exact locations of the sugarcane crops were unknown. For this reason, we decided to use the coordinates of the mills to obtain historical weather data relevant to their locations. Figure 2.5 shows the dataset's structure when using weather variables.

Then, considering that fieldwork spans 12 months (from November to October) and cane processing at the mill can vary but typically lasts around 7 months (from November to May), we adjusted the datasets by eliminating the first ten months

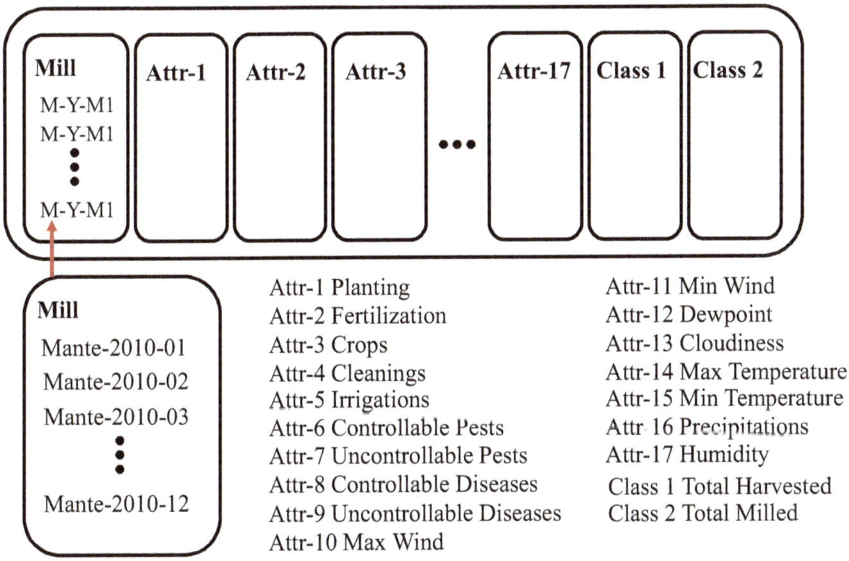

Fig. 2.5 Dataset structure with weather information

of 2010 and the last seven months of 2022. We then grouped them into three-dimensional lists with the structure (Batches x Length x Features), meaning the number of samples by the number of input (or output) data by the number of attributes. The input data will cover 12 months, while the output data will cover 7 months, but we sum them to obtain only one value for the total harvested and milled. Consequently, the dataset has less information; thus, we divided the dataset into 70% for training and 15% for validation and 15% testing, respectively. Figure 2.6 shows the structure for each division.

Before applying any technique, normalized data between values of −1 and 1 in the dataset to stabilize the model's training process and scale the attributes, excluding reference columns, which are the harvested and milled sugarcane quantities.

Table 2.6 presents partial information on the dataset before normalization; in this case, it shows the twelve steps of work and the nine agricultural activities (features) of the Jose Maria Morelos sugar mill. Table 2.7 presents the same data after applying min-max normalization. Normalization was performed feature-wise across the training dataset to ensure consistent input ranges and improve techniques' convergence behavior.

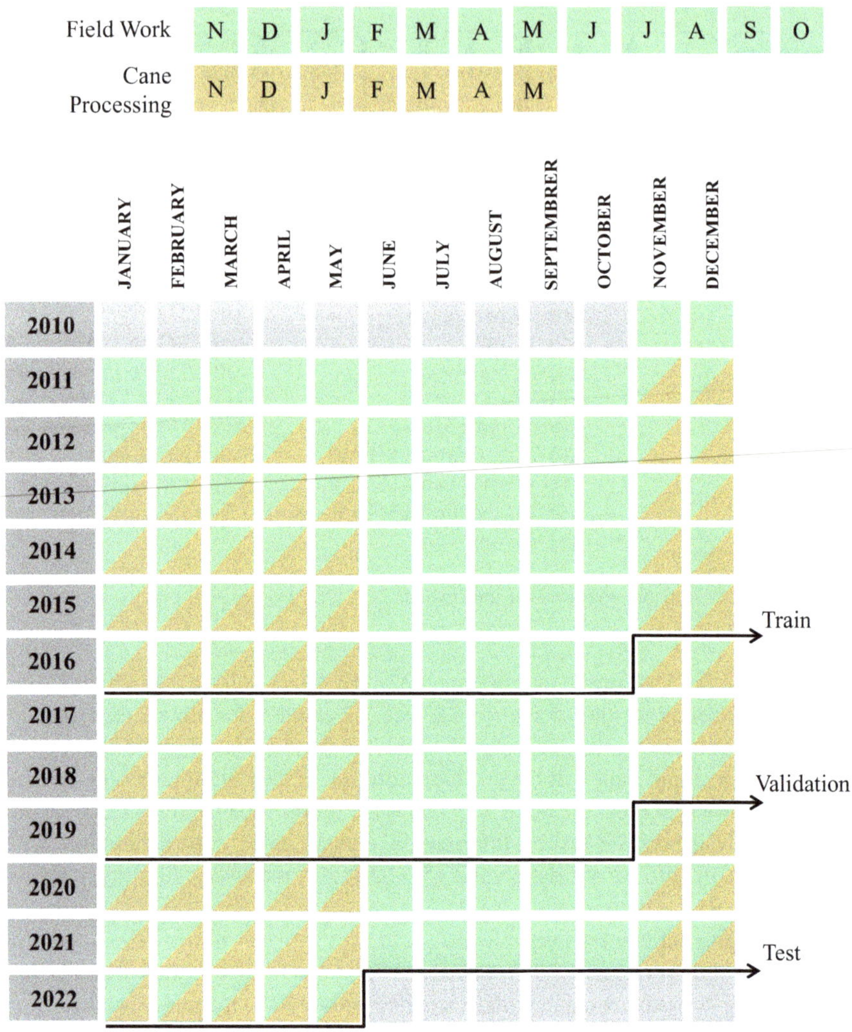

Fig. 2.6 Dataset division

2.5 Model Training

Subsequently, we divided the dataset into 70% for training and 15% for validation and testing. Normalization applies the same parameters across all subsets to ensure consistency. Thus, we proceed with LSTM and Random Forest.

Table 2.6 Data instance of Jose Maria Morelos sugar mill

Time	F1	F2	F3	F4	F5	F6	F7	F8	F9
T0	0.0	0.0	0.0	807.7	0.0	41906.5	36128.8	0.0	0.0
T1	0.0	0.0	0.0	447.4	0.0	47569.75	36962.8	0.0	0.0
T2	0.0	628.60	0.3	691.05	20.9	42336.07	32882.48	0.0	0.0
T3	41.95	20.45	0.8	106.9	0.0	43144.92	32882.48	0.0	0.0
T4	177.75	46.8	0.0	0.0	0.0	21579.16	16441.24	0.0	0.0
T5	301.35	23.4	12.7	7.6	0.0	46935.53	35842.72	0.0	0.0
T6	156.45	0.0	544.75	132.2	136.55	60243.25	46203.5	0.0	0.0
T7	38.8	0.0	741	287	142.05	51405.5	36962.8	0.0	0.0
T8	7.9	0.0	753.3	558.65	434.15	64357.15	36962.8	0.0	0.0
T9	15.7	0.0	1035.4	1048.05	992.5	98443.95	46203.5	0.0	0.0
T10	0.0	0.0	580.15	484.75	992.0	88957.65	44803.4	0.0	0.0
T11	0.0	1670.05	520.7	1871.8	991.5	33103.32	32882.48	0.0	0.0

Table 2.7 Normalization of data instance of Jose Maria Morelos sugar mill

Time	F1	F2	F3	F4	F5	F6	F7	F8	F9
T0	−1.0	−1.0	−1.0	−0.57	−1.0	0.87	0.75	−1.0	−1.0
T1	−1.0	−1.0	−1.0	−0.44	−1.0	1.0	1.0	−1.0	−1.0
T2	−1.0	0.93	−0.99	0.97	−0.59	0.75	0.77	−1.0	−1.0
T3	−0.87	−0.87	−0.40	1.0	−1.0	0.79	0.77	−1.0	−1.0
T4	−0.35	−0.65	−1.0	−1.0	−1.0	−0.28	−0.28	−1.0	−1.0
T5	1.0	−0.60	−0.96	−0.96	−1.0	1.0	1.0	−1.0	−1.0
T6	0.50	−1.0	−0.22	−0.72	−0.81	1.0	1.0	−1.0	−1.0
T7	0.69	−1.0	0.27	−0.08	−0.54	1.0	1.0	−1.0	−1.0
T8	−0.82	−1.0	0.39	1.0	0.04	1.0	1.0	−1.0	−1.0
T9	1.0	−1.0	1.0	1.0	1.0	1.0	1.0	−1.0	−1.0
T10	−1.0	−1.0	1.0	1.0	1.0	1.0	1.0	−1.0	−1.0
T11	−1.0	−0.41	−0.51	−0.21	−0.89	0.37	0.60	−1.0	−1.0

2.5.1 Long Short-Term Memory (LSTM)

Proposed in 1997 by Sepp Hochreiter and Jürgen Schmidhuber, LSTM neural network structure is a derivative of a recurrent neural network (RNN) that researchers commonly use for sequence-processing tasks, such as speech recognition, machine translation, and time-series analysis [22]. It aims to overcome the limitations of standard RNNs by learning long-term dependencies in sequences [1, 19]. The architecture of LSTMs features a *gate* mechanism composed of three stages that regulate the flow of information. First, it evaluates whether the information from the previous state is significant and should be retained or irrelevant, and the system discards it. Next, it incorporates new information from the current input and transmits the updated information from the current state to the next time step. As shown in Fig. 2.7.

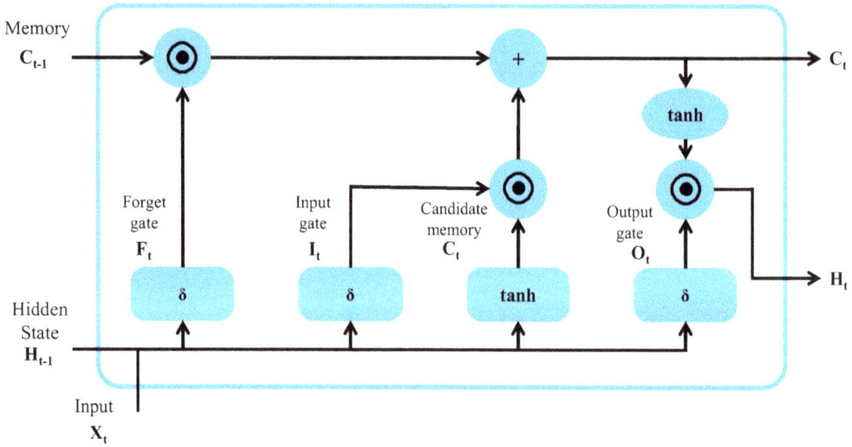

Fig. 2.7 LSTM's basic architecture

The model proposed in the investigation uses a Long Short-Term Memory (LSTM) recurrent neural network designed to capture long-term temporal dependencies in multivariate time series. The overall architecture includes:

- Four layers with N units correspond to the hidden state and the memory cell.
- Different activation functions were strategically assigned across the architecture to explore their individual impact:
  - ReLU (Rectified Linear Unit) was applied in the initial layers due to its computational efficiency and ability to mitigate the vanishing gradient problem commonly found in deep networks.
  - Tanh (hyperbolic tangent) was incorporated in other layers to better model nonlinear, zero-centered relationships, particularly beneficial when working with normalized data in the range $[-1, 1]$.
  - In certain experimental configurations, Softmax was used in the final layer—not typical in regression tasks—but was tested to explore its potential for stabilizing outputs or constraining them within a probabilistic-like distribution. Its inclusion highlights an avenue for future research into adaptive activation strategies across deep recurrent layers.
- Dropout layers, with rates ranging between 0.3 and 0.5, were added after each LSTM layer to reduce overfitting by randomly deactivating units during training.
- L2 regularization (ridge penalty) was applied to the LSTM weights to penalize overly large parameters, thus encouraging simpler and more generalizable models.
- The output layer is Dense employs a linear activation function, aligning with the continuous nature of the target variable in regression-based forecasting.

Table 2.8 Hyperparameter ranges explored and selected values for the LSTM model

Hyperparameter	Explored range	Selected value
Number of LSTM units	{50, 100, ..., 250}	80
Learning rate (Adam)	{0.0001, 0.001, 0.01, 0.1}	0.001
Dropout rate	{0.3, 0.4, 0.5}	0.5
L2 regularization factor	{0.0001, 0.001, 0.01, 0.1}	0.001
Batch size	{100, 150, 200, 250, 300}	100
Epochs	{500, 700, 1000}	1000
Early stopping patience	15% of total epochs	150

Algorithm 1 LSTM

for each unit in N_UNITS **do**
    for each rate in LR **do**
        LSTM ← **CreateModel**(N_UNITS, LR, Dropout, L2)
        $LSTM_{trained}$ ← **TrainModel**(LSTM, Adam, Train, Val, ES = true)
    end for
end for
eval ← **getPrediction**($LSTM_{trained}$, Test)

Using the Adam optimizer, we trained the model, chosen for its ability to individually adapt learning rates for each parameter, which accelerates convergence on non-convex problems such as deep networks. Besides, learning rates were explored within the range [0.0001, 0.1].

Early stopping was employed, with patience equal to 15% of the total number of epochs (patience = 0.15 * EPOCHS), monitoring validation loss. This technique avoids training beyond the improvement point, reducing the risk of overfitting.

Further, the hyperparameters were selected by combining grid search and cross-validation, considering the RMSE metric as a performance criterion. Table 2.8 summarizes the explored ranges and selected values. We use the trained model and the test set for prediction, as shown in Algorithm 1.

2.5.2 *Random Forest (RF)*

A Random Forest was implemented to complement the LSTM approach to provide a non-sequential baseline for multivariate time series forecasting.

Based on ensemble learning, Random Forest (RF) is a ML algorithm for classification and regression problems. Introduced by Leo Breiman in 2001 [5]. The RF goal is to combine multiple decision trees to obtain the prediction of a value from each one and then average each prediction, generating more robust models than individual decision trees. Each decision tree in the forest trains indcpendently using a random sample of the training data and a random subset of features or parameters. A decision tree is constructed using the previously created sample dataset, with a maximum depth and several attributes to consider at each tree split [3, 9].

Table 2.9 Hyperparameter ranges explored and selected values for the Random Forest model

Hyperparameter	Explored range	Selected value
Number of estimators	{10, 50, 100, 110, 150}	110
Min. samples to split	{5, 10, 15, 20, 40}	40
Maximum tree depth	{5, 10, 15, 25}	15
Max. features per split	{0.1, 0.4, 0.9, 10}	10
Min. samples per leaf	{5, 15, 20, 25}	20
Bootstrap sampling	{True, False}	True

Algorithm 2 Random forest

```
1: for each n_trees do
2:     RF ← CreateModel(n_trees, sampleSplit, deepthTree, maxFeatures, sampleLeaf)
3:     RF_trained ← TrainModel(RF, Train, Val)
4: end for
5: EvalRF ← getPrediction(RF_trained, Test)
```

Using RandomForestRegressor class from the scikit-learn library to train the model and hyperparameters selected through a combination of grid search and empirical evaluation on a validation dataset. Table 2.8 summarizes the explored ranges and selected values (Table 2.9).

The following key hyperparameters were tuned:

- **n_estimators**: The number of trees in the ensemble. A larger number improves stability but increases training costs.
- **min_samples_split**: Minimum number of samples required to split a node. Larger values help prevent overfitting by enforcing broader generalization.
- **max_depth**: Limits the depth of each tree, controlling model complexity.
- **max_features**: The fraction or number of features considered when splitting a node. Smaller values increase model variance but reduce correlation among trees.
- **min_samples_leaf**: Minimum samples required in a leaf. Higher values promote smoother predictions and prevent learning from noise.
- **bootstrap**: Whether each tree is trained on a bootstrap sample. Setting this to True is typical and promotes tree diversity.

Algorithm 2 shows the model construction using hyperparameters such as the number of trees (n trees), minimum samples to split the nodes (sampleSplit), maximum tree depth (depthTree), number of features (maxFeatures), and samples required to be in a node (sampleLeaf). These parameters control the model's complexity and ability to fit the data.

2.6 Experimental Results

The experimentation involved using the Python 3.11.9 programming language, employing the PyCharm IDE for implementation, on a laptop with an Apple M3

Pro processor running macOS Sonoma 14.5. The laptop features an 11-core CPU, a 14-core GPU, and a 16-core Neural Engine chip. It also has 18 GB of RAM and a 512 GB SSD.

This study proposes a model to generate forecasts based on the activities carried out in the agricultural field recorded by the active mills in Mexico. Therefore, we worked with the LSTM and Random Forest algorithms, and to observe the behavior of the results, we used an additional technique, the dense neural network. We used performance metrics to evaluate the model's performance: root mean square error (RMSE), mean absolute percentage error (MAPE), and coefficient of determination (R^2).

2.6.1 Experiment Results of LSTM

Considering that a dataset is complete (i.e., fit to apply the technique) when it contains 139 samples based on the previous splits shown in Fig. 2.6. And that a dataset represents a sugar mill with 13 sample years (2010–2022), covering 12 months per year, excluding certain months. We found that sugar mills with less than 50 samples completely at zeros achieved better results during testing. As a result, we discarded the data sets that did not meet this condition, leaving us with 40 sugar mills. Figures 2.8 and 2.9 present the results when we include weather data and exclude them. Tables 2.10 and 2.11 show the RMSE, MAPE, and (R^2) results for both cases in the LSTM tests.

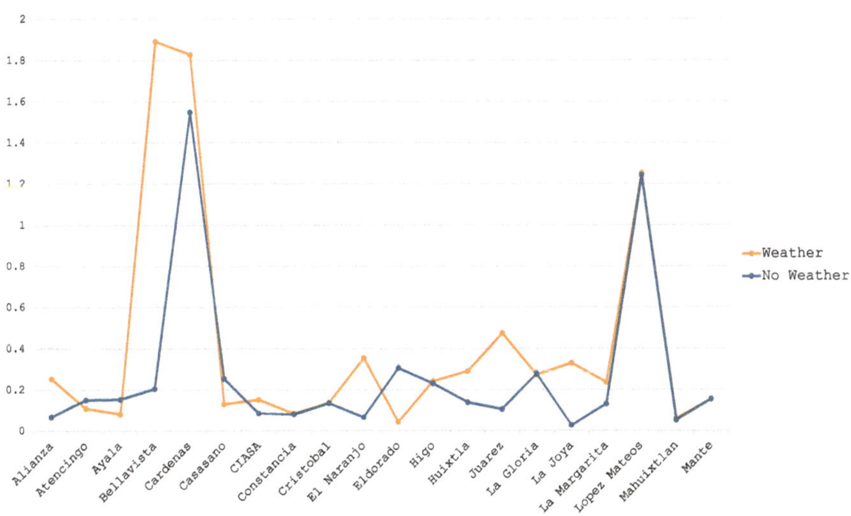

Fig. 2.8 MAPE results in LSTM

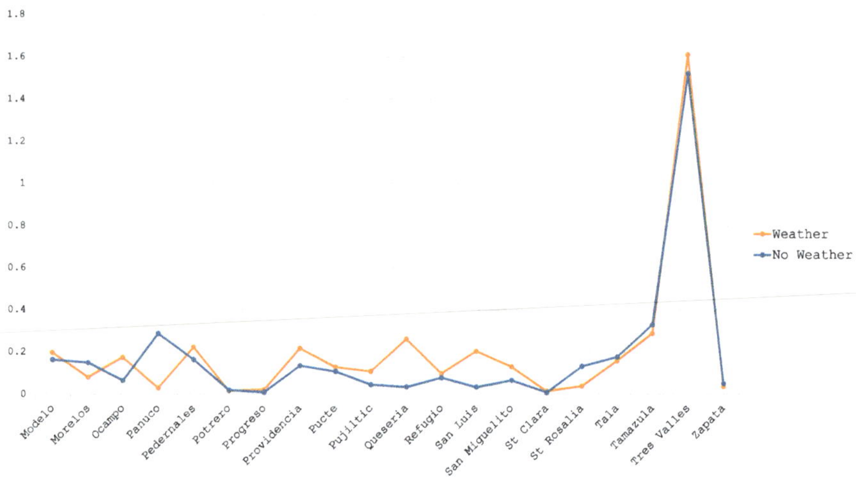

Fig. 2.9 MAPE results in LSTM

Figures 2.10 and 2.11 show the harvested values (Harvested Real) and their respective predictions, whether weather information or not. Twenty-seven sugar mills performed below or close to the actual value. Mahuixtlan and Santa Clara showed balanced results, while Eldorado and Santa Rosalia performed better with weather information, whereas P. de Ayala and Atencingo achieved better results without it. Then, Figs. 2.12 and 2.13 show the sugarcane milled values (Milled Real) and their respective predictions when using or not using weather information. In this case, Mahuixtlan, Santa Clara, Progreso, and Potrero had the closest values to the actual data with both datasets. Meanwhile, El Mante, E. Zapata, El Refugio, and Panuco performed better using climate information. The remaining mills either had very low values or exceeded the expected value.

2.6.2 Experiment Results of Random Forest and Dense Neural Network

Tables 2.12 and 2.13 present the results obtained by applying Random Forest. Figures 2.14 and 2.15 shows that, on average, whether using weather data or not, the error has only a few decimal differences. Tables 2.11 and 2.12 show the differences between the actual and predicted values.

In the case of the Dense network, Figs. 2.16 and 2.17 shows the minimum difference when using weather information or not, while Tables 2.14 and 2.15 present the results broken down by the mill.

The results show that the LSTM generates less loss than the other techniques. This neural network achieved better results when it did not use weather data, with

Table 2.10 Results of the dataset with weather data—LSTM

No.	Mills	RMSE	MAPE	R^2
1	Alianza	0.360	25.4	0.343
2	Atencingo	0.921	10.9	0.883
3	Ayala	0.101	8.2	0.937
4	Bellavista	1.157	189.3	−45.09
5	Cardenas	0.195	182.9	−22.0
6	Casasano	1.348	13.1	0.750
7	CIASA	1.098	15.2	0.767
8	Constancia	0.178	8.5	0.927
9	Cristobal	0.422	13.8	0.806
10	El Naranjo	0.456	35.5	−0.742
11	Eldorado	0.153	4.4	0.981
12	Higo	0.575	24.3	0.526
13	Huixtla	0.686	29.1	0.175
14	Juarez	0.806	47.5	−2.380
15	La Gloria	0.636	27.2	0.258
16	La Joya	0.833	33.0	0.008
17	La Margarita	0.506	23.7	0.607
18	Lopez Mateos	0.738	125.2	−14.04
19	Mahuixtlan	0.805	6.0	0.962
20	Mante	0.526	15.6	0.533
21	Modelo	0.448	19.5	0.624
22	Morelos	0.157	7.8	0.959
23	Ocampo	0.442	17.2	0.715
24	Panuco	0.241	2.8	0.994
25	Pedernales	0.153	22.1	0.274
26	Potrero	0.353	1.6	0.997
27	Progreso	0.354	2.1	0.993
28	Providencia	0.678	21.6	0.194
29	Pucte	0.201	12.7	0.837
30	Pujiltic	0.571	10.7	0.907
31	Queseria	0.330	26.0	0.325
32	Refugio	0.304	9.7	0.923
33	San Luis	0.419	20.3	0.595
34	San Miguelito	0.774	13.0	0.812
35	St Clara	0.264	1.5	0.996
36	St Rosalia	0.367	3.9	0.983
37	Tala	0.256	15.8	0.697
38	Tamazula	0.517	28.8	0.338
39	Tres Valles	1.309	161.0	−24.83
40	Zapata	0.885	3.6	0.984

Table 2.11 Results of the dataset no weather data—LSTM

No.	Mills	RMSE	MAPE	R^2
1	Alianza	0.496	6.7	0.953
2	Atencingo	0.818	15.1	0.813
3	Ayala	0.380	15.4	0.339
4	Bellavista	0.324	20.5	0.409
5	Cardenas	0.137	155.0	−14.66
6	Casasano	1.227	25.4	0.230
7	CIASA	1.362	8.5	0.964
8	Constancia	0.177	8.0	0.936
9	Cristobal	0.533	13.5	0.737
10	El Naranjo	0.447	6.6	0.940
11	Eldorado	0.562	30.5	0.072
12	Higo	0.692	23.0	0.473
13	Huixtla	0.550	13.9	0.815
14	Juarez	0.679	10.6	0.830
15	La Gloria	0.495	27.7	0.228
16	La Joya	0.995	2.8	0.994
17	La Margarita	0.766	13.1	0.823
18	Lopez Mateos	0.731	124.1	−13.78
19	Mahuixtlan	0.956	5.3	0.920
20	Mante	0.539	15.4	0.552
21	Modelo	0.438	16.2	0.747
22	Morelos	0.168	14.8	0.849
23	Ocampo	0.569	6.4	0.930
24	Panuco	0.403	28.6	0.119
25	Pedernales	0.473	16.3	0.464
26	Potrero	0.361	1.9	0.996
27	Progreso	0.375	0.9	0.997
28	Providencia	0.642	13.5	0.788
29	Pucte	0.165	10.7	0.886
30	Pujiltic	0.606	4.5	0.805
31	Queseria	0.343	3.4	0.970
32	Refugio	0.093	7.8	0.974
33	San Luis	0.461	3.4	0.982
34	San Miguelito	0.420	6.6	0.963
35	St Clara	0.204	0.9	0.911
36	St Rosalia	0.573	13.3	0.603
37	Tala	0.284	17.7	0.624
38	Tamazula	0.586	32.9	0.063
39	Tres Valles	1.184	151.9	−20.39
40	Zapata	0.529	5.1	0.972

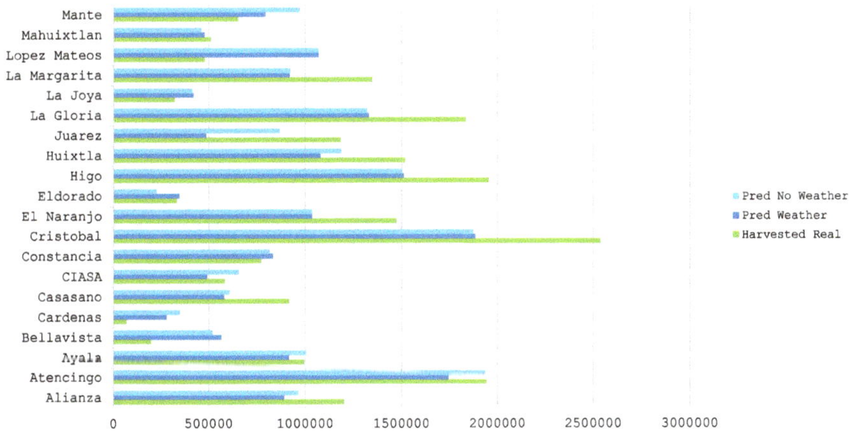

Fig. 2.10 Harvested results—LSTM

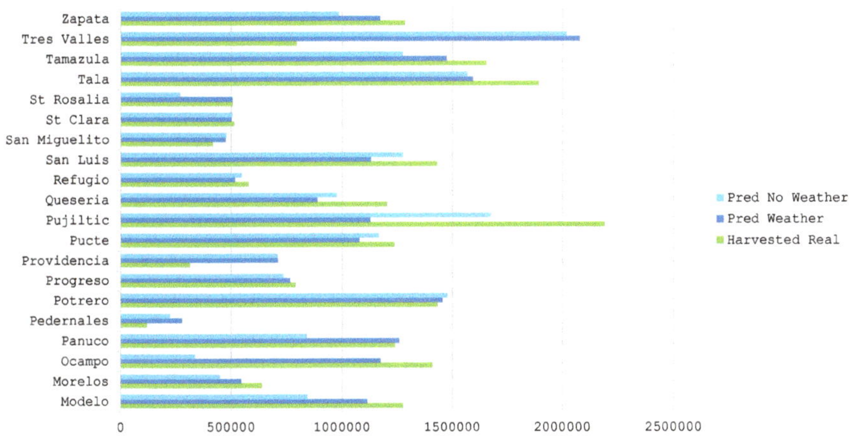

Fig. 2.11 Harvested results—LSTM

an average of 0.224 compared to 0.310. Meanwhile, the RF and Dense networks achieved better results when they used the data, but their values are very high, averaging 48.1 and 16.3, respectively. Of the 20 data sets tested, 12 and 14 values of the total harvested and milled production were close, with a less than 35% loss. Figures 2.18 and 2.19 show the difference between the MAPE and the techniques used.

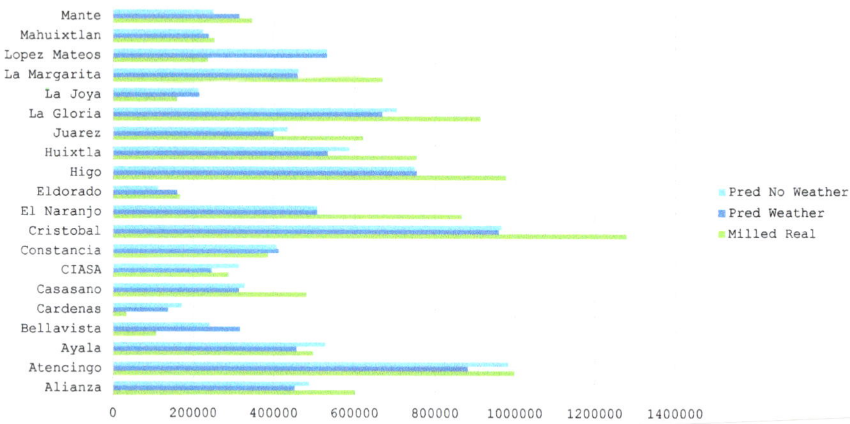

Fig. 2.12 Milled results—LSTM

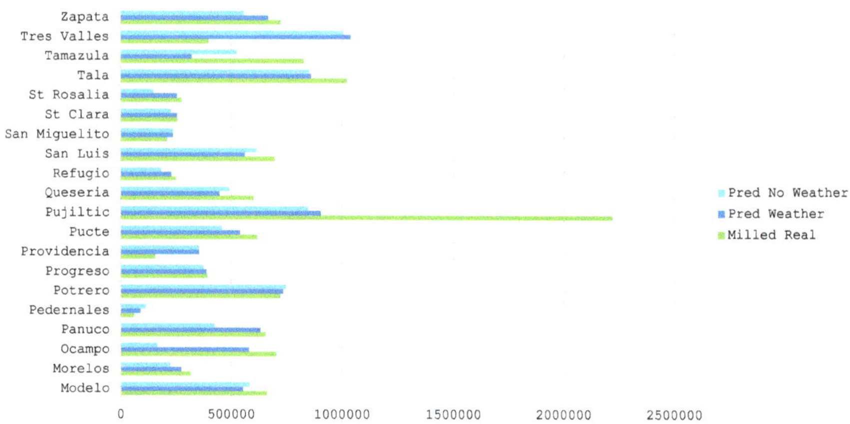

Fig. 2.13 Milled results—LSTM

2.6.3 Explaining RMSE, MAPE, and R^2 Variability Between Mills

lls is attributed to multiple data-driven and contextual factors. Notably, the RMSE, MAPE, and R^2 can behave differently depending on each mill's input data's characteristics and quality.

Firstly, mills with more irregular or volatile production data—caused by inconsistent environmental conditions, operational disruptions, or changes in agricultural practices—tend to yield higher RMSE and MAPE values. In contrast, mills with more stable and seasonally patterned data allow the LSTM to capture trends more effectively, resulting in lower forecasting errors. In this context, high RMSE values

Table 2.12 Results of the dataset with weather data—RF

No.	Mills	RMSE	MAPE	R^2
1	Alianza P.	0.393	28.307	0.202
2	Atencingo	0.903	11.231	0.876
3	Bellavista	1.2	198.21	−46.4
4	CIASA	1.11	14.026	0.801
5	Casasano	1.445	35.618	−0.429
6	Constancia	0.189	10.39	0.892
7	Eldorado	1.19	62.66	−2.927
8	E. Zapata	0.549	22.687	0.285
9	Gloria	0.489	27.424	0.247
10	Higo	0.689	22.396	0.502
11	Huixtla	0.674	28.476	0.201
12	Juarez	0.837	64.640	−3.632
13	La Joya	0.833	33.055	−0.021
14	Lopez Mateos	0.737	125.429	−14.2
15	Mahuixtlan	0.808	5.68	0.966
16	Mante	0.534	15.439	0.507
17	Margarita	0.766	30.659	0.08
18	Modelo	0.434	16.37	0.716
19	Naranjo	0.452	36.145	−0.86
20	Morelos	0.79	45.77	−1.013
21	Ocampo	0.904	55.772	−2.088
22	Panuco	0.191	2.768	0.994
23	Pedernales	1.179	138.353	−18.14
24	Potrero	0.348	1.403	0.999
25	Progreso	0.351	2.243	0.993
26	Providencia	0.762	128.277	−15.93
27	P. Ayala	0.101	8.23	0.936
28	P. San Luis	0.419	20.30	0.597
29	Pucte	0.222	13.886	0.807
30	Pujiltic	0.672	43.328	−1.012
31	Queseria	0.302	23.821	0.405
32	Refugio	0.305	9.456	0.937
33	S. Clara	0.269	1.133	0.998
34	L. Cardenas	0.736	306.113	−93.45
35	S.Cristobal	0.586	24.073	0.406
36	S. Rosalia	0.583	45.993	−1.616
37	S. Miguelito	0.706	9.934	0.903
38	Tala	0.283	17.578	0.621
39	Tamazula	0.877	49.75	−1.283
40	Tres Valles	1.37	167.42	−26.98

Table 2.13 Results of the dataset no weather data—RF

No.	Mills	RMSE	MAPE	R^2
1	Alianza P.	0.318	22.195	0.51
2	Atencingo	0.893	11.845	0.862
3	Bellavista	1.193	196.788	−45.5
4	CIASA	1.105	14.523	0.787
5	Casasano	1.483	36.663	−0.512
6	Constancia	0.177	8.076	0.935
7	Eldorado	1.128	59.542	−2.545
8	E. Zapata	0.536	21.614	0.351
9	Gloria	0.491	27.537	0.241
10	Higo	0.687	22.078	0.517
11	Huixtla	0.687	29.116	0.165
12	Juarez	0.864	66.329	−3.883
12	La Joya	0.841	31.140	0.095
14	Lopez Mateos	0.726	123.118	−13.7
15	Mahuixtlan	0.818	5.147	0.972
16	Mante	0.532	15.847	0.478
17	Margarita	0.785	31.897	0.004
18	Modelo	0.434	16.462	0.713
19	Naranjo	0.451	35.978	−0.836
20	Morelos	0.82	47.15	−1.139
21	Ocampo	0.912	56.181	−2.133
22	Panuco	0.181	2.711	0.993
23	Pedernales	1.217	144.75	−19.9
24	Potrero	0.392	3.337	0.991
25	Progreso	0.355	2.06	0.994
26	Providencia	0.729	123.161	−14.6
27	P. Ayala	0.105	8.963	0.925
28	P. San Luis	0.403	19.351	0.633
29	Pucte	0.184	11.808	0.86
30	Pujiltic	0.669	43.15	−0.994
31	Queseria	0.388	30.037	0.066
32	Refugio	0.304	9.025	0.943
33	S. Clara	0.278	0.398	1
34	L. Cardenas	0.733	304.917	−92.85
35	S.Cristobal	0.639	25.938	0.311
36	S. Rosalia	0.585	46.074	−1.623
37	S. Miguelito	0.684	8.939	0.922
38	Tala	0.296	18.348	0.586
39	Tamazula	0.879	49.82	−1.28
40	Tres Valles	1.367	167.071	−26.8

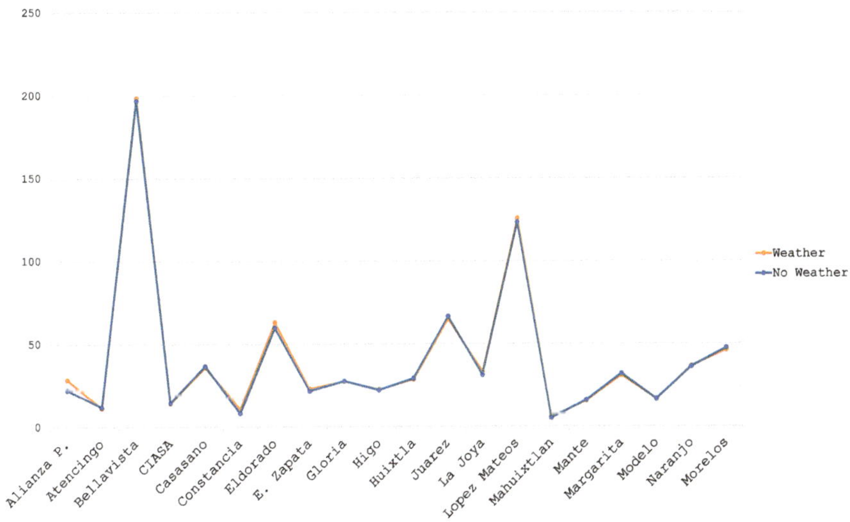

Fig. 2.14 MAPE results in random forest

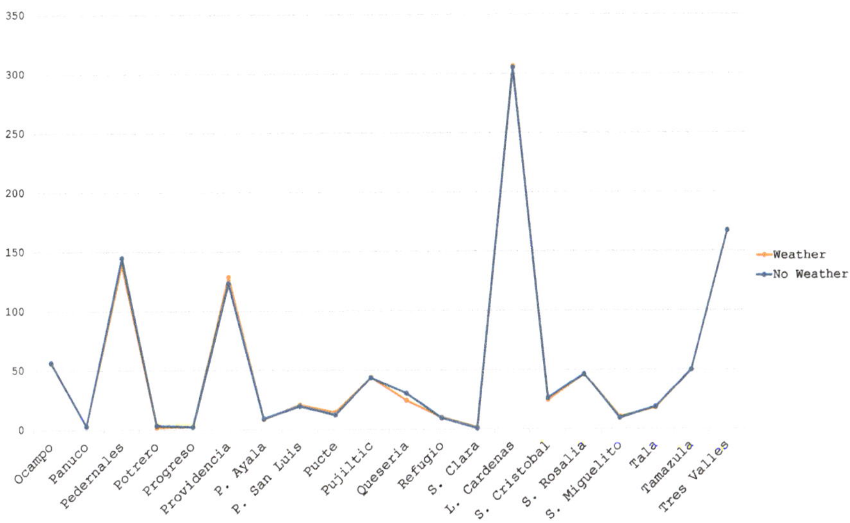

Fig. 2.15 MAPE results in random forest

may indicate the presence of extreme values or greater variance in production quantities, making the model's task more complex.

Secondly, the interpretability of R^2 requires special attention. In some cases, a negative R^2 appears, which suggests that the model performs worse than a simple mean-based predictor. This problem typically occurs when the model fails to generalize due to limited training data, poor-quality inputs, or when the target variables (e.g., cane harvested and milled) have weak correlations with the input features. Such cases may also reflect overfitting or underfitting behavior, especially if the LSTM architecture or training parameters are not well-tuned to the characteristics of a specific mill.

Additionally, it is common to observe scenarios where the RMSE appears relatively low, yet the MAPE remains high. Generally, this happens when the predicted values are numerically close to the actual values in absolute terms (hence low RMSE). However, some true values are small in magnitude, making the percentage of errors disproportionately large. The study highlights the sensitivity of MAPE to low denominators and justifies using multiple evaluation metrics to provide a comprehensive assessment of model accuracy.

Furthermore, the observed discrepancies in model performance are not limited to a single approach. In several cases, a given sugar mill achieved relatively accurate forecasts using the LSTM model, while the same data yielded weaker results under a Random Forest (RF) regression approach. The study highlights that not all modeling strategies generalize equally well across different data distributions and temporal patterns. For instance, mills with pronounced seasonality and sequential dependencies benefited from the temporal memory structure of LSTM, which is designed to capture time-based trends. In contrast, RF, which lacks temporal awareness, may struggle to learn such patterns effectively, particularly in the presence of long-range dependencies or lagged effects.

Some intermediate-performing mills showed favorable results using a Dense Neural Network model. These cases suggest that a feedforward architecture may provide a better balance between model complexity and generalization for specific data configurations—perhaps where relationships are nonlinear but not strongly sequential. The varying outcomes across mills and models reinforce the necessity of comparative evaluation rather than relying on a single predictive paradigm.

This modeling contrast provides important insights into the influence of data structure and model architecture on forecasting performance. It also underlines the potential of ensemble or hybrid strategies that combine the strengths of different algorithms to improve robustness across diverse operational contexts.

2.6.4 Statistical Analysis of Performance

Due to the nature of the data and the limited sample size, we conducted non-parametric statistical tests to evaluate the performance differences among the LSTM neural network, Random Forest, and a dense neural network. Specifically, we applied Friedman and Wilcoxon tests to compare the techniques and determine which performed best statistically. These methods were implemented on time-series datasets to forecast the amount of sugarcane yield after performing agricultural various activities in the field.

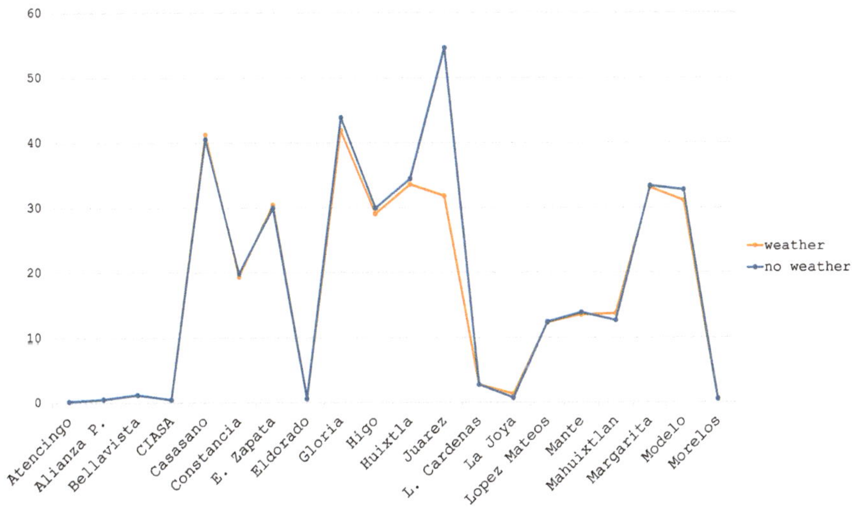

Fig. 2.16 MAPE results in ANN dense

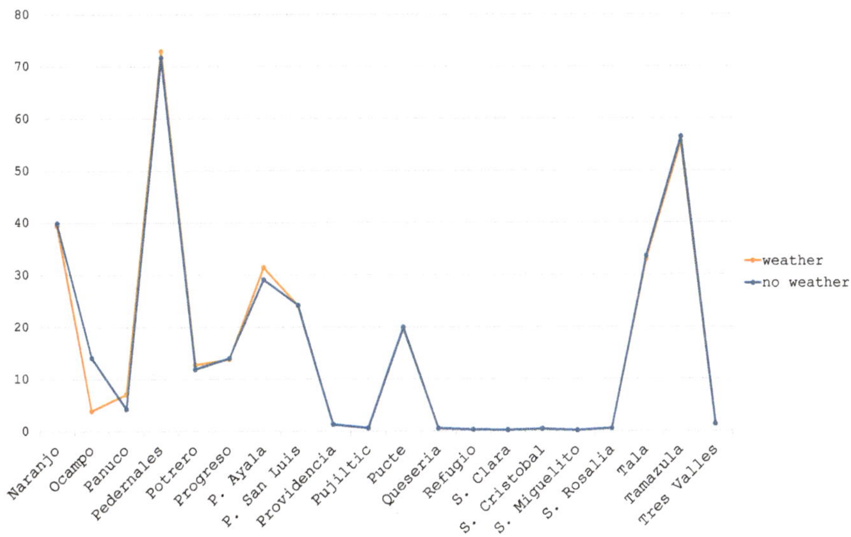

Fig. 2.17 MAPE results in ANN dense

2.6.4.1 Friedman Test

We conducted the Friedman test to determine whether there were significant differences in the performance of the three techniques across the proposed scenarios: when the datasets included weather information (*W*) and when they did not (*NW*). This analysis allowed us to assess the impact of weather data on the accuracy of

LSTM, Random Forest, and Dense Neural Networks, providing a statistical basis for selecting the most effective approach for time-series forecasting in sugarcane yield prediction.

The Friedman test results showed a X^2 value of 132.56, with 5 degrees of freedom and a p-value lower than 0.001, indicating statistically significant differences in the algorithms' performance across the different scenarios.

Table 2.16 presents the ranking mean obtained for each algorithm and scenario.

Given the test results, the analysis confirms significant differences. The rankings show that LSTM without climate data (1.35) and LSTM with climate data (1.85) achieve the best performance, followed by the dense neural network. Random Forest ranks the lowest in both scenarios, reinforcing that it does not provide the best solution for this problem.

These findings suggest that LSTM is the most effective algorithm for forecasting sugarcane yield, performing better when excluding climate data. These results suggest that climate variables may introduce noise or inconsistencies that negatively impact the model's predictions. Conversely, Random Forest performs the worst in both cases, indicating that it struggles to capture the complex patterns necessary for accurate time-series forecasting.

Furthermore, climate information plays a crucial role in model performance, as the ranking fluctuates depending on its inclusion or exclusion. These highlight the importance of carefully evaluating the relevance and impact of external variables in predictive modeling. Since the Friedman test confirmed significant differences among the algorithms, a post hoc test is necessary to pinpoint which specific pairs exhibit meaningful variations in performance. To accomplish this, researchers can conduct the Wilcoxon test with paired comparisons, allowing for a deeper statistical analysis to determine whether the observed differences remain consistent across multiple scenarios.

2.6.4.2 Wilcoxon Test

The Wilcoxon test is a non-parametric method for comparing two related samples. The Friedman test primarily identifies overall differences across a broader set of models. If significant differences are detected, the Wilcoxon tests compare specific model pairs with LSTM models with and without weather data. Generally, the Friedman test determines whether there is a global difference in performance among configurations, but it does not specify which model performs better or worse. For that, Wilcoxon tests make specific comparisons. Researchers use it when data do not follow a normal distribution, making it a robust alternative to the paired t-test. This test relies on three key ranking values: Negative ranks apply to cases where the first condition is lower than the second. Positive ranks apply to cases where the first condition is higher than the second. Ties are cases where both conditions have the same value.

Additionally, the Wilcoxon Z-statistic is derived from these ranks and helps compute the p-value (asymp. sig.), determining whether the difference between

Table 2.14 Results of the dataset with weather data—Dense

No.	Mills	RMSE	MAPE	R^2
1	Atencingo	1.024	0.134	0.819
2	Alianza P.	0.879	0.481	−1.33
3	Bellavista	0.978	1.17	−16.5
4	CIASA	1.067	0.422	−0.752
5	Casasano	1.817	41.2	−0.962
6	Constancia	0.484	19.3	0.628
7	E. Zapata	0.836	30.4	−0.265
8	Eldorado	1.17	0.539	−1.91
9	Gloria	0.932	41.8	−0.738
10	Higo	0.907	29	0.165
11	Huixtla	0.939	33.5	−0.099
12	Juarez	54.079	31.748	−12425
13	L. Cardenas	0.788	2.65	−69.9
14	La Joya	0.495	1.214	−8.095
15	Lopez Mateos	12.762	12.240	−1018
16	Mante	0.679	13.4	0.826
17	Mahuixtlan	0.846	13.6	0.813
18	Margarita	0.956	33	−0.074
19	Modelo	0.748	31	0.006
20	Morelos	0.999	0.484	−1.264
21	Naranjo	0.653	39.3	−1.174
22	Ocampo	8494	3831	−1.9
23	Panuco	0.315	7	0.959
24	Pedernales	0.953	72.9	−4.3
25	Potrero	0.416	12.6	0.838
26	Progreso	0.451	13.7	0.804
27	P. Ayala	0.542	31.3	0.026
28	P. San Luis	0.647	24	0.437
29	Providencia	0.863	1.201	−13.5
30	Pujiltic	0.938	0.502	−1.683
31	Pucte	0.492	19.7	0.61
32	Queseria	0.707	0.406	−0.647
33	Refugio	0.627	0.2	0.685
34	S. Clara	0.368	0.153	0.756
35	S.Cristobal	1.073	0.352	−0.273
36	S. Miguelito	0.595	0.043	0.986
37	S. Rosalia	0.683	0.424	−1.2
38	Tala	0.734	32.8	−0.292
39	Tamazula	1.145	55.4	−1.9
40	Tres Valles	1.146	1.259	−14.8

Table 2.15 Results of the dataset no weather data—dense

No.	Mills	RMSE	MAPE	R^2
1	Atencingo	1.024	0.134	0.819
2	Alianza P.	0.879	0.481	−1.33
3	Bellavista	0.978	1.17	−16.5
4	CIASA	1.067	0.422	−0.752
5	Casasano	1.78	40.5	−0.892
6	Constancia	0.489	19.8	0.609
7	E. Zapata	0.821	29.9	−0.223
8	Eldorado	1.17	0.539	−1.91
9	Gloria	0.973	43.8	−0.909
10	Higo	0.914	29.9	0.111
11	Huixtla	0.954	34.4	−0.158
12	Juarez	102.9	54.56	−6.2
13	L. Cardenas	0.788	2.65	−69.9
14	La Joya	0.711	0.662	−1.777
15	Lopez Mateos	14.801	12.350	−1063
16	Mante	0.668	13.8	0.795
17	Mahuixtlan	0.851	12.6	0.84
18	Margarita	0.955	33.3	−0.093
19	Modelo	0.776	32.7	−0.11
20	Morelos	0.999	0.484	−1.264
21	Naranjo	0.655	39.8	−1.227
22	Ocampo	3075.6	1400.7	−2.40
23	Panuco	0.315	4.2	0.986
24	Pedernales	0.945	71.6	−4.132
25	Potrero	0.401	11.8	0.857
26	Progreso	0.447	13.9	0.799
27	P. Ayala	0.502	29	0.168
28	P. San Luis	0.639	24.1	0.437
29	Providencia	0.863	1.201	−13.5
30	Pujiltic	0.938	0.502	−1.683
31	Pucte	0.491	19.9	0.603
32	Queseria	0.707	0.406	−0.647
33	Refugio	0.627	0.2	0.685
34	S. Clara	0.368	0.153	0.756
35	S.Cristobal	1.073	0.352	−0.273
36	S. Miguelito	0.595	0.043	0.986
37	S. Rosalia	0.683	0.424	−1.2
38	Tala	0.745	33.5	−0.351
39	Tamazula	1.155	56.4	−2.007
40	Tres Valles	1.146	1.259	−14.8

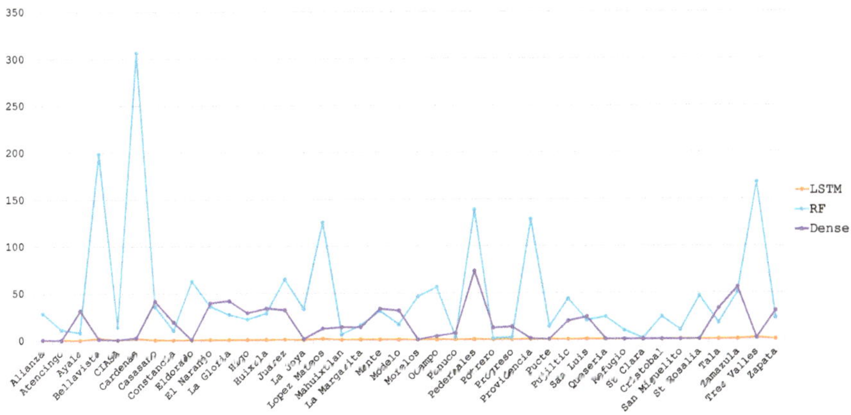

Fig. 2.18 Difference in MAPE between the techniques—weather data

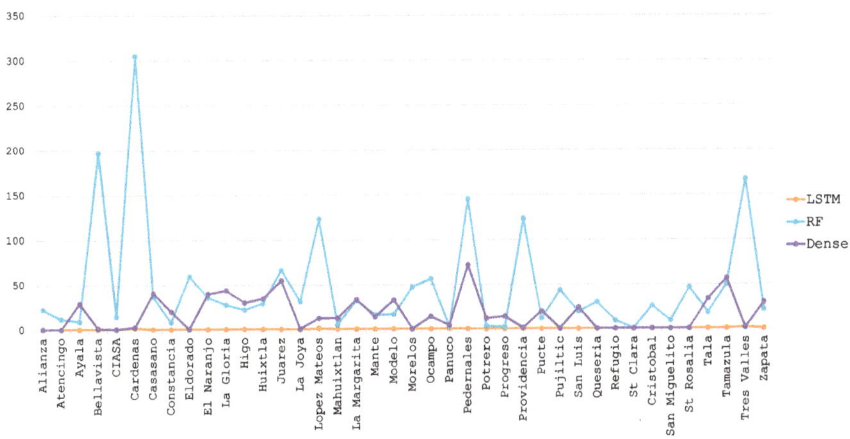

Fig. 2.19 Difference in MAPE between the techniques—no weather data

Table 2.16 Ranking of Friedman test results

Technique	Mean rank
LSTMNW	1.35
LSTMW	1.85
ANNDW	4.00
ANNDNW	4.20
RFNW	4.78
RFW	4.83

conditions is statistically significant. We applied the Wilcoxon test after the Friedman test to conduct pairwise comparisons and identify which models exhibited significant performance differences. This step is crucial in validating the ranking

Table 2.17 Wilcoxon test results to compare LSTM with weather and without use

Metric	Mean rank
Negative Ranks	N = 12, Mean Rank = 19.29, Sum of Ranks = 231.50
Positive Ranks	N = 28, Mean Rank = 21.02, Sum of Ranks = 588.50
Ties	0
Z	−2.40
Asymp. Sig. (2-tailed)	0.016

Table 2.18 Wilcoxon test results to compare LSTM and dense neural network

Metric	Mean rank
Negative ranks	N = 1, Mean Rank = 3.00, Sum of Ranks = 3.00
Positive ranks	N = 39, Mean Rank = 20.95, Sum of Ranks = 817.00
Ties	0
Z	−5.47
Asymp. sig. (2-tailed)	0.000

results and ensuring that any observed superiority of one model over another is not due to random variations but rather a consistent trend across multiple scenarios.

In this way, two tests compared the top three results: LSTM with climate data versus LSTM without weather data and LSTM without climate data versus the Dense Neural Network. Tables 2.17 and 2.18 show the results.

In the first test, results show that in 28 cases, the LSTM model without weather data outperformed the version that included it, while in 12 cases, the opposite occurred. The p-value of 0.016 indicates a statistically significant difference between both LSTM versions, confirming that the LSTM model without weather data performs significantly differently than when it incorporates weather data.

In the second test, results show that in 39 cases, the LSTM model without weather data achieved better performance, while in only 1 case, the Dense Neural Network with weather data performed better. The p-value was less than 0.001, indicating a significant difference between both models.

2.6.4.3 Explanation of Statistical Results

The LSTM model proves to be the most effective, mainly when the model does not use weather data. While climate information can improve LSTM's performance, its performance remains strong even without it. The Dense Neural Network falls short, performing worse than LSTM even when the model incorporates weather data.

The Friedman and Wilcoxon test results consistently support the conclusion that LSTM outperforms other models, including the Dense Neural Network. The Wilcoxon test, in particular, showed significant differences, confirming that the observed results are not random but reflect the true impact of the algorithm type and the data utilized.

The dense neural network consistently yielded worse results than LSTM without climate data when using climate data. This highlights LSTM as a more suitable model for forecasting in this specific context. LSTM consistently outperforms both the climate-dependent version and the dense neural network. Furthermore, these findings suggest that including climate data may introduce unnecessary complexities or noise, which could hinder the model's performance rather than enhance it. In summary, LSTM emerges as the superior algorithm for this problem, with or without climate data, while the Dense Neural Network fails to match its performance. These emphasize the importance of selecting the right model and considering the potential drawbacks of incorporating certain data types.

2.7 Discussion and Future Work

The proposed method trains the data for a whole year and forecasts the amount of cane that will harvested and the amount that the mill will receive. Knowing these values allows better management of field activities. Experiments showed that missing records (the presence of zeros) are irrelevant as long as the data sets accurately reflect actual production levels without being altered. The model successfully identifies patterns in the data despite the difficulties and was designed with this approach to cope with the lack of reliable data since focusing on the data otherwise would have been infeasible. Therefore, we minimized the data as much as possible to obtain reliable forecasts, considering that the production is 7 months (November-May) and thus generate an overall forecast, considering various factors that could influence the harvest. For example, the amount of rainfall in a particular month could significantly affect production levels, providing a more realistic view of the data and, consequently, improving the accuracy of the predictions.

Furthermore, we decided to use a more extended aggregation period to smooth out fluctuations that may occur due to irregularities in the weather or other external factors. By incorporating multiple months' data, the model can better account for seasonality and trends, leading to a more robust forecast. Understanding this is particularly crucial in agricultural contexts, where production can be highly variable due to unpredictable environmental factors.

While the results showed that the presence of zeros in the data did not significantly affect the model's accuracy, refining and preparing the dataset is important in ensuring that forecasts remain realistic and reliable. Further exploring how specific factors such as weather patterns, soil conditions, and farming practices influence production could enhance the method's performance in future work. Additionally, expanding the dataset by incorporating more diverse sources of information could help improve the model's generalizability and accuracy, ultimately contributing to more effective decision-making in agricultural planning and resource management.

The results from our LSTM model demonstrate its ability to provide predictions closer to the desired values compared to the Random Forest and Dense Neural Network models, which tend to generate higher values. Regarding the differences between predicted and actual values, we observed that most mills achieved highly accurate results with the LSTM model. Only a few mills managed to get closer values with the other models, further reinforcing the effectiveness of the LSTM.

Additionally, when analyzing the MAPE, the LSTM showed the lowest percentage error compared to the other models, standing out as the model with the highest prediction accuracy. These results confirm that the LSTM adjusts better to agricultural production data than the other models tested. We selected the LSTM model due to its ability to handle temporal sequences and complex patterns in agricultural data. Since the agricultural process follows a continuous cycle influenced by seasonal, climatic, and management factors, the LSTM demonstrated superior performance as it adapts well to fluctuations occurring in the harvest processes. This characteristic makes it the best choice for predicting the amount of sugarcane entering the mill based on available historical data. Furthermore, using recurrent neural networks, the LSTM can capture temporal dependencies in the data, providing more accurate and reliable forecasts.

Collecting these detailed data will allow producers to obtain precise information about the amount of sugarcane they harvest each season, facilitating strategic decision-making and improving supply chain efficiency in the sugarcane industry. The results obtained with the LSTM model open an opportunity to expand this approach to predict sugarcane production and integrate and optimize other parts of the agro-industrial cycle. Including data from different stages of the crop cycle and considering additional factors such as water usage, fertilizers, and agricultural management techniques could enrich the prediction process and improve results. For example, this extended model could predict the amount of sugarcane harvested and optimize resource use and production planning throughout the agricultural cycle.

2.8 Conclusion

In this article, we present an innovative model designed to predict the tons of sugarcane that will enter the mill for processing, utilizing advanced machine learning techniques. The main objective is to provide mill managers with accurate information regarding the future entry of sugarcane from the field so that they can appropriately plan the processing of the expected sugarcane. This model represents a valuable opportunity in Mexico, where, until now, little research has been done, unlike in other countries, where researchers have developed models to optimize sugarcane yield using well-established data sets. The present work serves as a fundamental basis for creating new agricultural research, working with data from sugar mills.

As mentioned above, this type of model would be of benefit to sugarcane growers, who will be able to know in advance how much cane they will harvest, the problem is that in Mexico, there are no data sets detailing the activities carried out individually, so we work with the data that was available to the public. The available data come mainly from the Sugar Mill Information System (SIIBA), collected by the mills, mainly through the field department, for statistical control purposes. However, the data show the activities carried out in the field globally, which limits the ability to perform deep and meaningful analysis, excluding factors such as optimal temperature ranges, crop management, and cane variety, which can significantly influence production and could provide an opportunity to analyze variables and consider more factors, requiring more detailed field activity records to understand production behavior better. This approach helps optimize agricultural yields, improve field management, optimize available resources, and ensure sustainable production. With more detailed information, producers could better adapt their strategies to changing environmental conditions and market demands, thus contributing to the growth of the sugarcane industry in Mexico and other Latin American countries. In future work, it would be advisable to conduct tests that propose scenarios considering the climatic factors that affect sugarcane production. In addition, working with broader historical data and going deeper into the different crop cycles (Planta, Soca, and Resoca) would allow producers to observe variations in production and make decisions on the need to replant. Also, keeping detailed records of farming activities would be crucial to improving the model's accuracy, as it would allow the model to learn and generate more accurate forecasts over time.

Although this type of model is not a novelty in the field of agricultural prediction, it represents an innovative contribution within the Mexican context. Farmers and agricultural managers in data-driven countries already use such systems to support planning and resource optimization. However, in Mexico, only medium and large-scale producers typically have access to detailed monitoring systems, while smallholders often rely on experience and manual records. This work sets a precedent for future research efforts aimed at strengthening the agricultural sector through data-based tools, promoting broader accessibility, and encouraging more evidence-based decision-making in sugarcane production.

Appendix

Some portions of this manuscript were enhanced with the assistance of generative AI tools. The authors carefully reviewed, edited and approved all AI-generated content. The authors take full responsibility for the accuracy and integrity of the entire chapter.

References

1. Agarwal, S., Tarar, S.: A hybrid approach for crop yield prediction using machine learning and deep learning algorithms. In: Journal of Physics: Conference Series, vol. 1714 (2021). https://doi.org/10.1088/1742-6596/1714/1/012012

2. Alavi, M., Albaji, M., Golabi, M., Naseri, A.A., Homayouni, S.: Estimation of sugar cane evapotranspiration from remote sensing and limited meteorological variables using machine learning models. J. Hydrol. **629**, 130605 (2024). https://doi.org/10.1016/j.jhydrol.2023.130605. https://linkinghub.elsevier.com/retrieve/pii/S0022169423015470
3. Berk, R.A.: Random forests, pp. 233–295 (2020). https://link.springer.com/chapter/10.1007/978-3-030-40189-4_5
4. Bhimavarapu, U., Battineni, G., Chintalapudi, N.: Improved optimization algorithm in LSTM to predict crop yield. Computers **12**, 10 (2023). https://doi.org/10.3390/computers12010010.
5. Breiman, L.: Random forests. Mach. Learn. **45**, 5–32 (2001). https://doi.org/10.1023/A:1010933404324
6. Canales, J.M.V.: El sector agroalimentario mexicano y las nuevas tecnologías. e-Agronegocios **8**, 89–113 (2022). https://doi.org/10.18845/ea.v8i2.6156. https://revistas.tec.ac.cr/index.php/eagronegocios/article/view/6156
7. Canata, T.F., Júnior, M.R.B., de Oliveira, R.P., Furlani, C.E.A., da Silva, R.P.: Ai-driven prediction of sugarcane quality attributes using satellite imagery. Sugar Tech. **26**, 741–751 (2024). https://doi.org/10.1007/s12355-024-01399-9. https://link.springer.com/10.1007/s12355-024-01399-9
8. Charoen-Ung, P., Mittrapiyanuruk, P.: Sugarcane Yield Grade Prediction using Random Forest and Gradient Boosting Tree Techniques. IEEE, Piscataway (2018)
9. Chaudhary, M.: Random forest algorithm - how it works and why it is so effective (2024). https://www.turing.com/kb/random-forest-algorithm
10. CONADESUCA: 11° informe estadístico del sector agroindustrial de la caña de azúcar en méxico — comité nacional para el desarrollo sustentable de la caña de azúcar — gobierno — gob.mx (2024). https://www.gob.mx/conadesuca/articulos/11-informe-estadistico-del-sector-agroindustrial-de-la-cana-de-azucar-en-mexico
11. CONADESUCA: Dieproc estimados y pronósticos de producción — comité nacional para el desarrollo sustentable de la caña de azúcar — gobierno — gob.mx (2025). https://www.gob.mx/conadesuca/documentos/infocana-estimados-y-pronosticos-de-produccion
12. CONADESUCA, SADER: Directorio de ingenios azucareros de méxico, zafra 2024/2025 (2025). www.gob.mx/conadesuca
13. CONADESUCA, SADER: Sistema infocaña (2025). https://www.siiba.conadesuca.gob.mx/infocana/
14. Datta, P., Faroughi, S.A.: A multihead lstm technique for prognostic prediction of soil moisture. Geoderma **433**, 116452 (2023). https://doi.org/10.1016/j.geoderma.2023.116452
15. de Almeida, G.M., Pereira, G.T., de Souza Bahia, A.S.R., Fernandes, K., Júnior, J.M.: Machine learning in the prediction of sugarcane production environments. Comput. Electron. Agric. **190**, 106452 (2021). https://doi.org/10.1016/J.COMPAG.2021.106452
16. de México, G.: La construcción de un trapiche para el impulso de la producción de azúcar en michoacán — archivo general de la nación — gobierno (2022). https://www.gob.mx/agn/articulos/la-construccion-de-un-trapiche-para-el-impulso-de-la-produccion-de-azucar-en-michoacan?idiom=es
17. da Silva, M.L., da Silva Alves, L.M., Roque, A., Pohren, D., Silva, A.S.D., de Freitas, E.P., Balbinot, A.: Prediction of sugarcane yield production by sensor fusion applying machine learning methods (2024). https://doi.org/10.2139/ssrn.4819150
18. DOF: Ley de desarrollo sustentable de la caña de azúcar (2005). https://www.fao.org/faolex/results/details/es/c/LEX-FAOC069602/
19. Espinosa-Zúñiga, J.J.: Aplicación de algoritmos random forest y xgboost en una base de solicitudes de tarjetas de crédito. Ingeniería, investigaciín y tecnología **21**, 1–16 (2020). https://doi.org/10.22201/FI.25940732E.2020.21.3.022
20. García-Arteaga, J.J., Zambrano-Zambrano, J.J., Alcivar-Cevallos, R., Zambrano-Romero, W.D.: Agricultural crop yield prediction using machine learning. Rev. Arbit. Interdiscip. Koinonía **5**, 144 (2020). https://doi.org/10.35381/r.k.v5i2.1013

21. Gupta, S.K., Agarwal, A.P.: Predicting total sugar production using multivariable linear regression. In: Proceedings - IEEE 2021 International Conference on Computing, Communication, and Intelligent Systems, ICCCIS 2021, pp. 465–469. Institute of Electrical and Electronics Engineers, Piscataway (2021). https://doi.org/10.1109/ICCCIS51004.2021.9397078. https://ieeexplore.ieee.org/document/9397078/authors#authors
22. Hochreiter, S., Schmidhuber, J.: Long short-term memory. Neural Comput. **9**, 1735–1780 (1997). https://doi.org/10.1162/neco.1997.9.8.1735
23. İnik, O., Özkan İ., Öztaş, T., Yüksel, A.: Soil temperature prediction with long short term memory (LSTM). Turkish J. Agric. Nat. Sci. **9**, 779–785 (2022). https://doi.org/10.30910/turkjans.1101753
24. Kulasekara, S., Kumarasiri, K., Sirimanna, T., Dissanayake, D., Karunasena, A., Pemadasa, N.: Machine learning based solution for improving the efficiency of sugar production in Sri Lanka. In: 2022 13th International Conference on Computing Communication and Networking Technologies (ICCCNT), pp. 1–7. IEEE, Piscataway (2022). https://doi.org/10.1109/ICCCNT54827.2022.9984496. https://ieeexplore.ieee.org/document/9984496/
25. Lárraga-Altamirano, H.R., Hernández-López, D.R., Piedad-Rubio, A.M., Blanco-Martínez, J.R.: Machine-learning model for estimating sugarcane production at crop level. J. Technol. Innov. **11**, 1–13 (2024). https://doi.org/10.35429/jti.2024.28.11.1.13
26. Lee, H., Kim, B.Y., Sang, M.J., Seo, W., Baek, M.C., Yang, J.K., Mo, J.E., Development, C., Park, S.H., Lee, B.Y., Kim, M.J., Sang, W., Seo, M.C., Baek, J.K., Yang, J.E., Mo, C.: Development of a soil moisture prediction model based on recurrent neural network long short-term memory (RNN-LSTM) in soybean cultivation. Sensors **23**, 1976 (2023). https://doi.org/10.3390/S23041976
27. Liakos, K.G., Busato, P., Moshou, D., Pearson, S., Bochtis, D.: Machine learning in agriculture: a review. Sensors **18**, 2674 (2018). https://doi.org/10.3390/S18082674
28. Ma, P., Zhang, N., Yang, Y., Wang, Z., Li, G., Fu, Z.: Forecast of sugarcane yield in chongzuo, guangxi—LSTM model based on fusion of trend yield and meteorological yield. Agronomy **14**, 2512 (2024). https://doi.org/10.3390/agronomy14112512
29. Narvekar, C., Rao, M.: Sugarcane production modeling using machine learning in Western Maharashtra. Int. J. Appl. Sci. Smart Technol. **4**, 123–130 (2022)
30. Nihar, A., Patel, N.R., Danodia, A.: Machine-learning-based regional yield forecasting for sugarcane crop in Uttar Pradesh, India. J. Indian Soc. Remote Sens. **50**, 1519–1530 (2022). https://doi.org/10.1007/s12524-022-01549-0
31. Panigrahi, B., Kathala, K.C.R., Sujatha, M.: A machine learning-based comparative approach to predict the crop yield using supervised learning with regression models. Proc. Comput. Sci. **218**, 2684–2693 (2023). https://doi.org/10.1016/j.procs.2023.01.241
32. Pignéde, E., Roudier, P., Diedhiou, A., Bi, V.H.N., Kobea, A.T., Konaté, D., Péné, C.B.: Sugarcane yield forecast in ivory coast (West Africa) based on weather and vegetation index data. Atmosphere **12**, 1459 (2021). https://doi.org/10.3390/atmos12111459
33. Ramadhan, A.J., Priya, S.R., Pavithra, V., Mishra, P., Dash, A., Abotaleb, M., Alkattan, H., Albadran, Z.: Machine learning techniques for sugarcane yield prediction using weather variables. In: BIO Web of Conferences, vol. 97. EDP Sciences, Les Ulis (2024). https://doi.org/10.1051/bioconf/20249700157
34. Rodríguez, K.A.F., García, A.M.T.G., Moreno, Y.M., Rosas, F.H., Sandoval, B.F.: Factores que explican el rendimiento de caÑa de azúcar a nivel municipal en México. Rev. Mexicana Cienc. Agríc. **6**, 1345–1358 (2015)
35. SADER: Etapas del cultivo de caÑa — secretaría de agricultura y desarrollo rural — gobierno — gob.mx (2019). https://www.gob.mx/agricultura/articulos/etapas-del-cultivo-de-cana
36. SADER: Caña de azúcar, una dulce producción — secretaría de agricultura y desarrollo rural — gobierno — gob.mx (2020). https://www.gob.mx/agricultura/es/articulos/cana-de-azucar-una-dulce-produccion-237168
37. SADER: Dulce caña de azúcar — secretaría de agricultura y desarrollo rural — gobierno — gob.mx (2021). https://www.gob.mx/agricultura/articulos/dulce-cana-de-azucar

38. SADER: México cuenta con disponibilidad suficiente de azúcar para atender el abasto nacional y exportaciones | secretaría de agricultura y desarrollo rural | gobierno (2022). https://www.gob.mx/agricultura/prensa/mexico-cuenta-con-disponibilidad-suficiente-de-azucar-para-atender-el-abasto-nacional-y-exportaciones
39. SADER: Secretaría de agricultura y desarrollo rural | gobierno | gob.mx (2022). https://www.gob.mx/agricultura/que-hacemos
40. SADER, CONADESUCA: Ficha técnica del cultivo de la caña de azúcar (2015)
41. Singla, S.K., Garg, R.D., Dubey, O.P.: Ensemble machine learning methods to estimate the sugarcane yield based on remote sensing information. Rev. d'Intell. Artif. **34**, 731–743 (2020). https://doi.org/10.18280/RIA.340607
42. Solidaridad, P.: Manual de Buenas Prácticas Agrícolas para el Cultivo de Caña de Azúcar (2021)
43. Som-Ard, J., Atzberger, C., Izquierdo-Verdiguier, E., Vuolo, F., Immitzer, M.: Remote sensing applications in sugarcane cultivation: a review. Remote Sens. **13**, 4040 (2021). https://doi.org/10.3390/RS13204040
44. Sridhara, S., Soumya, B.R., Kashyap, G.R.: Multistage sugarcane yield prediction using machine learning algorithms. J. Agrometeorol. **26**, 37–44 (2024). https://doi.org/10.54386/jam.v26i1.2411
45. Zhou, Y., Pan, M., Guan, W., Fu, C., Su, T.: Predicting sugarcane yield via the use of an improved least squares support vector machine and water cycle optimization model. Agriculture **13**, 2115 (2023). https://doi.org/10.3390/agriculture13112115
46. Zulu, N.S., Sibanda, M., Tlali, B.S.: Factors affecting sugarcane production by small-scale growers in ndwedwe local unicipality, South Africa. Agriculture **9**, 170 (2019). https://doi.org/10.3390/AGRICULTURE9080170

Chapter 3
LSTM Recurrent Neural Network for Streamflow Forecasting in the Upper Basin of the Guayalejo-Tamesí River, Mexico

Estefanía Román-Rangel, Jesús David Terán-Villanueva, Tomás Alejandro Peña-Alonso, Salvador Ibarra-Martínez, and Mirna Patricia Ponce-Flores

Abstract Water scarcity has emerged as a critical issue in Mexico, with water quality and management undergoing constant changes due to natural phenomena and human activities. Recent advancements in machine learning have demonstrated promising results in forecasting climatological and hydrological variables. This study employs a Long Short-Term Memory (LSTM) recurrent neural network to correlate 52 years of historical rainfall and streamflow data from the Upper Basin of the Guayalejo-Tamesi River in Tamaulipas, Mexico, aiming to provide accurate flow forecasts through the rainfall historical data. The results show that LSTM produced a Root Mean Squared Error (RMSE) of 110.44. We acknowledge various factors like data cleaning processes such as imputation and outlier removal, the accuracy of historical data, and unconsidered climatological variables as potential influences on the network's performance.

Keywords Long-short term memory · RNN · Streamflow forecasting · Basin · Time series

3.1 Introduction

Water resources in Mexico have become a central topic of research due to their importance and the need for better resource management. In 2022, in certain regions of Tamaulipas, such as Jaumave, communal landowners and farmers formed self-defense groups to protect the water of the Guayalejo River from the abuse of water

E. Román-Rangel · J. D. Terán-Villanueva · T. A. Peña-Alonso · S. Ibarra-Martínez (✉)
M. P. Ponce-Flores
Centro Universitario Tampico Madero, Universidad Autónoma de Tamaulipas (UAT), Tampico Tamaulipas, México
e-mail: sibarram@uat.edu.mx

G. Rivera et al. (eds.), *Recent Advances in Artificial Neural Networks*, Intelligent Systems Reference Library 283, https://doi.org/10.1007/978-3-032-13497-4_3

use by producers in that region [1]. In April 2024, the National Water Commission (CONAGUA) applied sanctions to 20 concessionaires for the clandestine use of water from the Guayalejo-Tamesí Basin [2]. Therefore, it is now more important to have certainty about water availability for proper management. One way to achieve this is by using machine learning, a tool that helps us interpret the information collected, such as rainfall or flow, and to build forecast models. For this purpose, two methods are commonly used [3]: (a) physical-conceptual models and (b) data-driven models.

Physical-conceptual models are based on physical and spatial concepts [4], while data-driven models use the relationship between inputs and outputs (historical data) [5]. In the case of forecasts with time series, it is usual to use some data-driven model, especially in studies related to hydrology [6]. Within this range are the LSTM recurrent neural networks, which are widely used for rainfall-flow models, where the forecast accuracy is linked to the quantity and quality of the data [7]. The particularity of LSTM networks is that they use their short and long-term memory to process variable-length input sequences [8]; they can decide which information is best to store and which to discard [9].

In this paper, we propose a data-driven model for forecasting a basin's Streamflow rate (in m^3/s) using historical rainfall and streamflow data. We begin by preprocessing the data and then conducting a hyperparameter search to enhance the model's accuracy, which we evaluate using the Root Mean Square Error (RMSE) metric. We compare two search methods and their results against the actual data. Additionally, we compare our model to an ARIMA model, where the LSTM network achieved better accuracy. The structure of this paper is as follows: Sect. 3.2 presents the background on hydrological time series forecasting. Section 3.3 describes the characteristics of the study area, the imputation of missing data, and the data cleaning processes. In Sect. 3.4, we detail the development of the forecasting model and the optimal hyperparameter settings. Section 3.5 presents the model's results and a discussion. In Sect. 3.6, we present the conclusions of the study.

3.2 Background

Man et al. [10], introduce in their study an enhanced LSTM model for daily runoff prediction in the upper Huai River Basin in China. They aim to improve the accuracy of runoff prediction. The enhanced LSTM model integrates a feature extractor and employs novel loss functions to emphasize peak flow prediction. The study compares this improved model with other data-driven and hydrological models to demonstrate its superior performance. They obtained daily precipitation and runoff data from 1951 to 2016 from the China Meteorological Data Service Center and the Henan Hydrology Bureau, respectively. They chose six meteorological stations in and around the basin to capture spatial heterogeneity. Due to missing precipitation data from some stations between 1951 and 1959, the model used runoff data from 1960 to 2016. For training, they used data from January 1960 to October 2005 and November 2005 to December 2016 to test the model. The enhanced LSTM model consists of four layers: an input layer, a feature extractor, a predictor, and an output layer. The feature extractor comprises three LSTM networks, one for each meteorological station, to extract temporal features of the input data. It uses precipitation (P), conditional cumulative precipitation (CCP), and historical runoff data as inputs. The authors designed two novel loss functions to improve peak flow prediction: Peak Error Tanh (PET) uses a tanh function alongside the mean squared error (MSE) to amplify more significant errors and reduce the smaller ones. Peak Error Swish (PES) combines MSE with a sigmoid function to better address the errors. Their model uses a mini-batch training technique with a batch size of 64 and epoch size of 200, along with the Adam optimizer, and normalized the input data in a 0-15 range. Regarding daily runoff prediction, the LSTM model outperformed all the comparative models, including data-driven models (SVR, ANN, GRU) and lumped hydrological models (AWBM, Sacramento, SimHyd, Tank Model). The LSTM model using PET as the loss function achieved the highest NSE of 0.924 for overall runoff prediction and using PES as the loss function performed best on extreme runoff prediction, achieving a mean NSE for floods of 0.873 and a QR of 92.3% Precipitation at meteorological stations with higher altitudes contributed more to runoff prediction than those at lower altitudes. The LSTM model, with its feature extractor and novel loss functions, is an effective tool for daily runoff prediction.

In 2021, in a study focusing on the Kentucky River basin in the eastern United States, Rahimzad et al. [3] compared the performance of several data-driven techniques for daily streamflow forecasting. The study employed Linear Regression (LR), Multilayer Perceptron (MLP), Support Vector Machine (SVM), and Long Short-Term Memory (LSTM) network models. The main objective was to compare the accuracy of these four data-driven techniques by using three scenarios based on historical precipitation and streamflow data over 26 years. They evaluated the performance of each model using a range of statistical criteria, including the coefficient of correlation (R), Nash-Sutcliff coefficient of efficiency (E), Nash-Sutcliff for High flow (EH), Nash-Sutcliff for Low flow (EL), normalized root mean square error (NRMSE), relative error in estimating maximum flow (REmax), threshold statistics (TS), and average absolute relative error (AARE). The study results showed that the LSTM network outperformed the other models in forecasting daily streamflow, with the lowest NRMSE and the highest EH, EL, and R values under all scenarios. These findings suggest that LSTM is a robust data-driven technique for characterizing time series behaviors in hydrological modeling applications. The study used daily streamflow data from the Kentucky River at a specific location and daily rainfall data from five rain gauges in the region. The data covered 26 years from 1960 to 1989, with some missing data in certain years. The authors considered three scenarios in the study: streamflow forecasted based on previous streamflow and precipitation, streamflow forecasted based on previous streamflow and current precipitation, and streamflow forecasted based on previous streamflow only. The study also looked at the effect of different input combinations on model performance by using cross correlation to choose the optimum inputs for the designed scenarios. In addition, they performed an outlier detection to ensure the results were valid. We can say that this study demonstrated the superior performance of LSTM over traditional machine learning algorithms like LR, MLP, and SVM for daily streamflow forecasting in the Kentucky River basin.

In 2018, Kratzert et al. [11] proposed a novel data-driven approach for rainfall-runoff modeling using Long Short-Term Memory (LSTM) networks. The authors explored the potential of LSTMs to model the rainfall-runoff behavior of many differently complex catchments at a daily timescale. To test their approach, they used the freely available CAMELS dataset, which includes meteorological data and observed discharge for 671 catchments across the contiguous United States. The study compared the LSTM results to the well-known Sacramento Soil Moisture Accounting Model (SAC-SMA) coupled with the Snow-17 snow routine. The study focused on 241 catchments from 4 out of 18 hydrological units (HUCs) to cover a wide range of hydrological conditions. The LSTM network was able to predict runoff from meteorological observations with accuracies comparable to (or slightly better than) the well-established SAC-SMA + Snow-17 model. The mean Nash-Sutcliffe Efficiency (NSE) was 0.63 for the LSTM model compared to 0.58 for the SAC-SMA + Snow-17 model. The study highlighted that traditional Recurrent Neural Networks (RNNs) are unsuitable for daily streamflow prediction because they cannot learn long-term dependencies. LSTM's ability to learn and store long-term dependencies is crucial for accurately modeling storage effects in catchments. The study suggests that 15 years of daily data is a lower bound for calibrating the LSTM model and they used various metrics to evaluate model performance as Nash-Sutcliffe Efficiency (NSE), Correlation Coefficient (r), Variance Bias (α), Total Volume Bias (β), Flow Duration Curve (FDC) and Mean Squared Error (MSE) as an objective criterion to train the LSTM. For experiment the median NSE difference between the LSTM and the benchmark model was -0.03, meaning the benchmark model performed slightly better in the median, though the LSTM outperformed the benchmark when considering the mean NSE. In experiment two, the regional models showed a more balanced under and overestimating discharge than models trained on individual catchments. In experiment three, they obtained almost the same median NSE as the reference model for the validation period (0.72 and 0.71 respectively), but the mean was approximately 15% higher for the models in experiment three. In summary, this study demonstrated the potential of LSTMs for rainfall-runoff modeling, showing their ability to learn complex relationships from data, their advantages over traditional RNNs, and the benefits of pre-training and fine-tuning strategies. The study also highlighted how the LSTMs performed comparably to a well-established hydrological model.

Hu et al. [12], focused on using a Long Short-Term Memory (LSTM) deep learning model for streamflow forecasting in small rivers, acknowledging the challenges of rapid and unpredictable flow changes. They compared the LSTM model's performance against traditional machine learning methods, specifically Support Vector Regression (SVR) and Multilayer Perceptron (MLP). The authors collected streamflow data from a hydrologic station in Tunxi, China, and rainfall data from 11 surrounding precipitation stations from 1981 to 2003. The goal was to predict streamflow volume 6 hours into the future using the past 12 hours of streamflow and precipitation data. They pre-processed data into a time series format and normalized it. The dataset was split into training (70%) and testing (30%), and the LSTM model was trained using the Keras library with specific parameters: 64 hidden layer nodes, Adam optimizer, a batch size of 72, and 30 epochs. Hu et al. compared the performance of the LSTM model with two other models: SVR, a traditional machine learning model, and MLP, a deep learning model15. The study used Root Mean Square Error (RMSE), Mean Absolute Error (MAE), and the coefficient of determination (R2) as evaluation metrics. The results showed that the LSTM model outperformed SVR and MLP across all evaluation metrics. Specifically, the LSTM model achieved an RMSE of 82.007, an MAE of 27.752, and an R2 of 0.9701. In addition, the LSTM predictions closely matched the actual streamflow data without the fluctuations seen in SVR and MLP predictions. Prediction accuracy decreased as the time step increased, and models with an encoder time step between 12 and 14 hours had the best forecast accuracy. The study concluded that the LSTM model is an effective method for streamflow forecasting due to its ability to handle non-linear time series data. The LSTM model demonstrated better stability, reliability, and accuracy in capturing the data features than the SVR and MLP models.

Yan et al. [13], introduce an LSTM network for streamflow prediction in small watersheds, emphasizing its ability to capture long-term pre-information from hydrological data. The authors note that this is the first LSTM architecture applied for streamflow prediction in small watersheds, which is reasonable, given that it helps discover and analyze inherent patterns between some hydrological data features and river flow. They point out that the flood mechanisms in small watersheds are often too complex to build a physical hydrology model, making data driven approaches like LSTM particularly useful. The research aims to evaluate the performance of LSTM in forecasting hydrologic time series data from a small watershed in China and compare it with traditional methods such as Support Vector Machines (SVM). The authors describe the study area, Tunxi (TX) catchment, a small watershed in southeast China with complex hydrogeological features. They collected around 19,000 data points from 1991 to 1999 during the flood period and divided them into training and testing sets. The model predicts the river flow one to six hours into the future, using past hydrology, past weather, and weather forecast data as input. The authors implemented their LSTM model using TensorFlow, while the SVM model was in Matlab. The study evaluates model performance using Root Mean Square Error (RMSE), Median Absolute Error (MedAE), and the coefficient of determination (R2). The results show that the LSTM model has better accuracy and robustness than the SVM model. The LSTM model, with its memory cells and gates, is better at capturing the temporal dependencies in the data and performs better in forecasting the time of flood peak occurrence and the values of flow peak. The authors conclude that the proposed LSTM model is effective and provides a valuable tool for predicting river flow in small watersheds. They suggest that future research could explore combining the LSTM model with other techniques.

In a study, Hunt et al. [14], explore the effectiveness of using a long short-term memory (LSTM) neural network to enhance river streamflow forecasts across the western United States. The research team trained the LSTM model using catchment-mean meteorological and hydrological variables from the ERA5 and GloFAS-ERA5 and historical streamflow data from ten river gauge stations. They compared the LSTM's performance against raw and bias-corrected outputs from the Copernicus Emergency Management Service (CEMS) physics-based GloFAS. The study included a testing phase, where the models used ERA5 data, and an operational phase, fed forecast variables from the European Centre for Medium-Range Weather Forecasts (ECMWF) Integrated Forecasting System (IFS). The results demonstrate the LSTM model's capability to provide skillful streamflow forecasts at up to 10 days lead times, even outperforming established physics-based models. The study also defines a hybrid forecasting system that combines physically based and statistical or machine learning methods. Using the ECMWF forecasts to feed an LSTM puts the study into hybrid forecasting. The LSTM was trained separately for each of the ten basins, and the mean of the five best-performing models was used for validation. This is the first time an LSTM has been used in a hybrid system to create a medium-range streamflow forecast.

A study by Sahoo et al. [15], investigate the suitability of a long short-term memory recurrent neural network (LSTM-RNN) for low-flow hydrological time series forecasting. The study utilizes daily discharge data from the Basantapur gauging station located in the Mahanadi River basin in India. The LSTM-RNN is an artificial intelligence (AI) technique that works on a sequential framework, considering all predecessor data. The authors compare the LSTM-RNN's performance to a standard recurrent neural network (RNN) model and a naïve forecasting method. The research indicates that LSTM-RNN outperforms both RNN and the naive method in low-flow forecasting. The LSTM-RNN model achieved a correlation coefficient (R) of 0.943, a Nash-Sutcliffe efficiency (ENS) of 0.878, and a root-mean-square error (RMSE) of 0.487, compared to the RNN model's R of 0.935, ENS of 0.843, and RMSE of 0.516, and the naive method's R of 0.866, ENS of 0.704, and RMSE of 0.793. This suggests that LSTM-RNN can be a reliable AI technique for low-flow forecasting. The study also emphasizes that AI techniques are increasingly popular in hydrological time series modeling because of their ability to analyze long series, handle large datasets, and recognize hidden patterns in historical data. The LSTM-RNN model includes memory cells, which help capture information over long training data sequences and have an advantage over traditional RNNs. The specific architecture used in this study is a many-to-one model for one-step-ahead forecasting, which uses all previous monthly low-flow values to predict the current month's value. The study also utilizes the Q75 discharge as a low-flow threshold and includes data pre-processing, such as logarithmic transformation, to stabilize time series variance. The model development process consists of three phases: training (70%), validation (15%), and testing (15%). The best model configuration is determined based on the Root Mean Square Error (RMSE) and involves varying the number of neurons in the hidden layers (1 to 5) and different time lags (1, 3, 6,

9, and 12). The results support the claim that LSTM-RNN is a robust method for modeling low-flow hydrological time series.

A paper by Ni et al. [16], introduces two hybrid models, WLSTM and CLSTM, to improve the accuracy of streamflow and rainfall forecasting using long short-term memory (LSTM) networks. The researchers coupled LSTM with wavelet transform (WT) in the WLSTM model and convolutional neural networks (CNN) in the CLSTM model. The goal was to investigate the potential of LSTM in forecasting these hydrometeorological variables and to enhance forecast accuracy by combining LSTM with data preprocessing techniques. The study uses two streamflow datasets and two rainfall datasets to evaluate the performance of the models, comparing them with a standard multilayer perceptron (MLP) and a standard LSTM model. The results show that while LSTM was suitable for time series prediction, WLSTM and CLSTM provided superior performance. WLSTM uses a trous algorithm of wavelet transform to decompose time series, while CLSTM uses convolutional layers to extract temporal features. Researchers evaluated the models using monthly streamflow volume data from the Cuntan and Hankou stations and monthly rainfall data from the Jinan and Wenjiang stations. CLSTM generally had the best forecast accuracy in most cases. Still, when the explanatory variables were few, and an accurate forecast of peak values was expected, WLSTM was a better alternative. The study also highlights that as the forecasting time steps increase, the performance of MLP decreases, while LSTM-based models show more stable performance. The LSTM-based models were able to capture peak values better, while MLP often underestimated or overestimated these values. This study indicates that using wavelet transform and convolutional layers as feature extractors improves the prediction accuracy of LSTM.

An investigation led by Yosuff et al. [17], focuses on using Long Short-Term Memory (LSTM) models to predict streamflow during peak rainfall events in the Piney River watershed. The research uses 20 years of daily data (2000–2019) from two USGS stations, including streamflow, water level, and rainfall. The study explores the effectiveness of LSTM models in predicting streamflow during a major flooding event on August 24, 2021, at Piney River Resort. The LSTM model is a type of recurrent neural network designed to address the vanishing gradient problem using backpropagation, which makes it effective in learning long-term dependencies in complex time-series data. The LSTM model's architecture includes a cell state, input gate, output gate, and forget gate, which controls information flow within the network. This study employed a single-layered LSTM model. The model uses a sequence of 14 days to predict streamflow values. The researchers tested four different LSTM models: one using only rainfall data (LSTMR), one using only water level data (LSTMWL), one using only streamflow data (LSTMSF), and one using all three data sets (LSTMFULL). The LSTMFULL model performed best, achieving an optimal Nash-Sutcliffe Efficiency (NSE) of 0.68 and a Root Mean Square Error (RMSE) of 13.84 m^3/s. The final model predicted a peak flow of 94 m3/s with the lowest prediction error. The study also investigated the sensitivity of streamflow to rainfall using a non-parametric estimator, which indicated that a 1% change in annual precipitation results in a 2.3% change in annual streamflow. The study also used a grid search for hyperparameter tuning, finding that a 64-neuron architecture gave optimal performance and computational cost balance. The findings from the study suggest that LSTM models can be effective tools for predicting streamflow.

3.3 Dataset

For this research, we created a dataset containing rainfall and steamflow data from stations in the upper Guayalejo-Tamesí river basin, which includes nine rainfall stations and one streamflow station.

3.3.1 Study Area

The Upper Basin (UB) is an important area for water harvesting because it experiences constant rain or fog discharges due to the collision of humid winds from the East and Southeast and the cold winds from the North with the Sierra Madre Oriental [18]. The UB is the highest part of the Gauyalejo-Tamesí Basin about sea level. It has a territorial extension of approximately 3793 km^2. It begins in the Sierra Madre Oriental between Nuevo León and Tamaulipas [19], and it ends at the height of the Streamflow station La Encantada [18]. We based the forecast on data from La Encantada station and the precipitation recorded at nine rainfall stations in this area. See Fig. 3.1.

Table 3.1 shows the characteristics of the streamflow and rainfall measurement stations considered for the hydrological analysis. We list the altitude in meters, UTM coordinates (East and North), and the percentage of missing data for each station. We collected Rainfall data (mm) from the National Meteorological System's historical records [20] for nine weather stations in the UB. Similarly, we obtained Streamflow data from the National Water Commission's National Surface Data Bank [21] (Fig. 3.2). Streamflow data was complete for the selected Streamflow station, but many rainfall stations had more than 50% missing values. We used a 52-year time series, applying data preprocessing techniques.

3.3.2 Imputation and Data Cleaning

Data processing is sometimes necessary to enhance the quality of information. Historical data often contains missing values or outliers, requiring techniques to minimize their impact on forecasts. We have employed two techniques in our study: data imputation and outliers removal.

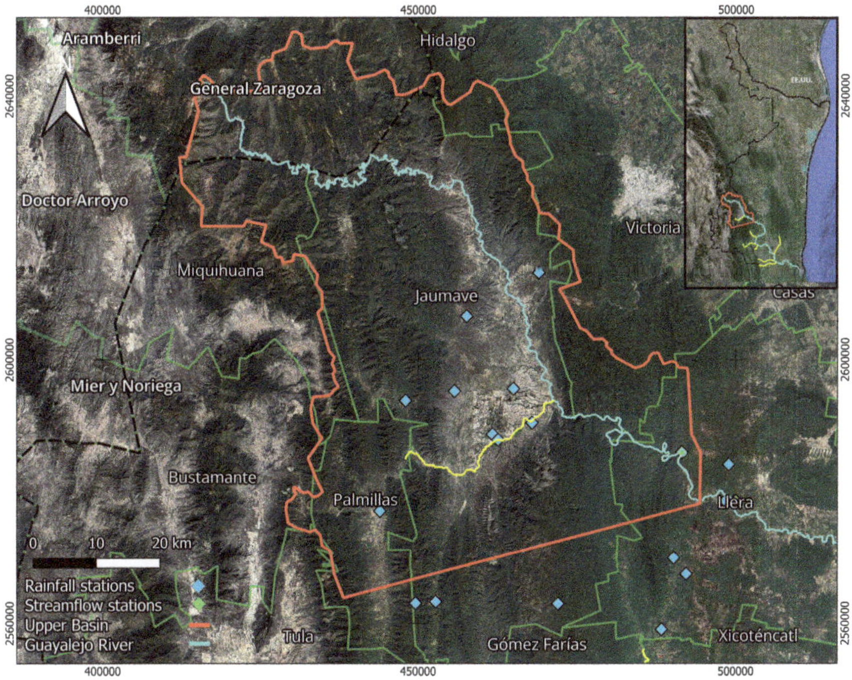

Fig. 3.1 Location of rainfall and Streamflow stations in the upper basin

Table 3.1 Features of streamflow and rainfall stations

	Station	Altitude (m)	UTMX	UTMY	Missing %
Streamflow	La Encantada	327	491484.17	2585958.13	0
Rainfall	Palmillas	1219	443931.28	2577082.41	1.92
	Magdaleno Aguilar	1205	447995.30	2593827.77	40.87
	La Vega	872	455745.21	2595247.82	66.99
	Plan de Ayala	833	457650.31	2606589.98	5.77
	La Reforma	795	468946.48	2613204.21	39.58
	Jaumave DGE	733	461657.16	2588680.93	4.49
	Jaumave SMN	705	462534.86	2587848.37	64.58
	Francisco Zorrilla	699	464937.63	2595622.78	16.19
	San Vicente	634	467876.55	2590418.89	8.49

3.3.2.1 Data Imputation

Using the PyCharm development environment and Python, we created a dataset from nine rainfall stations and one streamflow gauge from January 1961 to December 2012. Recognizing the importance of data completeness, we strategically employed a data imputation technique to address missing data in the dataset. This process involves estimating missing values based on existing data. A corre-

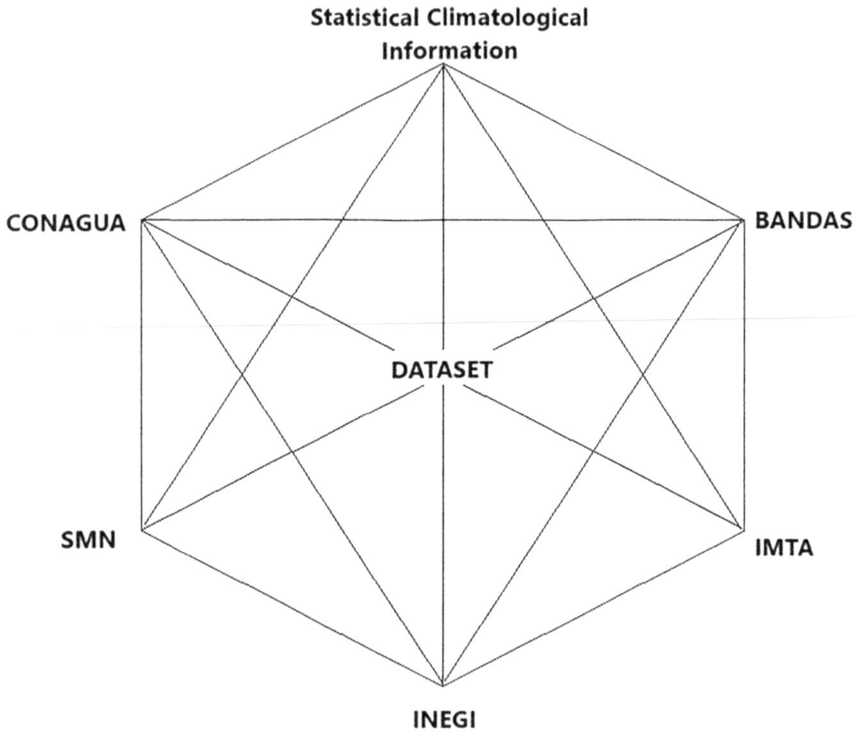

Fig. 3.2 Sources of information

lation matrix helped analyze the relationships among stations. Intermediate values (between −1 and 1) indicate the strength of the correlation. The closer the value is to 1, the stronger the linear relationship. Most of the values in the matrix were positive and above 0.5, indicating strong correlations (Fig. 3.3). Zeros signified a lack of data overlap between stations.

To impute missing values, we utilized a weighted average method using the correlation between stations. This involved calculating the global average precipitation for each station. For each missing value, three important calculations are executed:

1. Proportion Calculation: The proportion of the missing station's global average relative to the global average of other stations is determined.

$$proportion_{i,j} = \frac{avg(x_i)}{avg(x_j)} \tag{3.1}$$

   Where $avg(x_i)$ is the missing station's global average and $avg(x_j)$ represents the global average of each station with data.
2. Weighted Correlation Product: The proportion is multiplied by the station's value with data and then by the corresponding correlation matrix value.

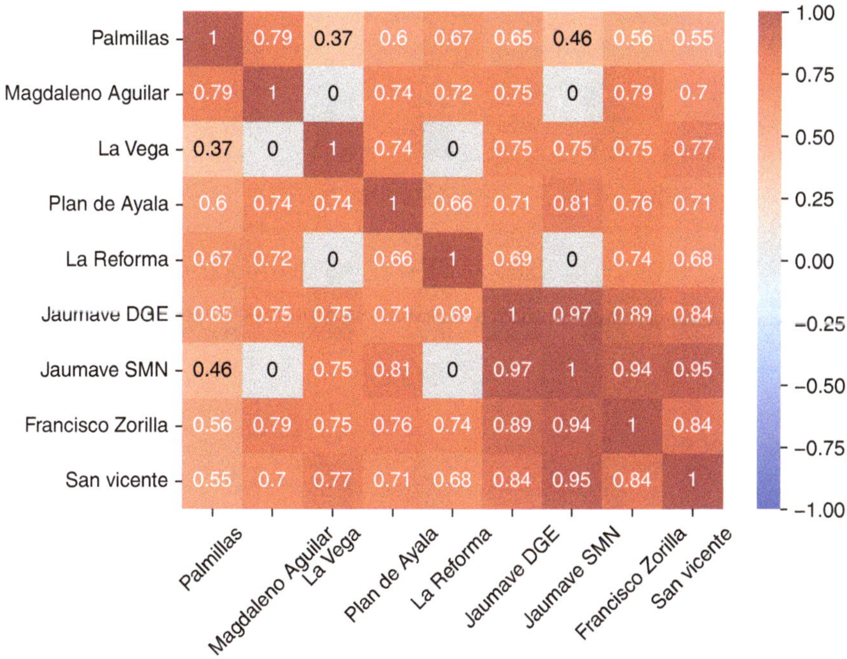

Fig. 3.3 Matrix correlation between rainfall stations

$$wcp_{i,j} = proportion_{i,j} * x_j * corr(x_j, x_i) \tag{3.2}$$

x_j represents stations with data, and x_i represents stations with missing values. $corr(xj, xi)$ is the correlation coefficient between the precipitation data from station j and station i, obtained directly from the correlation matrix shown in Fig. 3.3. This value quantifies the historical relationship of precipitation between these two stations.

3. Weighted Average: The weighted Correlation Product from all stations with data was summed and divided by the sum of the correlation matrix values.

$$x_{imp} = \frac{\sum wcp_{i,j}}{\sum corr(x_j, x_i)} \tag{3.3}$$

We expected the original dataset to contain 5616 values, but there were only 4063. This means that 27.66% of the rainfall data were imputed. In Fig. 3.4, we can see data distribution before and after imputation. They share a similar tendency, given that the highest and lowest values are in the same stations in both datasets. Furthermore, there appears to be a similar dispersion of data between them, indicating similar variability in rainfall values at each station.

This imputation method was appropriate since it did not significantly alter the data's variability.

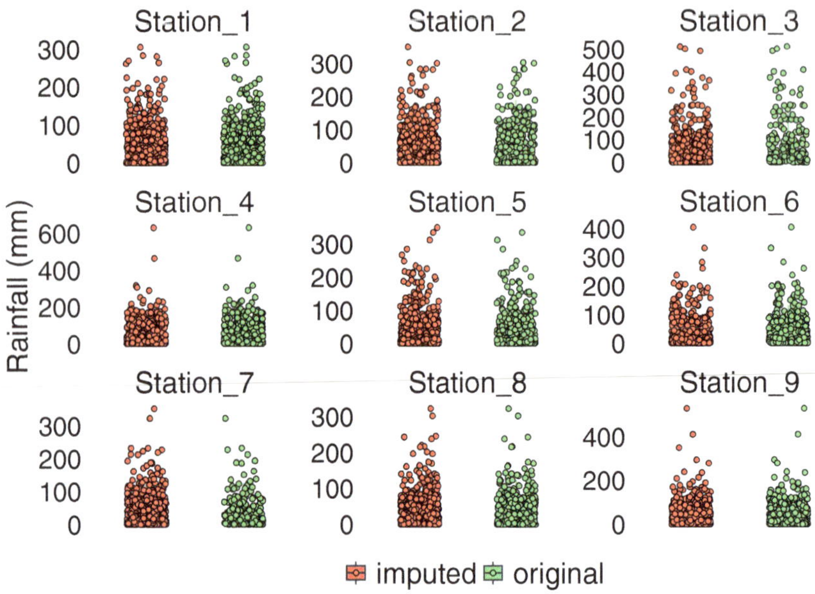

Fig. 3.4 Distribution of original and imputed rainfall values

3.3.2.2 Outliers

During our analysis of streamflow data, we encountered outliers (values that deviated significantly from the expected range without any apparent cause). To ensure data accuracy and reliability, we removed these outliers, comprising 12% of the total dataset.

We calculated the quartiles of the data to determine the interquartile range (IQR). Using the IQR to identify potential outliers, we established upper and lower limits and excluded any values that fell outside these limits (as illustrated in Fig. 3.5) from our forecasting model.

3.4 LSTM Forecast Model Development

In this study, we developed an LSTM model to forecast monthly streamflow. We divided the dataset into training (80%), validation (10%), and testing (10%) sets and normalized it to a scale of −1 to 1. The model's input was precipitation data from rainfall stations, and the output was the flow from the streamflow gauge. We use a sliding window strategy with a window size of 2 months to make one-step-ahead forecasts (1-month forecast horizon) of the monthly streamflow. We focused on tuning hyperparameters, particularly the sliding Window and number of epochs, as these parameters significantly impacted the model's performance (Table 3.2).

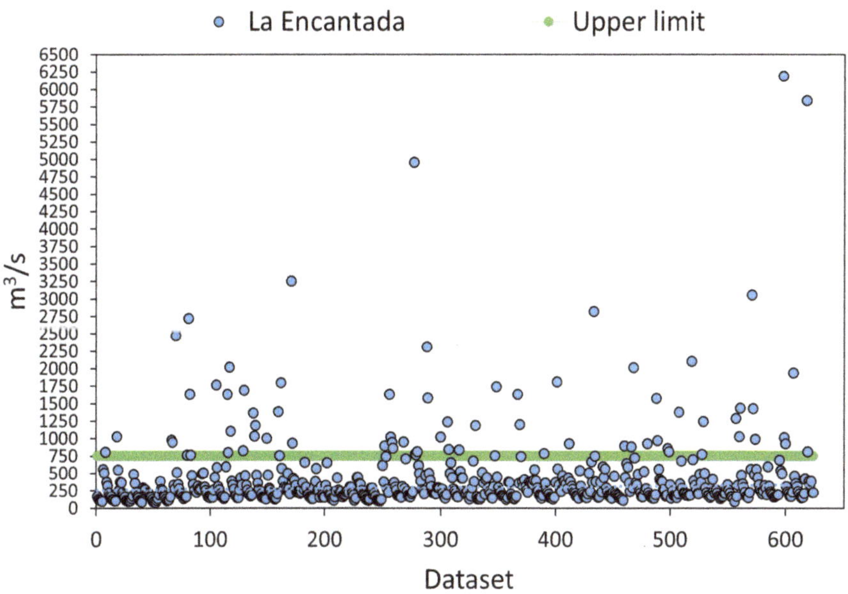

Fig. 3.5 Identification of outliers from the Streamflow station

Table 3.2 The most important hyperparameter settings of the LSTM model for monthly forecasting

	Monthly forecast											
Epochs	10	20	30	40	50	60	70	80	90	100		
Sliding window length	1	2	3	4	5	6	7	8	9	10	11	12

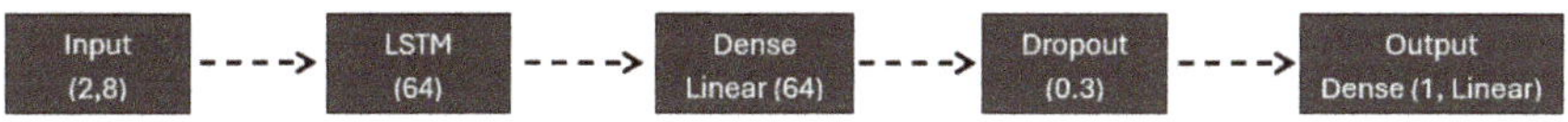

Fig. 3.6 Architecture of the LSTM neural network

We constructed a sequential model using an LSTM layer with 64 nodes, providing better results than 32 and 128. Each LSTM unit has an internal state that includes a memory cell (a vector of 64 learned weights) for long-term information retention and a hidden state (a vector of 64 learned weights) that carries over to the next layer. Additionally, we included a dense layer with linear activation and a Dropout layer for regularization. Then, we configure the output of the network with a Dense layer (See architecture in Fig. 3.6).

Through experimentation, we adjusted some hyperparameters to optimize performance. The details of the tested hyperparameters are provided in Table 3.3.

Table 3.3 Hyperparameters and values tested in LSTM

Hyperparameter	Example of values tested
Batch size	15, 25, 35, 45, 55, 65, 75, 85, 95
Optimizer	RMSprop, Adam, SGD
Memory cell	32, 64, 128
Hidden state	32, 64, 128
Activation	tanh, linear, relu
Dropout	0.1, 0.2, 0.3, 0.4, 0.5, 0.6, 0.7, 0.8, 0.9
Learning rate	1e-3, 1e-4, 1e-5, 1e-6, 1e-7, 1e-8, 1e-9
Epochs	10, 20, 30, 40, 50, 60, 70, 80, 90, 100

Table 3.4 Optimal hyperparameters settings for LSTM network test set through manual search

	Sliding window (months)	Optimization	Learning rate	Epochs	Batch size	RMSE (m^3/s)	MAPE
LSTM	2	RMSprop	1e-4	10	45	110.44	28.22

3.5 Results and Discussion

Table 3.4 displays the set of hyperparameters that resulted in the best performance during the preliminary test through manual hyperparameter search. The highest level of accuracy achieved is a Root Mean Squared Error (RMSE) of 110.44 and a Mean Absolute Percentage Error (MAPE) of 28.22. In simpler terms, our LSTM model had an error rate of 28.22%.

Window and Epochs were the hyperparameters that had the most significant impact on the results. Figures 3.7 and 3.8 show that increasing window and epochs leads to a higher RMSE value. Typically, more training leads to better results, but in our model, this improvement plateaus around number 10 and then enters a state of overfitting. In this state, the model is trained excessively on the current data, leading to poor predictions for future data.

A Sliding Window from 5 to 9 months does not help to forecast the next month, although a year-long perspective offers some improvement. Rainfall in a given month will likely be reflected within that month or the next, as factors like soil infiltration, evapotranspiration, water gradient, and proximity to the river may influence [22]. The forecast data varies on average by $110.44m^3/s$ from the actual values.

It is important to note that we did not test every possible combination of hyperparameters because of the significant search space. However, we did search for the optimal hyperparameters through Bayesian optimization and we obtained a RMSE of 123.91 (Table 3.5) on the test set. The hyperparameter settings identified by this process sought to minimize the error on the validation set, achieving an RMSE of 124.80 as we can see in Table 3.6. This approach is based on finding a configuration that generalizes well to unseen data during training and optimization. In contrast, the manual hyperparameter search had a slightly higher RMSE on the validation set of 131.34. This suggests that while this manual configuration

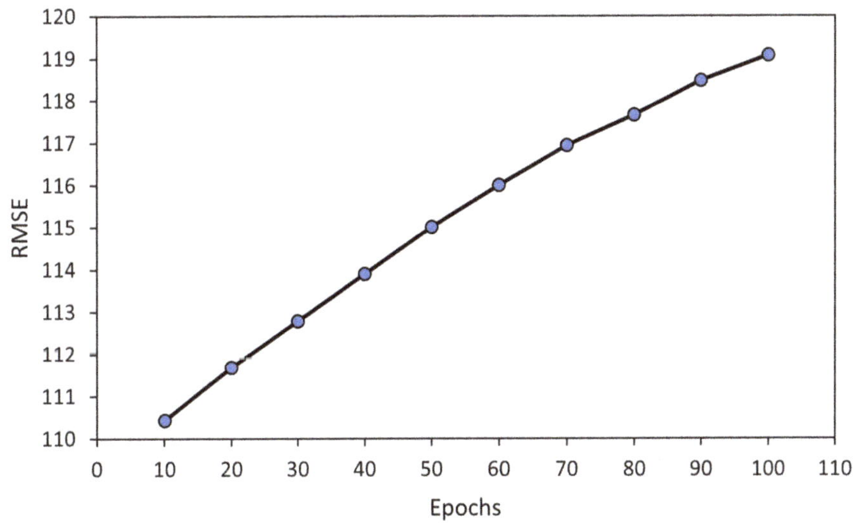

Fig. 3.7 RMSE results for different epochs testing a window of 2 months

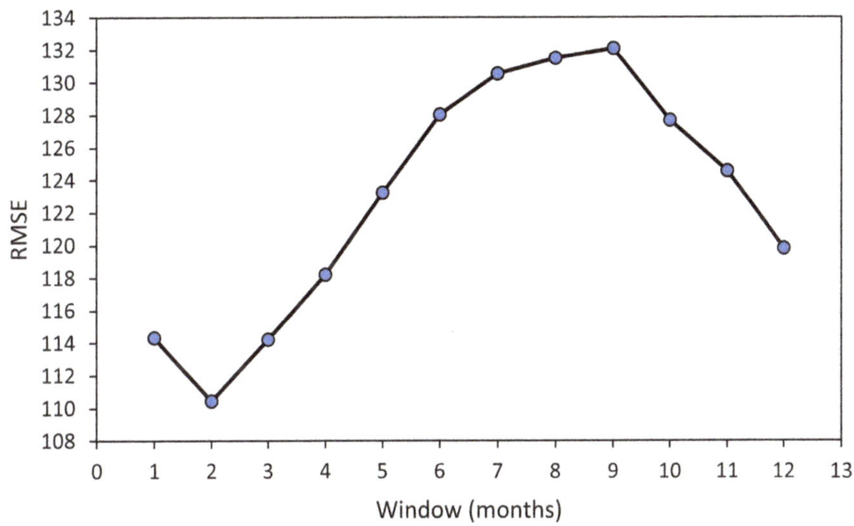

Fig. 3.8 Best RMSE performance testing a window of 2 months and 10 epochs

better fits the specific test set, it could be less robust to new unseen data than the configuration obtained through Bayesian optimization. The observed difference underscores the importance of the optimization strategy and the target metric. While Bayesian optimization focused on generalization evaluated on the validation set, the manual search explored a hyperparameter space that, coincidentally, resulted in better performance on the specific test set. However, considering the test set

Table 3.5 Optimal hyperparameters settings for LSTM network test set using Bayesian optimization

	Time lag (months)	Optimization	Learning rate	Epochs	Batch size	RMSE (m^3/s)	MAPE
LSTM	2	RMSprop	1e-3	50	85	123.91	34.45

Table 3.6 Comparison of RMSE obtained by manual search and Bayesian optimization

	RMSE test	RMSE validation
Manual search	110.44	131.34
Bayesian optimization	123.91	124.80

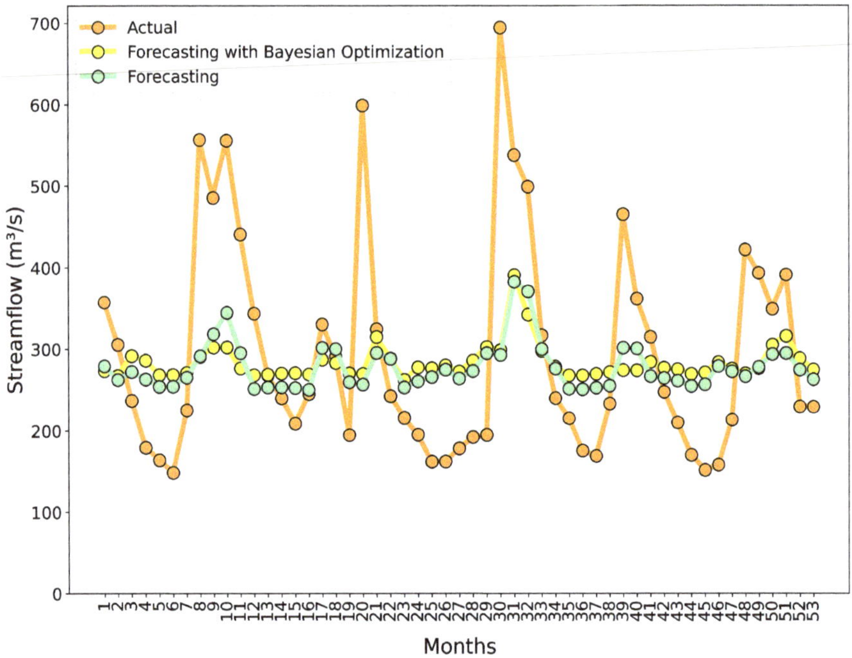

Fig. 3.9 Differences and trends between actual and forecast data

completely independent of the optimization process provides a more reliable estimate of the model's future performance.

In this paper, the root mean square error (RMSE) obtained through manual hyperparameter search on the test set was 110.44, which could be considered relatively high in absolute terms. However, for the specific objectives of this application problem, and as illustrated in Fig. 3.9, the overall trend of the predicted data (green line) follows the dynamics of the real data (orange line) over time. This ability to capture the underlying trend is an acceptable result for the current stage of this study. This level of error is sufficient for the initial objectives for several reasons. First, the quality and granularity of the available historical data can be limiting factors in the achievable accuracy. Refinement of data quality or incorporating larger and more accurate data sets in future work could reduce the error. Second, due to its exploratory nature and limited search space, manual hyperparameter search may not have identified the optimal configuration. Implementing more comprehensive hyperparameter search methods, such as Bayesian optimization (whose results we also present in this study), which explore a much larger space of variations, can significantly improve forecast accuracy in future research. Finally, the primary objective of this particular study is to demonstrate the feasibility of the methodology and analyze general trends in forecasts obtained with different approaches. While error reduction is an important long-term goal, the ability of the manual search model to replicate general streamflow dynamics, despite the relatively high RMSE, provides a useful basis for future improvements and comparisons with other methods.

3.5.1 *LSTM vs ARIMA*

We developed an ARIMA model to compare the LSTM forecasting with the results of another technique. When plotting results, at first glance, the LSTM results show a more similar trend to the actual data compared to ARIMA, which shows a softer line (Fig. 3.10). In addition, the ARIMA model yields an error of 128.32, more significant than that evaluated in the LSTM model. In concrete numbers, LSTM

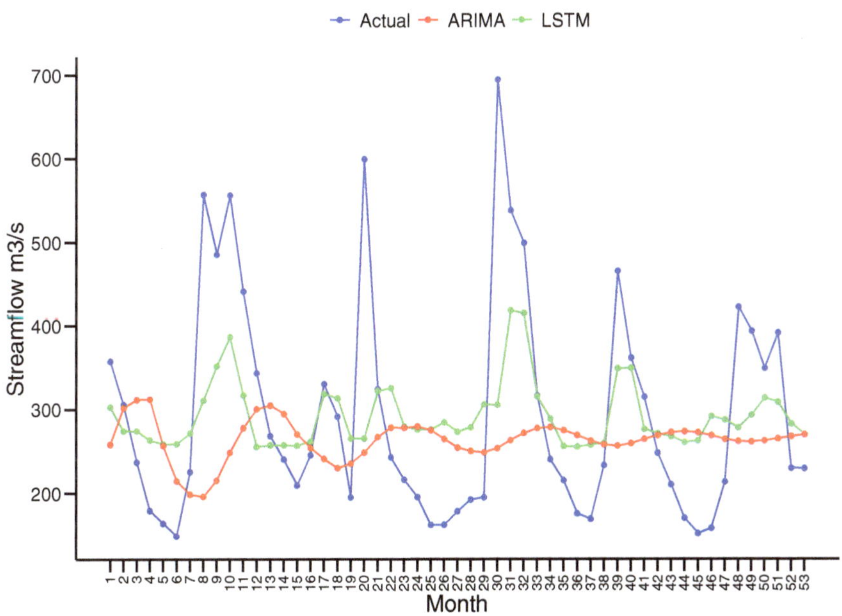

Fig. 3.10 Results among LSTM and ARIMA models

was 31 months closer to the actual values compared to the 22 months ARIMA gave for the 53-month forecast.

The quality of input data in a neural network is essential for accurate forecasting in hydrology, particularly when measuring precipitation [23]. While commonly used historical datasets, such as CAMELS and ERA5, can be employed to evaluate an LSTM neural network, we have opted to utilize regional information to set a precedent and further enrich this study. However, historical data in Mexico often lacks quality due to various factors, including device miscalibration, reading errors caused by obstructions, human mistakes in data handling, or the remote locations of monitoring stations. Therefore, we must make the most of the available resources when faced with these challenges.

3.6 Conclusions

In this study, we develop a Long-Short-Termemory (LSTM) forecasting model for monthly streamflow using historical rainfall and streamflow data. We implement some data cleaning: weighted averages as a treatment for data imputation and outlier removal using the interquartile range (IQR) to improve the quality and help the model work better. The graphical results indicate that the forecasted data closely follow the trend of the actual data; however, the error of 110.44 m^3/s can be minimized by addressing several factors that may impact the model, such as the amount of imputed data, the presence of outliers, and the sources of information used. When comparing the LSTM model with the ARIMA forecasting model, the LSTM consistently produces better results, with a trend that aligns more closely with the actual values than the predictions made by ARIMA.

For future work, we recommended improving the data preprocessing by using several imputation methods, a recollection of more precise data, and some techniques to cover every possible combination of hyperparameters.

AI Use Statement: Some portions of this manuscript were enhanced with the assistance of generative AI tools. The authors carefully reviewed, edited and approved all AI-generated content. The authors take full responsibility for the accuracy and integrity of the entire chapter.

References

1. Rivera, C.: Se levantan en armas autodefensas del agua en jaumave; llevan 5 años sin sembrar (2022). https://www.hoytamaulipas.net/notas/492009/Tamaulipas-Se-levantan-en-armas-autodefensas-del-agua-en-Jaumave-llevan-5-aos-sin-sembrar.html
2. Tapia, J.L.: Detectan tomas ilegales de agua en la cuenca guayalejo-tamesí (2024). https://www.elsoldetampico.com.mx/local/detectan-tomas-ilegales-de-agua-en-la-cuenca-guayalejo-tamesi-11739966.html

3. Rahimzad, M., Nia, A.M., Zolfonoon, H., Soltani, J., Mehr, A.D., Kwon, H.H.: Performance comparison of an LSTM-based deep learning model versus conventional machine learning algorithms for streamflow forecasting. Water Resour. Manag. **35**, 4167–4187 (2021)
4. Knight, D., Shamseldin, A.: River Basin Modelling for Flood Risk Mitigation. Taylor & Francis, Milton Park (2006)
5. Lees, T., Buechel, M., Anderson, B., Slater, L., Reece, S., Coxon, G., & Dadson, S.J.: Benchmarking data-driven rainfall-runoff models in great britain: a comparison of long short-term memory (LSTM)-based models with four lumped conceptual models. Hydrol. Earth Syst. Sci. **25**, 5517–5534 (2021)
6. C.L. Wu, K.W. Chau, Data-driven models for monthly streamflow time series prediction. Eng. Appl. Artif. Intell. **23**, 1350–1367 (2010)
7. Boulmaiz, T., Guermoui, M., Boutaghane, H.: Impact of training data size on the LSTM performances for rainfall–runoff modeling. In: Modeling Earth Systems and Environment, vol. 6, pp. 2153–2164. Springer, Berlin (2020)
8. Ramírez-Gil, C.M., Ramírez Gil W.A.: Artificial Intelligence Programming: A Practical Course. RA-MA Editorial, Madrid (2023)
9. Chumino, V., Rodriguez, R., Stalker, A.: Recurrent Neural Networks Applied to Indoor Localization Systems in WLAN. Thesis, Universidad de la Repuública, Montevideo (2021)
10. Man, Y., Yang, Q., Shao, J., Wang, G., Bai, L., Xue, Y.: Enhanced LSTM model for daily runoff prediction in the Upper Huai River Basin, China. Engineering **24**, 229–238 (2023)
11. Kratzert, F., Klotz, D., Brenner, C., Schulz, K., Herrnegger, M.: Rainfall-runoff modelling using long short-term memory (LSTM) networks. Hydrol. Earth Syst. Sci. **22**, 6005–6022 (2018)
12. Hu, Y., Yan, L., Hang, T., Feng, J.: Stream-flow forecasting of small rivers based on LSTM. arXiv:2001.05681 (2020)
13. Yan, L., Feng, J., Hang, T.: Small Watershed Stream-Flow Forecasting Based on LSTM, vol. 935, pp. 1006–1014. Springer, Berlin (2019)
14. Hunt, K.M., Matthews, G.R., Pappenberger, F., Prudhomme, C.: Using a long short-term memory (LSTM) neural network to boost river streamflow forecasts over the western united states. Hydrol. Earth Syst. Sci. **26**, 5449–5472 (2022)
15. Sahoo, B.B., Jha, R., Singh, A., Kumar, D.: Long short-term memory (LSTM) recurrent neural network for low-flow hydrological time series forecasting. Acta Geophys. **67**, 1471–1481 (2019)
16. Ni, L., Wang, D., Singh, V.P., Wu, J., Wang, Y., Tao, Y., Zhang, J.: Streamflow and rainfall forecasting by two long short-term memory-based models. J. Hydrol. 583, 124296 (2020)
17. Kareem, K.Y., Seong, Y., Jung, Y.: LSTM prediction of streamflow during peak rainfall of piney river LSTM piney river. J. Korean Soc. Disaster Secur. **14**, 17–27 (2021)
18. R. Vera-Vázquez, La cuenca del río Guayalejo-Tamesí, situación actual, políticas públicas y perspectivas. El Colegio de Tamaulipas (2004)
19. Proyecto de Integración y Desarrollo de Mesoamérica, Atlas of Natural Hazards in the Municipalities of Tampico, Madero and Altamira, Tamaulipas (2009). https://www.tamaulipas.gob.mx/proteccioncivil/wp-content/uploads/sites/36/2011/12/ATLAS-DE-RIESGOS-MPIOS-TAMPICO-MADERO-Y-ALTAMIRA.pdf
20. Nacional, S.M.: Información estadística climatológica (2018). https://smn.conagua.gob.mx/es/climatologia/informacion-climatologica/informacion-estadistica-climatologica
21. del Agua, C.N.: Bases de datos hidrométricas (2016). https://app.conagua.gob.mx/bandas/bases.html
22. Tarbuck, E.J., Lutgens, F.K.: Ciencias de la Tierra: una introducción a la geología física. Pearson-Prentice-Hall, Saddle River (2005)
23. Sikorska, A., Seibert, J.: Importance of precipitation data quality for streamflow predictions. Geophys. Res. Abstracts **17**, 2015–13369 (2015)

Chapter 4
Leveraging Language Models for Location-Specific Disease Mortality Rate Prediction from Large-Scale Social Data

Reshma Unnikrishnan, Sowmya Kamath, and V. S. Ananthanarayana

Abstract Social media platforms significantly influence users' outlook, shaping and molding their likes, dislikes, and thought processes, and even affecting mood and well-being. The content posted by social media users can potentially reveal signs of their emotions, indicating underlying mental and emotional health concerns. Given the rise in mental health disorders and the lack of access to effective healthcare, there is a growing need to explore research avenues that could benefit users experiencing anxiety-related disorders and depression, two of the most common conditions. While these conditions are explicitly studied, the interconnection between mental health and other stress-related chronic diseases like heart ailments and diabetes is also on the rise. In this article, an approach that leverages semantic representation techniques and deep neural models to analyze social media data, particularly community-level tweets, for predicting Atherosclerotic Heart Disease (AHD) mortality rates in vulnerable communities is presented. Transformer-based Masked Language Models (MLMs) were utilized to represent tweet data contextually, and Transfer learning methods were adopted for further optimization. Experiments revealed that tweets correlate with community-level AHD mortality rates, and finetuning experiments on open benchmark datasets highlighted the efficacy of finetuned MLM models, which outperformed all other language models considered for the study.

Keywords Social media analysis · Regional-level analysis · Heart disease prediction · Contextual embedding representation · Language modeling · Transfer learning

R. Unnikrishnan (✉) · S. Kamath · V. S. Ananthanarayana
Department of Information Technology, National Institute of Technology Karnataka, Mangalore, India
e-mail: reshma.anata@gmail.com; sowmyakamath@nitk.edu.in; anvs@nitk.edu.in

G. Rivera et al. (eds.), *Recent Advances in Artificial Neural Networks*, Intelligent Systems Reference Library 283, https://doi.org/10.1007/978-3-032-13497-4_4

4.1 Introduction

With the exponential increase in digital data, systems that utilize intelligent analytics to support various applications have also seen significant growth. Given their role in personal and business communication, networking, and various interactions, social media platforms are key contributors to this data surge. Since emotions are closely tied to an individual's actions, the language used in social media communications can provide profound insights into a person's mindset and opinions. Numerous research studies have explored applications such as the link between social media images and loneliness [49], predicting vaccine hesitancy in communities [35], personality assessment [3, 48], consumer purchase intentions [23], election result predictions [66, 67], the relationship between sentiments and sharing behavior [6, 63], understanding loneliness through sentiment analysis [59], estimating county health statistics [16, 51, 68], and predicting disease transmission [35, 55]. Over the past decade, researchers have increasingly leveraged social data to generate actionable insights for real-world applications, such as supporting epidemic outbreak detection [15], predicting political alignment [14], forecasting crime rates [25], and even predicting personality traits related to entrepreneurship and business growth [47].

Twitter, recently rebranded as X, continues to be a key platform for analyzing social behavior across various domains, including public health. Tweets have been utilized to compute happiness indices across geographical regions, helping to identify depressive and anti-depressive communities [53]. Additionally, intensity-level classifications and sentiment analyses have been conducted to understand emotional patterns [2, 8, 36], with studies also exploring global sentiments [60, 61] and assessing mental health status [54, 56]. Social media platforms are now extensively used for public health surveillance, particularly through tweet analysis. However, differentiating tweets that provide meaningful health insights from general noise remains a significant challenge. Efforts have been made to address this issue, including the development of an influenza infection detection algorithm capable of performing analysis across different geographical granularities, which was reported to closely align with data from the CDC and the Department of Health and Mental Hygiene of New York City [10]. Beyond public health surveillance, Twitter data has also been used to gauge community-level social well-being, or "life satisfaction rate". Studies suggest that positive terms related to outdoor activities, exercise, and fitness are associated with higher life satisfaction, while terms reflecting disengagement, such as boredom and tiredness, are linked to lower satisfaction [58]. Predictive models for life satisfaction have been developed using phone survey data, with demographic controls drawn from the U.S. Census Bureau.

A growing body of research has leveraged social media data to investigate correlations between linguistic patterns and health outcomes [31, 32, 42]. For instance, it has been shown that the use of future tense in tweets has a low correlation with human immunodeficiency virus (HIV) prevalence, with approaches such as dictionary-based and topic modeling being employed for analysis [30]. The pandemic and the resulting infodemic have intensified interest in analyzing social data, leading to numerous studies examining public health impacts and individual well-being. One line of research has focused on understanding public health beliefs, using machine learning techniques to analyze the perceived susceptibility, severity, benefits, and barriers of COVID-19 cases through the language used on social media platforms [69]. Other studies have explored the effects of physical distancing, applying deep learning techniques to measure its influence on health behaviors [50]. In addition to mental health and emotional responses during the pandemic [4, 43, 73], further investigations have examined the impact of vaccinations [37], lockdowns [39], and perceptions of healthcare workers [70]. While these studies have made significant contributions, many still rely on conventional data analysis techniques, with limited contextual representations and advanced data-centric methods in understanding social data for health predictions.

This work focuses on maximizing the potential of social media data for predicting locality-specific mortality rates. Social media posts are represented in a contextual vectorized format to identify locations with high heart disease mortality rates, using correlations with public health entities. Experiments are conducted to evaluate the effectiveness of various contextual embedding techniques in representing social media posts. By excluding conventional socio-economic, demographic, and health features, the aim is to leverage Twitter data exclusively for heart disease mortality rate prediction. The generated insights could be extended to predict the prevalence of other diseases and outbreaks. The contributions of this work are threefold:

1. Contextual vector representation of community-level tweet data using MLM and non-MLM models.
2. Fine-tuning of MLMs using Transfer learning techniques for descriptive and predictive analysis of AHD mortality rate prediction.
3. Benchmarking with existing state-of-the-art works for descriptive and predictive analysis.

The remainder of this chapter is organized as follows: Sect. 4.2 discusses existing works aligned with the current study. The proposed methodology and processes are presented in detail in Sect. 4.3, highlighting the representation techniques designed for the semantic representation of the data. The experimental evaluation and results are detailed in Sect. 4.4. A discussion of the results obtained is presented in Sect. 4.5, followed by the conclusion and potential future work directions.

4.2 Related Work

The widespread dissemination of social health data has led to diverse analyses and examinations of various public health perspectives. A recent survey [20] on using Twitter in public health research identified six broad domains: surveillance, event detection, pharmacovigilance, forecasting, disease tracking, and geographic identification. The survey also noted that many studies still rely on supervised techniques and labeled data, which limits the scope of larger-scale analysis. Examples of Twitter's diverse applications include the analysis of the rare Monkeypox disease outbreak in 2022 [64], classification of audience engagement with celebrity tweets based on content [17], predicting the spread of new tweets [5], identifying cyberbullying [7], and even forecasting morning traffic in Pittsburgh using overnight tweet data [75]. These studies demonstrate the platform's versatility in population-centric studies in varied domains.

In the context of population health research, Nguyen et al. [45] utilized geo-tagged tweet data as predictors to measure the correlation of various health variables with the mortality rate obtained from the Centre for Disease Control and Prevention (CDC). Using the MALLET toolkit, sentiment analysis was performed on the tweet data, revealing that communities with more tweets about food and physical activity had fewer chances of premature mortality and obesity. Similarly, positive attitudes and happiness were closely related to beneficial health outcomes. Communities with excessive alcohol consumption were more prone to higher accidental mortality rates. Another study [9] showed how negative emotions in tweet language were closely correlated with Trump's victory in the 2016 elections, with his supporters primarily from counties expressing negativity and anger over tweets. This research utilised open-source data from Eichstaedt et al. [21], covering 1347 counties in the U.S. Dredze et al. [19] highlighted the advantages of using social media data for monitoring health-related events. They proposed the Ailment Topic Aspect Model (ATAM), a probabilistic graphical model that predicts diseases from raw tweets, identifying correlations between dominant features in disease prediction. Their findings included positive correlations between smoking and cancer, negative correlations between exercise and obesity, and links between good health and diseases. The study emphasised the value of social media data, particularly Twitter, in uncovering private thoughts people may hesitate to share in real life, such as medication usage or insomnia management, to derive meaningful insights for societal benefit.

Another work by Tufts et al. [65] used tweets exclusively from Pennsylvania to assess overall health conditions and perspectives on health and medical diagnoses in the region. They selected 14 common health conditions and examined the correlation between word usage and each disease. Nguyen et al. [46] developed a graph-based approach that utilized location-tagged Twitter data for population health analysis, employing the Radial basis method and k-means clustering to explore inter-feature relationships and understand the inter-tweet graph. Zhang et al. [76] modeled the experience of patients undergoing chemotherapy through tweets from both personal and organizational accounts, performing content and sentiment analysis to understand topics relevant to arthritis during the pandemic. The mental state of individuals staying at home, focusing on vulnerable communities, was investigated by Xia et al. [71] using regression analysis. They used the open-source Twitter Sentiment Geographical Index (TSGI) dataset, labeled with Transformer-based models, and found that stay-at-home orders were directly correlated with the well-being of elderly people and small and medium enterprises.

Eichstaedt et al. [21] utilized Twitter data alongside socioeconomic, demographic, and health variables to explore the correlation between emotions and Atherosclerotic Heart Disease (AHD) mortality rates. Their findings indicated that positive words had a protective effect, while negative emotions were strongly correlated with higher AHD mortality rates. While these studies relied on Twitter data for analysis, Brown et al. [11] conducted a reanalysis that contradicted Eichstaedt et al. [21]. Brown's study used suicide as a predictor variable and found no connection between negative words and suicide rates, concluding that Twitter language was not reliable for predicting heart disease mortality rates. In response, Eichstaedt et al. [22] replicated Brown's study, incorporating elevation and rurality as confounding variables. They found that disengagement had a positive correlation with suicide, supporting their original findings. This demonstrates that Twitter language reflecting negativity and disengagement correlated with an increased risk of AHD mortality, whereas positivity and engagement did not. Their regression model incorporated ten socioeconomic, demographic, and health risk factors alongside Twitter data, with the AHD mortality rate as the dependent variable. The correlation between these variables and AHD mortality was stronger when Twitter data was included, compared to using only traditional variables. Their results were validated by comparing predicted outcomes with the Center for Disease Control and Prevention (CDC) data.

In this work, the focus is on capturing the contextual representation of community-level tweets, applied exclusively for predicting community-level Atherosclerotic Heart Disease (AHD) mortality rates. A significant limitation of many researchers' work is their focus on manually creating tweet dictionaries for topic identification. The dictionary-based approach lacks boundaries in limiting word tokens or having a fixed representation standard, while topic modeling becomes increasingly complex as the data size grows [56]. This challenge, known as the curse of dimensionality, can hinder effective modeling when processing large datasets [44]. To address these issues, manual efforts in creating dictionaries and topics are reduced by decreasing the time spent on descriptive analysis, while tweet representation is improved by incorporating embedding mechanisms that carry contextual information. Pre-processing techniques are applied to assess the impact of embedding methods on tweet formatting. Two transfer learning approaches, inductive and transductive, are used to fine-tune models for better tweet data representation. Dimensionality reduction techniques are also employed to derive insights from a lower-dimensional perspective, enhancing correlation analysis.

4.3 Materials and Methods

The overall workflow of the proposed approach is depicted in Fig. 4.1. Tweets from each county were considered for the evaluation (Details in Sect. 4.3.1) were first cleaned and aggregated in terms of communities and their AHD Mortality rate. Using different foundation language models, the community tweets were

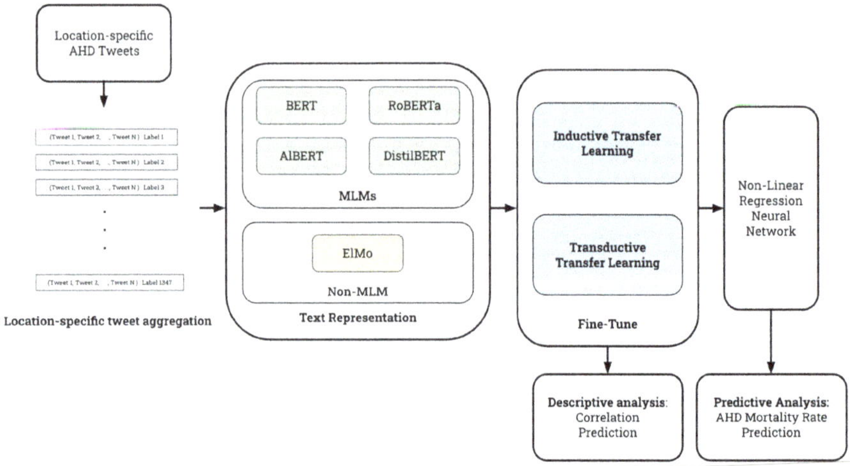

Fig. 4.1 Proposed methodology

represented as a contextual vector that depicts the latent features. These vectors and their labels are fed to the neural network-based regression architecture to perform the 10-fold cross-validation, and the correlation of the predicted values with ground truth values is computed. The best-performing foundation language model was then selected for further optimization using Transfer learning techniques. Here, the model is fine-tuned to adapt the domain knowledge through the Inductive and Transductive transfer learning methods [52].

4.3.1 Data Specifics

For experimental validation, a standard dataset [21] was used, and the self-reported locations of users were considered to map tweets to communities. The dataset consists of 148 million tweets from 1347 communities, covering approximately 88% of the U.S. population. County-level age-adjusted mortality rates for AHD were obtained from the CDC, represented by code `I25.1` for a population size typically per 100,000 or 1000. Demographic and health risk variables, including the percentage of married residents, high school and college graduation rates, percentage of female, Black, and Hispanic residents, as well as self-reported diabetes, obesity, smoking, and hypertension, were sourced from the U.S. Census Bureau and the CDC. The distribution of AHD mortality rates across U.S. states shows significant imbalance across different locations. This unbalanced data distribution is illustrated in Fig. 4.2, with the U.S. states on the X-axis and the AHD mortality rate on the Y-axis.

Descriptive analytics experiments were conducted to correlate commonly considered categorical features—such as socio-economic, demographic, and health

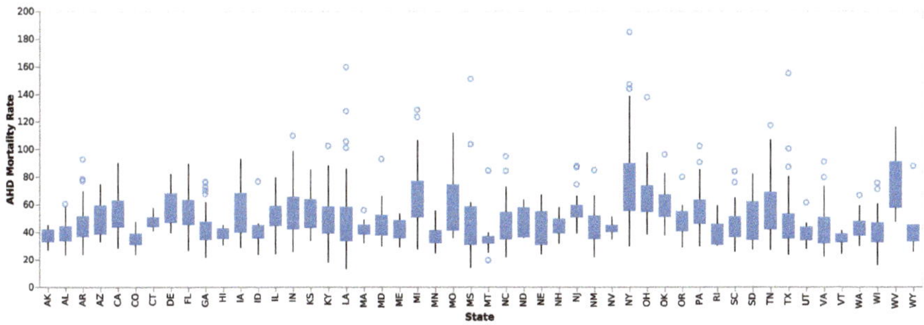

Fig. 4.2 Distribution of AHD mortality rate across US states

variables—with AHD mortality rates to assess their effectiveness. Demographic features (population, percentage of married residents, and percentage of female, Black, and Hispanic residents) were correlated with AHD mortality rates, with results presented in Table 4.1. The data indicates that nearly all demographic features are negatively correlated with AHD mortality rates. Socioeconomic features (county-level median income, unemployment rate, and high school and college graduation rates) exhibited similar results, with most features showing no significant correlation with mortality rates, except for unemployment, as detailed in Table 4.1. In contrast, health variables (self-reported diabetes, obesity, smoking, and hypertension) positively correlated with mortality rates, as shown in Table 4.1. These results suggest that socioeconomic and demographic features contribute less to the prediction task, whereas health variables show a stronger correlation.

4.3.2 Data Preparation

Tweets, being unstructured text data, often contain short forms, URLs, emojis, and other elements that require effective pre-processing to extract relevant insights. The pre-processing involved normalizing URLs, numbers, hashtags, mentions, reserved words (e.g., RT, FAV), emojis, and smileys using the tweet preprocessor tool.[1] The representation of tweets was extracted from raw data following a series of steps, including the removal of stop words, numeric values, and retention of only positive words, as well as both positive and negative words, using an opinion lexicon [28] for reference. The impact of these pre-processing steps is evident from Fig. 4.3, which highlights significant changes in tweet length (number of tokens) after each step. To avoid bias in the model due to variations in the number of words in each community's tweets, the number of words and their length were correlated with the AHD mortality rate and the results are presented in Fig. 4.4, which indicate

[1] http://pypi.org/project/tweet-preprocessor/.

Table 4.1 Common features vs. AHD mortality rate

Data	Features	Correlation
Demographics	femalePOP165210D$10	−0.063877
	hispanicPOP405210D$10	−0.026607
	blackPOP255210D$10	−0.081384
	forgnbornHC03_VC134ACS3yr$10	−0.039765
	marriedaveHC03_AC3yr$10	−0.040272
	marriedmaleHC03_VC37ACS3yr$10	−0.052400
	marriedfemHC03_VC44ACS3yr$10	−0.026492
	pop2010	0.028583
Socioeconomics	hsgradHC03_VC93ACS3yr$10	−0.134773
	bachdegHC03_VC94ACS3yr$10	−0.274085
	incomeHC01_VC85ACS3yr$10	−0.137372
	hsbachgradHD03_ACS3$10	−0.229059
	logincomeHC01_VC85ACS3yr$10	−0.140894
	unemployAve_BLSLAUS$0910	0.159699
Health	smokerCHR13$0511	0.245563
	diabeticCHR13$09	0.192862
	obeseCHR13$09	0.202981
	fairpoorhealthCHR13$0511	0.212318
	physunhlthdaysCHR13$0511	0.242264
	mentalunhlthdaysCHR13$0511	0.235520
	hypertenmaleIHME$09	0.152085
	hypertenfemIHME$09	0.138734
	hypertenaveIHME$09	0.147178

no significant correlation. The preprocessed tweets were subsequently aggregated based on community information obtained from the CDC.

4.3.3 Semantic Representation Techniques

Our approach is built on encoder-decoder architectures, where the encoder maps the input sequence to a higher-dimensional space while the decoder translates the abstract vector into the desired output sequence. As the quality of text representation directly affects the comprehension of the contextual information embedded within the text tokens, we experiment with various Transformer architectures to optimize the representations. BERT (Bidirectional Encoder Representations from Transformers) [33], an encoder-only model, was employed here. BERT is an instance of contrastive self-supervised learning that processes input text bi-directionally using a masking technique to represent tweets mapped to communities. Additionally, the ELMo (Embeddings from Language Models) [34], a Bidirectional

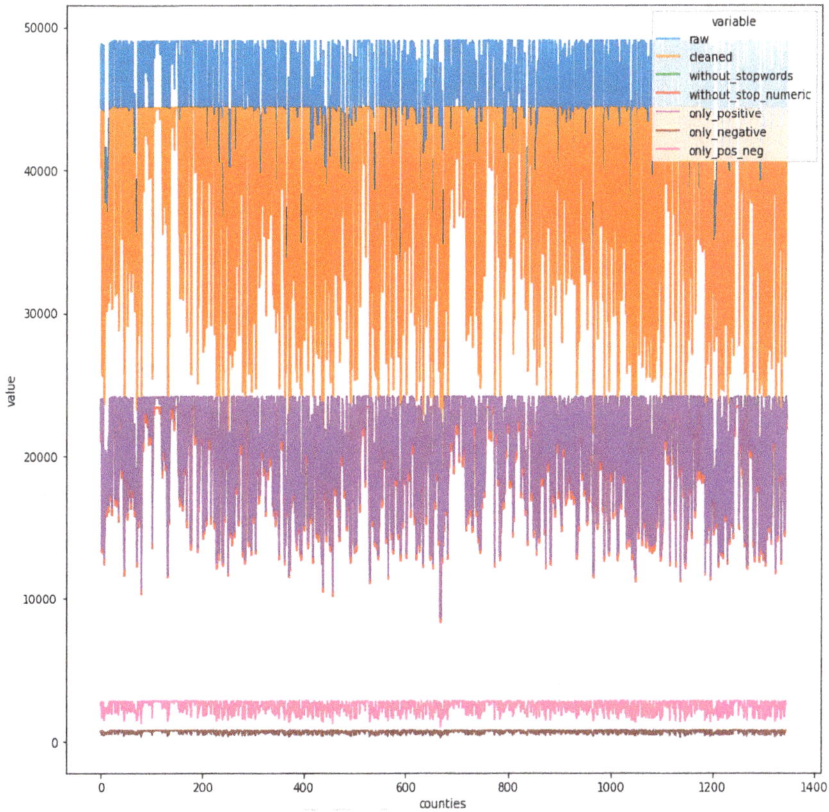

Fig. 4.3 Length of tweets in counties after preprocessing

LSTM-based model known for handling out-of-vocabulary (OOV) words due to its unique character-level embedding mechanism, was also experimented with for the contextual representation of community-mapped tweets data. These models replace the manual feature extraction pipelines used in earlier works, leveraging neural networks and embedding representation techniques to enhance predictive capabilities.

1. *ELMo:* a Bidirectional LSTM-based model that handles vocabulary words with its unique mechanism of character-level embedding for the contextual representation of the community-mapped tweets data. In ELMo, each word is computed to its lower-level representation using character embeddings that are good at representing words based on context and handling OOV words. This character-level representation is then used to train a convolutional neural network through several filters and a max-pooling layer. The computed representations are then sent through a two-layer highway network before being passed as input to the bidirectional LSTM layers. Bi-LSTM layers are run separately, forward

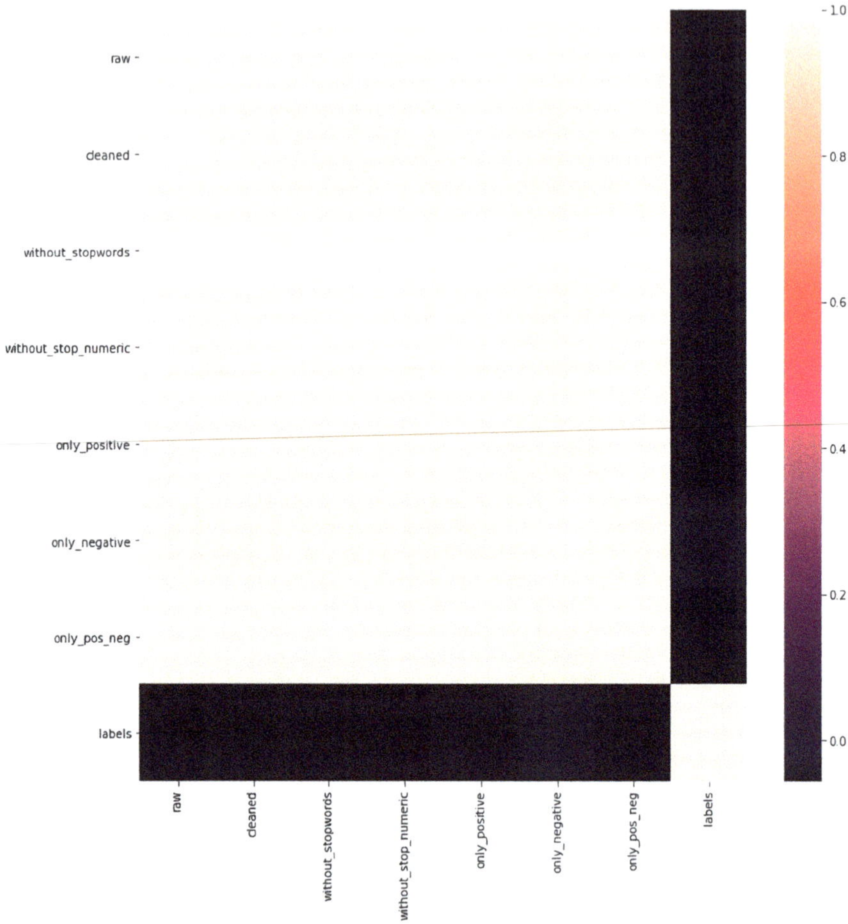

Fig. 4.4 Correlation between length of preprocessed tweets and AHD mortality rate

and backwards, then concatenated together at each layer, with the addition of residual connections between the two layers. The final word representation would be the combination of the generated word embeddings and 2 hidden layer representations from corresponding Bi-LSTM layers, as shown in Eq. (4.1). Assuming the input word i = '*happy*', s_i indicates hidden layer weights, and γ_j denotes the scaling factor while the '*task*' represents the ELMo model is flexible to be trained for task-specific requirements.

$$ELMo^{task} = \gamma_i(s_0^{task} * x_i + s_1^{task} * h_{1,i} + s_2^{task} * h_{2,i}) \tag{4.1}$$

ELMo can learn the embeddings from the unsupervised corpus in a supervised method so the model can be easily applied to different domains. The character embeddings built through Convolutional Neural Network (CNN), learns to discriminate between the morphemes, prefixes, and suffixes. Thus, it helps in finding the vector representation for the words which were not present in training (robust model) and also reduces vocabulary size (highly influential for morphologically rich languages) [34]. The Highway network in the architecture helps to carry the contributive features directly to the output layers, where other features undergo the affine transform followed by non-linearities to formulate the new features out of character embeddings [62]. Since the LSTM shows greater performance in language modelling tasks, the 2-layer Bi-LSTM included in this architecture is more suitable for the language understanding task [33, 74]. The residual connection between them helps solve any degradation problem in very deep networks, thereby allowing for successful network training [27].

2. *BERT:* well-suited for the task due to its distinctive feature of processing a given input sentence by conditioning each word based on both preceding and succeeding sequences, embedding contextual information into the produced vectors. Unlike existing contextual models that generate embeddings for a word based only on preceding sequences, BERT utilizes deep contextual information from surrounding words. The bidirectional mechanism is implemented using a masking technique, which ensures that the word of interest does not see itself during the random masking process. Using WordPiece tokenization technique the words are handled based on known subwords or characters and does not have a true OOV word list. This approach helps in handling unknown word list outside the predefined vocabulary size it is pretrained on and aids in unique representation to each token based on context. Inspired by the transformer architecture, BERT incorporates the encoder block, which includes self-attention and a Feed Forward Neural Network (FFNN) in each encoder layer. The self-attention layer resolves co-references by understanding which words are relevant for processing the current word. Additional features such as residual connections and normalization layers are included in the encoder to address issues like vanishing and exploding gradients and internal covariance shift between self-attention and FFNN [33].
3. *BERT Variants:* Four BERT variants were employed in this study—BERT-large-uncased [33], RoBERTa-large [40], ALBERT-large-v2 [38], and DistilBERT-base-uncased [57]. While RoBERTa-large uses Byte-Pair Encoding (BPE) to as tokenization method, the other BERT variants follow a similar approach to BERT. The BPE technique is similar to WordPiece tokenization except that it focuses on the most frequent pair of subwords rather than the maximum likelihood approach used in BERT. In this way, RoBERTa-large model gets generalized to unseen words by giving a robust representation to the words fed. The focus was on evaluating how different representation techniques, based on transformer architecture, handle tweet data and comparing the performance of these models in predicting AHD mortality rates. Although all these models are derived from the BERT framework, each offers distinct advantages, such as enhanced prediction mechanisms or reduced computational costs, when compared to BERT.

Empirical experiments revealed that RoBERTa [40] offers advantages over the BERT model but requires more training data, computational resources, and time. Performance improvements were noted with the use of a dynamic masking technique that varies during each training epoch to prevent the reuse of the same mask. Larger batch sizes and the inclusion of the full community tweet representation without Next Sentence Prediction (NSP) loss enhanced the training process. Byte Pair Encoding (BPE), with an expanded vocabulary of 50K, compared to BERT's 30K, effectively handled sub-word units without introducing unknown tokens. ALBERT [38], sharing the same architecture as BERT, was selected for its efficient use of parameters. While RoBERTa and ALBERT both omit NSP, ALBERT introduces Sentence Order Prediction (SOP) to improve context extraction by ensuring sentence segments are in order. ALBERT employs the same Masked Language Model (MLM) as BERT extended to 3 grams, and demonstrates improved parameter efficiency through shared parameters across the FFNN and self-attention layers. DistilBERT [57] aims to improve the inference time (4 times less than BERT) by training in only 16 GB data with only 66 million parameters compared to BERT base with 110 million and BERT large with 340 million parameters. With slight degradation in the performance of BERT, the method used approximates large neural networks into smaller ones by using the distillation technique.

4.3.4 Dimensionality Reduction

The foundation models are limited in handling smaller chunks of text (e.g., BERT is constrained to 512 tokens), necessitating an aggregation approach to summarise the representation of all chunks belonging to a county into a single vector. A factorisation method is preferred over averaging, as it preserves word order. To optimise computation and reduce latency, dimensionality reduction was applied to the tweet data. Singular Value Decomposition (SVD) [26], a proven data-driven generalisation of the Fourier Transform, was employed to extract the most relevant information. SVD was chosen over other matrix factorisation methods like Eigen Factorisation [1] and Non-Negative Matrix Factorisation [72], as it does not require the feature matrix to be positive-definite or square. Extracting insights from 148 million tweets requires a standardised method to infer features contributing to the correlation with AHD mortality rates. To examine this correlation, the feature matrix derived from different embedding techniques was decomposed using the top 20 singular values from Σ and singular vectors from U and V^T. The truncation from the economy SVD was computed in line with the Eckart-Young Theorem [13], minimising the difference between the original and approximated feature matrix in the Frobenius norm, resulting in the best rank for decomposition. Here, X is a matrix with $m = 1347$ rows (location-specific embedding vectors) and $n = 768/1024$ columns (depending on the dimensionality used by different models for each of these 1347 vectors), as shown in Eqs. (4.2) and (4.3), where, $x_{1347} \in \mathbb{R}^{1024}$. This is transformed as per Eqs. (4.4) and (4.5) during decomposition. The final decomposed matrix $\hat{X}_{1347\times 20} = U_{1347\times 20} \times \sigma_{20\times 20}$.

$$X = \begin{bmatrix} - & x_1 & - \\ - & x_2 & - \\ - & x_3 & - \\ & \vdots & \\ - & x_m & - \end{bmatrix}_{1347\times 1024} \tag{4.2}$$

$$U\Sigma V^T = \begin{bmatrix} | & | & & | \\ u_1 & u_2 & .. & u_m \\ | & | & & | \end{bmatrix} \begin{bmatrix} \sigma_1 & . & .. & . \\ . & \sigma_2 & .. & . \\ . & . & .. & . \\ . & . & .. & \sigma_n \\ & & & \end{bmatrix} \begin{bmatrix} - & v_1 & - \\ - & v_2 & - \\ - & v_3 & - \\ & \vdots & \\ - & v_n & - \end{bmatrix} \tag{4.3}$$

$$U = \begin{bmatrix} | & | & & | \\ u_1 & u_2 & .. & u_{1347} \\ | & | & & | \end{bmatrix}_{1347\times 20} \tag{4.4}$$

$$\sigma = \begin{bmatrix} \sigma_1 & . & . & . \\ . & \sigma_2 & . & . \\ . & . & . & . \\ . & . & . & \sigma_{20} \\ & & & \end{bmatrix}_{20\times 20} \tag{4.5}$$

$$\begin{bmatrix} -\hat{x}_1- \\ -\hat{x}_2- \\ -\hat{x}_3- \\ \vdots \\ -\hat{x}_m- \end{bmatrix} = \begin{bmatrix} | & | & & | \\ u_1 & u_2 & .. & u_m \\ | & | & & | \end{bmatrix} \begin{bmatrix} \sigma_1 & . & . & . \\ . & \sigma_2 & . & . \\ . & . & . & . \\ . & . & . & \sigma_{20} \end{bmatrix} \tag{4.6}$$

4.3.5 Sentiment Analysis

The language expressed through social media often reflects a person's emotions towards an issue, interest, product, event, or activity. Therefore, it is possible to infer certain psychological factors from shared tweets. Many previous studies have attempted to derive these factors using sentiment analysis-based approaches [24, 61, 63]. Sentiment analysis-based methods were employed to validate the sentiment conveyed in aggregated tweets and measure any correlation with the AHD mortality rate. Transfer learning techniques were used to perform both dictionary-based and deep learning-based sentiment analysis. Aggregated tweets from each county were analyzed using VADER (Valence Aware Dictionary and sEntiment Reasoner) [29], a dictionary-based approach that maps lexical features of a sentence or phrase to sentiment scores in a word-by-word manner to obtain the overall text score. This methodology was designed to handle unlabelled data and social media texts, including emoticons, punctuations, conjunctions, and slang. Similarly, tweet data was also analyzed using Sentiment Neuron, a model originally designed to predict the next word in a sentence. This model, with a linear combination of multiplicative LSTM structures and 4096 hidden units, was trained on 82 million Amazon reviews [18]. The computed sentiment classes and scores were then correlated with the AHD mortality rate to explore their relationship. The post-experimentation revealed that the dictionary-based approach showed almost zero correlation with the mortality rate, whereas the multiplicative LSTM-based model predicted a positive correlation of 0.268. This suggests that semantic representation approaches are more effective than dictionary-based lexical approaches.

4.4 Experimental Evaluation and Results

In this section, the various experiments designed to comprehensively study the effect of tweet embedding representations and the correlation of such representations with the AHD mortality rate by performing both descriptive and predictive analytics are discussed in detail. We also present in-depth analysis in a lower-dimensional space and the experiments conducted to understand the emotions inculcated within the tweet data from each of the 1347 communities.

4.4.1 Correlation Analysis

The pre-trained models of the BERT variants and the ELMo model were initially used without hyperparameter tuning or weight change to evaluate the latent contextual analysis of these embeddings in the aggregated raw community-specific tweet data. Table 4.2 summarizes the details of each model considered for the experiments. Each set of community-specific tweet data is first represented using the embedding techniques with the specified varying dimensions. After representation, the embedding features are correlated with the AHD mortality rate obtained from the CDC using Pearson's correlation measure, as employed in Eichstaedt et al.'s [21] study. We also report 95% Confidence Intervals here for consistency with Eichstaedt et al.'s [21] work. While most embeddings show a significant correlation with the AHD mortality rate, RoBERTa demonstrated correlations ranging from 0.17 to -0.17, closely aligning with the dictionary and topic modelling approaches

Table 4.2 Summary of model statistics

Model	Vocabulary size	Embedding dimension	Parameters
BERT-large-uncased	30,522	1024	336M
DistillBERT-base-uncased	30,522	768	66M
RoBERTa-large	50,265	1024	355M
ALBERT-large-v2	30,000	128	17M
ELMo	$\sim 800{,}000$	1024 (per token)	13.6M

Table 4.3 Descriptive analysis: benchmarking the experimented approach with state-of-the-art

Study	Approach	Variant	Max +ve corr	Max −ve corr
Eichstaedt et al. [21]	Topic modeling and manual analysis	Dictionary	0.17	−0.16
		Topics	0.27	−0.17
Brown et al. [11]		Dictionary	0.06	−0.17
Brown et al. [12]		Topics	0.21	−0.32
Eichstaedt et al. [22]		Dictionary	0.24	−0.20
Ours	Contextual representation—foundation model	bert-large-uncased	0.14	−0.15
		Roberta-large	0.17	−0.17
		Distilbert-base-uncased	0.16	−0.15
		Albert-large-v2	0.17	−0.16
		elmo	0.16	−0.15
	Transductive fine-tuning—RoBERTa	Raw text	**0.55**	**−0.53**
		Stop words removed	**0.55**	**−0.53**
		Stop words and numeric -value removed	0.48	−0.46
		Positive words	<u>0.50</u>	<u>−0.52</u>
		Positive and negative words	0.46	−0.48

Best in bold. <u>Second best</u> is underlined

reported by Eichstaedt et al. [21] in the descriptive analysis of tweet data. The descriptive evaluation results of the Contextual Embedding Models applied to raw tweet data are represented in Table 4.3. The results indicate that the contextual representation using pre-trained models that are not tuned for mortality prediction shows a promising correlation with the AHD mortality rate and could potentially be fine-tuned further.

4.4.2 Non-linear Regression Analysis

Eichstaedt et al.'s [21] model employs a Linear Regression classifier to predict the correlation between raw tweet data and the heart disease mortality rate. A similar evaluation approach was adopted, utilizing 10-fold cross-validation for better comparison. To eliminate manual feature engineering and adapt a data-driven model, an end-to-end prediction model was employed to remove outliers from the embedding features and use the most optimal features for prediction. The hypothesis that the number of positive and negative words in a tweet reflects the psychological factors of the respective community was considered, and normalization of embedding features was applied. With 1347 aggregated tweet embeddings in the processed corpus, the design focused on developing a generic framework to avoid over-fitting the model. Data augmentation was also explored by breaking down a particular county's tweets into chunks of the top 1000 frequent words and assigning the county's mortality rate to these chunks as labels. Although this reduced the data size to 0.244 million, it did not perform satisfactorily. These findings revealed that conventional linear regression does not adequately capture the inherent non-linear information of the respective county, even with a length of 1000 words.

Consequently, a Non-Linear Regression Network with one input layer, three hidden layers (256 neurons with ReLU activation function), and one output layer (1 neuron with a linear activation function) was employed to capture all the variations. The Mean Squared Error (MSE) loss function was employed as the performance metric to evaluate the Non-Linear Regression model's learning capability. For comparison, the dataset was split into 10 randomly sampled training and testing sets, and the average Pearson correlation value for the test sets was recorded. Each training and test set was fed into the regression network for 2000 epochs with early stopping, and the test sets were evaluated to obtain the Pearson correlation value corresponding to the model with the lowest MSE. This cross-validation process involved experimenting with different batch sizes for training, varying between 1 and 128, selected empirically.

4.4.3 Embedding Approaches

The contextual representation of tweet data for the five embedding models used in the experiments was carried out with varying dimensions: 1024 for ELMo, 1024 for *BERT-large-uncased*, 128 for *ALBERT-large-v2*, 1024 for *RoBERTa-large*, and 768 for *DistilBERT-base-uncased*. This vectorization was performed on the aggregated tweets corresponding to each 1347 counties and fed as input to the non-linear regression network to predict AHD mortality rates. Although the correlation obtained from these raw models was lower than the Pearson correlation reported by Eichstaedt et al. [21], both the ELMo and RoBERTa models still outperformed the others. The results, presented in Table 4.4, show that RoBERTa and ELMo were the top-performing models, achieving correlations of 0.21 and 0.17, respectively, after 2000 epochs. However, they appeared to struggle to capture the essential context from the tweet data, prompting further fine-tuning by experimenting with optimal preprocessing techniques.

4.4.4 Effect of Preprocessing

Preprocessing tweet data is essential due to the presence of plain texts along with variables such as URLs, emoticons, hashtags, acronyms, and more. Since the embedding models are trained on plain English texts, it is necessary to remove these variables to retain only the underlying context. As part of this process, stop words were removed at the first preprocessing stage, followed by filtering out unwanted numerals. Using the Opinion Lexicon as a base, the preprocessed tweets were further filtered to retain only positive and negative words to observe their impact on building the contextual embedding vector. A combination of positive and negative words was also created and added to different levels of preprocessing for experimentation. The RoBERTa model outperformed all others, and thus, a detailed analysis of the effect of various preprocessing techniques for this model is presented. After computing the embeddings for the preprocessed data, the embeddings were input into the non-linear regression network, trained over 2000 epochs with a batch size of 50. Four different levels of preprocessing were tested, and it was observed that retaining only positive and negative words from the tweet data helped better correlate with the reduced number of tokens to be processed by the foundation model (Refer Table 4.4).

4.4.5 Transfer Learning for Performance Optimization

The RoBERTa model is pre-trained on various datasets such as BookCorpus,[2] English Wikipedia,[3] CC-News,[4] OpenWebText[5] and Stories.[6] Using a pre-trained model without refinement might not always yield optimal performance. Retraining from scratch to incorporate domain-specific vocabulary for a given task can be challenging. However, fine-tuning such language models for specific downstream tasks can significantly enhance their performance and facilitate easier reuse. Based on the

[2] yknzhu.wixsite.com/mbweb.

[3] en.wikipedia.org/wiki/English_Wikipedia.

[4] commoncrawl.org/2016/10/news-dataset-available.

[5] github.com/jcpeterson/openwebtext.

[6] arxiv.org/abs/1806.02847.

Table 4.4 Predictive analysis: benchmarking the experimented approach with state-of-the-art

Study	Approach	Variant	Pearson (r) CV-10
Eichstaedt et al. [21]	Topic modeling and manual analysis	Only twitter	0.42
		All predictors	0.36
		Twitter + all predictors	0.42
Brown et al. [11]		Only twitter	0.24
Brown et al. [12]			
Eichstaedt et al. [22]		Only twitter All predictors	0.42 0.26
Ours	Contextual representation—foundation model	Bert-large-uncased	0.15
		Roberta-large	0.21
		Distilbert-base-uncased	0.14
		Albert-large-v2	0.14
		Elmo	0.17
	Contextual representation—RoBERTa	Raw	0.21
		Stop words removed	0.18
		Stop words and numeric value -removed	0.20
		Positive words	0.23
		Positive and Negative words	0.26
	Inductive fine-tuning—RoBERTa	Raw	0.19
		Stop words removed	0.19
		Stop words and numeric value -removed	0.17
		Positive words	0.17
		Positive and Negative words	0.17
	Transductive fine-tuning—RoBERTa	Raw	0.22
		Stop words removed	0.24
		Stop words and numeric value -removed	0.33
		Positive words	0.34
		Positive and Negative words	**0.48**

Best in bold

results of previous experiments and the need for fine-tuning, further analysis was conducted by incorporating Inductive and Transductive transfer learning methods.

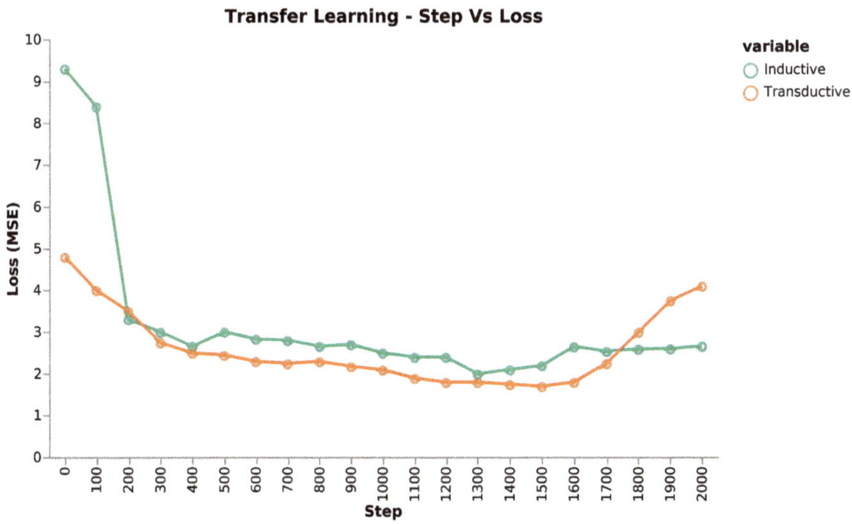

Fig. 4.5 Training loss of inductive and transductive transfer learning

4.4.5.1 Inductive Transfer Learning

As part of inductive transfer learning, the base RoBERTa model was fine-tuned using a sequence classification task. The source task involves language modelling, while the target task predicts the AHD mortality rate, a regression problem. This transformer-based regression setup includes a fully connected layer on the transformer model. The hidden layer consists of a single output neuron with a linear activation function to predict the target. In this sequence classification task, the RoBERTa model was trained using aggregated tweets as input, the mortality rate as the target, and mean square error (MSE) as the loss function, while all other parameters remained unchanged.

As shown in Fig. 4.5, it can be observed that after a certain point in training, the MSE loss reaches saturation, with no further significant reductions. During the initial iterations, the training loss rapidly decreases from a high value of 9.5 to approximately 3.2 indicating that the model quickly learns and improves. The training loss starts around 9.5 and drops to approximately 3.2 within the first 200 iterations. After this rapid decline, the loss gets down to 2.8 at step 400 and suddenly increases to 3 during the 500th iteration suggesting steady training. Although there are some fluctuations, the loss remains in this stable region, with minimal further decrease, suggesting the model is nearing convergence. Beyond step 1600 the curve smoothed out to a straight line, implying that the model settled into a stable state after overcoming midrange challenges.

After representing the aggregated tweets as embeddings using a fine-tuned embedding model, the embedding features were correlated with the AHD mortality rate. Since the model was tuned specifically for the regression task of predicting the mortality rate, the embeddings appeared to be overfitted to the target labels, with resulting descriptive outcomes ranging between − 0.99 and 0.99. The aggregated tweet embeddings and their variations, computed using the tuned embedding model from the inductive transfer learning method, were then fed into the Non-Linear Regression network with a batch size of 50 to predict the mortality rate. Although the embeddings were fine-tuned with the mortality rate as the final task, it can be seen that the model failed to generalize the embedding representations or exhibit meaningful characteristics for AHD mortality rate prediction (Refer Table 4.4).

4.4.5.2 Transductive Transfer Learning

In the transductive transfer learning method, RoBERTa was fine-tuned using the same task on which the base model was initially trained, specifically the masked language modelling task, which predicts the words surrounding a given word based on context. The model was trained with 148 million raw tweets as input, and dynamic masking was applied at each training step instead of static masking. The model inferred variations in the text sequences by applying masks to different word positions during training. This approach resulted in slower convergence due to dynamic masking, which is acceptable since it prevents the model from overfitting even with fewer training epochs. Additionally, padding was applied dynamically, with the maximum pad length based on batches instead of being fixed across the entire training dataset, a more suitable approach for aggregated tweet data from each county.

Similar to the inductive method, both descriptive and predictive analysis were performed using the transductively fine-tuned model. Although the model was not specifically tuned for the mortality rate, it successfully generalized the embedding representations, unlike the inductive method. This conclusion is further supported by Fig. 4.5. The training loss starts around 4.9 and drops to approximately 1.9 by learning continuously for the first 1500 iterations. The steady decline in loss shows that the model is consistently learning at a slower rate. Beyond this step, there is a sudden elevation in loss which turned the smooth line into a rising curve. Consequently, training is carried out for only a few epochs beyond this point.

The results for the best-performing nonlinear regression model trained using only positive and negative words from raw tweets, with the RoBERTa model fine-tuned through the transductive learning method, is tabulated in Table 4.4. Figure 4.6 shows the training and validation loss of the nonlinear regression network with a batch size of 50 during the training phase, where no early stopping parameter was applied. Although the training continued for 2000 epochs, the MSE loss did not change significantly after a certain point, underscoring the necessity of an early stopping parameter. Figure 4.7 presents the top embedding features that contributed most to the nonlinear regression network's predictions of AHD mortality rates. This visualization is instrumental in identifying the key features driving the model's performance. SHAP values (SHapley Additive exPlanations) [41] were used to explain the machine learning model's outputs, assigning each feature an importance

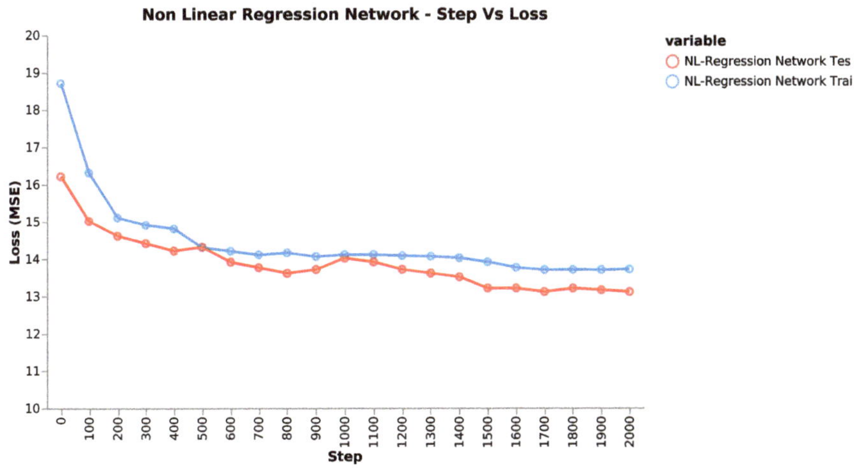

Fig. 4.6 Training and validation loss of non-linear regression network without early stopping

score based on its contribution to the prediction. Some features dominated the model's decision-making process, likely containing crucial information for accurate predictions (e.g., 767 and 768). Some features offer additional context or secondary information, influencing the model outputs (e.g. 378 and 729). Most other features played a very minor role, potentially containing redundant or less relevant data.

To assess the quality of embeddings in lower dimensions, the embeddings were reduced from higher dimensions to a space size of 20 using Singular Value Decomposition (SVD). These top 20 feature vectors were then fed into the Non-Linear Regression model to predict the AHD mortality rate. A notable correlation of 0.31 was observed between the lower-dimensional embeddings and the mortality rate. Figure 4.8 illustrates a significant improvement in speed (measured in seconds) during the training and evaluation phases when using lower-dimensional embeddings. The model trained in higher dimensions took approximately 5 seconds per epoch, compared to 3 seconds for lower-dimensional spaces, due to the increased complexity and volume of data in higher dimensions. However, the evaluation time remained similar, around 1 second, regardless of dimensionality, indicating that the model's evaluation process does not demand extensive computational resources once trained. The results were compared with those of Eichstaedt et al. [21] and are presented in Tables 4.3 and 4.4. The descriptive analysis in Table 4.3 shows that the RoBERTa model, fine-tuned using the transductive learning method, achieved a 28% higher correlation than the results reported by Eichstaedt et al. [21]. The predictive analytics results are tabulated in Table 4.4, which shows a 6% more significant correlation measure than Eichstaedt et al.'s work.

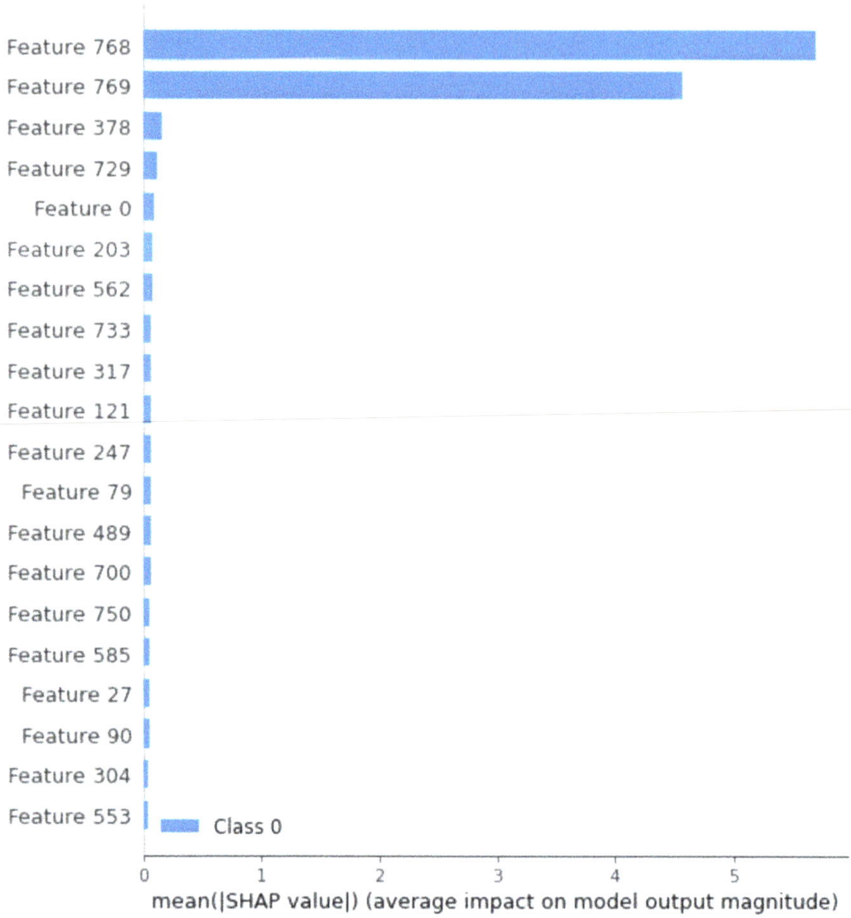

Fig. 4.7 Feature importance learned by Non-linear regression model

4.5 Discussion

In our work, experiments with models employing both inductive and learning methods for predicting community-specific AHD mortality rates using contextual tweet representations and transfer learning techniques were considered. The primary objective was to emphasize the social context reflected in tweet data rather than relying on traditional demographic, socioeconomic, and health variables, which have typically served as default parameters for such prediction tasks. Various preprocessing techniques and embedding representation models (both pre-trained and fine-tuned) were explored to assess the potential of social communication in capturing the characteristics of a community, which in turn mirrors the general behaviour of the population. Experiments with Transformer-based language

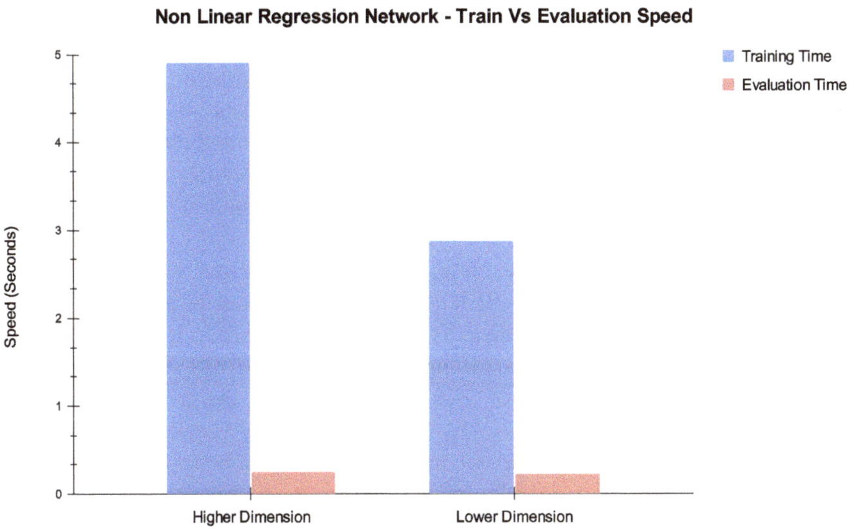

Fig. 4.8 Train and evaluation speed of non-linear regression network

models and the Bi-LSTM model were also undertaken to generate aggregated tweet embeddings corresponding to a community. Embeddings computed using BERT, RoBERTa, ALBERT, DistilBERT, and ELMo from raw tweet data were correlated with mortality rates obtained from the CDC. Unlike previous human-centred approaches that utilized dictionaries and topic modelling, this study applied descriptive analysis based on contextual vectorization and correlation. The vectorized inputs from the five models were fed into a Linear Regression classifier for comparison. As anticipated, the results were unsatisfactory due to the high dimensionality of the input feature vectors. This challenge was addressed by implementing a Non-Linear Regression Network classifier, which improved the correlation to 0.21 using the RoBERTa model. Based on this improvement, further experimentation was concentrated on the RoBERTa model.

Next, the impact of four pre-processing techniques—stop-word removal, numeric value removal, retaining positive words, and retaining both positive and negative words—was examined. Using the RoBERTa model in combination with the Non-Linear Regression Network classifier, the highest Pearson's correlation value of 0.26 was achieved when retaining both positive and negative words for each community. While the RoBERTa model effectively captured the contextual meaning of tweets, a key limitation was the reduced training data size caused by tweet aggregation at the community level. This observation led to further fine-tuning of the RoBERTa model using inductive and transductive transfer learning techniques.

In the inductive learning approach, the model was trained on 148 million tweets, employing a regression model with a single output neuron to predict the mortality rate based on tweet content. Once trained, the embedding from the last hidden state was utilized as the computed representation for each community and was then input into the Non-Linear Regression Network for predicting AHD mortality rates. The predictive model achieved a maximum correlation of only 0.19, while the descriptive model exhibited a significantly higher correlation, with a positive maximum of 0.99 and a negative maximum of −0.99. It is understood that the disparity between the high correlation in descriptive analysis and the low correlation in the predictive task arises from the model's failure to generalize effectively when learning from tweet representations, particularly when the mortality rate was set as the output parameter.

In the case of transductive learning, the approach utilized was analogous to the underlying mechanism of RoBERTa, employing masked language modelling (MLM) techniques on the tweet data. The model was trained to learn surrounding words dynamically by randomly shuffling the masks at each time step. This method yielded a predictive correlation of 0.48, which surpasses the predictive measure obtained by Eichstaedt et al. [21], who formed embedding vectors using only the positive and negative words relevant to a community. A better correlation was observed in the descriptive analysis, with a maximum positive correlation of 0.55 and a maximum negative correlation of -0.53. This significant improvement in both predictive and descriptive tasks may be attributed to the training mechanisms adopted.

Another aspect of the analysis involved studying the effect of representing tweets in lower dimensions and understanding the emotions associated with each community. To achieve this, Singular Value Decomposition (SVD) and sentiment analysis techniques were employed. The contextual embedding vector representation, derived from the best-performing model in transductive learning, was projected from higher to lower dimensions by considering the top twenty features. When this learned representation was fed into the Non-Linear Regression Network classifier, it predicted a correlation of 0.31, despite the reduced dimensionality. Sentiment analysis was conducted using both dictionary-based learning and a multiplicative LSTM model, leveraging raw tweet data to extract emotions associated with community-specific tweets. The dictionary-based approach demonstrated minimal correlation, while the multiplicative LSTM model achieved a correlation of 0.268. This finding reinforces the notion that semantic representation of data can yield better results than dictionary-based look-ups, as it effectively captures contextual information from the tweet data. Although this study did not further pursue sentiment analysis, plans are in place to extend this research to other diseases and languages.

Furthermore, masked and non-masked language models were also considered and selected based on recent breakthroughs in NLP driven by transformer-based architectures. While the MLM-based RoBERTa emerged as the best-performing model, the non-MLM-based ELMo demonstrated comparable performance with other models. This aspect warrants further study and experimentation for extension to other diseases and languages. Overall, our work demonstrates that fine-tuning pre-trained embedding representations using transfer learning techniques can effectively extract contextual information from complex, unstructured natural language data, facilitating the generation of meaningful insights and enabling advanced predictive applications.

4.6 Conclusion and Future work

In this article, a location-specific disease mortality rate prediction approach based on social data was presented. Tweet data is contextually represented using various language modelling techniques, including BERT, RoBERTa, ALBERT, DistilBERT, and ELMo. Descriptive and predictive analyses of mortality rate predictions are conducted based on raw aggregated tweet data for each community. A series of optimizations were implemented, including preprocessing the tweet data and fine-tuning the best-performing model (RoBERTa) using inductive and transductive transfer learning techniques, with the Non-Linear Regression Neural Network model serving as the classifier. Results from the transductive transfer learning indicated a maximum positive correlation of 0.55 and a maximum negative correlation of − 0.53, while the predictive analysis achieved a Pearson correlation value of 0.48. The best-performing embedding was then projected to lower dimensions using the SVD technique to evaluate its correlation in a reduced space. The correlation value of 0.31 demonstrated that Twitter data, when represented using contextual embedding techniques, can yield valuable insights even in lower-dimensional spaces. Additionally, both dictionary-based and semantic-based sentiment analyses were performed on the raw tweet data. As anticipated, the multiplicative LSTM-based sentiment model outperformed the dictionary approach, yielding a correlation of 0.268. This reinforces the importance of semantic representation in text data analysis.

The extensive experimentation conducted in this study provided significant insights into how language expressed in tweets reflects the psychological behaviour of a community and its effects on AHD mortality rates. A predictive correlation of 0.48 and a descriptive correlation ranging between 0.55 and − 0.53 were observed when using MLM techniques. Though this study seems to give a better correlation and signifies the importance of social text as a health indicator, it still lags behind the generalizability of using different datasets and diseases. This motivates us to extend this research to other datasets and diseases associated with psychological effects that contribute to their outbreaks. The feasibility of such extensions is supported by the unsupervised nature of the MLM models. The key focus will be testing these techniques on low-resource languages and less-explored events.

References

1. Abdi, H.: The eigen-decomposition: eigenvalues and eigenvectors. In: Encyclopedia of Measurement and Statistics, pp. 304–308 (2007)
2. Abuelenin, S., Elmougy, S., Naguib, E.: Twitter sentiment analysis for arabic tweets. In: Proceedings of the International Conference on Advanced Intelligent Systems and Informatics 2017, pp. 467–476. Springer, Berlin (2018). https://doi.org/10.1007/978-3-319-64861-3_44
3. Agrawal, A., George, R.A., Ravi, S.S., Kamath, S., Kumar, A.: Leveraging multimodal behavioral analytics for automated job interview performance assessment and feedback. In: Second Grand-Challenge and Workshop on Multimodal Language (Challenge-HML), pp. 46–54 (2020)

4. Al-Khalifa, K.S., Bakhurji, E.A., Halawany, H.S., Alabdurubalnabi, E.M., Nasser, W.W., Shetty, A.C., Sadaf, S.: Pattern of dental needs and advice on twitter during the covid-19 pandemic in Saudi Arabia. BMC Oral Health **21**, 456 (2021). https://api.semanticscholar.org/CorpusID:237551527
5. Anwar, M.M., Li, J., Liu, C.: Predicting the spread of a new tweet in twitter. In: Australasian Database Conference (2015). https://api.semanticscholar.org/CorpusID:13143353
6. Ashok, M.V., Rajanna, S., Joshi, P.V., Sowmya Kamath, S.: A personalized recommender system using machine learning based sentiment analysis over social data. In: 2016 IEEE Students' Conference on Electrical, Electronics and Computer Science (SCEECS), pp. 1–6 (2016). https://api.semanticscholar.org/CorpusID:14189299
7. Balet, T., Vo, Q., Salem, O., Mehaoua, A.: Cyberbullying detection on tweets from twitter using machine learning algorithms. In: 2023 International Conference on Intelligent Computing, Communication, Networking and Services (ICCNS), pp. 177–182 (2023). https://api.semanticscholar.org/CorpusID:260388108
8. Berkovic, D., Ackerman, I.N., Briggs, A.M., Ayton, D.: Tweets by people with arthritis during the covid-19 pandemic: content and sentiment analysis. J. Med. Internet Res. **22**(12), e24550 (2020). https://doi.org/10.2196/24550
9. Bernecker, K., Wenzler, M., Sassenberg, K.: Tweeted anger predicts county-level results of the 2016 united states presidential election. Int. Rev. Soc. Psychol. **32**, 6 (2019). https://api.semanticscholar.org/CorpusID:155374833
10. Broniatowski, D.A., Paul, M.J., Dredze, M.: National and local influenza surveillance through twitter: an analysis of the 2012–2013 influenza epidemic. PloS One **8**(12), e83672 (2013). https://doi.org/10.1371/journal.pone.0083672
11. Brown, N.J., Coyne, J.: No evidence that twitter language reliably predicts heart disease: a reanalysis of eichstaedt et al. (2015a) (2018)
12. Brown, N.J., Coyne, J.C.: Does twitter language reliably predict heart disease? A commentary on eichstaedt et al. (2015a). Peer J. **6**, e5656 (2018). https://doi.org/10.7717/peerj.5656
13. Brunton, S.L., Kutz, J.N.: Data-Driven Science and Engineering: Machine Learning, Dynamical Systems, and Control. Cambridge University Press, Cambridge (2022). https://doi.org/10.1017/9781108380690.002
14. Conover, M.D., Gonçalves, B., Ratkiewicz, J., Flammini, A., Menczer, F.: Predicting the political alignment of twitter users. In: 2011 IEEE Third International Conference on Privacy, Security, Risk and Trust and 2011 IEEE Third International Conference on Social Computing, pp. 192–199. IEEE, Piscataway (2011). https://doi.org/10.1109/PASSAT/SocialCom.2011.34
15. Culotta, A.: Towards detecting influenza epidemics by analyzing twitter messages. In: Proceedings of the First Workshop on Social Media Analytics, pp. 115–122 (2010). https://doi.org/10.1145/1964858.1964874
16. Culotta, A.: Estimating county health statistics with twitter. In: Proceedings of the SIGCHI Conference on Human Factors in Computing Systems, pp. 1335–1344 (2014). https://doi.org/10.1145/2556288.2557139
17. Das, S., Goard, J., Murray, D.S.: How celebrities feed tweeples with personal and promotional tweets: celebrity twitter use and audience engagement. In: Proceedings of the 8th International Conference on Social Media & Society (2017). https://api.semanticscholar.org/CorpusID:24835458
18. Donnelly, J., Roegiest, A.: On interpretability and feature representations: an analysis of the sentiment neuron. In: Advances in Information Retrieval: 41st European Conference on IR Research, ECIR 2019, Cologne, Germany, April 14–18, 2019, Proceedings, Part I 41, pp. 795–802. Springer, Berlin (2019). https://doi.org/10.1007/978-3-030-15712-8_55
19. Dredze, M.: How social media will change public health. IEEE Intell. Syst. **27**(4), 81–84 (2012). https://doi.org/10.1109/MIS.2012.76
20. Edo-Osagie, O., de la Iglesia, B., Lake, I.R., Edeghere, O.: A scoping review of the use of twitter for public health research. Comput. Biol. Med. **122**, 103770 (2020). https://api.semanticscholar.org/CorpusID:218656958

21. Eichstaedt, J.C., Schwartz, H.A., Kern, M.L., Park, G., Labarthe, D.R., Merchant, R.M., Jha, S., Agrawal, M., Dziurzynski, L.A., Sap, M., et al.: Psychological language on twitter predicts county-level heart disease mortality. Psychol. Sci. **26**(2), 159–169 (2015). https://doi.org/10.1177/0956797614557867
22. Eichstaedt, J.C., Schwartz, H.A., Giorgi, S., Kern, M.L., Park, G., Sap, M., Labarthe, D.R., Larson, E.E., Seligman, M.E., Ungar, L.H.: More evidence that twitter language predicts heart disease: a response and replication. PsyArXiv Preprints **10** (2018). https://doi.org/10.31234/osf.io/p75ku
23. Erkan, I., Evans, C.: The influence of ewom in social media on consumers' purchase intentions: an extended approach to information adoption. Comput. Hum. Behav. **61**, 47–55 (2016). https://doi.org/10.1016/j.chb.2016.03.003
24. Gaspar, R., Pedro, C., Panagiotopoulos, P., Seibt, B.: Beyond positive or negative: qualitative sentiment analysis of social media reactions to unexpected stressful events. Comput. Hum. Behav. **56**, 179–191 (2016). https://doi.org/10.1016/j.chb.2015.11.040
25. Gerber, M.S.: Predicting crime using twitter and kernel density estimation. Decis. Support Syst. **61**, 115–125 (2014). https://doi.org/10.1016/j.dss.2014.02.003
26. Golub, G.H., Reinsch, C.: Singular value decomposition and least squares solutions. In: Handbook for Automatic Computation: Volume II: Linear Algebra, pp. 134–151. Springer, Berlin (1971). https://doi.org/10.1007/BF02163027
27. He, K., Zhang, X., Ren, S., Sun, J.: Deep residual learning for image recognition. In: Proceedings of the IEEE Conference on Computer Vision and Pattern Recognition, pp. 770–778 (2016). https://doi.org/10.1109/CVPR.2016.90
28. Hu, M., Liu, B.: Mining and summarizing customer reviews. In: Proceedings of the Tenth ACM SIGKDD International Conference on Knowledge Discovery and Data Mining, pp. 168–177 (2004). https://doi.org/10.1145/1014052.1014073
29. Hutto, C., Gilbert, E.: Vader: a parsimonious rule-based model for sentiment analysis of social media text. In: Proceedings of the International AAAI Conference on Web and Social Media, vol. 8, pp. 216–225 (2014). https://doi.org/10.1609/icwsm.v8i1.14550
30. Ireland, M.E., Schwartz, H.A., Chen, Q., Ungar, L.H., Albarracín, D.: Future-oriented tweets predict lower county-level hiv prevalence in the united states. Health Psychol. **34**(S), 1252 (2015). https://doi.org/10.1037/hea0000279
31. Joshi, S., Sowmya Kamath, S.: Network science based predictive analysis on social media data. In: 2023 14th International Conference on Computing Communication and Networking Technologies (ICCCNT), pp. 1–6 (2023). https://api.semanticscholar.org/CorpusID:265406674
32. Joshi, S., Sowmya Kamath, S.: Social network science approaches for disease named entity recognition and extraction. In: 2024 International Conference on Information Networking (ICOIN), pp. 96–101. IEEE, Piscataway (2024). https://doi.org/10.1109/icoin59985.2024.10572092
33. Kenton, J.D.M.W.C., Toutanova, L.K.: Bert: pre-training of deep bidirectional transformers for language understanding. In: Proceedings of naacL-HLT, vol. 1, p. 2 (2019). https://doi.org/10.48550/arXiv.1810.04805
34. Kim, Y., Jernite, Y., Sontag, D., Rush, A.: Character-aware neural language models. In: Proceedings of the AAAI Conference on Artificial Intelligence, vol. 30 (2016). https://doi.org/10.1609/aaai.v30i1.10362
35. Krishnan, G.S., Sowmya Kamath, S., Sugumaran, V.: Predicting vaccine hesitancy and vaccine sentiment using topic modeling and evolutionary optimization. In: International Conference on Applications of Natural Language to Information Systems, pp. 255–263. Springer, Berlin (2021)
36. Kumar, E.R., Rao, A., Nayak, S.R.: Emotional level classification and prediction of tweets in twitter (2020). https://api.semanticscholar.org/CorpusID:226657168
37. Kwok, S.W.H., Vadde, S.K., Wang, G.: Tweet topics and sentiments relating to covid-19 vaccination among Australian twitter users: machine learning analysis. J. Med. Internet Res. **23**(5), e26953 (2021). https://api.semanticscholar.org/CorpusID:235197293

38. Lan, Z.: Albert: a lite bert for self-supervised learning of language representations (2019). arXiv:1909.11942. https://doi.org/10.48550/arXiv.1909.11942
39. Li, M., Hua, Y., Liao, Y., Zhou, L., Li, X., Wang, L., Yang, J.: Tracking the impact of covid-19 and lockdown policies on public mental health using social media: infoveillance study. J. Med. Internet Res. **24**(10), e39676 (2022). https://api.semanticscholar.org/CorpusID:252693938
40. Liu, Y.: Roberta: a robustly optimized bert pretraining approach (2019). arXiv:1907.11692. https://doi.org/10.48550/arXiv.1907.11692
41. Lundberg, S.: A unified approach to interpreting model predictions (2017). arXiv:1705.07874. 10.48550/arXiv.1705.07874
42. Mahajan, R., Kamath, S.: Binary classification of english tweets reporting children's medical disorders. In: Proceedings of The 9th Social Media Mining for Health Research and Applications (SMM4H 2024) Workshop and Shared Tasks, pp. 133–135 (2024). https://aclanthology.org/2024.smm4h-1.31
43. Massell, J., Lieb, R., Meyer, A.H., Mayor, E.: Fluctuations of psychological states on twitter before and during covid-19. PLOS ONE **17**(12), e0278018 (2022). https://api.semanticscholar.org/CorpusID:254684447
44. Murali, S., Sowmya, S., Supreetha, R.: Remag-kr: retrieval and medically assisted generation with knowledge reduction for medical question answering. In: Proceedings of the 62nd Annual Meeting of the Association for Computational Linguistics (Volume 4: Student Research Workshop), pp. 62–67 (2024)
45. Nguyen, Q.C., McCullough, M., Meng, H.W., Paul, D., Li, D., Kath, S., Loomis, G., Nsoesie, E.O., Wen, M., Smith, K.R., et al.: Geotagged us tweets as predictors of county-level health outcomes, 2015–2016. Am. J. Public Health **107**(11), 1776–1782 (2017). https://doi.org/10.2105/AJPH.2017.303993
46. Nguyen, H., Nguyen, T., Nguyen, D.T.: A graph-based approach for population health analysis using geo-tagged tweets. Multimedia Tools Appl. **80**, 7187–7204 (2020). https://api.semanticscholar.org/CorpusID:225070770
47. Obschonka, M., Lee, N., Rodríguez-Pose, A., Eichstaedt, J.C., Ebert, T.: Big data methods, social media, and the psychology of entrepreneurial regions: capturing cross-county personality traits and their impact on entrepreneurship in the USA. Small Bus. Econ. **55**, 567–588 (2020). https://doi.org/10.1007/s11187-019-11187-019
48. Park, G., Schwartz, H.A., Eichstaedt, J.C., Kern, M.L., Kosinski, M., Stillwell, D.J., Ungar, L.H., Seligman, M.E.: Automatic personality assessment through social media language. J. Person. Soc. Psychol. **108**(6), 934 (2015). https://doi.org/10.1037/pspp0000020
49. Pittman, M., Reich, B.: Social media and loneliness: why an instagram picture may be worth more than a thousand twitter words. Comput. Hum. Behav. **62**, 155–167 (2016). https://doi.org/10.1016/j.chb.2016.03.084
50. Raamkumar, A.S., Tan, S.G., Wee, H.L.: Use of health belief model–based deep learning classifiers for covid-19 social media content to examine public perceptions of physical distancing: model development and case study. JMIR Public Health Surveill. **6**(3), e20493 (2020). https://api.semanticscholar.org/CorpusID:220647762
51. Reshma, U., Kamath, S.S., Ananthanarayana, V.: Population-centric profiling with social data for large-scale epidemiological studies. In: Proceedings of the 5th Joint International Conference on Data Science & Management of Data (9th ACM IKDD CODS and 27th COMAD), pp. 302–303 (2022). https://doi.org/10.1145/3493700.3493752
52. Ruder, S., Peters, M.E., Swayamdipta, S., Wolf, T.: Transfer learning in natural language processing. In: Proceedings of the 2019 Conference of the North American Chapter of the Association for Computational Linguistics: Tutorials, pp. 15–18 (2019). https://doi.org/10.18653/v1/N19-5004
53. Sadasivuni, S.T., Zhang, Y.: An outcome-based method for computing happiness index from mental health related tweets of twitter users. 2021 IEEE International Conference on Intelligent Systems, Smart and Green Technologies (ICISSGT) pp. 99–102 (2021). https://api.semanticscholar.org/CorpusID:247192057

54. Sadasivuni, S.T., Zhang, Y.: Timestamp analysis of mental health tweets of twitter users along with covid-19 confirmed cases. In: 13th ACM International Conference on Bioinformatics, Computational Biology and Health Informatics, pp. 1–6 (2022). https://doi.org/10.1145/3535508.3545543
55. Sadilek, A., Kautz, H., Silenzio, V.: Predicting disease transmission from geo-tagged micro-blog data. In: Proceedings of the AAAI Conference on Artificial Intelligence, vol. 26, pp. 136–142 (2012). https://doi.org/10.1609/aaai.v26i1.8103
56. Saini, G., Yadav, N., Sowmya Kamath, S.: Ensemble neural models for depressive tendency prediction based on social media activity of twitter users. In: Security, Privacy and Data Analytics: Select Proceedings of ISPDA 2021, pp. 211–226. Springer, Berlin (2022)
57. Sanh, V.: Distilbert, a distilled version of bert: smaller, faster, cheaper and lighter (2019). arXiv:1910.01108. https://doi.org/10.48550/arXiv.1910.01108
58. Schwartz, H., Eichstaedt, J., Kern, M., Dziurzynski, L., Lucas, R., Agrawal, M., Park, G., Lakshmikanth, S., Jha, S., Seligman, M., et al.: Characterizing geographic variation in well-being using tweets. In: Proceedings of the International AAAI Conference on Web and Social Media, vol. 7, pp. 583–591 (2013). https://doi.org/10.1609/icwsm.v7i1.14442
59. Shah, H.A., Agus, M., Househ, M.: Sentiment visualization of correlation of loneliness mapped through social intelligence analysis. Comput. Methods Programs Biomed. Update **5**, 100144 (2024). https://doi.org/0.1016/j.cmpbup.2024.100144
60. Siva Rama Rao, A.V., Kuchu, S.S., Thota, D., Chennam, V.S., Yantrapragada, H.: Sentiment analysis: twitter tweets classification using machine learning approaches. In: Information and Communication Technology for Competitive Strategies (ICTCS 2021) ICT: Applications and Social Interfaces, pp. 411–419. Springer, Berlin (2022). https://doi.org/10.1109/IC3.2018.8530517
61. Sowmya Kamath, S., Bagalkotkar, A., Kandelwal, A., Pandey, S., Poornima, K.: Sentiment analysis based approaches for understanding user context in web content. In: 2013 International Conference on Communication Systems and Network Technologies, pp. 607–611. IEEE, Piscataway (2013)
62. Srivastava, R.K., Greff, K., Schmidhuber, J.: Highway networks (2015). arXiv:1505.00387. https://doi.org/10.48550/arXiv.1505.00387
63. Stieglitz, S., Dang-Xuan, L.: Emotions and information diffusion in social media—sentiment of microblogs and sharing behavior. J. Manag. Inf. Syst. **29**(4), 217–248 (2013). https://doi.org/10.2753/MIS0742-1222290408
64. Thakur, N.: Monkeypox2022tweets: a large-scale twitter dataset on the 2022 monkeypox outbreak, findings from analysis of tweets, and open research questions. Infect. Disease Rep. **14**, 855–883 (2022)
65. Tufts, C., Polsky, D.E., Volpp, K.G.M., Groeneveld, P.W., Ungar, L.H., Merchant, R.M., Pelullo, A.P.: Characterizing tweet volume and content about common health conditions across pennsylvania: Retrospective analysis. JMIR Publ. Health Surveil. **4**(4), e10834 (2018). https://api.semanticscholar.org/CorpusID:13796063
66. Tumasjan, A., Sprenger, T., Sandner, P., Welpe, I.: Predicting elections with twitter: what 140 characters reveal about political sentiment. In: Proceedings of the International AAAI Conference on Web and Social Media, vol. 4, pp. 178–185 (2010). https://doi.org/10.1609/icwsm.v4i1.14009
67. Tumasjan, A., Sprenger, T.O., Sandner, P.G., Welpe, I.M.: Election forecasts with twitter: how 140 characters reflect the political landscape. Soc. Sci. Comput. Rev. **29**(4), 402–418 (2011). https://doi.org/10.1177/0894439310386557
68. Unnikrishnan, R., Kamath, S., Ananthanarayana, V.: Adaptive learning model for effective detection, extraction and normalization of adverse drug events from social media data. In: Proceedings of The Seventh Workshop on Social Media Mining for Health Applications, Workshop & Shared Task, pp. 95–97 (2022). https://aclanthology.org/2022.smm4h-1.27
69. Wang, H., Li, Y., Hutch, M., Naidech, A., Luo, Y.: Using tweets to understand how covid-19–related health beliefs are affected in the age of social media: twitter data analysis study. J. Med. Internet Res. **23**(2), e26302 (2021). https://doi.org/10.2196/26302

70. Xavier, T., Lambert, J.: Sentiment and emotion trends in nurses' tweets about the covid-19 pandemic. J. Nurs. Scholar. **54**(5), 613–622 (2022). https://doi.org/10.1111/jnu.12775
71. Xia, X., Zhang, Y., Jiang, W., Wu, C.Y.H.: Staying home, tweeting hope: mixed methods study of twitter sentiment geographical index during us stay-at-home orders. J. Med. Internet Res. **25**, e45757 (2023). https://api.semanticscholar.org/CorpusID:259403172
72. Xu, W., Liu, X., Gong, Y.: Document clustering based on non-negative matrix factorization. In: Proceedings of the 26th Annual International ACM SIGIR Conference on Research and Development in Informaion Retrieval, pp. 267–273 (2003). https://doi.org/10.1145/860435.860485
73. Xue, J., Chen, J., Hu, R., Chen, C., Zheng, C., Su, Y., Zhu, T.: Twitter discussions and emotions about the covid-19 pandemic: machine learning approach. J. Med. Internet Res. **22**, e20550 (2020). https://api.semanticscholar.org/CorpusID:219955648
74. Yang, Z.: Xlnet: Generalized autoregressive pretraining for language understanding (2019). arXiv:1906.08237. https://doi.org/10.48550/arXiv.1906.08237
75. Yao, W., Qian, S.: From twitter to traffic predictor: next-day morning traffic prediction using social media data (2020). ArXiv **abs/2009.13794**. https://api.semanticscholar.org/CorpusID:221996013
76. Zhang, L., Hall, M., Bastola, D.: Utilizing twitter data for analysis of chemotherapy. Int. J. Med. Inf. **120**, 92–100 (2018). https://doi.org/10.1016/j.ijmedinf.2018.10.002

Chapter 5
SZ-Net: A Hybrid CNN-BiLSTM-BiGRU Framework for Schizophrenia Detection and Classification Using Electroencephalogram Signals

Arghyasree Saha and Pawan Kumar Singh

Abstract Schizophrenia (SZ), an intricate mental health condition, is defined by a variety of symptoms, such as delusions, disorganized thinking, hallucinations and cognitive impairments. Its heterogeneous nature, which varies significantly across individuals in terms of symptom severity, progression, and response to treatment, makes diagnosis particularly challenging. The electroencephalogram (EEG), which records brain electrical activity through scalp electrodes, plays a vital role in SZ research. Its capacity to detect nuanced variations in neural activity using high temporal resolution offers valuable insights into brain function. Various machine learning (ML) and deep learning (DL) models have been designed to improve the accuracy of SZ detection, aiming for early diagnosis and better patient outcomes. This research introduces a hybrid DL model (SZ-Net), combining Convolutional Neural Networks (CNN), Bidirectional Long Short-Term Memory (BiLSTM), and Bidirectional Gated Recurrent Units (BiGRU) with attention mechanisms to analyze EEG signals for SZ detection. The model is evaluated on two benchmark datasets—the EEG dataset obtained from M.V. Lomonosov Moscow State University (MSU), comprising 39 healthy and 45 SZ patients, and the Kaggle EEG dataset from a basic sensory task in SZ, comprising 32 healthy and 49 SZ patients. Our proposed framework achieves accuracy rates of 98.21% and 99.27%, respectively. It outperforms pioneering results, showing its exceptional capability in EEG-based SZ diagnosis. These results highlight the framework's capability as a dependable solution for early diagnosis and intervention. The hybrid architecture effectively captures both spatial and temporal patterns in EEG signals, enabling a more comprehensive analysis. Additionally, the study opens avenues for future research focused on enhancing model efficiency, real-time implementation, and interpretability, with the goal of providing accessible and accurate diagnostic solutions for SZ.

A. Saha · P. K. Singh (✉)
Department of Information Technology, Jadavpur University, Jadavpur University Salt Lake Campus, Kolkata, West Bengal, India

G. Rivera et al. (eds.), *Recent Advances in Artificial Neural Networks*, Intelligent Systems Reference Library 283, https://doi.org/10.1007/978-3-032-13497-4_5

Keywords SZ-Net · Schizophrenia detection · Hybrid deep learning model · Electroencephalogram signals · Machine learning · Convolutional neural networks · Attention mechanisms · Recurrent neural networks

5.1 Introduction

Schizophrenia (SZ), being a multifaceted as well as chronic mental health condition, impacts nearly 1% of people worldwide [1, 2]. It is defined by a diverse range of symptoms, which can generally be grouped into positive, negative, and cognitive categories [3, 4]. Positive symptoms encompass disorganized thinking, delusions, and hallucinations; negative symptoms comprise social isolation, emotional flatness as well as apathy [4]. Cognitive deficits, like difficulty with attention, memory, and executive functioning, are also widespread, adding to the disorder's overall severity [5]. These symptoms frequently impair the individual's ability to function in daily life, posing substantial obstacles in maintaining relationships, employment, and independent living [6]. The categorized representation of SZ symptoms into positive, negative and cognitive domains is shown in Fig. 5.1.

A timely as well as precise identification of SZ is essential for commencing efficient therapy interventions, which usually involve antipsychotic medicines and psychotherapy [7]. Early detection and management have been demonstrated to improve long-term outcomes, reduce symptom severity, and boost an individual's quality of life [8]. However, identifying SZ remains difficult due to the varied character of its symptoms, which coincide with those of other psychiatric diseases [4]. Furthermore, there is no conclusive biological test to validate the diagnosis, therefore the approach is mostly based on subjective clinical assessments [9].

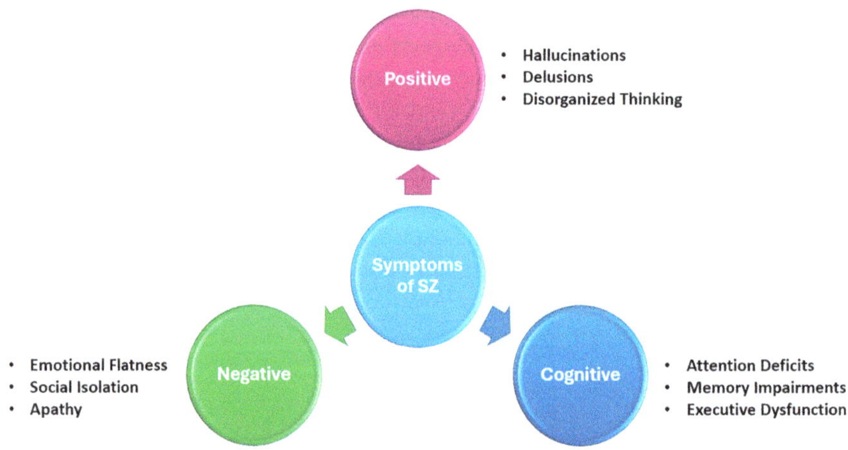

Fig. 5.1 Overview of SZ symptoms

Currently, the gold standard for diagnosing SZ is based on clinical interviews and behavioral evaluations, which are frequently guided through established diagnostic guidelines, like those specified in the Diagnostic and Statistical Manual of Mental Disorders (DSM) [10]. While these procedures are necessary, they can be influenced by the clinician's prior experience, biases, and the patient's willingness to participate [11]. Furthermore, the examination is mostly qualitative, which allows for errors and variation in diagnosis between clinicians and situations [4]. This subjectivity can lead to mis-diagnosis or delayed diagnosis, resulting in ineffective therapies or a lack of early intervention [12].

The shortcomings of conventional diagnostic methods underscore the critical need for more objective, data-driven approaches to SZ detection and categorization [13]. Recent advances in electrophysiological techniques, such as Electroencephalography (EEG), provide a promising path to more accurate and trustworthy diagnosis [14–16]. EEG captures the dynamic temporal aspects of brain activity, offering valuable insights into the functional abnormalities that are considered to underpin SZ. A visual representation of EEG signal acquisition, showcasing the placement of electrodes on the scalp to capture brain activity [17] is provided in Fig. 5.2. However, the complexity and volume of EEG data necessitate the

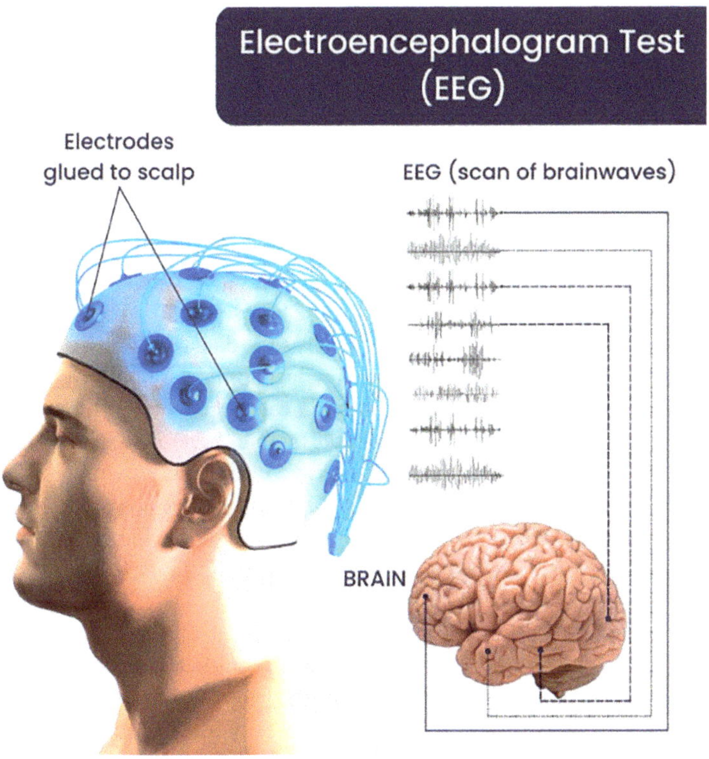

Fig. 5.2 Schematic representation of EEG acquisition and its role in SZ diagnosis [17]

development of advanced computational methods, including machine learning (ML) and deep learning (DL) algorithms, to efficiently process and interpret the data for therapeutic use [18].

DL has transformed medical data analysis due to its capacity to automatically discover complicated patterns from vast datasets [19]. This feature is particularly essential in sectors like SZ identification, where traditional approaches frequently fail to capture the subtle patterns found in multimodal data [20]. In this study, we propose a unified Convolutional Neural Network, Bidirectional Long Short-Term Memory as well as Bidirectional Gated Recurrent Unit (CNN-BiLSTM-BiGRU) framework, to address the challenge of analyzing EEG signals for the classification of healthy and SZ individuals. The model leverages the strengths of CNN and recurrent neural networks (RNN), complemented by attention mechanisms, to achieve high diagnostic accuracy [18, 21].

CNNs are highly effective in spatial feature extraction, enabling the model to automatically discover meaningful spatial patterns in raw EEG data [22]. This capability is particularly valuable for identifying abnormalities in brain activity associated with SZ. By learning directly from the raw signals, CNNs eliminate the need for manual feature engineering, providing a robust and adaptive method for feature extraction [23]. RNNs, particularly BiLSTM and BiGRU architectures, excel in processing sequential or time-series data, which is inherent to EEG signals [24, 25]. EEG captures the dynamic temporal patterns of brain activity over time, and alterations in these patterns can provide critical insights into the onset and progression of SZ [18]. BiLSTMs and BiGRUs enhance the temporal modeling of EEG data by processing information in both forward and backward directions. Unlike unidirectional architectures, which capture dependencies only from past to future, bidirectional networks account for both past and future contexts simultaneously [26].

This bidirectional capability allows the model to uncover subtle temporal dependencies that might otherwise go unnoticed, such as intricate changes in brain wave patterns indicative of SZ [24, 26]. These architectures are particularly advantageous for EEG analysis, where the sequential nature of the data often involves long-range dependencies and overlapping patterns that reflect complex brain activity dynamics [18]. When combined with CNNs, these recurrent architectures form a powerful hybrid model that excels in both spatial and temporal feature extraction [22, 23]. CNNs are adept at identifying spatial patterns and local features within EEG signals, while BiLSTM and BiGRU networks analyze how these features evolve over time [25]. This synergistic combination ensures that the model effectively captures the comprehensive information embedded in EEG signals, encompassing both static and dynamic aspects [18, 21].

The incorporation of attention mechanisms further enhances the model's effectiveness by enabling it to concentrate specifically on the most important features [21]. In EEG data, not all spatial or temporal patterns contribute equally to the diagnosis of SZ. Attention mechanisms dynamically assign higher importance to critical spatial features and temporal sequences that are most indicative of the condition [27]. This prioritization allows the model to filter out noise and irrelevant information, ensuring that its predictions are based on the most diagnostically significant patterns [28]. By emphasizing these key features, attention mechanisms enhance the interpretability and accuracy of the model, making it a highly effective tool for SZ detection and classification [18].

This study highlights the potential of a hybrid DL framework (named as SZ-Net)—combining CNN-BiLSTM-BiGRU with attention mechanisms, to deliver robust and accurate analysis of EEG signals for SZ detection. By streamlining the analysis pipeline and leveraging advanced network architectures, this approach provides a powerful tool for understanding and classifying brain activity patterns in SZ.

The key highlights of this paper are outlined below:

1. Development of a hybrid DL framework (SZ-Net), a CNN-BiLSTM-BiGRU model with attention mechanisms, to analyze EEG data for SZ detection and classification, enhancing the model's focus on critical spatial and temporal features for improved diagnostic accuracy.
2. Rigorous evaluation of the proposed framework using benchmark EEG datasets, demonstrating their adaptability and effectiveness in handling complex time-series data.
3. Highlighting the potential of unified models to streamline EEG data analysis pipelines, providing robust and consistent performance while achieving state-of-the-art classification accuracy.

The paper is organized in the following manner: A thorough review of existing literature (Sect. 5.2), an extensive description concerning the dataset employed in this research (Sect. 5.3), the proposed methodology, encompassing data pre-processing, model framework, implementation, and results accompanied by an in-depth discussion (Sect. 5.4), a comparative analysis with existing studies carried out to assess the effectiveness of the suggested approach (Sect. 5.5), conclusion as well as potential avenues for subsequent studies (Sect. 5.6).

5.2 Literature Analysis

SZ detection has garnered significant scientific interest due to the challenges of diagnosis and the growing demand for objective, data-driven methods. Recent research has increasingly focused on using electrophysiological data, particularly EEG signals, to identify biomarkers associated with the disorder [29, 30]. EEG provides valuable insights into the brain's electrical activity, capturing dynamic patterns and temporal abnormalities that are often linked to SZ, making it a promising tool for advancing diagnostic accuracy [31, 32].

Several researchers have used ML and DL models to assess EEG characteristics for diagnosis [16, 33–35]. Phang et al. [36] utilized the MSU dataset [37] to classify SZ using EEG signals utilizing deep neural network designed with a deep belief network (DNN-DBN) structure. Their method achieved 95% accuracy within theta and beta frequency ranges. Using the same dataset, Rajesh et al. [38] introduced a method for SZ diagnosis utilizing symmetrically weighted local binary patterns (SLBPs). They developed an automated system where histogram features derived from SLBP were refined using correlation-driven feature selection method. The optimized set of features was then analyzed using a LogitBoost classification algorithm, achieving an accuracy of 91.66% in distinguishing SZ patients from healthy controls (HCs). Sobahi et al. [39] introduced an innovative approach for SZ detection employing EEG data by mapping signals to images. They encoded rhythmic patterns from EEG data utilizing one-dimensional local binary patterns (1D LBP). For enhancing the dataset, they employed autoencoders (AE) based on extreme learning machines (ELM). Advanced deep transfer learning techniques were then applied to differentiate SZ patients from HCs, resulting in an impressive accuracy of 97.7%. Aslan et al. [40] performed an investigation utilizing DL methods for SZ diagnosis. EEG data were converted into two-dimensional representations through continuous wavelet transform (CWT) to extract time-frequency patterns. In contrast to conventional ML methods, the model automatically extracted features directly from the EEG data for training. They achieved classification accuracies of 98% as well as 99.5% across two distinct datasets with people from different age groups. The explainability of the results, represented through multiple approaches, revealed a clear association amid the frequency elements in HCs and individuals with SZ.

Using Kaggle EEG dataset [41], Lei Zhang [42] created a ML-based classifier to differentiate between individuals with SZ and HCs by extracting features from electroencephalogram (EEG) signals, specifically event-related potentials (ERPs). The EEG signals were pre-processed, and given the small dataset size, the random forest technique was used for classification utilizing several feature sets. The study found that including specific EEG-derived variables considerably enhanced classification accuracy, with a result of 81.1%. Using the same dataset, Siuly et al. [43] sought to develop an automated SZ detection system based on electroencephalogram (EEG) data in order to address existing issues as well as assist healthcare professionals and investigators. They used the empirical mode decomposition (EMD) technique to efficiently regulate the nonlinear and dynamic nature of EEG signals. Every EEG signal was broken down into intrinsic mode functions (IMFs) using EMD technique, along with twenty-two quantitative attributes were retrieved. The Kruskal–Wallis test was applied to determine five important traits. Among the classifiers tested, the ensemble bagged tree (EBT) classifier performed the best, achieving a classification accuracy of 93.21% for SZ as well as an overall performance of 89.59% using IMF 2. Lei Zhang [44] created an artificial neural network (ANN) model in which EEG data were captured during tasks related to event-related potentials (ERPs), specifically key presses and sound tone presentations. The model utilized five time-domain features derived from the EEG data, along with two demographic variables, to train and evaluate the neural network model. After applying a wavelet transformation to the ERP EEG data, additional time-frequency characteristics were extracted for classification. They achieved an accuracy exceeding 98.5%. These results highlight the considerable promise of creating a precise and objective

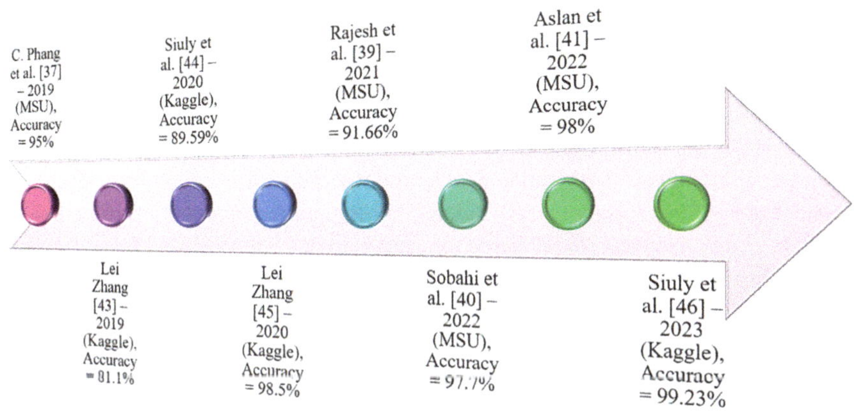

Fig. 5.3 Timeline of studies utilizing MSU [37] and Kaggle EEG [41] data for SZ classification

diagnostic instrument for SZ using EEG signals. Siuly et al. [45] explored the application of deep residual networks (ResNet) for the automated detection of SZ from EEG data. The method involved three key steps: preprocessing EEG data with an average filtering technique to eliminate noise, extracting features through a deep ResNet model that autonomously identified important patterns from the EEG signals, and categorizing the extracted features using a Softmax activation layer. Additionally, various ML techniques, such as support vector machine (SVM), were employed on the same set of features. The suggested approach attained a classification accuracy of 99.23% with the SVM classifier, outperforming both the ResNet classifier and other existing methods.

Our proposed study utilized both the MSU [37] and Kaggle EEG [41] datasets for evaluation. A chronological representation highlighting key studies that have employed MSU [37] and Kaggle EEG [41] data for SZ classification is depicted (Fig. 5.3).

The studies reviewed demonstrate significant progress in classifying SZ using EEG data through a variety of methodologies [18]. While these approaches have shown promising results in accurately identifying the disorder, particularly by leveraging a combination of features, several challenges persist. These include the requirement for bigger and more varied datasets, reducing processing times, developing robust and efficient models, and seamlessly integrating these models into medical practices [29]. Tackling these obstacles is crucial to assist healthcare professionals in delivering accurate diagnoses and, ultimately, improving patient results.

Motivation for the Proposed Study Previous research has extensively explored various ML approaches and CNN-based models for EEG-based SZ detection [18, 30]. While studies have implemented combinations such as CNN-LSTM or CNN-GRU, the integration of CNN, LSTM, and GRU within a single framework has not been explored. Building upon this, we initially developed a CNN-LSTM-GRU model and later advanced it to a CNN-BiLSTM-BiGRU model, which demonstrates superior accuracy, marking a novel contribution to this domain. We have combined the strengths of three powerful models into a single cohesive framework. Initially, EEG signals are pre-processed to ensure optimal data quality, after which they are classified using the CNN-BiLSTM-BiGRU model enhanced with a channel attention mechanism. This channel attention mechanism enables the model to concentrate on the most significant spatial and temporal elements of EEG signals, improving the accuracy of SZ detection. Our unified approach is tested on two distinct EEG datasets, demonstrating its robustness and effectiveness in advancing SZ detection and classification.

5.3 Dataset Description

In this study, we have employed two publicly available EEG datasets to evaluate the performance of the proposed SZ-Net model for SZ detection: MSU [37] dataset and Kaggle EEG [41] dataset. These datasets contain EEG recordings from both HCs and individuals diagnosed with SZ, offering a diverse representation of neural activity patterns associated with the disorder. In the following subsections, we provide a detailed description of each dataset, including the acquisition protocol, channel configuration, and the preprocessing steps applied.

5.3.1 MSU [37] Dataset

In our research work, we have utilized the openly available EEG dataset obtained from M.V. Lomonosov Moscow State University (MSU) [37] to investigate SZ detection. The dataset consists of EEG recordings from adolescents who participated in a comprehensive psychiatric screening conducted by experienced psychiatrists. Based on the results of these screenings, the participants are categorized into two distinct sections: one category consisting of 39 healthy individuals, and the other consisting of 45 individuals, exhibiting symptoms indicative of SZ. The EEG recordings are obtained during various cognitive tasks and resting states, providing a rich source of temporal and spatial brain activity data. By analyzing these EEG signals, we aim to identify distinctive patterns and features that differentiate between healthy individuals and those with SZ, contributing to the development of more effective diagnostic models.

Each EEG file in the dataset corresponds to a single subject's EEG recording, containing detailed data captured across various electrode positions. Within each file, there is a dedicated column for the samples recorded from 16 distinct electrode positions on the scalp. Each value in these columns represents the EEG amplitude (in microvolts, μV) at a specific sample point. The data is organized sequentially, with the first 7680 samples corresponding to the first channel (F7), followed by the samples for the second channel (F3), and so on, up to the 16th channel. A one-minute EEG recording is composed of 7680 samples, having a sampling frequency of 128 Hz. The 16 electrode channels used for recording are positioned across various regions of the scalp, specifically: T6, T5, F8, Pz, Cz, T3, F7, P3, O1, F4, C4, T4, C3, P4, F3, and O2.

5.3.2 Kaggle EEG [41] Dataset

The EEG recordings from a fundamental sensory activity in SZ [41] comprises event-related potentials (ERPs) triggered by button presses and auditory tones. The dataset includes recordings from 81 human subjects, consisting of 49 individuals diagnosed as SZ and 32 as HC. The study population includes 67 males as well as 14 females, having a mean age of 39 years, with an average schooling duration of 14.5 years, providing a diverse yet controlled sample.

The EEG data has been continuously captured with a high sampling frequency of 1024 Hz, ensuring precise temporal resolution. The data is offline-referenced to averaged earlobe electrodes, a common method to reduce common noise and enhance signal quality. This reference method improves the accuracy of the EEG recordings, particularly in identifying subtle brain activity associated with SZ. The EEG signals are captured from a total of 64 scalp sites and 8 external sites, ensuring wide coverage of the brain, using the BioSemi Active Two system [46], which is known for its high signal fidelity and versatility in brain wave monitoring. Three experimental conditions are employed during the task to elicit specific sensory responses—(a) a tone generated immediately after pressing the button, (b) passive listening with a consistent tone, as well as (c) pressing a button without producing a tone.

These experimental conditions are designed to study both sensory (auditory) and motor (button press) processes in the brain, providing insights into the neural mechanisms that may differ between individuals with SZ and HCs. The detailed recording of brain activity from multiple sites across the scalp allows researchers to explore the complex neural dynamics involved in SZ, providing a foundation for further studies on diagnostic biomarkers or therapeutic interventions. The distribution of EEG data across HC and SZ patients in the utilized datasets, highlighting the class-wise sample counts for each dataset is illustrated (Fig. 5.4).

5.4 Proposed Methodology

The overall workflow of the proposed strategy for EEG-based SZ detection is shown (Fig. 5.5). The process begins with raw EEG signals as input, which undergo data pre-processing to prepare the signals for analysis. The pre-processed data is then fed into the SZ-Net model, a hybrid DL framework specifically designed for SZ

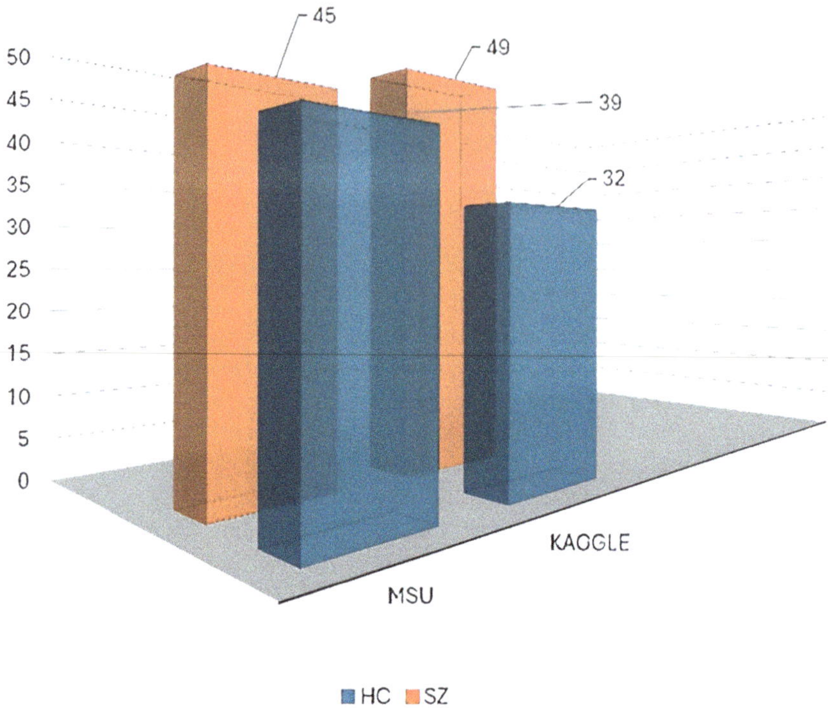

Fig. 5.4 Data statistics of MSU [33] and Kaggle EEG [41] datasets

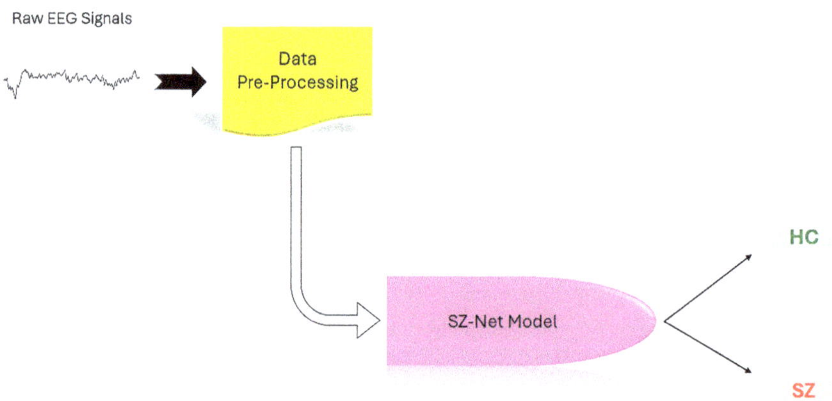

Fig. 5.5 Overall pipeline of our proposed approach for EEG-based SZ diagnosis

detection. Finally, the model outputs a classification into two categories: HC or SZ. This streamlined workflow ensures an effective and accurate pipeline for leveraging EEG data in SZ diagnosis.

5.4.1 Data Pre-Processing

The preprocessing of EEG data is a critical phase in ensuring the success of the classification task. This sub-section describes the detailed pipeline of both the datasets, starting from loading raw EEG recordings to obtaining normalized feature vectors ready for ML. Each step is carefully designed to maximize signal fidelity and enhance feature relevance for SZ detection.

5.4.1.1 Pre-Processing of MSU [37] Dataset

The MSU [37] dataset includes EEG recordings in .edf format for each class—healthy and SZ. Using the mne library, each .edf file is loaded with EEG channels selected explicitly via a pick_types function. Loading the data this way sets up the recordings for systematic preprocessing.

Raw EEG data contains low-frequency drifts, high-frequency noise, and other artifacts irrelevant to the task at hand. To address this, a bandpass filter is applied to retain frequencies between 1 and 40 Hz, the range most relevant to brain activity. This frequency range encompasses critical EEG frequency bands – delta, theta, alpha and beta. The filtering process is implemented using the FIR (Finite Impulse Response) method with the firwin design, which provides precise control over the frequency response while minimizing phase distortion. Filtering is applied uniformly across all recordings to standardize the dataset and eliminate artifacts such as power-line interference or motion-related noise.

To analyze the temporal dynamics of EEG signals, the continuous recordings are segmented into smaller overlapping windows. Each segment is 2 s long, equivalent to 256 samples at a sampling frequency of 128 Hz. A 55% overlap between consecutive windows is introduced, resulting in a step size of 115 samples (approximately 45% of the window size). This overlapping scheme ensures effective data utilization by preventing the loss of significant temporal information.

For each segmented window, features are extracted from all 16 EEG channels individually. Specifically, statistical features (mean amplitude and standard deviation) and spectral features (spectral entropy) are computed for each channel. This results in a 48-dimensional feature vector per window (3 features $\times$ 16 channels).

To construct the final feature set, the feature vectors from all channels are concatenated into a single, unified vector representing the entire window. For each EEG recording, approximately 65 overlapping segments are generated. Each segment yields a 48-dimensional feature vector, resulting in 65 feature vectors per recording. The final dataset comprises all feature vectors extracted across the EEG recordings, where each feature vector is directly associated with its respective class label (0 for healthy, 1 for SZ).

To ensure consistency across features and avoid bias due to differing scales, normalization is performed using StandardScaler. This transformation standardizes each feature to possess:

- Zero mean: Making sure that all features are centered.
- Unit variance: Ensuring that features play an equal role in the model's training process.

Normalization is applied globally across the dataset to align feature scales and enhance the interpretability of learned patterns.

After preprocessing, the data is organized into two key components:

- Feature Matrix (X): A 2D array where each row corresponds to a normalized feature vector for a specific segment.
- Label Array (Y): A 1D array of labels corresponding to the class of each feature vector.

The healthy and SZ data are combined to form a single dataset, ready for training and evaluation. The final dataset preserves temporal, statistical, and spectral information, enabling robust classification of SZ from EEG signals.

5.4.1.2 Pre-Processing of Kaggle EEG [41] Dataset

The Kaggle EEG [41] dataset is loaded from .csv files and located in two specified directories (dataset_path1 and dataset_path2). As each file is processed, it checks the file name, which contains the subject ID, and maps it to the corresponding label using a dictionary (diagnosis_dict) derived from the demographic data. The demographic data, stored in a separate CSV file (demographic.csv), provides the subject IDs and their corresponding diagnosis labels (SZ or HC). The correct label is assigned to each EEG file during the loading process.

Similar to the previous dataset, the first key step in preprocessing is the application of a bandpass filter to the EEG signals. This is achieved using the bandpass_filter function, which applies a Butterworth filter to remove unwanted noise and retain the frequency range of interest. Specifically, a lower cutoff frequency set to 1 Hz and an upper cutoff frequency of 40 Hz are used. This filtering step ensures that the data primarily contains brainwave signals while removing low-frequency drift and high-frequency noise that can distort the analysis.

Once the data is filtered, the next step is to segment the continuous EEG data into smaller, overlapping windows using the segment_data function. The window size is set to 2 s (i.e., $2 \times 1024 = 2048$ samples), and the overlap between consecutive windows is set to 30%, resulting in a step size of 1434 samples (approximately 70% of the window size). This segmentation allows the model to extract the time-varying patterns of EEG data by analyzing smaller chunks of data rather than the entire continuous signal. The function slides a window across the data, extracting overlapping segments for further analysis.

After segmentation, the extract_features function is used to calculate statistical features from each EEG window. Specifically, for every segment, the mean and standard deviation of the EEG signal across all channels are computed, resulting in a 128-dimensional feature vector per window (2 features per channel for 64 channels). To construct the final feature set, the feature vectors from all channels are concatenated into a single, unified vector representing the entire window. For each EEG recording, only one segment is generated, resulting in a single 128-dimensional feature vector per recording. The final dataset comprises all feature vectors extracted across the EEG recordings, where each feature vector is directly associated with its respective class label (0 for healthy, 1 for SZ).

Then, the data is normalized using StandardScaler. Normalization guarantees that every feature has a mean of zero and a variance of one, which is important for ML models to perform optimally. This process prevents features with larger magnitudes from overpowering the learning procedure and aids the model to converge more efficiently in the course of training.

The datasets are partitioned into training and testing subsets, with 80% allocated for training and 20% for evaluation. The partitioning is performed at the subject level to prevent data leakage, ensuring the reliability of the reported results. Specifically, the EEG recordings from each subject are treated as independent units during the train-test split process. This ensures that the same subject's data does not appear in both the training and testing sets, preserving the model's ability to generalize to unseen subjects. Stratification is applied to guarantee that the training and test sets are carefully balanced, maintaining an equal distribution of both classes (SZ and HC). This step is crucial to ensure that the model is trained and evaluated on a sample that accurately represents the data distribution. Hence, the data is cleaned, standardized, and partitioned appropriately, making it ready for use in predictive modeling tasks to effectively classify SZ versus HC subjects.

5.4.2 Model Framework and Implementation

In this study, we propose SZ-Net, a hybrid DL framework designed for EEG-based SZ detection. The combination of convolutional and recurrent layers allows SZ-Net to extract complex features and temporal patterns, while the channel attention mechanism enhances the model's sensitivity to discriminative EEG channels. In the following subsection, we describe the detailed architecture of SZ-Net, including the layer configuration, activation functions, and regularization strategies.

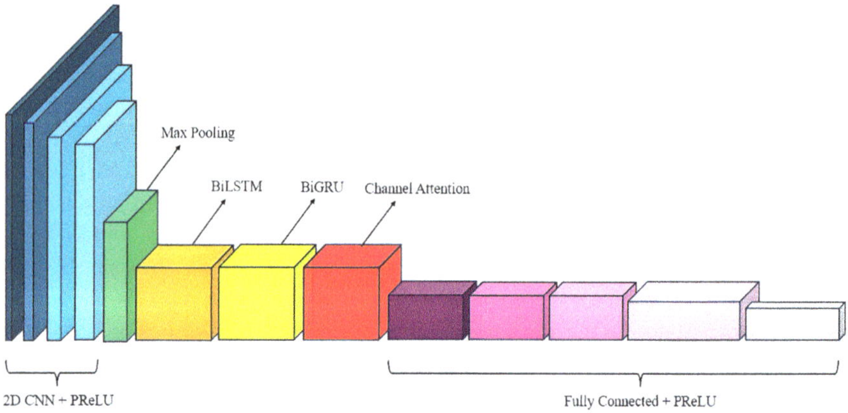

Fig. 5.6 The proposed SZ-Net model framework

5.4.2.1 Proposed SZ-Net Architecture

The SZ-Net model is specifically designed for the binary classification task of distinguishing between healthy and SZ subjects. It integrates CNN layers, RNNs (BiLSTM and BiGRU), and a custom channel attention mechanism for robust feature extraction and classification. The use of Parametric Rectified Linear Unit (PreLU) activations and dropout layers ensures effective learning and generalization. The architecture of the proposed SZ-Net model for EEG-based SZ detection is illustrated (Fig. 5.6).

The model uses the extracted feature vectors as input, generated through the pre-processing steps detailed in Sects. 5.4.1.1 and 5.4.1.2, respectively, for the MSU [37] and Kaggle EEG [41] datasets. Regarding the input format, the extracted feature vectors are in 2D tensor format of shape (timepoints, features) and are directly fed into the model. The model architecture is designed to accept 2D feature representations, which aligns with the multi-dimensional feature vectors extracted during pre-processing. Additionally, 16 EEG channels (for the MSU [37] dataset) and 64 EEG channels (for the Kaggle EEG [41] dataset) are used in the model.

The model begins with a series of four 2D CNN layers. Each layer is configured with 128 filters and a kernel dimension of (3, 1) to focus on localized spatial and temporal relationships in the input data. A PReLU activation function is employed after each CNN layer. This activation function, due to its learnable parameters, adapts to the variability in the EEG data, thereby enhancing feature learning. Stacking multiple CNN layers allows the model to capture hierarchical features ranging from simple oscillatory patterns to more complex characteristics indicative of SZ. Following the CNN layers, a MaxPooling2D operation reduces the spatial size of the feature maps, which helps lower computational complexity while also retains the most salient features of the input.

Once the spatial features are extracted, the output is flattened and reshaped into a 3D tensor suitable for sequential processing by RNN layers. The initial RNN layer in the model consists of a BiLSTM layer with 256 units as well as tanh activation function. BiLSTM networks are particularly effective for capturing bidirectional temporal dependencies, enabling the model to process information from both past and future contexts in the EEG signals. This is followed by a BiGRU layer, also with 256 units. BiGRU layers provide computational efficiency while retaining the ability to effectively model bidirectional temporal patterns, complementing the BiLSTM in this hybrid architecture. Both layers are configured with return_sequences = True to ensure that the sequential outputs are passed to subsequent layers for further processing.

To enhance the model's focus on critical features, a custom Channel Attention mechanism is proposed. This approach adaptively modifies the significance of each feature channel, enabling the model to emphasize channels that contribute significantly to distinguishing between healthy and SZ EEG patterns. The Channel Attention mechanism integrates global average pooling as well as global max pooling across the feature maps, succeeded by two dense layers to compute attention weights. The weighted features are then combined and applied back to the input feature maps, selectively amplifying relevant channels.

After applying the Channel Attention mechanism, the data undergoes further processing through a series of dense layers with increasing complexity. Each dense layer uses PReLU activation, ensuring that the model adapts to the characteristics of the EEG signals. The count of neurons in these layers progressively increases, with the final dense layers consolidating the extracted features. To avoid overfitting, a dropout layer with a 0.5 rate is introduced prior to the final classification stage.

The model ends with a fully connected layer equipped with a sigmoid activation function, which outputs the likelihood that the input belongs to either HC or SZ. The loss function employed is binary cross-entropy, which is suitable for binary classification tasks. The Adam optimizer is utilized to adjust the learning rates dynamically, promoting efficient convergence throughout the training process. To further enhance the training process, the model employs EarlyStopping and ReduceLROnPlateau callbacks. EarlyStopping halts training once the validation performance stagnates, while ReduceLROnPlateau reduces the learning rate dynamically to overcome plateaus in model improvement.

This architecture effectively combines the strengths of CNN, RNN, and attention mechanisms, making it well-suited for analyzing complex EEG data and achieving robust classification performance in the detection of SZ.

5.4.3 Results and Discussion

We have employed the Kaggle platform, utilizing Python programming and dual NVIDIA T4 GPUs with 15 GB memory to optimize our computational resources for EEG signal processing. The performance of our DL-based classification model is evaluated using several key metrics: true positives (TP), true negatives (TN), false positives (FP), as well as false negatives (FN), which together form the Confusion Matrix. The following formulas are used to calculate the primary evaluation metrics:

Accuracy (Acc): This metric indicates the ratio of accurate predictions (both positive and negative) made by the model relative to the total number of predictions.

$$\text{Acc} = \frac{T_P + T_N}{T_P + T_N + F_P + F_N}$$

Precision (Prec): This measures the model's reliability in identifying SZ cases correctly among all positive predictions.

$$\text{Prec} = \frac{T_P}{T_P + F_P}$$

Recall (Rec): Also known as sensitivity or true positive rate, this measures the model's capacity to accurately identify all actual SZ cases.

$$\text{Rec} = \frac{T_P}{T_P + F_N}$$

F1-Score (F1): This provides a harmonic mean of precision and recall, balancing the trade-offs between these metrics.

$$F_1 = 2 \times \frac{\text{Prec} \times \text{Rec}}{\text{Prec} + \text{Rec}}$$

These evaluation metrics comprehensively reflect the predictive power and generalizability of the proposed framework, offering a robust measure of its diagnostic performance.

5.4.3.1 Results of MSU [37] Dataset

To evaluate its performance in EEG-based SZ detection, the proposed SZ-Net model is rigorously tested using fivefold cross-validation. During training, the model demonstrates consistent improvements in both accuracy and loss across epochs, with minimal signs of overfitting. For instance, in Fold 1, the accuracy during training begins at 73.44% in the initial epoch and quickly increases to 99.93% by epoch 21. The corresponding validation accuracy increases from 94.39% in the first epoch to a maximum of 98.31%, indicating the model's capability to perform effectively on new, unseen data. The training loss decreases steadily, reflecting effective optimization of the network parameters. Similar trends are observed across all folds, showcasing the robustness of the proposed model. The loss and accuracy curves across folds reveal that the model maintains a balance between training and validation performance. Early stopping is employed based on validation loss, ensuring efficient training without overfitting. The validation accuracy remains high, typically above 98%, highlighting the effectiveness of the SZ-Net architecture in capturing temporal and spatial dependencies in EEG data.

The model achieves outstanding performance on the test dataset, with an average classification accuracy of 98.21% across folds, while precision, recall, F1-score are found to be 98.36%, 98.43%, 98.39%, correspondingly. These metrics indicate that the model effectively distinguishes between healthy and SZ EEG signals with minimal misclassifications. The elevated precision and recall values further highlight the model's capability to accurately detect both true positives as well as true negatives accurately, ensuring reliability in clinical scenarios.

Graphical analysis provides additional understanding of the model's training dynamics as well as classification performance. The plot depicts the evolution of training and validation loss throughout the model training process with early stopping (Fig. 5.7a). The training loss exhibits a steady decline, reflecting effective learning from the data, while the validation loss initially decreases, indicating good generalization. Minor fluctuations in validation loss after the tenth epoch are observed, but overall, it remains low, demonstrating robustness. Early stopping ensures that training halts automatically at the optimal point when validation loss ceases to improve, preventing overfitting, reducing unnecessary training time, and achieving a balance between bias and variance. This highlights the model's performance and capacity to adapt effectively to new, unseen data.

Similarly, the accuracy progression of the model for the data used for both training and validation is shown (Fig. 5.7b). The training accuracy increases steadily, approaching 100%, demonstrating that the model effectively acquires knowledge and adapts to the data used for training. The accuracy on validation data also rises rapidly during the initial epochs and stabilizes close to the training accuracy, demonstrating strong generalization to new, unseen data. The alignment between the accuracy curves for training and validation reflects the absence of overfitting as well as underscores the model's capability to maintain balanced performance across datasets. The overall high validation accuracy confirms that the model captures the underlying patterns effectively and is likely to perform well on new data.

The class-wise metrics offers a thorough assessment of the model's effectiveness for both healthy and SZ classes (Fig. 5.8). The accuracy for both classes is found to be identical at 98.37%, indicating the model's high overall reliability. Precision is slightly higher for the SZ class (99.03%) compared to the healthy class (97.57%), demonstrating the model's superior capacity to prevent incorrect positive predictions for SZ cases. Conversely, recall is higher for the healthy class (98.80%) than the SZ class (98.03%), suggesting improved sensitivity in identifying healthy cases. The F1-score, which harmonizes precision and recall, is comparable for both classes, with the SZ class achieving 98.53% and the healthy class 98.18%. These results reveal a well-balanced performance across both classes, highlighting the model's robustness and fairness, which are critical for ensuring equitable and accurate medical diagnoses in SZ detection.

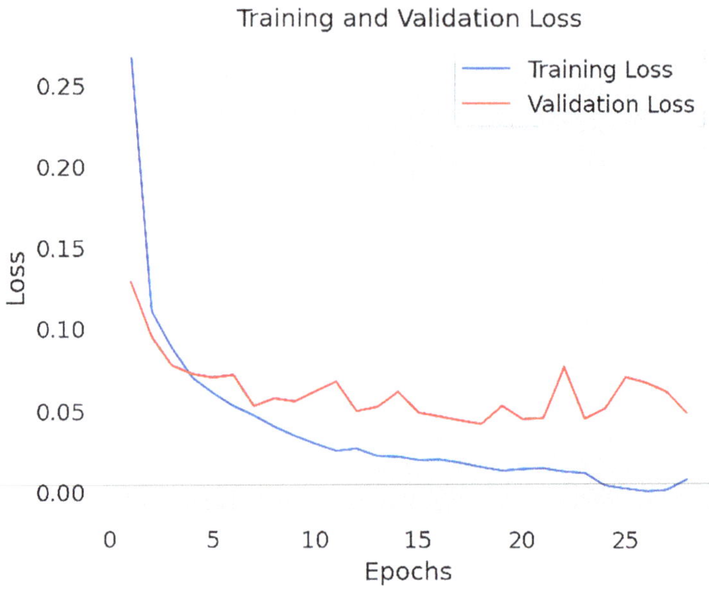

(a)

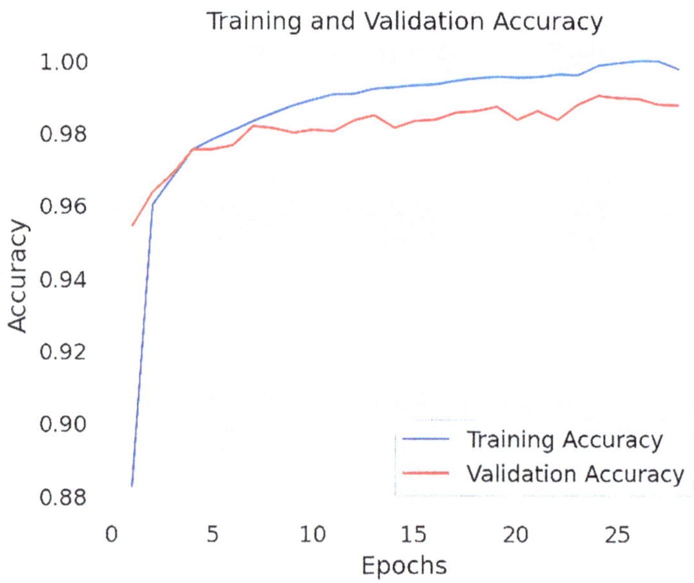

(b)

Fig. 5.7 **(a)** Curves for training and validation loss **(b)** Curves for training and validation accuracy, for the MSU [37] dataset

Fig. 5.8 Class-wise performance metrics for HC and SZ patients using the proposed SZ-Net framework for the MSU [37] dataset

The confusion matrix for the binary classification task provides a detailed overview of the performance outcomes (Fig. 5.9). Majority of the samples are correctly classified, with very few instances of false positives or false negatives. This reflects the model's robustness in handling class imbalances and noise in the EEG signals.

5.4.3.2 Results of Kaggle EEG [41] Dataset

For the second dataset, we have applied a fivefold cross-validation approach, consistent with the method used for the first dataset, to ensure reliable and generalizable performance evaluation. The model demonstrates impressive accuracy improvement across epochs. The accuracy during training improves from 79.81% in the first epoch to 99.98% by the 31st epoch. Likewise, validation accuracy sees a steady improvement from 94.28% to 99.51%, indicating the model's robust generalization capability. The loss metrics shows a similar trend, with the training loss dropping from 0.3993 to 7.1796×10^{-4}, and validation loss improving from 0.1464 to 0.0281 by the final epoch. Similar trends are observed across all folds, further reinforcing the model's consistency and generalizability across different data splits.

The model achieves a remarkable average classification accuracy of 99.27%, reflecting its effective capability to distinguish between healthy and SZ subjects. Precision is 99.43%, suggesting the model is highly accurate in predicting positive

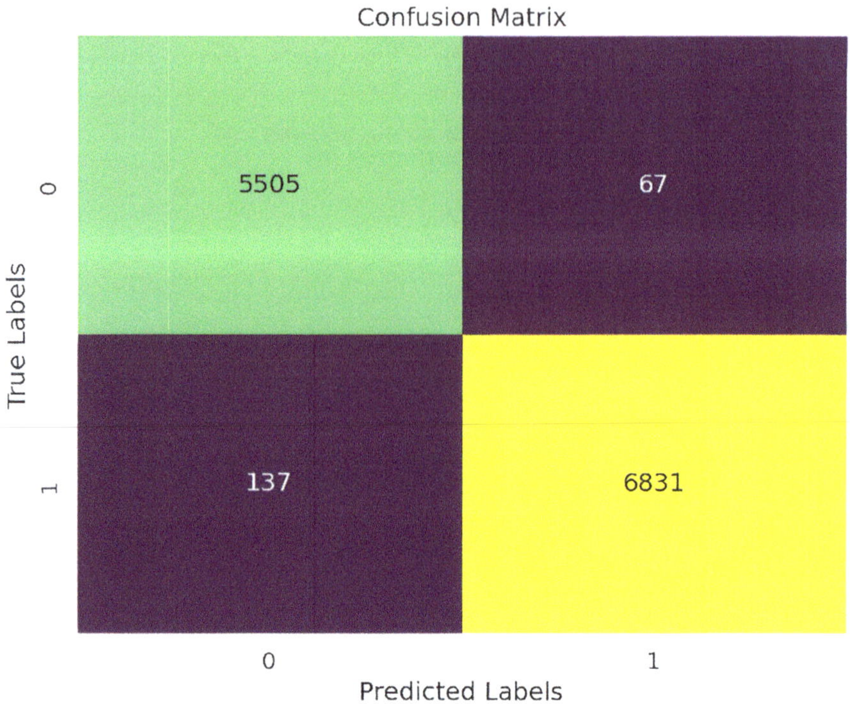

Fig. 5.9 Confusion matrix of the proposed SZ-Net framework for the MSU [37] dataset

SZ cases. Recall is 99.37%, indicating the model's excellent sensitivity in detecting actual SZ patients. These high values are critical in medical contexts where false negatives (missed diagnoses) need to be minimized. The F1-score of 99.4% indicates that the model effectively balances precision and recall, which is particularly important for diagnosing SZ, where both false positives as well as false negatives must be reduced to a minimum.

The trend of training as well as validation error throughout the process for the second dataset is shown (Fig. 5.10a). The training loss drops significantly during the first epochs and continues to fall consistently, indicating effective learning as the model minimizes the discrepancy in the training dataset. The validation error shows an equivalent downward pattern at the start, showing that the model performs effectively on new, unseen data at the initial stages of training. However, after about the fifth epoch, the validation loss starts to fluctuate significantly, but it remains modest and consistent generally. The persistent difference between training and validation loss is small, suggesting that the model did not overfit to the data used for training. This balance emphasizes the model's stability and great generalization abilities on the second dataset.

The plot depicts the advancement of the training as well as validation accuracy on the second dataset (Fig. 5.10b). Training accuracy grows significantly in the

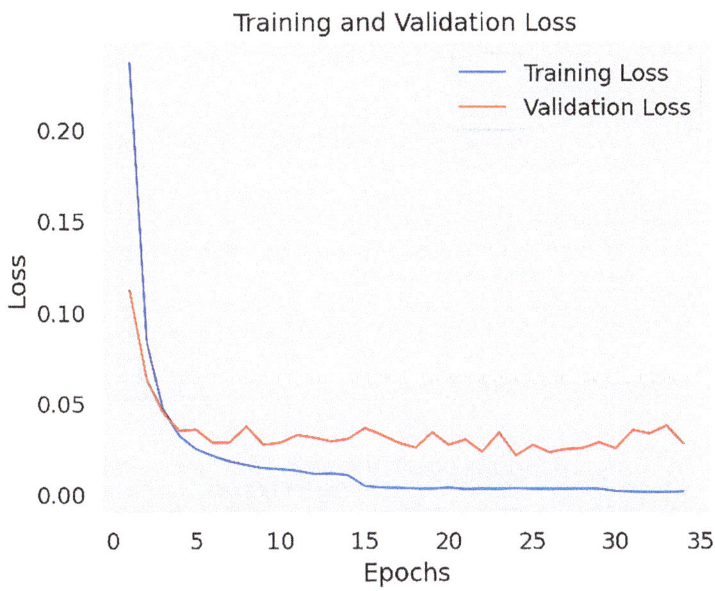

(a)

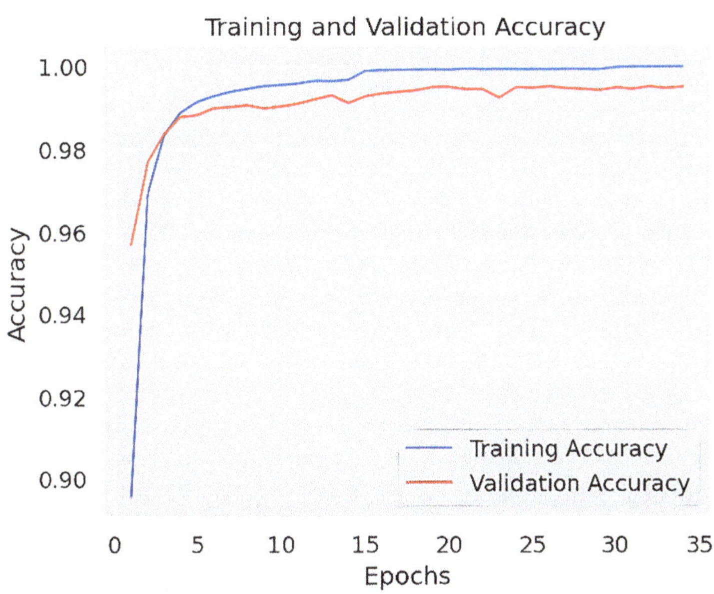

(b)

Fig. 5.10 **(a)** Curves for training and validation loss **(b)** Curves for training and validation accuracy, for the Kaggle EEG [41] dataset

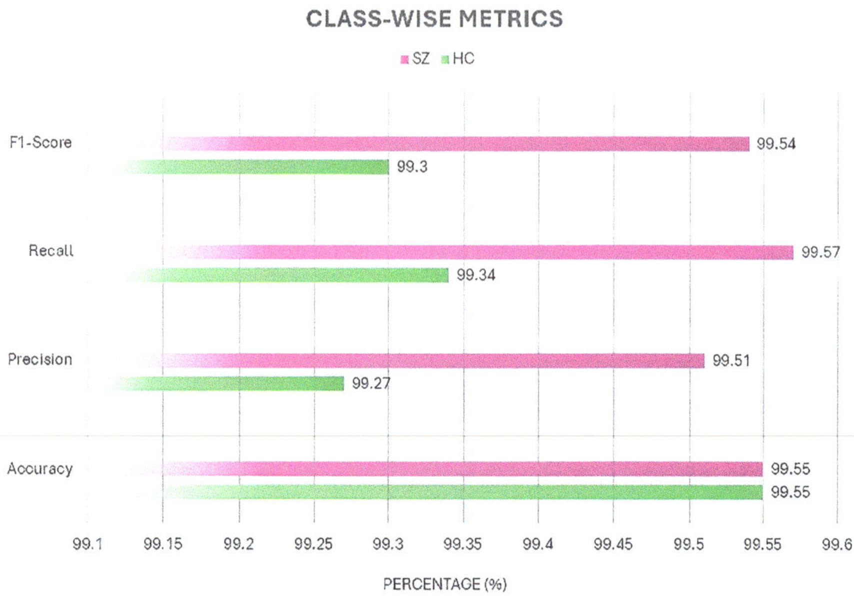

Fig. 5.11 Class-wise performance metrics for HC and SZ patients using the proposed SZ-Net framework for the Kaggle EEG [41] dataset

initial epochs, suggesting that the model rapidly learns to accurately categorize the training data. It steadily improves, eventually reaching near-perfect accuracy as training proceeds. The validation accuracy also increases dramatically in the beginning, indicating significant generalization early in training. After a few epochs, the validation accuracy stabilizes slightly lower than the training accuracy, with minor variations during the remaining epochs. The minor gap between training and validation accuracy suggests that the model adapts effectively to unseen data without significant overfitting. This trend, combined with the usage of early stopping, demonstrates that the model delivers high performance efficiently, ensuring robust accuracy on the second dataset.

The performance metrics of the model for both healthy and SZ classes is illustrated (Fig. 5.11). The model earns an excellent 99.55% accuracy in both classes, indicating its overall reliability. In terms of precision, the SZ class marginally surpasses the healthy class, with values of 99.51% and 99.27%, respectively, demonstrating the model's capacity to reduce incorrectly classified positives for SZ. Similarly, the SZ class has a higher recall (99.57%) than the healthy class (99.34%), showing greater sensitivity in recognizing SZ cases. The F1-scores are also high and close, with 99.54% for SZ and 99.30% for healthy, indicating a trade-off between precision and recall. These findings validate the model's constant and robust capability, indicating its efficacy in detecting SZ.

The confusion matrix for the binary classification task offers a thorough evaluation of the model's effectiveness (Fig. 5.12). This matrix highlights the model's

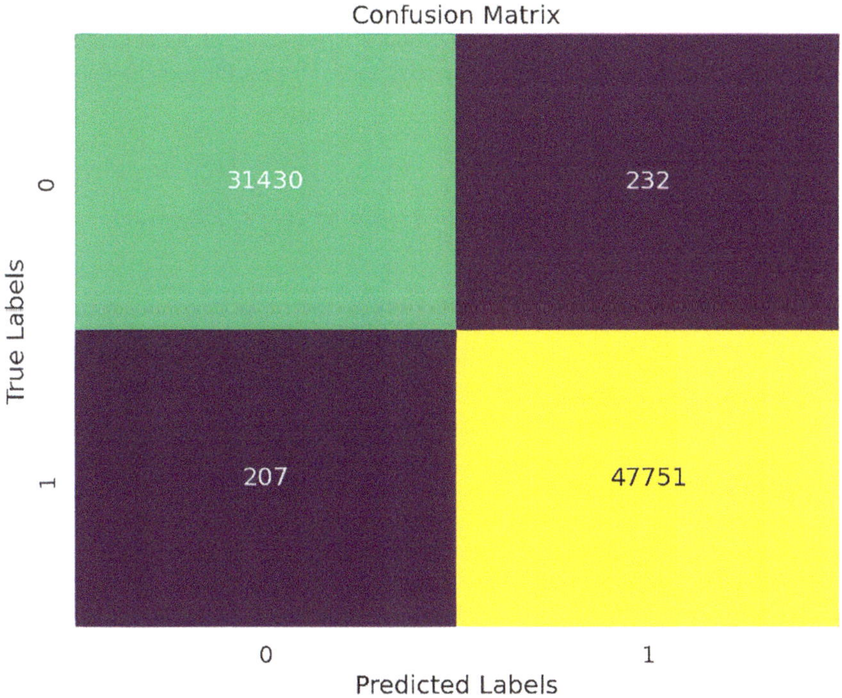

Fig. 5.12 Confusion matrix of the proposed SZ-Net framework for the Kaggle EEG [41] dataset

Table 5.1 Performance metrics of our proposed SZ-Net framework on the utilized datasets

Dataset	Accuracy (%)	Precision (%)	Recall (%)	F1-score (%)
MSU [37]	98.21	98.36	98.43	98.39
Kaggle EEG [41]	99.27	99.43	99.37	99.4

efficiency in managing class imbalances and noise present in the EEG signals. The classification outcomes of our suggested SZ-Net framework on both MSU [37] and Kaggle EEG [41] datasets are tabulated (Table 5.1).

5.5 Comparative Analysis with Existing Studies

In this section, we perform a comprehensive comparison between our approach and the methods used in existing studies within the field of SZ classification using EEG signals. Our primary goal is to gauge the model's effectiveness relative to previously published works that have utilized the same datasets, allowing for a clear assessment of the advancements and improvements our approach offers. This comparison allows us to not only confirm the robustness of our model but also identify areas where it outperforms existing methods or introduces novel improvements. Through this analysis, we aim to position our work in the larger scope of the field and add to the ongoing development of more accurate and reliable classification models for SZ detection.

5.5.1 Performance Comparison Using MSU [37] Dataset

Using the MSU dataset [37], Rajesh et al. [38] obtained 91.66% accuracy using SLBP-derived histogram features with a LogitBoost Classifier, emphasizing the role of texture-based analysis and ensemble methods. C. Phang et al. [36] achieved 95% accuracy by combining Direct Connectivity and Complex Network (DC-CN) features with a DNN-DBN model, highlighting the importance of advanced network-based features in classification tasks. Sobahi et al. [39] reported 97.7% accuracy by utilizing Time-Frequency features with an ELM-based Autoencoder (AE), effectively capturing temporal and spectral information for accurate classification. Aslan et al. [40] utilized the Short-Time Fourier Transform (STFT) to convert EEG data into scalogram images, which were then classified using a Convolutional Neural Network (CNN). They secured a high accuracy rate of 98% on the MSU [37] dataset, demonstrating the effectiveness of their approach in classifying SZ patients and healthy individuals.

In comparison, our framework achieves a superior performance of 98.21% accuracy by utilizing a hybrid CNN-BiLSTM-BiGRU architecture. This advanced model is designed to capture long-range dependencies in EEG signals and emphasize important features through channel attention, resulting in more precise classification of SZ patients and healthy individuals. This significant improvement demonstrates the effectiveness of our approach in addressing the complexities inherent in EEG signal analysis for SZ detection. The detailed comparison of our framework with existing approaches using MSU [37] dataset is shown (Table 5.2).

Table 5.2 Comparison of our proposed SZ-Net framework with existing models on the MSU [37] dataset

Work	Technique	Accuracy (%)
Rajesh et al. [38]—2021	SLBP	91.66
Phang et al. [36]—2019	DNN-DBN	95
Sobahi et al. [39]—2022	ELM-driven AE	97.7
Aslan et al. [40]—2022	CNN	98
Proposed SZ-Net Framework—2025	**CNN-BiLSTM-BiGRU**	**98.21**

Table 5.3 Comparison of our proposed SZ-Net framework with existing models on the Kaggle EEG [41] dataset

Work	Technique	Accuracy (%)
Lei Zhang [42]—2019	RF	81.1
Siuly et al. [43]—2020	EBT	89.59
Lei Zhang [44]—2020	ANN	98.5
Siuly et al. [45]—2023	Deep ResNet	99.23
Proposed SZ-Net Framework—2025	**CNN-BiLSTM-BiGRU**	**99.27**

5.5.2 *Performance Comparison Using Kaggle EEG [41] Dataset*

Using Kaggle EEG [41] dataset, Lei Zhang [42] achieved 81.1% classification accuracy using the Random Forest (RF) algorithm to differentiate between SZ and HCs according to EEG-derived features. The accuracy was enhanced by incorporating specific feature sets from the EEG signals. Siuly et al. [43] developed an automated approach for SZ diagnosis utilizing EEG data, employing EMD and EBT classifiers. Their approach achieved an overall accuracy of 89.59%, effectively distinguishing SZ patients from HCs. Later, Zhang [44] achieved over 98.5% classification accuracy for SZ detection using EEG signals with ANN. This was after applying wavelet transforms and extracting time-frequency features from the event-related potentials (ERP). Also, Siuly et al. [45] achieved 99.23% accuracy in SZ detection from EEG signals using a deep residual network (ResNet) for feature extraction, with classification via an SVM. This approach outperformed the ResNet classifier and other existing methods.

In comparison, our framework achieves an impressive 99.27% accuracy by integrating a hybrid CNN-BiLSTM-BiGRU architecture. By combining BiLSTM and BiGRU layers with convolutional layers, the model enhances feature extraction and outperforms simpler models in classifying EEG data. This combined architecture optimizes both spatial and temporal information, making the model highly effective in classifying EEG data with remarkable accuracy. A comprehensive comparison of our framework with existing methods using the Kaggle EEG [41] dataset is shown (Table 5.3).

5.6 Conclusion and Upcoming Avenues

This research showcases the efficiency of the CNN-BiLSTM-BiGRU model with channel attention mechanisms in analyzing EEG data for SZ detection. The SZ-Net model's architecture combines the advantages of Convolutional Neural Networks (CNNs) for robust extraction of features, Bidirectional Long Short-Term Memory (BiLSTM) networks for acquiring both past and future temporal dependencies, and

Bidirectional Gated Recurrent Units (BiGRU) for enhancing the model's capacity to derive insights from sequential data. By using these advanced techniques in tandem, the model efficiently classifies EEG signals into healthy and SZ categories, achieving high accuracy and offering a promising approach for early diagnosis.

Incorporating attention mechanisms into the model further strengthens its performance by allowing the model to emphasize the most significant aspects of the EEG signal, enhancing classification accuracy. These mechanisms contribute to the interpretability of the model, enabling us to better understand which elements of the signal are critical for SZ detection. The combination of these techniques makes the SZ-Net model a powerful tool for advancing the usage of DL for diagnosing as well as understanding of neurological disorders, especially in the context of SZ. This approach marks a major advancement in utilizing EEG as a diagnostic tool, addressing challenges in both accuracy and interpretability, and paving the way for more effective and efficient clinical applications.

To further enhance the reliability of the proposed model, we have performed cross-dataset validation by evaluating it on two independent EEG datasets: the MSU and Kaggle EEG datasets. These datasets differ significantly in terms of sampling rates, population demographics, and the number of EEG channels, making them suitable for assessing the model's robustness. By testing the model on datasets with varying characteristics, we demonstrate its ability to generalize across different EEG acquisition protocols. Furthermore, the use of stratified fivefold cross-validation on both datasets ensures a thorough evaluation, reducing the likelihood of overfitting and confirming the model's effectiveness in accurately classifying SZ and HC subjects. This cross-dataset validation strengthens the reliability and potential applicability of the proposed model in real-world SZ detection scenarios.

Upcoming studies may investigate the use of the SZ-Net model on larger and more varied EEG datasets, incorporating multi-center data to evaluate the model's generalizability across different populations and environments. Incorporating other modalities, such as fMRI or genetic information alongside EEG may enhance the model's diagnostic capabilities, leading to more robust and comprehensive SZ detection frameworks. As with many DL models, the interpretability of results is a key challenge. Subsequent studies should concentrate on making the model more interpretable, helping clinicians understand the problem-solving approach of the model, which is crucial for clinical adoption. Although the model has shown promising results, further work can focus on improving its computational efficiency, making it feasible for use in live applications and on systems with restricted processing capabilities. Continued progress in these areas will not only enhance the diagnostic capabilities of EEG-based systems but also contribute to a deeper understanding of the neurological mechanisms underlying SZ, ultimately improving patient outcomes.

Authors Contributions Conceptualization, A.S. and P.K.S.; methodology, P.K.S.; software, P.K.S.; validation, A.S. and P.K.S.; formal analysis, A.S.; investigation, A.S.; resources, P.K.S.; data curation, A.S.; writing—original draft preparation, A.S.; writing—review and editing, P.K.S.;

visualization, P.K.S., and A.S.; supervision, P.K.S.; project administration, P.K.S. Both authors have read and agreed to the published version of the manuscript.

Conflict of Interest The authors declare no conflict of interest.

Data Availability The datasets utilized in the current research are available at:

1. MSU Dataset [37]: http://brain.bio.msu.ru/eeg_schizophrenia.htm
2. Kaggle EEG Dataset [41]**:** https://www.kaggle.com/datasets/broach/button-tone-sz

Source Code Availability The source codes related to this research are available at: https://github.com/Arghyasree/SZ-Net.

References

1. Saha, S., Chant, D., Welham, J., McGrath, J.: A systematic review of the prevalence of schizophrenia. PLoS Med. **2**(5), e141 (2005). https://doi.org/10.1371/journal.pmed.0020141
2. Saha, A., Park, S., Geem, Z.W., Singh, P.K.: Schizophrenia detection and classification: a systematic review of the last decade. Diagnostics. **14**(23), 2698 (2024). https://doi.org/10.3390/diagnostics14232698
3. Andreasen, N.: Symptoms, signs, and diagnosis of schizophrenia. Lancet. **346**(8973), 477–481 (1995). https://doi.org/10.1016/S0140-6736(95)91325-4
4. Tandon, R., Nasrallah, H.A., Keshavan, M.S.: Schizophrenia, "just the facts" 4. Clinical features and conceptualization. Schizophr. Res. **110**(1–3), 1–23 (2009). https://doi.org/10.1016/j.schres.2009.03.005
5. Green, M.F.: What are the functional consequences of neurocognitive deficits in schizophrenia? Am. J. Psychiatry. **153**(3), 321–330 (1996). https://doi.org/10.1176/ajp.153.3.321
6. Burns, T., Patrick, D.: Social functioning as an outcome measure in schizophrenia studies. Acta Psychiatr. Scand. **116**(6), 403–418 (2007). https://doi.org/10.1111/j.1600-0447.2007.01108.x
7. Moreno-De-Luca, D., Sanders, S.J., Willsey, A.J., Mulle, J.G., Lowe, J.K., Geschwind, D.H., et al.: Using large clinical data sets to infer pathogenicity for rare copy number variants in autism cohorts. Mol. Psychiatry. **18**(10), 1090–1095 (2013). https://doi.org/10.1038/mp.2012.138
8. Marshall, M., Rathbone, J.: Early intervention for psychosis. Cochrane Database Syst. Rev. (6), CD004718 (2011). https://doi.org/10.1002/14651858.CD004718.pub3
9. Aich, U., Saha, A., Woźniak, M., Ijaz, M.F., Singh, P.K.: Schizophrenia detection from electroencephalogram signals using image encoding and wrapper-based deep feature selection approach. Scientific Reports, **15**(1), 21390 (2025). https://doi.org/10.1038/s41598-025-06121-7
10. Anderson, D., Lord, C., Risi, S., DiLavore, P., Shulman, C., Thurm, A., Pickles, A., American Psychiatric Association: Diagnostic and statistical manual of mental disorders. Washington, DC: Author. Linguist. Cogn. Effects Bilingual. Child. Autism Spectrum Disord. **21**, 175 (2013). https://doi.org/10.1176/appi.books.9780890425596
11. Ward, P.B., Firth, J., Rosenbaum, S., Samaras, K., Stubbs, B., Curtis, J.: Lifestyle interventions to reduce premature mortality in schizophrenia. Lancet Psychiatry. **4**(7), e14 (2017). https://doi.org/10.1016/S2215-0366(17)30235-3
12. Patel, K.R., Cherian, J., Gohil, K., Atkinson, D.: Schizophrenia: overview and treatment options. Pharm. Ther. **39**(9), 638 (2014) https://www.ncbi.nlm.nih.gov/pmc/articles/PMC4159061
13. Kapur, S., Phillips, A.G., Insel, T.R.: Why has it taken so long for biological psychiatry to develop clinical tests and what to do about it? Mol. Psychiatry. **17**(12), 1174–1179 (2012). https://doi.org/10.1038/mp.2012.105

14. Habets, P., Marcelis, M., Gronenschild, E., Drukker, M., Van Os, J.: Reduced cortical thickness as an outcome of differential sensitivity to environmental risks in schizophrenia. Biol. Psychiatry. **69**(5), 487–494 (2011). https://doi.org/10.1016/j.biopsych.2010.08.010
15. Bhadra, R., Singh, P.K., Mahmud, M.: HyEpiSeiD: a hybrid convolutional neural network and gated recurrent unit model for epileptic seizure detection from electroencephalogram signals. Brain Inf. **11**(1), 21 (2024). https://doi.org/10.1186/s40708-024-00234-x
16. Singh, A.K., Singh, P.K., Kaiser, M.S., Mahmud, M.: A deep learning-based framework for detecting depression from electroencephalogram signals. In: International Conference on Trends in Electronics and Health Informatics, pp. 3–16. Springer Nature Singapore, Singapore (2023). https://doi.org/10.1007/978-981-97-3937-0_1
17. Electroencephalogram Test (EEG). Available online: https://www.neuroinjuryspecialists.com/diagnostic-testing/electroencephalogram-test-eeg/
18. Craik, A., He, Y., Contreras-Vidal, J.L.: Deep learning for electroencephalogram (EEG) classification tasks: a review. J. Neural Eng. **16**(3), 031001 (2019). https://doi.org/10.1088/1741-2552/ab0ab5
19. LeCun, Y., Bengio, Y., Hinton, G.: Deep learning. Nature. **521**(7553), 436–444 (2015). https://doi.org/10.1038/nature14539
20. Liao, W., Fan, Y.S., Yang, S., Li, J., Duan, X., Cui, Q., Chen, H.: Preservation effect: cigarette smoking acts on the dynamic of influences among unifying neuropsychiatric triple networks in schizophrenia. Schizophr. Bull. **45**(6), 1242–1250 (2019). https://doi.org/10.1093/schbul/sby184
21. Vaswani, A., Shazeer, N., Parmar, N., Uszkoreit, J., Jones, L., Gomez, A.N., et al.: Attention is all you need. Adv. Neural Inf. Proces. Syst. **30** (2017). https://doi.org/10.48550/arXiv.1706.03762
22. Lawhern, V.J., Solon, A.J., Waytowich, N.R., Gordon, S.M., Hung, C.P., Lance, B.J.: EEGNet: a compact convolutional neural network for EEG-based brain–computer interfaces. J. Neural Eng. **15**(5), 056013 (2018). https://doi.org/10.1088/1741-2552/aace8c
23. Iyer, S.H., Matthews, S.A., Simeone, T.A., Maganti, R., Simeone, K.A.: Accumulation of rest deficiency precedes sudden death of epileptic Kv1. 1 knockout mice, a model of sudden unexpected death in epilepsy. Epilepsia. **59**(1), 92–105 (2018). https://doi.org/10.1111/epi.13953
24. Schuster, M., Paliwal, K.K.: Bidirectional recurrent neural networks. IEEE Trans. Signal Process. **45**(11), 2673–2681 (1997). https://doi.org/10.1109/78.650093
25. Cho, K., Van Merriënboer, B., Gulcehre, C., Bahdanau, D., Bougares, F., Schwenk, H., Bengio, Y.: Learning Phrase Representations Using RNN Encoder-Decoder for Statistical Machine Translation. arXiv preprint arXiv:1406.1078 (2014). https://doi.org/10.48550/arXiv.1406.1078
26. Graves, A., Schmidhuber, J.: Framewise phoneme classification with bidirectional LSTM and other neural network architectures. Neural Netw. **18**(5–6), 602–610 (2005). https://doi.org/10.1016/j.neunet.2005.06.042
27. Bahdanau, D. (2014). Neural Machine Translation by Jointly Learning to Align and Translate.. arXiv preprint arXiv:1409.0473. https://arxiv.org/abs/1409.0473
28. Luong, M.T.: Effective Approaches to Attention-Based Neural Machine Translation. arXiv preprint arXiv:1508.04025 (2015). https://doi.org/10.48550/arXiv.1508.04025
29. Uhlhaas, P.J., Singer, W.: Abnormal neural oscillations and synchrony in schizophrenia. Nat. Rev. Neurosci. **11**(2), 100–113 (2010). https://doi.org/10.1038/nrn2774
30. Ravì, D., Wong, C., Deligianni, F., Berthelot, M., Andreu-Perez, J., Lo, B., Yang, G.Z.: Deep learning for health informatics. IEEE J. Biomed. Health Inform. **21**(1), 4–21 (2016). https://doi.org/10.1109/JBHI.2016.2636665
31. Dhara, T., Singh, P.K., Mahmud, M.: A fuzzy ensemble-based deep learning model for EEG-based emotion recognition. Cogn. Comput. **16**(3), 1364–1378 (2024). https://doi.org/10.1007/s12559-023-10171-2
32. Dhara, T., Singh, P.K.: Emotion recognition from EEG data using hybrid deep learning approach. In: Frontiers of ICT in Healthcare: Proceedings of EAIT 2022, pp. 179–189. Springer Nature Singapore, Singapore (2023). https://doi.org/10.1007/978-981-19-5191-6_15

33. Dave, M., Mukherjee, S.K., Singh, P.K., Mahmud, M.: Deep ensemble learning approach for multimodal emotion recognition. In: International Conference on Trends in Electronics and Health Informatics, pp. 211–223. Springer Nature Singapore, Singapore (2023). https://doi.org/10.1007/978-981-97-3937-0_15
34. Mukherjee, S.K., Dave, M., Singh, P.K.: A transfer learning approach for the classification of human emotions using electroencephalogram signals. In: International Conference on Emerging Applications of Information Technology, pp. 219–228. Springer Nature Singapore, Singapore (2024). https://doi.org/10.1007/978-981-97-7532-3_17
35. Ghosh, S., Kim, S., Ijaz, M.F., Singh, P.K., Mahmud, M.: Classification of mental stress from wearable physiological sensors using image-encoding-based deep neural network. Biosensors. **12**(12), 1153 (2022). https://doi.org/10.3390/bios12121153
36. Phang, C.R., Ting, C.M., Samdin, S.B., Ombao, H.: Classification of EEG-based effective brain connectivity in schizophrenia using deep neural networks. In: 2019 9th International IEEE/EMBS Conference on Neural Engineering (NER), pp. 401–406. IEEE (2019). https://doi.org/10.1109/NER.2019.8717087
37. Gorbachevskaya, K., & Borisov, S.. EEG of Healthy Adolescents and Adolescents with Symptoms of Schizophrenia. (2019).. Available online: http://brain.bio.msu.ru/eeg_schizophrenia.htm
38. Rajesh, K.N., Kumar, T.S.: Schizophrenia detection in adolescents from EEG signals using symmetrically weighted local binary patterns. In: 2021 43rd Annual International Conference of the IEEE Engineering in Medicine and Biology Society (EMBC), pp. 963–966. IEEE (2021). https://doi.org/10.1109/EMBC46164.2021.9630232
39. Sobahi, N., Ari, B., Cakar, H., Alcin, O.F., Sengur, A.: A new signal to image mapping procedure and convolutional neural networks for efficient schizophrenia detection in EEG recordings. IEEE Sensors J. **22**(8), 7913–7919 (2022). https://doi.org/10.1109/JSEN.2022.3151465
40. Aslan, Z., Akin, M.: A deep learning approach in automated detection of schizophrenia using scalogram images of EEG signals. Phy. Eng. Sci. Med. **45**(1), 83–96 (2022). https://doi.org/10.1007/s13246-021-01083-2
41. EEG Data from Basic Sensory Task in Schizophrenia. Available online: https://www.kaggle.com/datasets/broach/button-tone-sz
42. Zhang, L.: EEG signals classification using machine learning for the identification and diagnosis of schizophrenia. In: 2019 41st Annual International Conference of the IEEE Engineering in Medicine and Biology Society (EMBC), pp. 4521–4524. IEEE (2019). https://doi.org/10.1109/EMBC.2019.8857946
43. Siuly, S., Khare, S.K., Bajaj, V., Wang, H., Zhang, Y.: A computerized method for automatic detection of schizophrenia using EEG signals. IEEE Trans. Neural Syst. Rehabil. Eng. **28**(11), 2390–2400 (2020). https://doi.org/10.1109/TNSRE.2020.3022715
44. Zhang, L.: EEG signals feature extraction and artificial neural networks classification for the diagnosis of schizophrenia. In: 2020 IEEE 19th International Conference on Cognitive Informatics & Cognitive Computing (ICCI* CC), pp. 68–75. IEEE (2020). https://doi.org/10.1109/ICCICC50026.2020.9450257
45. Siuly, S., Guo, Y., Alcin, O.F., Li, Y., Wen, P., Wang, H.: Exploring deep residual network based features for automatic schizophrenia detection from EEG. Phys. Eng. Sci. Med. **46**(2), 561–574 (2023). https://doi.org/10.1007/s13246-023-01225-8
46. BioSemi Active Two system. Available online: https://www.biosemi.com/

Chapter 6
TD-Net: A Deep Learning Technique with Spatial and Channel Attention for Efficient Tuberculosis Detection from Chest X-Ray Scans

Debgandhar Ghosh and Pawan Kumar Singh

Abstract The disease, Tuberculosis (TB) is a bacterial infection that results from *Mycobacterium tuberculosis* and can be fatal if not recognized early. Chest X-rays are used to identify TB, thus requiring close supervision from radiologists and doctors. Tuberculosis (TB) can be discovered and recognized through chest X-rays employing deep learning methodologies, leveraging the capabilities of Convolutional Neural Networks (CNNs) to differentiate between X-rays of individuals with TB and those that are normal. In this work, we developed a new deep learning methodology, named as TD-Net, incorporating pre-trained MobileNetV2 architecture with spatial and channel attention to elevate the accuracy, F1 Score, Recall and AUC-ROC of MobileNetV2. The experiment shows that the proposed TD-Net methodology gains a test accuracy of 98 and 97% respectively and an AUC-ROC score of 0.99 and 0.98 respectively on two benchmark TB datasets namely, a publicly available dataset curated by Twasifur Rahman et al. and TBX11 dataset.

Keywords Tuberculosis detection · TD-Net · Chest X-Ray images · Ensemble deep learning · MobileNetV2 · Spatial attention · Channel attention · TBX11 dataset

6.1 Introduction

Tuberculosis (TB) is an infectious disease that is usually fatal, targeting the lungs though it may target the other parts of the human body. The World Health Organization (WHO) asserts that around 12.5 lakh individuals succumbed to tuberculosis in 2023. The disease is curable once detected early, but might prove fatal if proper medical attention is not sought. Chest X-rays are mostly employed for TB diagnosis

---

D. Ghosh · P. K. Singh (✉)
Department of Information Technology, Jadavpur University, Jadavpur University Salt Lake Campus, Kolkata, West Bengal, India

G. Rivera et al. (eds.), *Recent Advances in Artificial Neural Networks*, Intelligent Systems Reference Library 283, https://doi.org/10.1007/978-3-032-13497-4_6

of the lungs in pulmonary TB cases. Generally, a conformational sputum test is followed after consultation with a medical professional. Chest X-rays help in the quick and early detection of the disease so that appropriate actions and remedies can be taken. The procedure is non-invasive, that is it requires no insertion of any device or instrument in the patient's body nor requires any kind of fluid collection from the patient which makes it more comfortable for the patient. Deep learning algorithms have been used effectively to detect TB from chest X-rays in different studies. Convolutional Neural Networks (CNNs) present an excellent approach towards the classification task of the chest X-ray of patients affected with tubercular TB and other normal patients.

The challenges associated with properly annotated high-quality images of the chest X-ray often hinder the implementation of deep learning Approaches in TB Detection. Although a large number of publicly available datasets are present for TB detection, chest X-rays often vary from one another due to the differences in imaging equipment and the position of the patient while the X-ray image is acquired. Further annotation of chest X-rays requires supervision from a Radiologist and might be a time-consuming task. Often the class imbalance present in the datasets might cause the deep learning models to underfit. Though imbalance is a natural phenomenon and most TB cases are fewer than normal ones, owing to an insufficient amount of data, the models may fail to learn specific features. TB may present radiographic similarities to conditions such as lung cancer and Chronic Obstructive Pulmonary Disease (COPD), making it challenging to differentiate TB based solely on chest X-rays. In the initial stages, TB may be undetectable using chest X-rays, rendering its identification a formidable challenge (Fig. 6.1).

- In this study, we leverage the power of both spatial and channel attentions [1, 2] for efficient detection of TB. In the proposed TD-Net methodology, we have embedded these attention modules into a pre-trained MobileNetV2 [3] architecture.
- The MobileNetV2 [3] is trained on the ImageNet [4] dataset, which is a large-scale dataset containing 14 million labeled data distributed among 1000 classes.
- Incorporating both spatial and channel attentions into the MobileNetV2 architecture [3], improves detection performance by directing model emphasis where it is needed.
- The proposed TD-Net Architecture achieves an accuracy of 98% and ROC of 0.99 outperforming the state-of-the-art models on the Kaggle Dataset curated by Tawsifur Rahman et al. [5].
- Our TD-Net model also achieves a classification accuracy of 97% and AUC-ROC of 0.98 on TBX11 dataset [6].

MobileNet architecture provides high feature extraction capabilities without increasing the computational cost. Furthermore, the inference time of MobileNetV2 is less compared to either VGG or DenseNet architectures. This allows us to draw faster inferences from X-rays and build a fast and scalable system for TB detection. Additionally, the use of attention mechanisms helps us solve the issues related to class imbalance as the number of normal classes exceeds the TB-positive class by

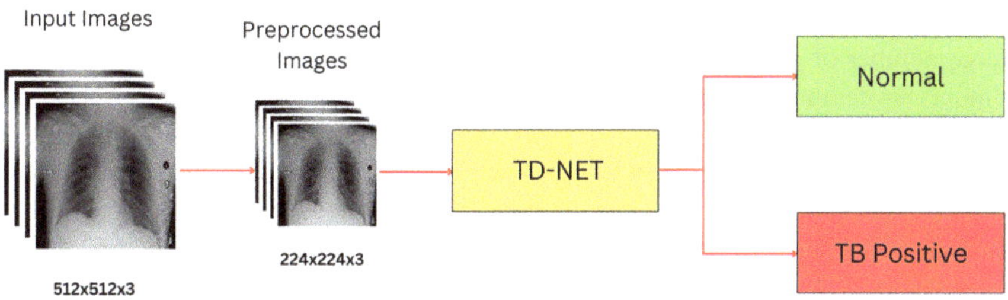

Fig. 6.1 Block diagram of our proposed TD-Net methodology

a vast number. Thus, the addition of an attention mechanism helps the model to focus on relevant regions for better TB detection, and separating it from diseases with similar radiographic features such as lung cancer or COPD, which the previous works lacked. This makes our proposed methodology ideal for early and accurate detection of TB.

6.2 Literature Review

The application of disease and anomaly detection from medical imaging using deep learning techniques has been one emerging space of research [7–9]. Several research have utilized the potential of CNNs for the effective identification of TB from chest X-rays. Lakhani and Sundaram [10] demonstrated the ensemble learning technique, making use of the weighted averages of probabilities coming from two Deep CNNs namely AlexNet [11] and GoogLeNet [12] with and managed to reach an AUC of 0.99. The ensemble technique of potential models of medical imaging shows an application in the aspect of DCNN architecture, which managed to successfully increase the aspect of accuracy and robustness for diagnostics. The datasets that the models were trained on included two publicly available ones by the National Institutes of Health [13] of Montgomery County, Maryland, and Shenzhen, China respectively. Hwa et al. [14] proposed an ensemble learning method with canny edge detection for the classification of TB. They used two publicly available datasets, Montgomery and Shenzhen datasets [13]. The ensemble learning utilized two deep neural networks, VGG16 [15] and InceptionNetV3 [16]. The images were first resized to a shape of 250 X 250 to ensure that all the images were of the same size. This is followed by Canny Edge detection so that only the edges are processed from the Chest X-rays. During the evaluation, the ensemble approach produced a noteworthy accuracy of 89.77%, along with a sensitivity 90.91%, and a specificity 88.64%.

F. Pasa et al. [17] proposed a simple lightweight architecture for image classification-based detection and classification of TB from Chest X-Rays. The detection and classification of TB have been approached with methods such as

Support Vector Machines (SVM) [18] or CNN. The use of complex models with a large number of parameters often results in the problem of overfitting. Since these models have numerous parameters, they tend to be resource-intensive and frequently need a lot of data to train. In their work, the authors propose a lightweight CNN architecture with 5 convolution blocks, each consisting of residual connections and ending with a pooling layer. They employed this architecture on three datasets that they created from the NIH TB chest X-ray datasets [13], one for Maryland, one for Shenzhen and they created the last one by combining the first two. The experimental results of F. Pasa et al. demonstrate accuracies of 79, 84.4, and 86.2% respectively on the three datasets along with AUC of 0.811, 0.900, and 0.925 respectively. The proposed network was lightweight and ideal for resource-limited environments, however, their approach could not leverage the power of transfer learning as similar works in this domain previously. Ahmed T. Sahlol et al. [19] proposed a hybrid approach that combines the MobileNet [20] architecture and the Artificial Ecosystem-based Optimization (AEO) [21] to effectively classify tuberculosis from chest x-rays. The pre-trained MobileNet model acts as the feature extractor. The metaheuristic algorithm known as Artificial Ecosystem-Based Optimisation (AEO) was created with the environment in mind to address a variety of optimization issues. Natural ecosystems, which are intricate and dynamic relationships between organisms and their surroundings that enable them to survive and preserve the ecological system, are the foundation upon which the AEO algorithm is based. The foundation of AEO's methods for problem space search and optimization is these relationships, which include competition, collaboration, resource sharing, and recycling. Since each organism has a specific role in natural ecosystems, the level of system stability and productivity is dependent upon them. For instance, producers (plants) produce resources and energy, consumers (carnivores and herbivores) depend on these organisms for energy sustenance and decomposers (fungi and microorganisms) absorb organic pollutants and replenish the environment with nutrients. AEO abstracts and mimics how these functions and processes work in the real world to determine the best or approximately best solution to a mathematical optimization task. The algorithm generates, assesses, and refines potential solutions in a cycle. To strike a balance between exploration, which involves looking widely across the solution space, and exploitation, which focuses on potential areas of the solution space, the program uses concepts derived from ecosystem dynamics. The method can also identify globally optimal solutions in quite difficult high-dimensional or non-linear problem spaces because of this balance, which keeps it from becoming stuck in poor solutions. The authors used two datasets namely, the Shenzhen dataset [13] and another publicly available dataset [22]. Ahmed T. Sahlol et al. achieved classification accuracies of 90.2 and 94.1% respectively for the two datasets. The proposed methodology achieved specificity and sensitivity of 0.901 and 0.914 respectively on the first dataset and 0.97 and 0.872 respectively on the second dataset.

Muhammad Ayaz et al. [23] proposed an Ensemble Learning Technique that combines the Gabor Filter with Deep Learning Techniques for feature extraction and detection of TB. The authors used the Montgomery and Shenzhen datasets [13].

The proposed method uses seven Deep Convolutional Neural Networks: VGG16, VGG19, Inceptionv3 [24], InceptionResnetv2 [24], MobileNet, and XceptionNet [25]. These were pre-trained on the ImageNet dataset [4]. The pre-trained CNNs were used for feature extraction. Using the Gabor filter at different scales and orientations the method can capture multi-scale features and highlight the edges and texture that set the TB-affected regions of the lungs apart. The significant handcrafted elements enhance the integrated features obtained from the pre-trained CNNs. This makes the Ensemble Learning Technique more robust as it takes both hand-crafted features as well as deep learning-based features leading to improved capability to detect TB. The integration of Gabor filters with CNN features addresses the issues of variation in X-ray pictures regarding resolution, contrast, and noise. This way the system can have a better textural image analysis that might be missed by only deep learning, thus achieving high diagnostic accuracy especially when applied to datasets such as Shenzhen and Montgomery. The proposed methodology achieves ROC of 0.99 and 0.97 and accuracy of 93.47 and 90.6% on Shenzhen and Montgomery datasets respectively using K-Fold Cross Validation.

Kotei and Thirunavukarasu [26] demonstrated the use of U-Net [26] for extracting the region of interest from Chest X-rays and feeding it to deep convolutional networks for feature extraction. The dataset used in their work is a combination of two publicly available datasets, first the National Library of Medicine(NLM) [13] dataset and the second dataset is available in Kaggle [5], it is curated by authors from Qatar University, Doha, Qatar, and the University of Dhaka, Bangladesh and contributors from Malaysia. Chest X-ray images were fed to the U-Net architecture to get segmented images, which contain only the portions of the Chest X-rays that might be useful for the classification task. The segmented images were then fed to nine different pre-trained CNNs, namely VGG16 [15], VGG19 [15], Inceptionv3 [16], MobileNet [20], InceptionResNetv2 [24], XceptionNet [25], DenseNet201 [27], EfficientNetB1 [28], ResNet50 [29] along with a custom CNN model. This custom CNN model was primarily used for feature extraction. The models were trained using both segmented and unsegmented images. A Softmax activation layer [30] was added to the last layer of the pre-trained networks to get the class probabilities. A SVM [18] algorithm is applied to the feature space extracted by the custom CNN model to get predictions for the TB classification, i.e. Normal or TB Class. A stack algorithm-based ensemble learning is used to train a meta-model from the two outputs, one from the pre-trained model and the other from the custom CNN model. This metamodel was used for the final prediction. The ensemble technique proposed by E. Kotei and R. Thirunavukarasu attained 98.38% accuracy, 98.89% sensitivity, and 98.70% specificity.

Rajat Mehrrotraa et al. [31] employed ensemble learning to detect TB and COVID-19 from chest X-rays. The authors used some standard, well-known CNNs namely, DenseNet201 [27], InceptionResNetv2 [16], and XceptionNet [25], for feature extraction. The feature space was then subjected to ensemble techniques, bagging, boosting, and subspace K-NN. The resultant features are then subjected to two popular algorithms, the Support Vector Machine (SVM) [18] and the K-Nearest Neighbors (KNN) [32] for the final prediction. Their work uses three datasets namely TB X-Ray (TBX11K) dataset [6], the Augmented COVID-19 x-ray Image dataset [33], and the Shenzhen dataset [13]. Their proposed methodology achieves accuracies of 87.90 and 99.10% respectively and AUC of 0.94 and 1 respectively for classifying TB and COVID-19. Natarajan et al. [34] proposed a novel methodology, tbXpert for the detection of TB from a dataset containing two classes of chest x-rays normal and TB positive, having 3500 images in each class. The proposed methodology consists of three main steps, image augmentation, Linear Triangular Interpolation, and a pre-trained deep-learning model. The images were obtained from four different publicly available datasets from the National Library of Medicine (NLM) [13], Belarus [5], the National Institute of Allergy and Infectious Disease (NIAID) [5], and the Radiological Society of North America (RSNA) [35]. Each image was pre-processed and downscaled to a size of 299 x 299 pixels. The authors leveraged the power of the triangular interpolation method to upscale the images as they come from different sources and can offer variation in the distribution of data. This technique addresses the problem of images having variable image quality, contrast, and radiological imaging equipment. The authors used InceptionNet [12] based network with residual connections with Adam [36] optimizer. Their suggested methodology performs well on the test set, achieving 99.2% accuracy, 98.9% sensitivity, 99.6% specificity, 99.6% precision, and 99.4% AUC (Table 6.1).

A novel technique for the classification of TB was proposed by Tawsifur Rahman et al. [37] on a set of 40,000 chest X-ray images. That uses a pre-trained model with CheXNet, to act as a feature extractor, and then a Multilayer Perceptron classifier based on an operational neural network. It further uses the PCA [38] algorithm for feature dimensionality reduction of the features that have been extracted from the CheXNet-based [39] CNN encoder to project higher-dimensional data into a lower dimensional space that reduces the redundant data and increases the overall accuracy of the system. CheXNet [39] is the model based on a pre-trained DenseNet121 model [27], which used numerous chest X-rays for training purpose. Their approach attains an accuracy of 93.2%, recall of 93.38%, and F1 Score of 94.58%. A Score-CAM analysis was done to explain the model's decisions, and it revealed that the model focuses on relevant areas in the lung images that might align with abnormalities. Rakhi Wajgi et al. [40] employed a pre-trained VGG19 [15] which harnessed the power of Transfer learning on a database of 4200 images of TB [5]. The paper mentions the precision of TB detection by the VGG19 [15] model for different activation functions, dropout rates, optimizers, and pooling layers as part of the Hyperparameter Tuning process. Rakhi Wajgi et al. reached 98% accuracy along with F1 Score and Recall of 0.9808 and 0.9811 respectively.

The previous works done in the domain of tuberculosis detection from chest X-ray images using deep learning techniques utilized basic CNNs along with feature-based approaches like Canny edge detection and Gabor filters. This resulted in poor performance because, unlike models trained specifically for TB detection, they failed to concentrate on the areas of the chest X-ray images that are most diagnostic of patients with or without TB. Although deeper architectures such as DenseNet201, XceptionNet, and InceptionResNetV2 outperform other networks in

Table 6.1 Summary of recently proposed research works and their results reported by previous researchers for TB detection.

Study	Model(s) used	Dataset(s)	Accuracy (%)	AUC	Precision	Recall	Specificity
Lakhani and Sundaram [10]	AlexNet, GoogLeNet (Ensemble)	Montgomery, Shenzhen	–	0.99	–	–	–
Hwa et al. [14]	VGG16, InceptionNetV3	Montgomery, Shenzhen	89.77	–	–	0.90	0.92
F. Pasa et al. [17]	Custom Lightweight CNN	NIH (Montgomery, Shenzhen, Combined)	79 / 84.4 / 86.2	0.811/0.900/0.925	–	–	–
Ahmed T. Sahlol et al. [19]	MobileNet + AEO	Shenzhen, Public Dataset	90.2/94.1	–	–	0.914/0.872	0.901/0.97
Muhammad Ayaz et al. [23]	Gabor Filter + CNN Ensemble (VGG16, VGG19, InceptionV3, etc.)	Montgomery, Shenzhen	93.47/90.6	0.99/0.97	–	–	–
E. Kotei and R. Thirunavukarasu [26]	U-Net + CNN Ensemble (VGG16, VGG19, InceptionV3, etc.)	NLM, Kaggle Dataset	98.38	–	–	0.98	0.99
Rajat Mehrrotraa et al. [31]	Ensemble (DenseNet201, InceptionResNetV2, XceptionNet + KNN)	TBX11K, COVID-19 Augmented, Shenzhen	87.90 / 99.10	0.94 / 1.0	–	–	–
Natarajan et al. [34]	tbXpert	NLM, Belarus, NIAID, RSNA	99.2	0.99	0.99	0.98	0.99
Tawsifur Rahman et al. [37]	CheXNet + MLP	40,000 Chest X-Rays	93.2	–	–	0.93	–
Rakhi Wajgi et al. [40]	VGG19 (Transfer Learning)	4200 TB Images (Kaggle Dataset)	98	–	–	0.98	–
TD-Net	MobileNetV2 + Attention Mechanism	4200 TB Images (Kaggle Dataset), TBX11	98 / 97	0.99 / 0.98	0.98	0.99	–

terms of accuracy it has several drawbacks including a high tendency of overfitting especially when trained on limited datasets, and consume a lot of computational power, which can be a constraint to its deployment in regions where such resources are scarce. One of the biggest concerns in earlier approaches has been the uneven quality of X-ray images: the differences in the resolution, image noise, and contrast between Montgomery or Shenzhen datasets decrease the models' stability and generality across the groups.

These challenges have been overcome by the proposed MobileNetV2 architecture with spatial and channel attention mechanisms integrated. The incorporation of attention modules allows the architecture to improve the focus of certain areas of the X-ray images that are diagnostically relevant while allowing a balance to be made between global and local features to be extracted. This design enhances robustness in terms of varied datasets while preserving computational costs that can be easily implemented in real-world healthcare environments, which are often characterized by limited resources. Therefore, the proposed TD-Net approach provides an effective solution step that solves the problem of drawbacks in prior models through an optimal balance of model prediction accuracy, attention-based feature selection, and resource utilization.

6.3 Datasets Used

The diversity of huge medical imaging databases gradually improves the performance of necessary diagnosis systems. On the other hand, the model generalization is often limited by challenges like data imbalance, variability in image quality, and a lack of labeled datasets. For example, datasets for TB detection are often built using chest X-rays sourced from large populations containing high variation in terms of resolution, contrast, and anatomy. To permit generalization across various clinical situations, this variation emphasizes the significance of appropriate preprocessing and feature extraction techniques, which will be crucial during modeling. Furthermore, to facilitate models capable of accurate predictions in practice, it is essential to incorporate all pertinent labeled data and high-quality information.

In this study, we have used a public dataset curated by Tawsifur Rahman et al. [5], named as Dataset #1, available in Kaggle repository for training and testing purposes. Additionally, we have also used TBX11 [6] dataset, named as Dataset #2, for testing our TD-Net model (Fig. 6.2).

6.3.1 Dataset #1

The first dataset, curated by Twasifur Rahman et al. [5] comprises 4200 chest X-rays divided into two classes Normal and TB, in image format. Each image is in.png format and has a dimension of 512 X 512. The Normal class contains 3500 images

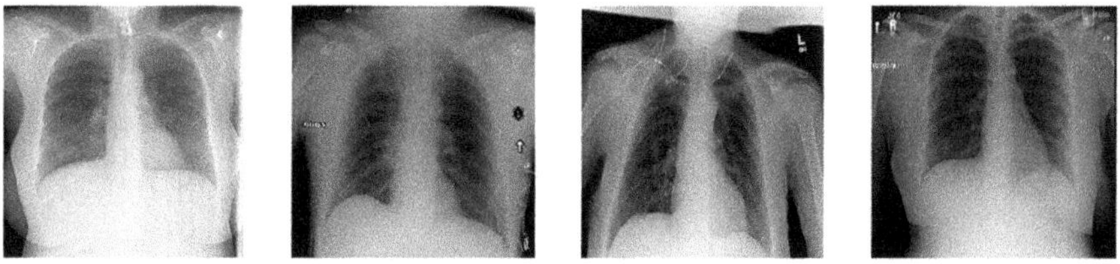

Fig. 6.2 Chest X-ray images: samples showing normal and abnormal (TB) cases used in the study for TB classification from Dataset #1

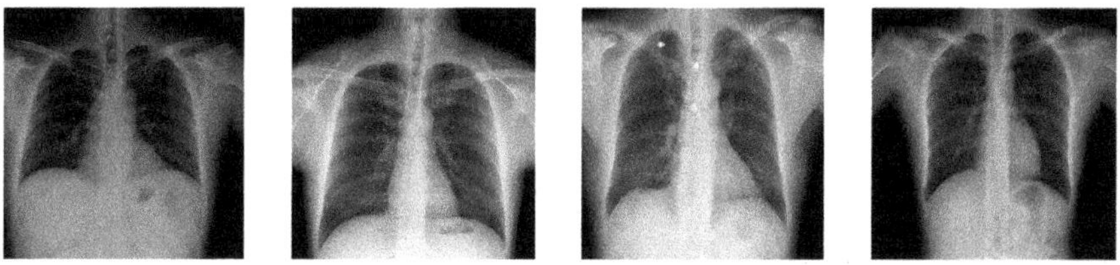

Fig. 6.3 Chest X-Ray images: samples showing normal and abnormal (TB) cases used in the study for TB classification from Dataset #2

and TB contains 700 images. Tawsifur Rahman et al. combined images from four well-known datasets such as NLM dataset, Belarus dataset, NIAID TB dataset, and RSNA CXR dataset (Fig. 6.3).

The images have been aggregated from four well-known publicly available datasets and hence can have variations in terms of image quality such as contrast, brightness, etc. This can introduce challenges for training the model, as it will introduce bias in the system. The model might not perform well in real-life scenarios due to variations in imaging equipment and demographics. The data distribution of the two classes in the dataset has a significant imbalance, the normal class contains 3500 images whereas the TB class contains 700 images. This imbalance can affect the learning of the model especially if proper augmentation is not applied. Furthermore, TB shares Radiological similarities with other lung diseases like pneumonia, lung cancer, etc., hence proper validation of the labeling is required by professionals.

6.3.2 *Dataset #2*

The second dataset, TBX11 [6] contains 3800 chest X-ray images belonging to the healthy class and 800 images belonging to the tb class. The other classes of the dataset #2 are not considered as the TD-Net model is originally trained for binary classification task. The images in the dataset are of dimensions 512 x 512 and each image is in .png format. The dataset has been formed by aggregating four datasets, namely DA and DB [41], Montgomery, and Shenzhen X-ray datasets [13]

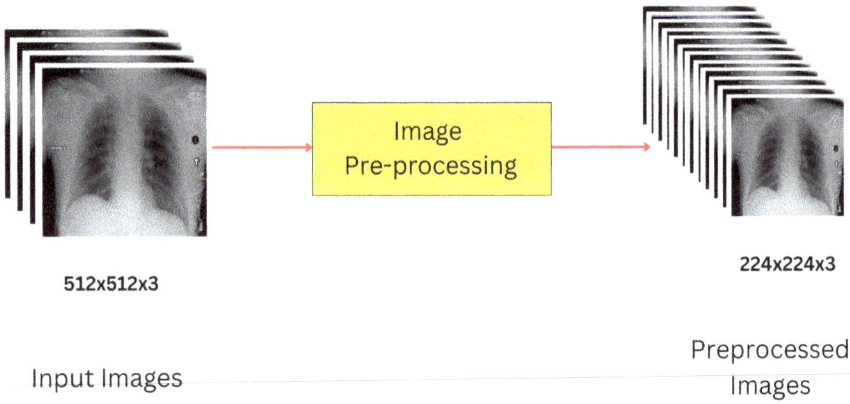

Fig. 6.4 Images of Dataset #1 before and after pre-processing

6.4 Pre-processing

The images in Dataset #1 are resized into 224 x 224 x 3 (Fig. 6.4) to verify results with MobileNet architectures. The MobileNet architecture requires the input image to be of either 224 x 224 pixels, 192 x 192 pixels, 160 x 160 pixels, or 128 x 128 pixels, hence before passing the image through the network it is resized to the appropriate shape for further processing. We plot the data distribution (shown in Fig. 6.5) across the two classes in the given Dataset #1 in order to observe the imbalance present in the data.

Imbalance in classes in data is a natural phenomenon and in real-life scenarios, the system trained on the data will face similar trends. The imbalance in the data can often cause hindrances in the model training if the minority class has a very small amount of data. The model will fail to capture the features of the minority class if an inadequate amount of data is present. To address this issue, we add augmented images to both the classes in train and validation sets, to address the class imbalance, however to ensure that the model is not overfitting to the data distribution, we evaluate the model on the same training set, once without applying augmentation and once applying augmentation.

Firstly, 60, 10, and 30% of the data are separated into train, validation, and test sets, respectively, so that a total of 2520 images are in the train set, 420 images are in the validation set, and 1260 images are in the test set.

Next, one image per image for the Normal Class and 9 images per image for the TB class are produced using image augmentation to address the class imbalance. The images are randomly chosen in batches of 4 and subjected to random rotation, flip, width shift, height shift, zoom, and shear. The train and validation set contains 7609 and 1346 images respectively after adding images using image augmentation. The test set after applying image augmentation contains 3919 images. Figure 6.6

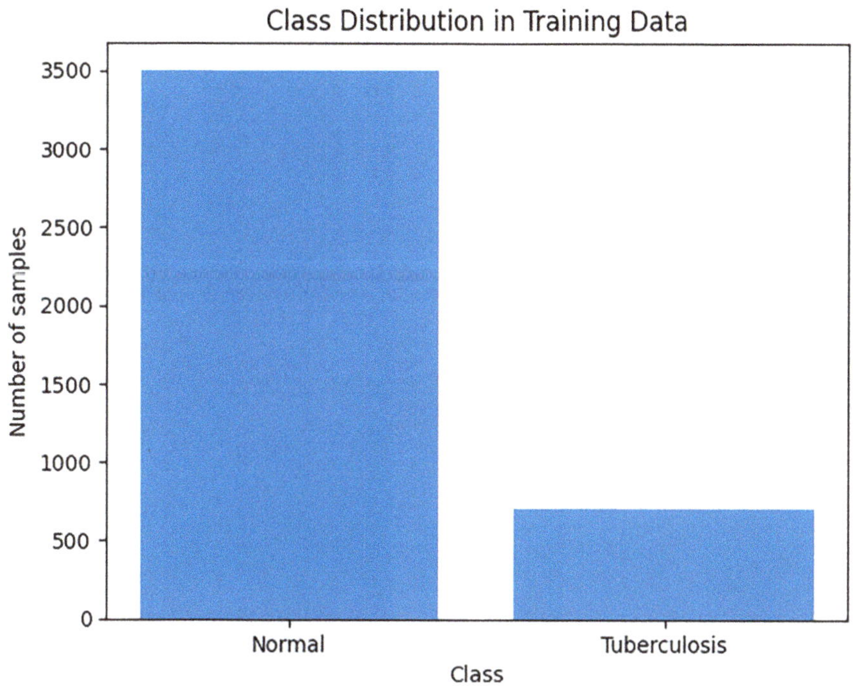

Fig. 6.5 Distribution of chest X-ray images across classes: normal and TB. This plot highlights the imbalance in Dataset #1, with 3500 images in the normal class and 700 in the TB class

shows the data distribution in the train, test and validation set after applying augmentation on Dataset #1.

For dataset #2, we do not apply any augmentation as it is used for testing purposes. The images are resized to 224 *X* 224 *X* 3 to ensure compatibility with the TD-Net architecture. The normalized images are then randomly taken in batches of 4 for the evaluation process.

6.5 Methodology

In this study, we have employed MobileNetV2 architecture [3] for accurate and real-time detection of TB from chest x-rays. The incorporation of spatial and channel attention helps in capturing the features from the chest X-rays in a more efficient way. These attention modules enhance the output of networks by focusing on specific spatial regions in the chest X-rays, leading to rich feature space and increasing the accuracy of the diagnosed disease. Furthermore, the use of MobileNetV2 makes the overall architecture lightweight and suitable for low-resource systems.

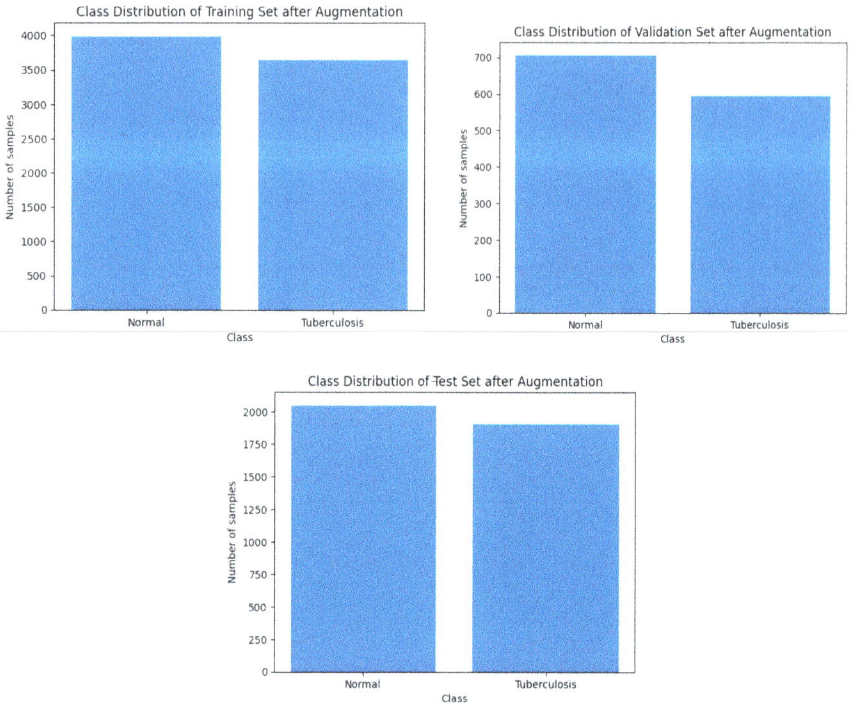

Fig. 6.6 Data distribution after image augmentation on dataset #1

6.5.1 MobileNetV2

MobileNetV2 [3] is a lightweight CNN that is best for mobile and embedded system applications. It is the successor of MobileNetV1 [20], incorporating several enhancements. The backbone of MobileNetV2 are inverted residual blocks and linear bottlenecks (Fig. 6.7), which helps in reducing the computational cost without compromising with accuracy of the overall system. The architecture comprises 53 layers in total. 17 bottle-necks further comprise pointwise and depthwise convolution layers. Like its predecessor MobileNetV1, MobileNetV2 makes use of Depthwise Separable Convolution, which breaks down traditional convolution operations for spatial filtering and channel-wise operations. The depthwise convolution operation and pointwise convolution operation together act as the building blocks of depthwise separable convolution. The decomposition of the convolution operation reduces the overall computational cost, thereby making MobileNet a suitable architecture for environments with resource constraints. Furthermore, the MobileNetV2 architecture incorporates ReLU6 [20][42] activation function, which contributes to the models robustness, by mitigating vanishing and exploding gradient problem. ReLU6 activation layer uses the following equation for computation:

$$\text{ReLU6}(x) = \min(\max(x, 0), 6) . \tag{6.1}$$

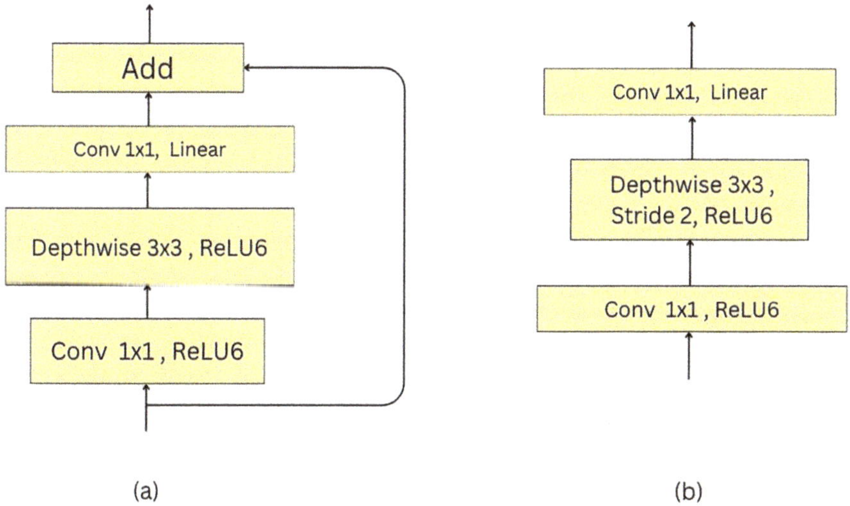

Fig. 6.7 MobileNet V2 architecture [3] (**a**) Stride=1 block (**b**) Stride=2 block

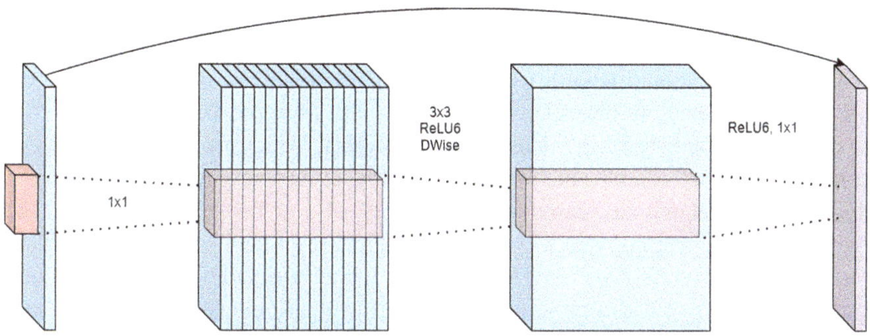

Fig. 6.8 Inverted residual blocks [3]

6.5.2 Inverted Residual Blocks

The inverted residual blocks are one of the key features of MobileNet architecture. It varies from traditional residual blocks in terms of approach. While traditional residual blocks tend to first compress the number of channels of input space and then expand them again, Inverted Residual Blocks increase the number of channels of the input space and later compress them with Depthwise Convolution. This approach reduces the computational cost of the overall process by making use of feature reuse thus making the architecture lightweight yet robust (Fig. 6.8).

Sankararaman et al. [43] in the year 2020 used the term Gradient confusion to refer to the problem that deep neural networks encounter when the gradients provided across various mini-batches of training data are not always aligned but instead point in different directions and frequently in opposite directions to one another. Mini-batches from one distribution have a negative correlation with mini-batches from another distribution, which is why this can occur. There is therefore ambiguity when one gradient update indicates that shifting the weights in a specific manner as dictated by a gradient is optimal, but another gradient indicates differently. The net movement towards the optimum is slowed down because these opposing forces are in balance. The network performs better when gradient confusion is minimized because it concentrates gradient updates in a single direction which is advantageous. This enables faster convergence towards a minimizer or closer to the area surrounding it. Due to the skip connections at the bottlenecks, gradient flow ambiguity becomes a design concern in the context of MobileNetV2's inverted residual blocks so by upscaling the bottleneck features to lower dimensions, the inverted residual block indirectly facilitates the gradient and information flow through the network, in contrast to the typical residual blocks. This transformation results in fewer gradients moving backward throughout the learning process and less information passing between the block's input and output. In particular, diminishing skip connections between layers makes it impossible for layers to maintain a sufficient information flow. Gradient propagation and training procedures become unsuccessful as a result.

This design decision presents difficulties because there may be less propagated gradients, which could lead to more cross-talk. Gradient dissimilarities generally serve as optimization obstacles since they make it more difficult to find the best updating path for a particular set of weights, which can cause a delay in convergence. Furthermore, the volume of gradients returned via the network may decrease due to the decreased dimensionality of the bottleneck structure. By reducing the control that deeper layers have on the training of shallower ones, this reduction modifies the learning dynamics and may ultimately limit the models' capacity to learn more complex clothing. However, MobileNetV2 overcomes its efficiency and gradient flow problems with its innovative architecture. By raising the corresponding feature dimensions in the intermediate layer and utilizing depthwise separable convolutions to facilitate gradient computation, the inverted residual block's partial restriction of gradient flow is circumvented. Even though there aren't many gradients that reach the final components, these methods allow them to have substantial meaning. This is advantageous for resource-constrained contexts since MobileNetV2 maintains a reasonable level of competitiveness while achieving a reduced computing load.

In contrast, MobileNetV2 overcomes its efficiency and gradient flow problems with its innovative architecture. By raising the corresponding feature dimensions in the intermediate layer and utilizing depthwise separable convolutions to facilitate gradient computation, the inverted residual block's partial restriction of gradient flow is circumvented. Even though there aren't many gradients that reach the final components, these methods allow them to have substantial meaning. This is advantageous for resource-constrained contexts since MobileNetV2 maintains a reasonable level of competitiveness while achieving a reduced computing load.

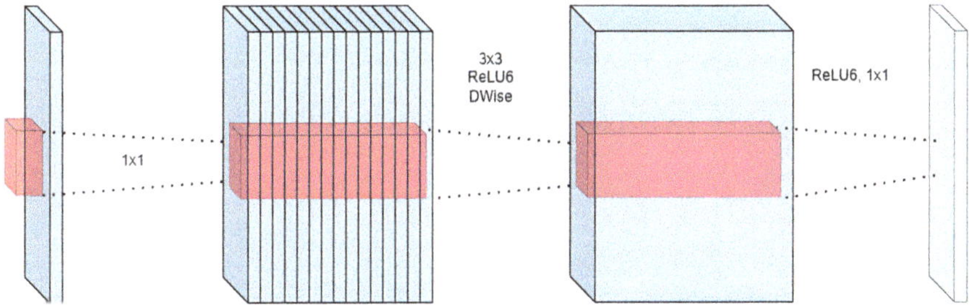

Fig. 6.9 Bottleneck with expansion layer used in our proposed TD-Net model [3]

6.5.3 Linear Bottleneck

The Linear Bottlenecks are one of the main characteristics of MobileNetV2 Architecture which is developed to solve some problems faced by deep neural networks mostly in resource-constrained environments. The key property of the linear bottleneck is its ability to maintain network efficiency while preventing information loss during feature compression. Traditional bottleneck layers perform a ReLU nonlinear activation function at each stage of the forward pass, whereas linear bottleneck layers do not. This architectural choice prevents the possibility of losing crucial features because ReLU obliterates all negative values. For lightweight networks, which rely on this for improved speed, the linearity in the bottleneck guarantees that any changes in the feature space are maintained. As one of the primary constituents of the structure of the inverted residual block, the linear bottleneck consists of the pointwise convolution that the input features undergo initially to bring them to a higher dimensional space. This extension makes it feasible for the network to carry out the feature extraction in a feature-rich space, in turn, improving its representational capacity (Fig. 6.9).

The depthwise separable convolutions decrease the computational cost by operating only on the channel-wise level. In the end, the last thinning of the bottleneck happens as the functional features, which were previously enlarged, are now compressed into a new space through pointwise convolutions operated by the linear activation function. This squeezing of the memory and the computation load of the network allows no crucial information to be lost during the projection and remain efficient.

One of the main advantages of the linear bottleneck is that a model is to make it even more expressive without any potential overhead costs. There is no room for quality loss on account of the overuse of non-linearities, and thus preservation of linear transformations through the bottleneck stage safeguards feature diversity. One of the salient points of building small models like MobileNetV2 is power consumption. Also, the linear bottleneck smoothens the use of skip connections in the inverted residual block, keeping the information and gradient flow among the layers. By doing so, it not only maintains the balance between compactness and

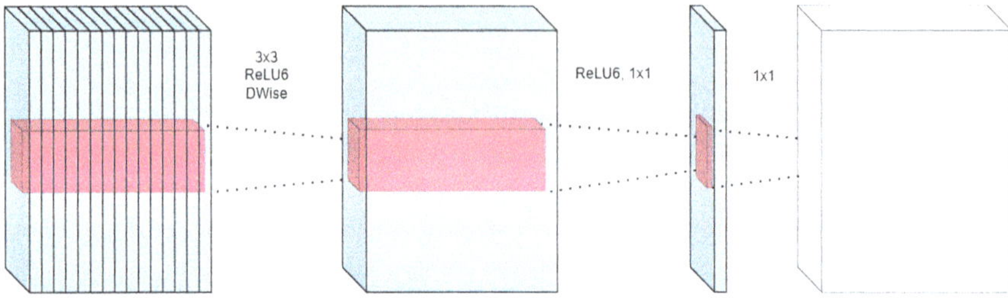

Fig. 6.10 Separable with linear bottleneck used in our proposed TD-Net model [3]

performance but also makes the network as accurate as it used to be despite being diminished in size. Linear bottlenecks can help to the general goal of MobileNetV2. The linear bottleneck is used to cut down the number of parameters and operations needed in the architecture, which makes the model perfect for deployment on mobile devices and embedded systems with limited hardware capabilities. The overfitting error is minimized by using this approach, which leads to the generalization capability of the network to be improved. This is implemented by performing various operations such as mean and max pooling, followed by transformations through dense blocks (Fig. 6.10).

6.5.4 Spatial Attention

Spatial attention is a mechanism that improves the result of models by helping them to concentrate on the important spatial regions in data. It is widely used to improve the representation of input space in a variety of domains, such as computer vision, natural language processing, etc. Spatial attention refers to the concept of adjusting input by applying feature-specific weights to each spatial position, indicating the significance of various areas within the image. These weights are determined following a comparison of each region to the task. By using a high-level strategy to visualize the data, such as regions, the systems that use spatial attention are better equipped to decide which data to use for decision-making (Fig. 6.11).

The creation of an attention map is the standard method for performing spatial attention. The summary of the input on specific dimensions, such as channels or time steps, appears to have been the most often used in numerous ways. The key features of the attention map are extracted by feature extraction techniques like average-pooling and max-pooling. An enhanced attention map is produced by concatenating and processing these descriptors. Fully connected layers or convolutional layers are utilized to improve the attention map. Since the model learns the pattern of local and global relations, the attention map is the one that most properly represents the relationships between the geographical regions. The attention map is then normalized, often using a sigmoid or Softmax function, to produce bounded values.

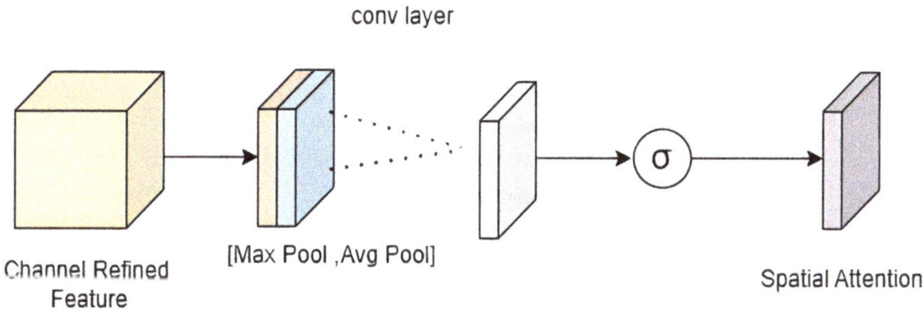

Fig. 6.11 Spatial attention used in our proposed TD-Net model [2]

Following its creation, the attention map is multiplied by the input, pixel by pixel. This increases the impact of the most significant but least relevant locations by making the original feature vectors grow or shrink based on their attention weight. As the feature extraction becomes more targeted, structured information is better acquired, and noise from the periphery is decreased.

Spatial attention is excellent for applications requiring the identification of confined patterns. In image recognition it allows models to ignore the backdrop and concentrate on the key objects of interest in a scene. Spatial attention in natural language processing aids the model in recognizing keywords or phrases in a document and, consequently, understanding its relationships and context. The versatility of the spatial attention mechanism is among its most notable characteristics. It can be utilized in various architectures that utilize different types of data making it adaptable. In context with this, spatial attention is a highly helpful tool that enables a model to improve its performance through interactions with the data it works with. Additionally, it is a lightweight, computationally efficient method that may be used in contexts with limited resources without causing significant system overheads.

6.5.5 Channel Attention

Channel attention is a technique created to improve the representational capabilities of models by highlighting the most useful channels and reducing the influence of unnecessary ones. It has also demonstrated a strong hold on fields like signal processing and computer vision, where creating reliable feature representations is the fundamental task. The fundamental idea behind channel attention is that different channels in a feature map identify distinct input patterns or characteristics. A feature map is then created by combining several channels. This is accomplished by feeding images and resampling outputs into the model after first resampling from the intermediate $1 \times 1 \times C$ tensor that contains the information on the edge mappings (where C represents the total number of channels).

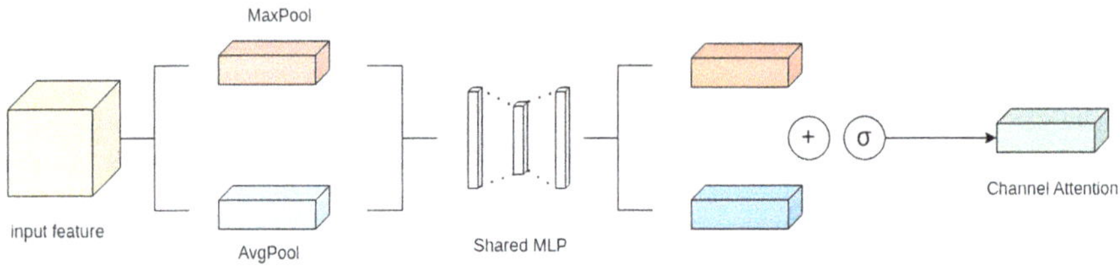

Fig. 6.12 Channel attention used in our proposed TD-Net model [2]

A feature map, typically an input spatial dimension aggregated along, is required for channel attention computation. Global average-pooling and global max-pooling are two well-liked functional approaches that have long been used to process feature maps. Separating the spatial data into tiny units, known as features, while maintaining the overall impact of each channel is one of the primary roles of Channel Attention. To learn the attention weights, the pooled descriptors are then fed into networks that undergo different transformations through convolutional layers and fully connected layers. After the attention weights are determined, element-wise multiplication is used to apply them to the input feature map (Fig. 6.12).

Channel attention is particularly noticeable in activities that require the recognition of hierarchical relationships or fine-grained patterns. To highlight the characteristics of object textures or colors that the model can differentiate between classes during image classification, channel attention can be used. The most important diagnostic characteristics in the medical image might also be highlighted, improving the readability of the outcome. Channel attention's adaptability to different model architectures is one of its greatest advantages. CNNs, transformers, and other deep learning frameworks can easily incorporate it. Additionally, it is an excellent fit for areas with limited resources because of its lightweight design, which ensures little to no computation. To improve the features in real-world scenarios, channel attention is frequently combined with spatial attention. The model can handle both channel-level and spatial-level features because of this linkage, which increases its accuracy and resilience in more challenging tasks.

6.5.6 Proposed TD-Net Methodology

In this study, we have incorporated channel and spatial attention to the MobileNetV2 architecture for building an enhanced yet lightweight model. While the traditional MobileNetV2 uses linear bottleneck and inverted residual blocks to attain high efficiency with low cost, we add spatial and channel attention to capture features from chest X-rays in a more refined manner. To make sure the network focuses on the most pertinent representations, Channel Attention selectively highlights the most important feature channels. The Global Average Pooling (GAP) and Global Max Pooling (GMP) techniques are combined in Channel Attention to extract the mean and prevailing responses of all channels. The pooled features are passed through two fully connected layers, with the first dense layer reducing the number of channels by a parameterized reduction ratio of 8, which also applies a ReLU activation function to introduce non-linearity. To generate attention weights, the second layer applies a sigmoid activation function to restrict the final outputs between 0 and 1 and resizes the features back to the channel dimension. The outputs (GAP and GMP) are multiplied element-wise, along with the original features, to direct the network's focus towards the most relevant feature channels. Mathematically channel attention can be represented as:

$$CA(F) = F \cdot \sigma \left(W_{\text{avg}} \left(\text{AvgPool}(F) \right) + W_{\text{max}} \left(\text{MaxPool}(F) \right) \right) \tag{6.2}$$

where $CA(F)$ is the channel attention output, F is the input feature map, σ denotes the Sigmoid activation, and W_{avg}, W_{max} are the dense transformations applied to average and max-pooled features, respectively.

The equation for Sigmoid is given as follows:

$$\text{Sigmoid}(x) = \frac{1}{1 + e^{-x}} \tag{6.3}$$

The equation for ReLU is given as follows:

$$\text{ReLU}(x) = \max(0, x) \tag{6.4}$$

Spatial attention improves the model's capacity to focus on important regions within an image, whereas Channel attention fine-tunes the choice of feature channels. Spatial attention uses Global Average Pooling and Global Max Pooling across the channel axis to extract both average and maximum spatial responses, much like the channel attention process. A 7×7 convolutional layer is used to learn spatial dependencies while maintaining local information once these pooled features have been concatenated. The most relevant parts of the input are highlighted by a spatial attention map created by a sigmoid activation function. Less relevant parts are subsequently suppressed by multiplying the spatial attention map by the original feature maps. By combining both attention methods, the model may dynamically modify its focus according to channel-wise and spatial significance, resulting in enhanced feature discrimination.

$$SA(F) = F \cdot \sigma \left(\text{Conv}_{7\times7} \left(\text{Concat} \left(\text{AvgPool}(F), \text{MaxPool}(F) \right) \right) \right) \tag{6.5}$$

where $\text{Conv}_{7\times7}$ indicates a convolution operation with a 7×7 kernel, and $SA(F)$ is the spatial attention output.

Following the feature extraction layer, the pre-trained MobileNetV2 is supplemented with the attention modules, namely the spatial attention and the channel attention. A dense layer having 128 neurons follows the attention modules, and dropout layer with a rate of 0.4 is added to avoid overfitting. The output of the final dense layer is subjected to sigmoid activation function (Fig. 6.13).

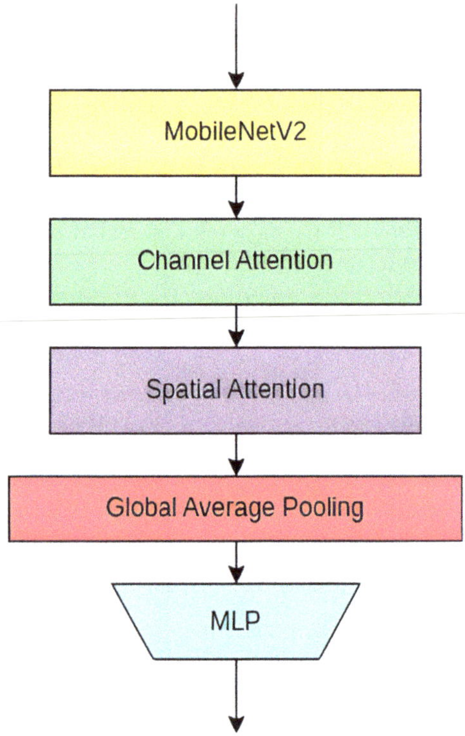

Fig. 6.13 Architecture of our proposed TD-Net methodology

To address the issue of class imbalance in the dataset, we have added class weights while training, to penalize the model to misclassify the TB positive class, as the data distribution has less number of samples in the TB class. The class weights are calculated as follows:

$$w_i = \frac{2 \cdot N_i}{N} \tag{6.6}$$

where, N is the total number of samples, and N_i *is the number of samples in class* i.

For our dataset w_0, the weight for the normal class or the TB negative class is calculated to be 0.94, and w_1, the weight for the TB positive class is calculated to be 1.07. This ensures that the model gives attention to the minority class during training. The weighted binary cross-entropy [44] loss function is given by the following formula:

$$\text{Loss} = -\sum_{i=1}^{N} w_{y_i} \cdot (y_i \log(p_i) + (1 - y_i)\log(1 - p_i)) \ , \tag{6.7}$$

where, y_i is the true label, p_i is the predicted probability, and w_{y_i} is the weight corresponding to the true label y_i.

The metrics chosen for the model evaluation are Accuracy, Precision, Recall, F1 Score and Area Under the Receiver Operating Characteristic Curve (AUC-ROC). The formulas used to evaluate the evaluation metrics can be found in our previous work [45].

RMSprop [46] is used as optimizer. The equation of RMSprop is given by :

$$E[g^2]_t = \beta E[g^2]_{t-1} + (1-\beta)g_t^2$$

$$\theta_{t+1} = \theta_t - \frac{\eta}{\sqrt{E[g^2]_t + \epsilon}} g_t$$

where:

- g_t is the gradient at time step t,
- $E[g^2]_t$ is the exponentially decaying average of past squared gradients,
- β is the decay rate ,
- η is the learning rate,
- ϵ is a small constant to prevent division by zero,
- θ_t are the parameters at time step t.

Furthermore, to prevent the model from overfitting we have added L1 regularization to the Dense Layers in the networks. The L_1 regularization penalty applied to the weights of the dense layers is given by:

$$L_1 = \lambda \sum_{i=1}^{n} |w_i| \tag{6.8}$$

where, λ is the regularization coefficient, w_i represents the weights of the dense layer, and n is the total number of weights.

6.5.7 Training TD-Net Model

This section describes the experiment conducted to train the TD-Net model. We select MobileNetV2 as our base model, as it is lightweight and ideal for fast and real-time inferences. We first find the optimal hyperparameters of the model i.e., optimizer, learning rate, etc. for 5 epochs. Once the optimal hyperparameters are found we compare the results of the model with other standard deep learning models such as EfficientNetB0 and DenseNet201 on the test set of the Chest X-Ray dataset. The models are trained on a system with Tesla T4 GPU, with a VRAM of 15, and 12.7 GB of RAM. The three models, pre-trained on the ImageNet dataset, are taken and after removing all the dense layers of the models, we add a global average pooling layer followed by two dense layers. Figure 6.14 illustrates the customized

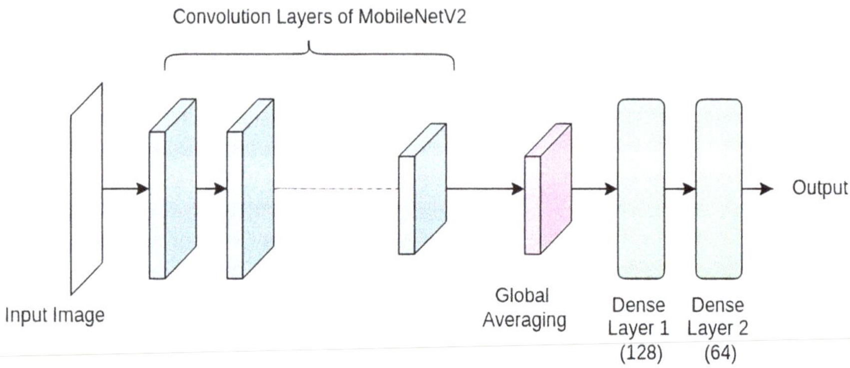

Fig. 6.14 Basic architecture of MobileNetV2 used in comparative study

MobileNetV2 architecture used for the comparison study. There are 128 neurons in the first layer and 64 in the second. Among the two dense layers, for the final layer, we used the Sigmoid activation function and for the other layers, we used ReLU [42] activation function.

To all the dense layers in the network, we add an L1 kernel regularizer with a strength of 0.01. During the training process, 7609 images resized to 224x224x3 with a batch size of 4 are taken. 1346 images are reserved for validation purposes. The models are trained utilizing the RMSprop optimizer with a learning rate of 0.0001 and a binary cross-entropy loss function for 20 epochs, incorporating early stopping to avoid overfitting. For testing purposes 1260 images without augmentation, the same set, increased to 3919 images after augmentation using the dataset #1 [5] and the 4600 images from the second dataset, dataset #2 [47] are used. To prove that the model is not overfitting on the data distribution, the test set of the dataset is taken before augmentation and after augmentation and the results are compared.

The early stopping criterion is configured to monitor the validation AUC-ROC, with a patience of 7 and baseline of 0.99, that is, early stopping will be employed whenever there is no improvement for 7 epochs or when the validation AUC-ROC reaches 0.99. The weights of the best-performing model are saved over epochs for further inference on the test set. Furthermore, class weights of 0.94 for normal and 1.07 for TB class is used to ensure the models are penalized for the minority class and generalize well for both classes. It is observed that although the models have good accuracy, they show poor F1 scores. On manual intervention, plotting the Grad-CAM [48] plot and checking the x-ray images we could conclude that the models are failing to focus on relevant regions. To address this issue, we add the attention modules to the best performing model Fig. 6.20.

The best weights of the models saved over the epochs are taken and loaded into the model. The inferences of the three models are recorded on the test images that are reserved. The test metrics, that is the accuracy, precision, recall, F1 Score and AUC-ROC of the models are evaluated. To our base model, that is

Table 6.2 Model comparison based on RAM usage and inference time on Dataset #1

Model name	Model size	Average RAM usage	Average inference time (s)
Attention-MobileNetV2	12.41 MB	2.61 MB	1.7631
MobileNetV2	9.24 MB	9.65 MB	0.9912
VGG16	56.38 MB	7.68 MB	2.9213
EfficientNetB0	16.07 MB	47.42 MB	2.6384
DenseNet121	27.35 MB	50.96 MB	10.3456
InceptionV3	84.17 MB	16.40 MB	5.2178

MobileNetV2, spatial and channel attention are added and three scenarios are tested, MobileNetV2 with spatial attention (A), MobileNetV2 with channel attention (B), and MobileNetV2 with spatial and channel attention (C). The same experimental setting that is used during training of the three previously mentioned models, MobileNetV2, DenseNet201, and EfficientNetB0 is used for the three models A, B, and C. The inferences are taken on the test data of the dataset #1 [5] without augmentation and the best-performing model is chosen based on accuracy, F1 Score, and recall.

We have finally evaluated our proposed TD-Net model and compared the same with five state-of-the-art pre-trained CNN models such as MobileNetV2, VGG16, EfficientNetB0, DensNet121, and InceptionNetV3. A set of 10 images for each candidate model are taken randomly from the test data of dataset #1 [5], without augmentation, and the model is evaluated on one image at a time. The average RAM usage and Inference time are recorded. Additionally, we considered the size of each model after training and the comparison is shown in Table 6.2

6.6 Results

We have used two datasets, one curated by Tawsifur Rahman et al. [5], known as Dataset #1, applied image augmentation, and Dataset #2 [47]. Dataset #1 is used for training, validation and testing purpose, while Dataset #2 is used for solely for testing purpose. This experimental setup is used to show the robustness of the model. The images are then re-sized to 224 X 224 X 3 dimensions. For selecting a base model, we have chosen MobileNetV2. We have found out the optimal hyperparameters of the model using hyperparameter tuning. For the optimal hyperparameter tuning, we have employed Hyperband algorithm [49] using Keras Tuner. A distinct set of hyperparameters is evaluated, and the optimal set is selected based on accuracy and F1 Score. Table 6.3 shows the various hyperparameters taken along with the corresponding accuracies and F1 Scores achieved by our proposed TD-Net model.

The MobileNetV2 model is trained on the training data for 20 epochs. It is compared with two more standard deep learning models trained under the same

Table 6.3 Hyperparameter configurations, corresponding classification accuracies, and F1 Scores on Dataset #1 (*Here * shows the best set of hyperparameters*)

#neurons in dense layer	Dropout rate	Optimizer	Learning rate	Accuracy	F1 score
64	0.3	Adam	0.0001	0.9780	0.8521
64	0.4	SGD	0.0010	0.9663	0.8082
128	0.2	RMSprop	0.0010	0.9512	0.7422
128	0.4	RMSprop	0.0001	0.9879*	0.8830*
96	0.4	SGD	0.0001	0.9398	0.7343
32	0.4	SGD	0.0001	0.9064	0.6575
128	0.3	Adam	0.0010	0.9510	0.7464

Table 6.4 Comparison of MobileNetV2 model with other standard deep learning models on Dataset #1*(Here "*" denotes the best performing model)*

Model	Accuracy	F1 score	Recall
EfficientNetB0	84%	0.67	0.77
DenseNet201	78%	0.53	0.79
MobileNetV2 *	94%	0.84	0.90

Table 6.5 Performance metrics for different models *(Here "*" denotes the best performing model)*

Model	Accuracy	F1 score	Recall
MobileNetV2 with spatial attention	98%	0.95	0.98
MobileNetV2 with channel attention	91%	0.63	0.46
TD-Net*	98%	0.98	0.98

experimental conditions. The accuracy, F1 Score, and recall of the three other pre-trained CNN models, i.e., EfficientNetB0, DenseNet201, and MobileNetV2 trained on the test set of the chest X-ray images are given in Table 6.4.

MobileNetV2 model demonstrates superior performance, attaining an accuracy of 94% and an F1 Score of 0.84 whereas the DenseNet201 is the worst performing model among the three, with an accuracy of 78% and an F1 score of 0.53. The EfficientNetB0 model, on the other hand, performs relatively poorly with an accuracy of 84% and an F1 score of 0.67.

To get a better understanding, we have done a Grad-CAM plot to find that our TD-Net model does not give importance to the relevant regions in the chest X-ray images (see Fig. 6.20 for more details). To improve the model's feature representation, we thereby include attention modules. As the MobileNetV2 model outperforms the other models, it is selected as the base model for the backbone of our architecture.

The classification accuracy, F1 Score, and recall of the base model after incorporation of spatial and channel attention can be seen in Table 6.5.

It is evident that the TD-Net model performs the best, with accuracy and F1 Score of 98% and 0.98, respectively. MobileNetV2 with spatial attention comes in second with 98% accuracy and 0.95 F1 Score. Despite having channel attention, the

MobileNetV2 model performs badly, with an accuracy of 91% and an F1 Score of 0.63.

The proposed methodology, TD-Net, achieves accuracy and AUC-ROC of 98% and 0.99 respectively, on the first test dataset without augmentation, 98% and 0.99 respectively on the first test dataset after augmentation, and 97% and 0.98 on Dataset #2 .

6.6.1 Results on Dataset #1

The training and validation accuracy, loss, F1-Score and ROC-AUC over epochs plot of our TD-Net model is given in Fig. 6.15. The accuracy, F1-Score and AUC-ROC show a rapid increase while the loss shows a rapid decrease in the first 5 epochs, after which the metrics change steadily. The confusion matrix plot for TD-Net on the Dataset #1 without augmentation is given in Fig. 6.16.

In the first test set, out of the 1260 images, our TD-Net model classifies images of 1029 positive class (Normal class) correctly, 205 images of negative class (TB

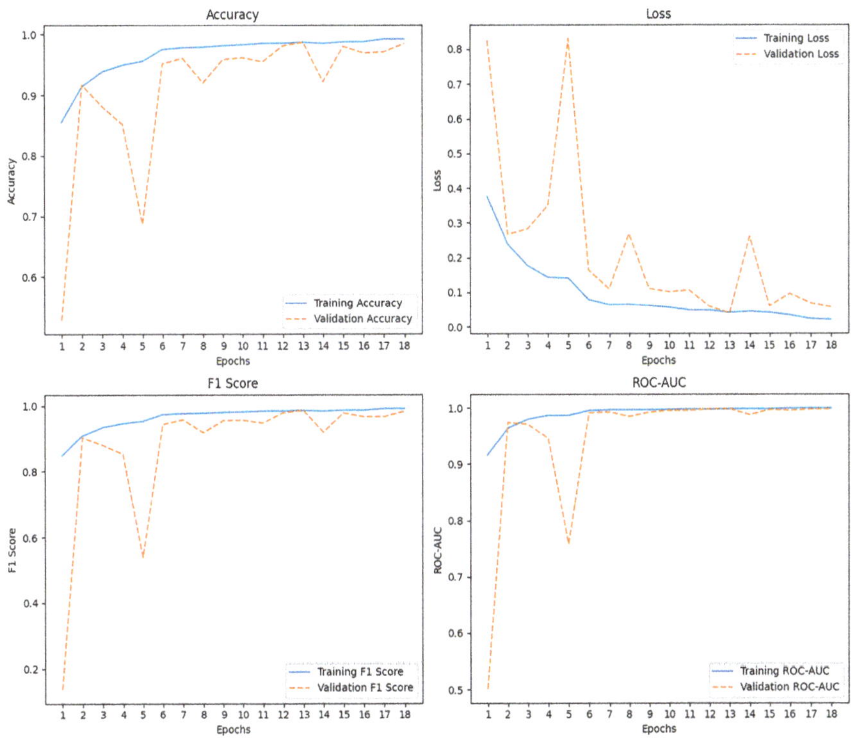

Fig. 6.15 Training metrics of MobileNetV2 with spatial and channel attention on Dataset #1

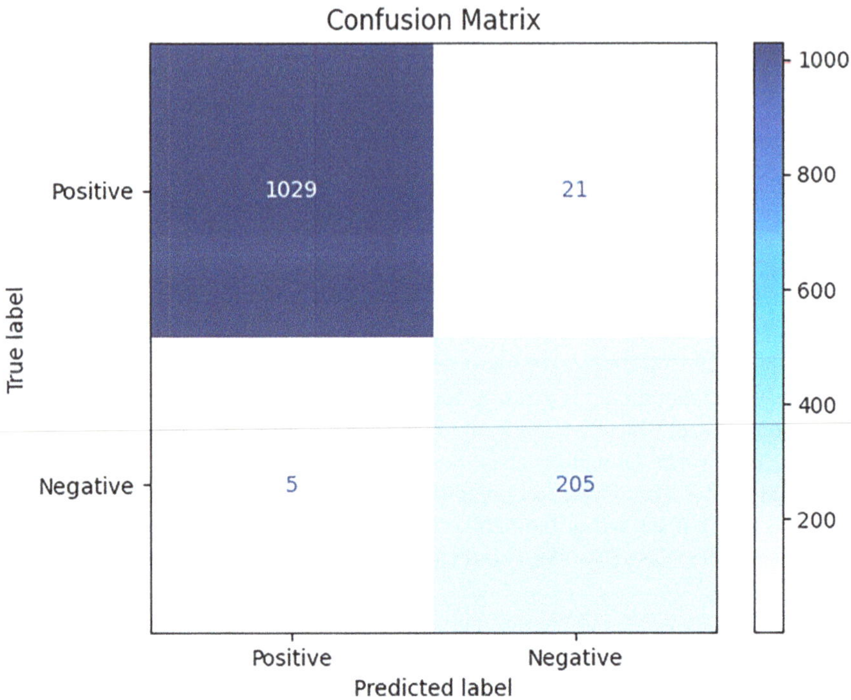

Fig. 6.16 Confusion matrix given by our proposed TD-Net model on Dataset #1 without augmentation

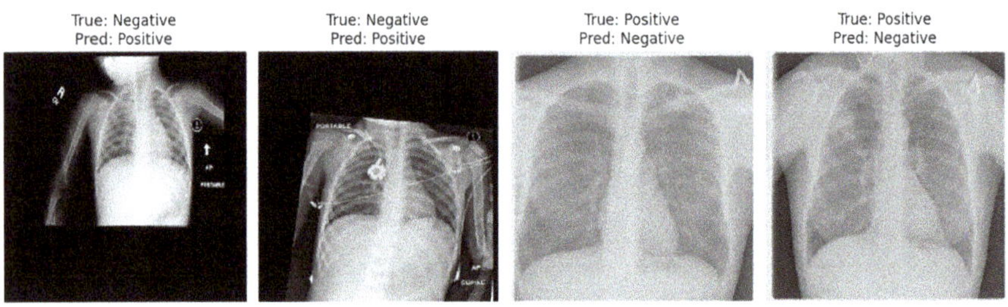

Fig. 6.17 Images misclassified by our proposed TD-Net model on Dataset #1

positive) correctly, however, it failed to classify 21 images of positive class and 5 images of negative class.

To get a better understanding of the model's performance, we have plotted the misclassified images. With manual inference, we can infer that the misclassified images are mostly noisy, which makes it difficult to distinguish between TB and Normal classes (Fig. 6.17). Furthermore, the lung regions found in the images are blurry, making it difficult for the model to capture the features of the images.

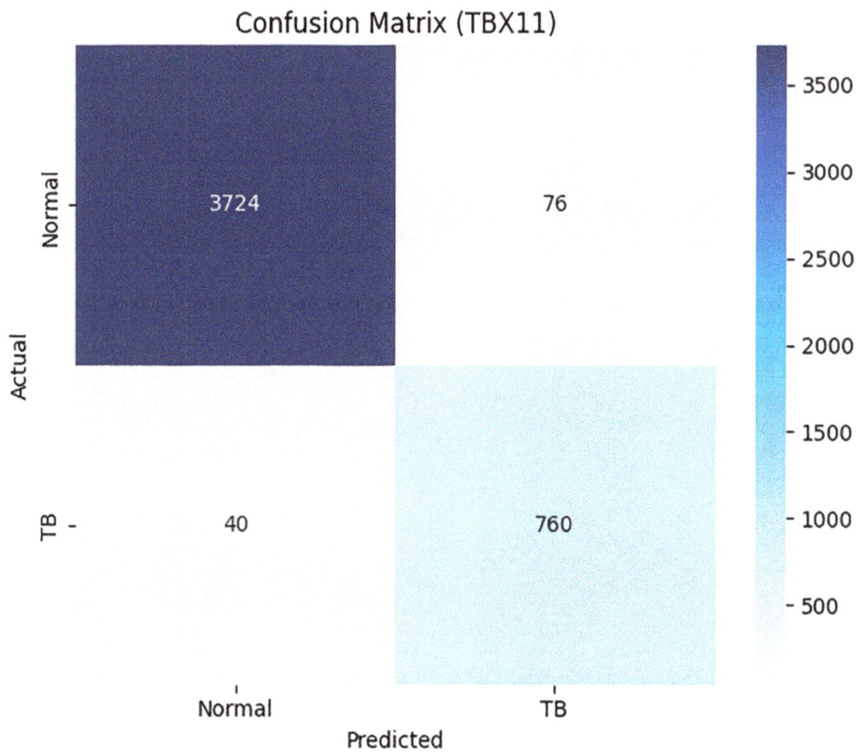

Fig. 6.18 Confusion matrix given by our proposed TD-Net model on Dataset #2

6.6.2 *Results on Dataset #2*

The confusion matrix plot for TD-Net model on the Dataset #2 is given in Fig. 6.18. In Dataset #2 out of the 4600 images, TD-Net classifies images of 3224 positive class (Normal class) correctly, 760 images of negative class (TB positive) correctly, however, it failed to classify 76 images of positive class and 40 images of negative class.

The images misclassified by our TD-Net model is shown in Fig. 6.19. With manual inference, we can infer that the misclas- sified images are mostly noisy, which makes it difficult to distinguish between TB and Normal classes.The majority of the misclassified images are noisy, which makes it challenging to differentiate between the TB and Normal classes.

6.6.3 *Grad-CAM Analysis*

Grad-weighted Class Activation Mapping, or Grad-CAM [48] is an effective visualization tool that elucidates the decision-making processes of CNNs. Grad-

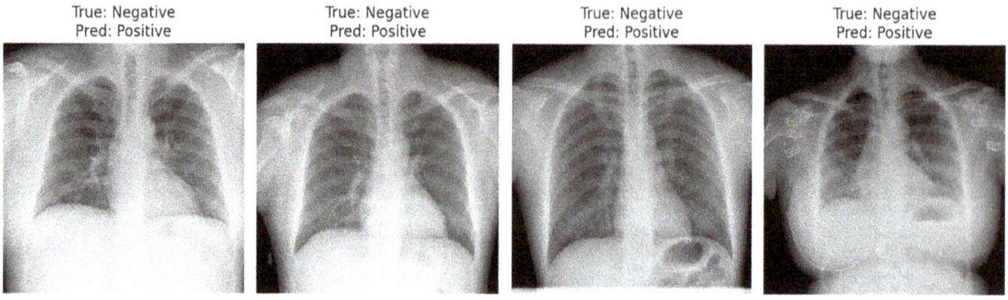

Fig. 6.19 Images misclassified by our proposed TD-Net model on Dataset #2

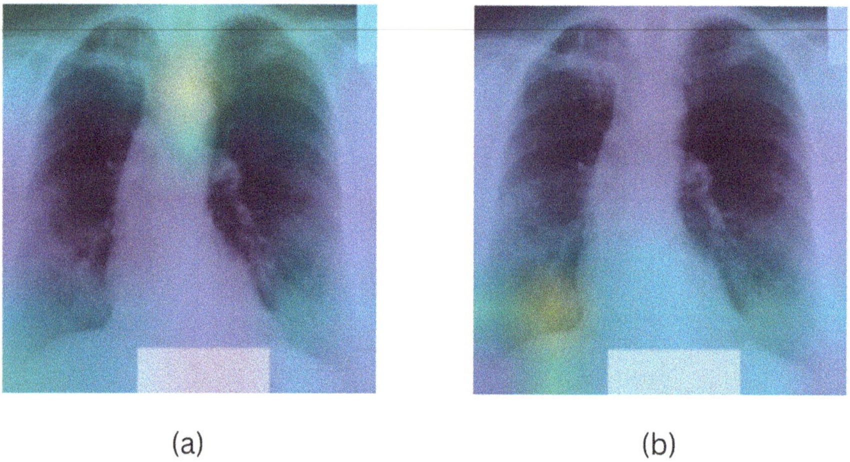

Fig. 6.20 Impact of using spatial and channel attention in MobileNetV2 on Dataset #1: (**a**) Grad-CAM visualization before adding attention modules (**b**) Grad-CAM visualization after adding attention modules

CAM highlights the areas of an input image that have the biggest impact on predictions, improving model performance and interpretability. This is crucial for TB identification from chest X-rays since it enables doctors and researchers to assess whether the model is concentrating on pertinent areas, such as lung fields, or is sidetracked by extraneous regions.

To get a better insight, we have plotted the Grad-CAM visualization of the model, MobileNetV2 before and after the addition of Spatial and Channel Attention. It is seen that the TD-Net model before the addition of the attention mechanism fails to focus on relevant regions, i.e., the regions in the X-rays that show infected parts of the lungs. This led us to add the attention mechanism to our model, to make its feature representation more rich. It is observed that upon adding the attention modules, the model emphasizes relevant regions as seen in Fig. 6.20.

The model's ability to better capture the features of chest X-ray images and prioritize infected lung regions is enhanced by the addition of spatial and channel attention. This makes the model reliable and ensures that it has generalized well.

Thus, using Grad-CAM visualization, we can detect the difference between the model before and after the addition of the attention modules, and the impact of the attention modules on the architecture.

6.7 Conclusion and Future Work

In this study, we performed TB detection from chest X-ray images with the help of TD-Net, a deep learning technique leveraging the power of transfer learning along with MobileNetV2 incorporating spatial and channel attention modules on two publicly available, standard Kaggle dataset curated by Tawsifur Rahman et al. [5] and TBX11 [6]. The results show that our proposed TD-Net methodology surpasses previous works on Dataset #1 with promising accuracy and AUC-ROC of 98% 0.99 respectively whereas on Dataset #2, the classification accuracy and AUC-ROC are found to be 97% and 0.98 respectively. By leveraging the power of transfer learning, we have achieved the desired results with comparatively less data. Furthermore, to decrease the chances of underfitting for the minority class, we applied image augmentation. Since we have used class weights while training the model, it ensures that the model generalizes well. The model can focus on specific areas of the X-ray images with the help of spatial and channel attention. Furthermore, the overall model is lightweight which decreases the inference time by a significant amount making the model well-suited for fast, scalable, and real-time applications.

The field of deep learning in healthcare sectors lacks good quality publicly available datasets, which hinders model training. Furthermore, gathering data becomes a problem due to patient data security issues. Hence, a large-scale quality dataset can be curated and made publicly available for researchers willing to work on the same. Furthermore, a zero-shot classification model can be prepared for medical imaging purposes. The model can prove to be ideal for healthcare sectors, especially with scenarios having data scarcity issues.

Data Availability

Dataset #1 can be found in the Kaggle repository at https://www.kaggle.com/datasets/tawsifurrahman/TB-tb-chest-xray-dataset

Dataset #2, TBX11 can be found in the Kaggle repository https://www.kaggle.com/datasets/usmanshams/tbx-11

Code Availability

The source codes for the architecture and training of TD-Net is made publicly available in the following GitHub repository : https://github.com/debg48/TD-Net. The code implementation uses Google's TensorFlow Library for building the architecture.

Acknowledgments We would like to acknowledge and extend our heartfelt gratitude towards Ms. Ongira Badyopadhyay, final year student of M.B.B.S at Medical College Kolkata for validating our results and guiding us with our initial idea about TB and its detection with the help of medical imaging.

References

1. Hu, J., Shen, L., Sun, G.: Squeeze-and-excitation networks (2017). CoRR, abs/1709.01507
2. Woo, S., Park, J., Lee, J.-Y., Kweon, I.S.: CBAM: convolutional block attention module. In Proceedings of the European Conference on Computer Vision (ECCV), pp. 3–19 (2018)
3. Sandler, M., Howard, A., Zhu, M., Zhmoginov, A., Chen, L.-C: MobileNet2: inverted residuals and linear bottlenecks (2019). arXiv:1801.04381
4. Deng, J., Dong, W., Socher, R., Li, L.-J., Li, K., Fei-Fei, L.: Imagenet: a large-scale hierarchical image database. In 2009 IEEE Conference on Computer Vision and Pattern Recognition, pp. 248–255 (2009)
5. Rahman, T., Khandakar, A., Kadir, M.A., Islam, K.R., Islam, K.F., Mazhar, R., Hamid, T., Islam, M.T., Kashem, S., Mahbub, Z.B., Ayari, M.A., Chowdhury, M.E.H.: Reliable tuberculosis detection using chest x-ray with deep learning, segmentation and visualization. IEEE Access **8**, 191586–191601 (2020)
6. Liu, Y., Wu, Y.-H., Ban, Y., Wang, H., Cheng, M.-M.: Rethinking computer-aided tuberculosis diagnosis. In: IEEE/CVF Conference on Computer Vision and Pattern Recognition, pp. 2646–2655 (2020)
7. Banerjee, A., Bhattacharya, R., Bhateja, V., Singh, P.K., Lay-Ekuakille, A., Sarkar, R.: Cofe-net: an ensemble strategy for computer-aided detection for covid-19. Measurement **187**, 110289 (2022)
8. Kundu, R., Singh, P., Ferrara, M., Ahmadian, A., Sarkar, R.: ET-NET: an ensemble of transfer learning models for prediction of covid-19 infection through chest CT-scan images. Multimedia Tools Appl. **81**, 31–50 (2022)
9. Dey, A., Chattopadhyay, S., Singh, P., Ahmadian, A., Ferrara, M., Senu, N., Sarkar, R.: Mrfgro: a hybrid meta-heuristic feature selection method for screening covid-19 using deep features. Sci. Rep. **11**, 24065 (2021)
10. Lakhani, P., Sundaram, B.: Deep learning at chest radiography: automated classification of pulmonary tuberculosis by using convolutional neural networks. Radiology **284**(2), 574–582 (2017)
11. Krizhevsky, A., Sutskever, I., Hinton, G.E.: Imagenet classification with deep convolutional neural networks. In Pereira, F., Burges, C.J., Bottou, L., Weinberger, K.Q. (eds.), Advances in Neural Information Processing Systems, vol. 25. Curran Associates, Red Hook (2012)
12. Szegedy, C., Liu, W., Jia, Y., Sermanet, P., Reed, S., Anguelov, D., Erhan, D., Vanhoucke, V., Rabinovich, A.: Going deeper with convolutions. In Proceedings of the IEEE Conference on Computer Vision and Pattern Recognition (CVPR), pp. 1–9 (2015). https://doi.org/10.1109/CVPR.2015.7298594

13. Jaeger, S., Candemir, S., Antani, S., Wáng, Y.X., Lu, P.X., Thoma, G.: Two public chest x-ray datasets for computer-aided screening of pulmonary diseases. Quant. Imag. Med. Surg. **4**(6), 475–477 (2014)
14. Hwa, S., Hijazi, M., Bade, A., Yaakob, R., Jeffree, M.: Ensemble deep learning for tuberculosis detection using chest x-ray and canny edge detected images. IAES Int. J. Artif. Intell. **8**, 429 (2019)
15. Simonyan, K., Zisserman, A.: Very deep convolutional networks for large-scale image recognition (2014). arXiv:1409.1556
16. Szegedy, C., Vanhoucke, V., Ioffe, S., Shlens, J., Wojna, Z.: Rethinking the inception architecture for computer vision. In Proceedings of the IEEE Conference on Computer Vision and Pattern Recognition, pp. 2818–2826 (2016)
17. Pasa, F., Golkov, V., Pfeiffer, F., Cremers, D., Pfeiffer, D.: Efficient deep network architectures for fast chest x-ray tuberculosis screening and visualization. Sci. Rep. **9**, 6268 (2019)
18. Cortes, C., Vapnik, V.: Support-vector networks. Mach. Learn. **20**(3), 273–297 (1995)
19. Sahlol, A.T., Abd Elaziz, M., Tariq Jamal, A., Damaševičius, R., Farouk Hassan, O.: A novel method for detection of tuberculosis in chest radiographs using artificial ecosystem-based optimisation of deep neural network features. Symmetry **12**(7), 1146 (2020)
20. Howard, A.G., Zhu, M., Chen, B., Kalenichenko, D., Wang, W., Weyand, T., Andreetto, M., Adam, H.: MobileNets: efficient convolutional neural networks for mobile vision applications (2017). https://arxiv.org/abs/1704.04861
21. Zhao, W., Wang, L., Zhang, Z.: Artificial ecosystem-based optimization: a novel nature-inspired meta-heuristic algorithm. Neural Comput. Appl. **32**, 1–43 (2020)
22. Kermany, D,S., Goldbaum, M., Cai, W., Valentim, C.C.S., Liang, H., Baxter, S.L., McKeown, A., Yang, G., Wu, X., Yan, F., et al.: Identifying medical diagnoses and treatable diseases by image-based deep learning. Cell **172**(5), 1122–1131 (2018)
23. Ayaz, M., Shaukat, F., Raja, G.: Ensemble learning based automatic detection of tuberculosis in chest X-ray images using hybrid feature descriptors. Phys. Eng. Sci. Med. **44**(1), 183–194 (2021)
24. Szegedy, C., Ioffe, S., Vanhoucke, V., Alemi, A.: Inception-v4, inception-resnet and the impact of residual connections on learning. In Proceedings of the AAAI Conference on Artificial Intelligence, vol. 31, no. 1 (2017)
25. Chollet, F.: Xception: deep learning with depthwise separable convolutions. In Proceedings of the IEEE Conference on Computer Vision and Pattern Recognition, pp. 1251–1258 (2017)
26. Kotei, E., Thirunavukarasu, R.: Ensemble technique coupled with deep transfer learning framework for automatic detection of tuberculosis from chest x-ray radiographs. Healthcare **10**(11), 2335 (2022)
27. Huang, G., Liu, Z., van der Maaten, L., Weinberger, K. Q.: Densely connected convolutional networks. In Proceedings of the IEEE Conference on Computer Vision and Pattern Recognition (CVPR), pp. 4700–4708 (2017). https://doi.org/10.1109/CVPR.2017.243
28. Tan, M., Le, Q.: Efficientnet: rethinking model scaling for convolutional neural networks. In International Conference on Machine Learning, pp. 6105–6114. PMLR (2019)
29. He, K., Zhang, X., Ren, S., Sun, J.: Deep residual learning for image recognition. In Proceedings of the IEEE Conference on Computer Vision and Pattern Recognition, pp. 770–778 (2016)
30. Bridle, J.S.: Probabilistic interpretation of feedforward classification network outputs, with relationships to statistical pattern recognition. In Neurocomputing: Algorithms, Architectures and Applications, pp. 227–236. Springer, Berlin (1990)
31. Mehrrotraa, R., Ansari, M.A., Agrawal, R., Tripathi, P., Bin Heyat, M.B., Al-Sarem, M., Muaad, A.Y.M., Nagmeldin, W.A.E., Abdelmaboud, A., Saeed, F.: Ensembling of efficient deep convolutional networks and machine learning algorithms for resource effective detection of tuberculosis using thoracic (chest) radiography. IEEE Access **10**, 85442–85458 (2022)
32. Fix, E., Hodges, J.L.: Discriminatory analysis, nonparametric discrimination: consistency properties. Technical Report 4, USAF School of Aviation Medicine, Randolph Field, Texas (1951)

33. Alqudah, A.M., Qazan, S., Alquran, H., Qasmieh, I.A., Alqudah, A.: COVID-19 detection from x-ray images using different artificial intelligence hybrid models. Jordan J. Electr. Eng. **6**(2), 168–178 (2020)
34. Natarajan, S., Sampath, P., Arunachalam, R., Shanmuganathan, V., Dhiman, G., Chakrabarti, P., Chakrabarti, T., Margala, M.: Early diagnosis and meta-agnostic model visualization of tuberculosis based on radiography images. Sci. Rep. **13**(1), 22803 (2023)
35. Shih, G., wu, C., Halabi, S., Kohli, M., Prevedello, L., Cook, T., Sharma, A., Amorosa, J., Arteaga, V., Galperin-Aizenberg, M., Gill, R., Godoy, M., Hobbs, S., Jeudy, J., Archana, T.A., Shah, P., Vummidi, D., Yaddanapudi, K., Stein, A.: Augmenting the national institutes of health chest radiograph dataset with expert annotations of possible pneumonia. Radiol. Artif. Intell. **1**, e180041 (2019)
36. Kingma, D.P., Ba, J.: Adam: a method for stochastic optimization (2014). arXiv:1412.6980
37. Rahman, T., Khandakar, A., Rahman, A., et al.: TB-CXRNet: tuberculosis and drug-resistant tuberculosis detection technique using chest x-ray images. Cognit. Comput. **16**, 1393–1412 (2024)
38. Karl Pearson, F.R.S.: Liii. on lines and planes of closest fit to systems of points in space. Lond. Edinburgh Dublin Philos. Mag. J. Sci. **2**(11), 559–572 (1901)
39. Rajpurkar, P., Irvin, J., Zhu, K., Yang, B., Mehta, H., Duan, T., Ding, D., Bagul, A., Langlotz, C., Shpanskaya, K., Lungren, M.P., Ng, A.Y.: CheXNet: Radiologist-level pneumonia detection on chest X-rays with deep learning (2017). https://arxiv.org/abs/1711.05225
40. Wajgi, R., Yenurkar, G., Nyangaresi, V.O., Wanjari, B., Verma, S., Deshmukh, A., Mallewar, S.: Optimized tuberculosis classification system for chest Xray images: fusing hyperparameter tuning with transfer learning approaches. Eng. Rep. **6**(11), e12906 (2024)
41. Chauhan, A., Chauhan, D., Rout, C.: Role of gist and phog features in computer-aided diagnosis of tuberculosis without segmentation. PloS One **9**(11), e112980 (2014)
42. Agarap, A.F.: Deep learning using rectified linear units (relu) (2018). arXiv:1803.08375
43. Sankararaman, K.A., De, S., Xu, Z., Huang, W.R., Goldstein, T.: The impact of neural network overparameterization on gradient confusion and stochastic gradient descent. In International Conference on Machine Learning, pp. 8469–8479. PMLR (2020)
44. Mao, A., Mohri, M., Zhong, Y.: Cross-entropy loss functions: theoretical analysis and applications (2023)
45. Halder, A., Gharami, S., Sadhu, P., Singh, P.K., Woźniak, M., Ijaz, M.F.: Implementing vision transformer for classifying 2d biomedical images. Sci. Rep. **14**(1), 12567 (2024)
46. Ruder, S.: An overview of gradient descent optimization algorithms (2016). arXiv:1609.04747
47. Pan, C., Zhao, G., Fang, J., Qi, B., Liu, J., Fang, C., Yu, Y.: Computer-aided tuberculosis diagnosis with attribute reasoning assistance. In: International Conference on Medical Image Computing and Computer-Assisted Intervention, pp. 623–633. Springer, Cham (2022)
48. Selvaraju, R.R., Cogswell, M., Das, A., Vedantam, R., Parikh, D., Batra, D.: Grad-CAM: visual explanations from deep networks via gradient-based localization. Int. J. Comput. Vis. **128**(2), 336–359 (2019)
49. Li, L., Jamieson, K., DeSalvo, G., Rostamizadeh, A., Talwalkar, A.: Hyperband: a novel bandit-based approach to hyperparameter optimization. J. Mach. Learn. Res. **18**(185), 1–52 (2018)

Chapter 7
Neural Networks Applied in Disease Prediction: A Systematic Review

Guadalupe E. Rivera-García, Juan C. Ramírez-Vázquez, Miriam J. Cervantes-López, Arturo LLanes-Castillo, and Irahan Otoniel José-Guzmán

Abstract Nowadays it is possible to use neural networks for medical applications, ranging from the prediction and diagnosis of diseases to the determination of the most appropriate treatment for patients. The aim of this article is to present a systematic review of scientific articles related to neural networks used for disease prediction, with the purpose of analyzing, by means of a systematic review, the models used and to demonstrate their prediction accuracy. PRISMA methodology was used to analyze 121 relevant articles, inclusion and exclusion criteria were applied and finally 38 remained. As results, 38 studies (2016–2024) using neural networks for prediction of 26 diseases are presented; CNNs prevail and, together with ANN, LSTM and hybrid models, achieve accuracies >95% (some cases ≈ 100%). 80% of the publications date from 2022 to 2024, 84% come from Asia and are disseminated in 28 journals. The SWOT points out as strengths: high throughput and early detection; weaknesses: black box, need for big data and ethical regulatory challenges, so it is concluded that the incorporation of neural networks brings significant benefits both in the prediction and treatment of diseases.

Keywords Artificial intelligence · Machine learning · Neural networks · Disease prediction

---

G. E. Rivera-García (✉) · J. C. Ramírez-Vázquez
TecNM. Instituto Tecnológico Superior de Pánuco, Pánuco, Veracruz, Mexico
e-mail: esmeralda.rivera@itspanuco.edu.mx; carlos.ramirez@itspanuco.edu.mx

M. J. Cervantes-López · A. LLanes-Castillo
Facultad de Medicina de Tampico "Dr. Alberto Romo Caballero" de la Universidad Autónoma de Tamaulipas, Facultad de Medicina de Tampico "Dr. Alberto Romo Caballero", Tampico, Tamaulipas, Mcxico
e-mail: mcervantes@docentes.uat.edu.mx; allanes@docentes.uat.edu.mx

I. O. José-Guzmán
TecNM. Instituto Tecnológico Superior de Misantla, Misantla, Veracruz, Mexico
e-mail: iojoseg@itsm.edu.mx

G. Rivera et al. (eds.), *Recent Advances in Artificial Neural Networks*, Intelligent Systems Reference Library 283, https://doi.org/10.1007/978-3-032-13497-4_7

7.1 Introduction

Making a medical diagnosis is a complex process that requires a thorough evaluation of several elements, such as the information provided by the patient about the symptomatology presented during the consultation, the clinical signs and the findings of the physical examination performed on the patient [1]. Nowadays, machine learning (ML), as a fundamental part of artificial intelligence (AI) [2], has driven significant advances. Within this field, neural networks have emerged as one of the most powerful tools in modern medicine. A neural network is a machine learning model or program that seeks to mimic the functioning of biological human neurons to identify phenomena, weigh options, and reach conclusions [3]. Such models are clear evidence of the remarkable ability to analyze complex data and identify patterns that might not be detected by traditional methods.

In recent decades, there has been a notable increase in efforts to incorporate predictive analysis into healthcare systems and, in turn, to develop machine learning systems that allow us to facilitate diagnosis [1]. This type of algorithms has been widely used in various disciplines, and have shown great progress and acceptance in the medical field, despite the complexity and risk associated with the preservation of human integrity, where any misdiagnosis can have serious consequences for the health of patients, even costing lives, which represents a great responsibility [4]. The use of neural networks in disease prediction has opened new horizons in medical diagnosis and prognosis, raising new opportunities on how to improve various factors such as efficiency and accuracy in medical care. There are several types of neural networks, the most common being Artificial Neural Networks (ANN), Convolutional Neural Networks (CNN) and Recurrent Neural Networks (RNN).

An artificial neural network (ANN) is a model of a computational algorithm inspired by the functioning of biological neurons [5]. Areas such as signal processing, pattern recognition, speech processing, handwriting recognition, time series prediction, data understanding and feature extraction are applications largely inspired by the functioning of the human brain. Such systems are commonly known for their ease of adaptation to specific problems and their remarkable ability to handle nonlinear data [6, 7]. The multiple layers of interconnected artificial neural neurons are responsible for forming artificial neural networks, which are responsible for processing data and making predictions based on patterns learned from previous data [8]. ANNs have proven to be highly efficient in the prediction of various diseases such as cancer, diabetes, as well as some cardiovascular diseases, all derived from the ability to model complex and nonlinear relationships in large clinical datasets [9].

It is important to keep in mind that the most basic neural networks commonly used are ANNs, in which all the neurons in one layer are connected to all the neurons in the next layer. These networks in the area of medicine are commonly focused on predicting diseases using tabular data such as medical records or clinical variables.

In recent years, convolutional neural networks (CNNs) have proven to be very useful in the analysis and processing of medical images, such as X-rays and magnetic resonance imaging (MRI). They are used to perform tasks such as image classification and image anomaly detection, and they are also ideal because they can help us detect certain spatial features in visual data [10]. Radiologists and other healthcare professionals have begun to use CNNs more frequently as a method of early detection of diseases such as breast cancer and lung cancer. CNNs allow us to perform more specialized analysis of grid-structured data, such as images. Convolutional filters are responsible for searching and detecting certain local features in the data by using clustering operations that seek to reduce dimensionality. They are widely used in medical image analysis.

The analysis of retinal images for the detection of eye diseases, the identification of lesions in MRI scans, and the detection of cancer in mammograms are some examples. In certain areas such as automated diagnostics, CNNs are frequently employed for image classification tasks, resulting in accuracy and efficiency that have been shown to be superior to traditional classification methods [11, 12].

Recurrent neural networks have an architecture created in which connections between neurons form directed graphs along a time sequence. This structure allows them to visualize dynamic behavior, retain an internal memory and store relevant information about events from previous time steps. Furthermore, it allows them to analyze time series data [13]. The processing of sequential and temporal data is an ideal process to be performed with RNNs. The cyclic connections of these networks help to facilitate the modeling of long-term dependencies and the prediction of future events in time series, such as genetic sequences or medical records. They are notoriously good at analyzing genetic patterns and predicting the progression of chronic diseases. Some variants that are included are Long Short-Term Memory Networks (LSTM), which are useful for predicting the development of chronic diseases. Gated Recurrent Units (GRU), which are similar to LSTMs but with a simpler structure, are suitable for sequential prediction and treatment response. Generative neural networks (GANs), although less common, generate synthetic data to train other models or extend data sets. The application of these neural network methodologies together with advanced data preprocessing and machine learning techniques has enabled researchers and healthcare professionals to achieve greater efficiency in the accuracy of disease prediction. This has allowed earlier diagnoses and personalized treatments for each patient [9].

In this context, there are previous literature reviews that have exploited the use of machine learning in the medical field. For example, Rajkomar et al. (2019) [14], which provided an overview of machine learning in medicine, while more recent research has focused on chronic or infectious diseases using artificial intelligence algorithms. However, until now, few systematic reviews have focused specifically on the use of neural network models for the prediction of a wide range of diseases. This paper seeks to fill that gap by providing a comprehensive and up-to-date perspective on the topic, and by highlighting the strengths and limitations of the existing literature. In this way, the present study adds value to previous work by including diverse medical fields and types of neural networks, identifying common trends and possibilities for improvement.

The objective of this research is to perform a systematic review of the scientific literature on the use of neural network models in disease prediction, in order to identify their accuracy and to highlight the advances, challenges and opportunities in the field of artificial intelligence based medical prediction. The sections are structured as follows: First, the materials and methods section are presented, detailing the application of the PRISMA methodology, and the collection of key data (author, year, pathology, architecture, data set and accuracy) for each investigation. Subsequently, comparative results and a bibliometric analysis describing temporal and geographic trends are presented. The discussion section of the findings is followed by an analysis of the comparison with other systematic reviews, as well as research gaps and future lines of work. Finally, the conclusions of the study are presented, addressing the benefits of neural networks in disease prediction and treatment.

7.2 Materials and Methods

7.2.1 Search Strategy

An exhaustive search of scientific articles was carried out between July and September 2024 in the PubMed, Scopus and Web of Science databases. We used key terms in Spanish and English combined with Boolean operators; for example: ("machine learning" OR machine learning) AND ("neural network" OR neural network) AND ("prediction" OR prediction) AND "disease". Search strategies were tailored to each database to maximize retrieval of relevant literature.

7.2.2 Inclusion and Exclusion Criteria

We included original studies (1) published in English, (2) within the last 10 years (2015–2024), (3) peer-reviewed, and (4) focused on the use of neural network models for human disease prediction (diagnosis or prognosis). We excluded papers that (a) did not focus on disease prediction (e.g., studies of artificial intelligence in other medical fields), (b) did not employ neural networks as the main tool, (c) were short communications or abstracts without detailed data, (d) did not have full text available, or (e) did not report performance metrics (accuracy or other) that would allow evaluation of model results.

7.2.3 *Selection of Studies*

Following the PRISMA guidelines [14], two reviewers independently screened the obtained titles and abstracts to discard clearly irrelevant studies. Subsequently, they evaluated the full texts of potentially eligible studies, applying the aforementioned criteria. Discrepancies were resolved by consensus, with the intervention of a third reviewer if necessary. The selection process is summarized in the flow chart in Fig. 7.1. Of 121 references initially identified, 2 were duplicates and 2 corresponded to studies in animal populations and were therefore discarded. After screening of titles/abstracts, 49 irrelevant studies and 1 without access to the full text were excluded. In the eligibility phase, 67 full texts were evaluated, eliminating 29 that did not report predictive model accuracy metrics. Finally, 38 studies were included in the qualitative synthesis.

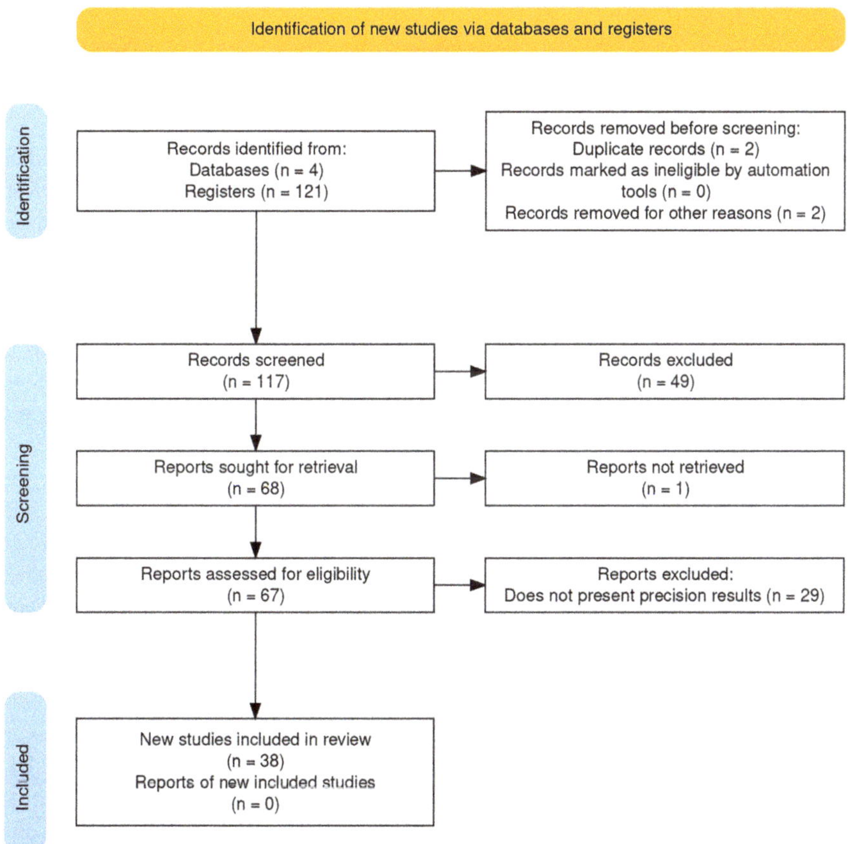

Fig. 7.1 PRISMA flow chart of the study selection process

7.2.4 Data Extraction and Analysis

The following data were extracted from each of the 38 included studies: authors, year, geographic region, target medical condition, type(s) of neural network(s) used, dataset or input characteristics used, and main performance outcomes (e.g., percentage accuracy, sensitivity, specificity, area under the curve). Extraction was performed independently by two authors, with cross-checking to ensure the accuracy of the information. The results were organized in tables and grouped by type of disease and type of neural network model to facilitate comparison. In addition, descriptive biblio-metric analyses (distribution of studies by year, geographic region and journal of publication) were carried out to contextualize research trends in this field.

7.3 Results

7.3.1 Selection of Scientific Articles

After carefully reviewing and evaluating 121 articles, 38 particularly relevant studies were selected. This selection covers a wide range of medical conditions: 1 study on stroke, 3 on Alzheimer's, 1 on arrhythmias, 1 on colon cancer, 1 on prostate cancer, 1 on mild cognitive impairment, 4 on diabetes, 1 on pregnancy, 2 on breast cancer, 1 on cardiovascular disease, 1 on malaria, 4 on heart disease, 1 on brain disease, 1 on kidney disease, 1 on autism spectrum, 2 on myocardial infarction, 1 on skin lesions, 1 on lymphoblastic leukemia, 1 on leukemia, 1 on meningiomas, 1 on osteoarthritis of the knee, 3 on brain tumors, 1 on ulcers and erosions, 1 on pneumonia, 1 on oral cancer and 1 on monkeypox. These studies provided valuable results on the application, impact and accuracy of neural networks in disease prediction.

Table 7.1 shows the research findings such as the authors of the articles, the neural network models used, the medical condition examined, the input data and the percentage of accuracy in disease prediction. It can be observed in the studies presented that CNN (Convolutional Neural Networks) is the most used model, especially in applications related to medical imaging, such as breast cancer, prostate cancer, myocardial infarction, brain tumors, among others. ANN (Artificial Neural Networks) and its variations are also frequently used, for example, to predict arrhythmias, heart disease and diabetes. Combined models such as CNN-LSTM and CNN-GRU have been shown to be effective in predicting heart disease, combining the strengths of different neural architectures. Some studies use more recent models such as LSTM, GAN, GCN, and ViT-CNN, showing the evolution in the application of neural networks to medicine. In some cases, such as GCN and GAN for dementia and Alzheimer's disease, they reached 100% accuracies.

In general, the models achieved high accuracy for example: prostate cancer (CNN 3D DenseNet-121) with 99.88% accuracy, brain tumors (CNN + cDCGAN),

Table 7.1 Results of the systematic review

Author (s)	Techniques and models used	Medical condition studied	Input data	Accuracy percentage
[15]	Random Forest, ExtraTrees, XGBoost, ANN and GANN	Strokes	Sex, age, various diseases, smoking, BMI, physical activity, hypertension, heart disease, lifestyle and others	ExtraTrees and AUC = 98.24% Random Forest and AUC = 98.03%
[16]	CNN	Alzheimer's	Biological samples (brain tissues)	PCA-CNN = 96.6% SVD-CNN = 97.08%
[17]	CNN and GCN	Dementia and Alzheimer's states	MRI images	Accuracy = 100%
[18]	CNN (MobileNet)	Alzheimer's	MRI images	Accuracy =96. 22%
[19]	CNN	Breast cancer	Breast masses	Accuracy = 96.39% Sensitivity = 97.73% Specificity = 94.8%
[20]	ANN	Arrhythmias	Electrocardiograms	Accuracy = 96.69%
[21]	LRAN	Oral cancer	Optical coherence tomography (OCT)	Accuracy = 91.62% Sensitivity = 91.6% Specificity = 92.58%
[22]	CNN 1-D), MLCP, (BiLSTM)	Colon cancer	RNA sequencing data from extracellular vesicles (EVs) from healthy and colon cancer individuals	Accuracy = 97.67%
[23]	CNN (3D DenseNet-121)	Prostate cancer	3D MRI images	Accuracy = 99.88%
[24]	CNN	Myocardial infarction	Electrocardiograms	Accuracy = 93.53% and 95.22% with noise and without noise respectively
[25]	CNN and CNN-LSTM	Diabetes	Electrocardiograms	Accuracy = 95.7%
[26]	LSTM (RNN)	Mild cognitive impairment	Digital biomarkers	Accuracy = 83.84%
[27]	LSTM	Diabetes	Medical history, clinical tests and lifestyle factors	Accuracy = 99.64%

(continued)

Table 7.1 (continued)

Author (s)	Techniques and models used	Medical condition studied	Input data	Accuracy percentage
[28]	ANN, (K-NN), (SVM), Naive Bayes (NB), Decision Tree, Random Forest (RF) and Linear Discriminant Analysis (LDA)	Diabetes mellitus type 2	Age, body mass index (BMI), blood pressure and glucose levels	Accuracy = 98.38%
[29]	CNN-LSTM	Diabetes	Electrocardiograms	Accuracy: 95.1%
[30]	ANN	Pregnancy	Hormonal readings	Precision = 96.5%
[31]	DBN-NN	Breast cancer	Biopsies, which may include measurements such as tumor size, shape, texture, and other histological factors	Accuracy = 99.68% Sensitivity = 100% Specificity = 99.47%
[32]	ANN	Cardiovascular disease	Kaggle Cardiovascular Diseases Dataset	Accuracy = 74.14%
[33]	CNN, MobileNetV2, ResNet-50	Malaria disease	Microscopic images of blood smear	Accuracy CNN = 95.90% MobileNetV2 = 97.06% ResNet-50 = 96.70%
[34]	ANN	Heart diseases	Electrocardiograms	Accuracy = 99%
[35]	CNN	Heart diseases	Electrocardiograms	Accuracy = 97.2%
[36]	CNN-LSTM and CNN-GRU	Heart diseases	Age, gender, blood pressure, cholesterol, heart rate, medical history, lifestyle habits, BMI, blood sugar level, electrocardiogram data	Accuracy = 98.41%
[37]	CNN and RNN	Heart diseases	MRI images	EfficientNetB0 = 77.1% Red móvil = 70.6% NASNetMóvil = 74.4% ResNet50V2 = 76.0%

(continued)

Table 7.1 (continued)

Author (s)	Techniques and models used	Medical condition studied	Input data	Accuracy percentage
[38]	GAN and GCN	Brain diseases	Neuroimaging	Accuracy = 97.76%
[39]	CNN	Kidney diseases	Ultrasounds	Accuracy = 99.23%
[40]	CNN	Autism spectrum	High-resolution facial images of children diagnosed with ASD	Accuracy = 95.0%
[41]	ViT-CNN	Leukemia	Images of leukemic B lymphoblastic cells (cancer cells) and B lymphoid precursors (normal cells)	Accuracy = 99.03%
[42]	CNN	Pneumonia	Chest x-ray	Accuracy = 97.92%
[43]	CNN	Myocardial infarction	Electrocardiograms	Accuracy = 99.0%
[44]	CNN	Skin lesions	Images of skin lesions	Accuracy HAM10000 = 98.66% ISIC 2018 = 95.96%
[45]	CNN	Lymphoblastic leukemia	Blood smear images	Precision = 96.0%
[46]	CNN	Meningiomas	Magnetic resonance	Accuracy NF-2 = 73.0%, S100 = 80.0%
[47]	CNN and SVM	Knee osteoarthritis	X-ray images of knees	Accuracy =90.8%
[48]	CNN	Brain tumor	MRI images	Accuracy =97.4%
[49]	CNN, cDCGAN	Brain tumors	MRI images	Accuracy = 99.0%
[50]	CNN	Brain tumors	MRI images	Accuracy = 100.0%
[51]	CNN	Ulcers and erosions	Enteroscopy images	Accuracy = 98.7%
[52]	ANN	Monkeypox	Epidemiological data	Accuracy = 99.0%

dementia and Alzheimer's disease (CNN and GCN) reaching 100% accuracy, leukemia (ViT-CNNN) with 99.03% accuracy, models such as CNN-LSTM for diabetes prediction achieve 95.1% accuracy, which shows their ability to process clinical data sequences. In the study of myocardial infarction, differences in accuracy are observed when adding noisy data 93.53% and noiseless data 95.22%, this finding highlights the importance of data quality to obtain more robust predictive models. Some models, such as EficientNetB0 and Mobile Network for heart disease, show lower accuracies (between 70.6% and 77.1%), suggesting possible limitations of these approaches for certain types of data or specific medical conditions.

Several studies report both sensitivity and specificity, in addition to overall accuracy. For example: breast cancer and oral cancer, these values are important to assess not only the accuracy in detecting the disease, but also their ability to avoid false positives.

Types of data used include: Medical imaging (MRI, ultra-sound, CT scans, X-ray images), clinical data such as biomarkers, hormone readings, genetic testing, and lifestyle records, electrocardiograms to predict heart disease and diabetes. In performance by medical area, it can be noted that breast, prostate and colon cancer have shown an accuracy of over 97% in most cases, suggesting that the current models are highly effective in these areas. Heart disease varies more in accuracy, with some studies reaching over 99%, while other models show more modest results, such as those based on EficientNet.

7.3.2 *Bibliometric Analysis*

The 38 included studies were published between 2016 and 2024, with a clear increasing trend in recent years. In the first years of the period (2016–2018), few studies per year were found, whereas from 2022 onwards, a strong increase in the number of relevant publications is observed. Figure 7.2 illustrates the distribution by year of publication of the articles included, showing how more than 80% of the studies are concentrated in the last three years (2022–2024), which reflects the current boom in the use of neural networks in this field.

Regarding geographic origin, most of the research was conducted in Asian countries (including the Middle East). In fact, approximately 84% of the studies

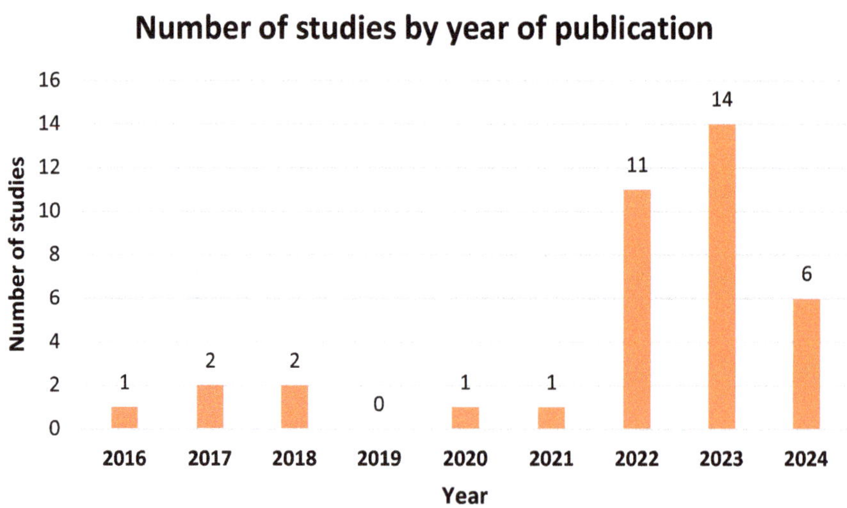

Fig. 7.2 Distribution of studies by year of publication

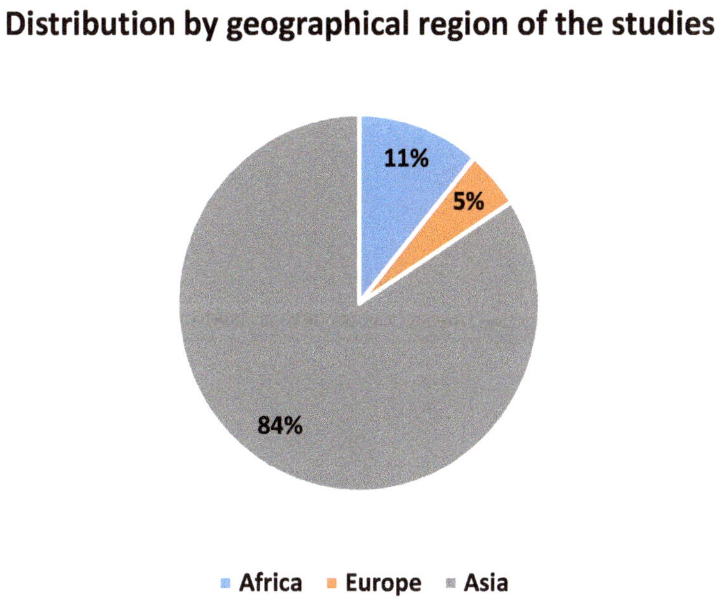

Fig. 7.3 Distribution of studies by geographic region

came from Asian institutions, while ~11% of contributions were identified from Africa and about 5% from Europe. Figure 7.3 shows this percentage distribution by geographic region. This concentration of papers in Asia can be attributed, in part, to the growing interest and development in medical artificial intelligence in countries such as China and India in recent years, as well as to international collaborations that include authors from the Middle East. It should be noted that the reasons for the absence of North American studies is mainly due to the fact that they did not meet the strict inclusion criteria of the present investigation (for example, they focused on machine learning in general and not specifically on neural networks, or they did not report accuracy metrics).

The 38 studies were distributed among some 28 journals, as shown in Fig. 7.4, with no dominant concentration: Diagnostics published 5 articles, Applied Sciences and Electronics 4, Sensors and Bioengineering 3 each; the remaining 53% (20 articles) appeared in isolation in other journals such as Biomedicines, Pattern Recognition or Expert Systems with Applications. This dispersion confirms the interdisciplinary nature of the subject, encompassing both medical journals and journals on biomedical engineering, medical informatics and artificial intelligence.

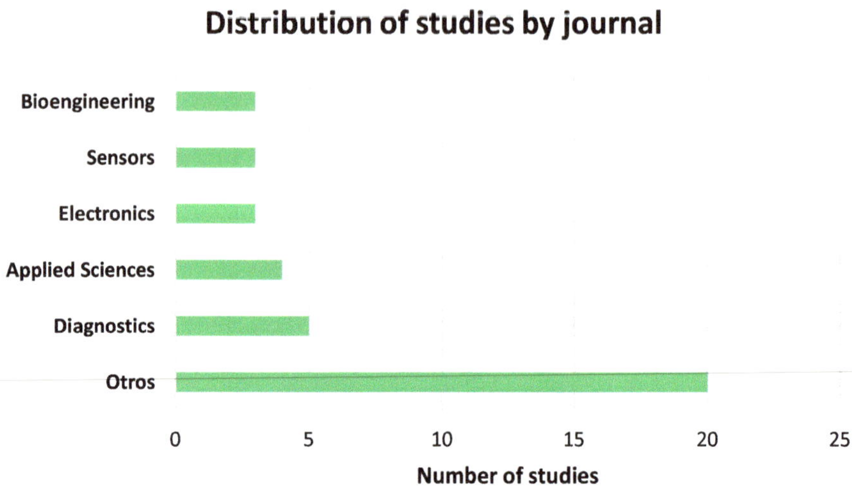

Fig. 7.4 Distribution of studies by journal

7.3.3 *Summary of Results by Medical Domains*

Neural networks excel in all the medical domains reviewed: in oncology, CNNs applied to images reach accuracies above 95% in the detection of breast, prostate, colon, brain tumors and leukemias; in neurology, combinations of CNNs, GCNs and hybrid models reach accuracy levels close to 100% for Alzheimer's and other dementias; in metabolic and chronic pathologies such as diabetes and cardiovascular diseases, ANNs and RNN/LSTMs analyzing tabular clinical data and electrocardiographic signals obtain between 90% and 99% accuracy; and in infectious diseases (malaria, pneumonia, monkeypox), CNNs on blood images, radiographs or skin lesions achieve accuracies of 95%–99%. Taken together, these results confirm that deep learning is a reliable tool for predicting the onset and progression of various diseases, with performances that in many cases exceed 95% accuracy.

7.3.4 *SWOT Analysis of the Neural Network-Based Approach*

The SWOT analysis of neural networks for disease prediction reveals as strengths the high accuracy and versatility for different types of data, early detection and objective clinical decision support. Weaknesses include poor interpretability ("black box"), need for large and varied data sets, risk of bias, high computational costs and heterogeneity of metrics. Neural networks open opportunities for further advances in explainable AI, multimodal data fusion, personalized medicine, integration into telemedicine and wearable devices, and greater access to medical records through interdisciplinary collaborations and open repositories. As threats, neural networks present ethical-legal challenges (privacy, liability, bias), resistance or lack of staff training, rapid technological obsolescence, upgrade costs, and risk of extrapolating models without extensive clinical validation.

7.4 Discussions

Derived from the opaque nature of many algorithms, lack of confidence is one of the main challenges currently facing healthcare-focused AI tools [53]. Specific diagnostic or patient-assigned treatment recommendations using "black box" methods, such as neural networks, random forests, and gradient boosting models, are difficult for clinicians and researchers to understand [54, 55], unlike when dealing with some "white box" algorithms such as logistic regression and decision trees that are more trans-parent [54]. Skepticism affects not only clinicians, but also patients, who are often less tolerant of errors produced by an AI than clinicians [56]. A key way to combat this challenge is to provide training for healthcare professionals to effectively integrate these complex systems into their medical practice [57]. In AI research, it is also of utmost importance to standardize metrics because outcomes are reported with different measures, which can make them difficult to compare. Furthermore, although models may be remarkable in theoretical studies, bringing these successes into clinical practice on a frequent basis can be challenging [58, 59].

There are a number of ethical issues related to the application of AI models, such as patient informed consent, security, transparency, human bias, data privacy, and legal implications such as data protection, cybersecurity, and intellectual property that represent a major challenge for these technologies [60, 61]. AI technologies in medicine present a number of challenges today, such as ensuring transparency and patient safety, standardizing information and financial aspects, training medical personnel, and establishing regulatory standards for evaluating the efficacy and safety of these emerging technologies [62]. It is vitally important to use diverse and representative data in all phases of training, evaluation and monitoring to avoid human bias and overfitting that can affect the accuracy of predictions [63].

7.4.1 Comparison with Other Systematic Reviews

Compared with three previous systematic reviews, our synthesis confirms the current dominance of deep neural networks, especially CNNs and ANNs, in disease prediction, often outperforming classical methods such as SVM or random forest reported as competitive in previous studies by Roobini 2024 and Delpino 2022, [64, 65]. We agree on its high performance in heart disease, Alzheimer's, cancer and chronic diseases (diabetes, cardiovascular), and extend the evidence to accuracies >95%. Both our work and that of Keshavamurthy 2022 [66], emphasize that data quality and heterogeneity are critical factors; without robust data, even the best algorithms fail. Overall, the reviews converge in recognizing neural networks as leaders in accuracy, but agree on the need to improve their interpretability and generalizability for full clinical adoption.

7.4.2 Research Gaps and Future Lines of Work

There are some main gaps in research on neural networks for disease prediction: (1) lack of standardization: few studies apply a uniform set of metrics or compare architectures under the same conditions, which requires common benchmarks; (2) poorly explored medical domains: areas such as autoimmune diseases, psychiatry or pediatrics hardly appear, compared to the predominance of oncology, cardiology and neurology; (3) low external validation: most models are not prospectively tested in real clinical settings, so their practical impact is unknown; (4) low interpretability: rare papers apply XAI techniques (LIME, SHAP, heat maps) or design explainable networks, limiting clinical confidence; and (5) insufficient data bias and clinical integration: few studies address demographic equity, creation of user-friendly interfaces, physician training or protocols for use. Addressing these gaps, such as systemic comparison, new domains, clinical trials, explainability and debiasing, are key opportunities for future research.

7.5 Conclusions

Neural network models have become powerful tools for disease prediction, which has allowed the research and medical community to present remarkable advances in the diagnosis and prediction of diseases. The accuracy in diagnoses and predictions issued by these models has shown significant improvements thanks to their ability to handle large volumes of data and detect complex patterns, performing their processes mainly through deep neural networks such as CNN and LSTM. These networks adapt to a variety of data types, such as genetic sequences, medical images and tabular clinical data, facilitating their use in various contexts. Such advances frequently outperform traditional methods [10, 14]. Neural networks can lead to earlier interventions and better patient outcomes [67]. By providing fast and accurate diagnoses, automation of data analysis using neural networks can also lighten the workload of healthcare professionals and reduce costs [68]. However, the quality and diversity of the data used determine the effectiveness of these models, and biases in the data can affect the accuracy and fairness of the results [9].

Neural network research continues to advance, with new approaches and techniques under development to improve interpretability [68], integrate multimodal data, and optimize training algorithms [69]. Neural network models have great potential to assist with personali-zed medicine by providing prognoses and treatment recommendations tailored to individual patient characteristics, thereby improving treatment efficacy and disease management [70]. However, to maximize their impact on healthcare, it is essential to address issues related to data quality, interpretation, and ethical considerations.

Research findings on the use of neural networks in disease prediction, highlighting that CNNs are the most widely used, especially for medical imaging related to cancer, myocardial infarction and brain tumors. Models such as ANN and its variants are also co-mune, and combinations such as CNN-LSTM have been shown to be effective for cardiac diseases. The accuracy of the models is remarkable, with examples reaching up to 100% for dementia and Alzheimer's disease, and 99.88% for prostate cancer, although variation in accuracy is observed for infarcts with noisy data. Some models, such as EficientNet, showed lower accuracies (between 70.6% and 77.1%). Not only accuracies, but also sensitivity and specificity are reported, which is crucial to avoid false positives. The data used include medical and clinical images, with outstanding performance in breast, prostate and colon cancer, which exceed 97% accuracy, while cardiac diseases show variability in results.

Neural network techniques, especially CNN and ANN, show outstanding performance in the prediction of various diseases, particularly in the field of medical imaging. Studies of cardiovascular diseases, cancer and neurological diseases have been the main beneficiaries of these techniques. The use of hybrid architectures such as CNN-LSTM and improved data cleaning can significantly increase the accuracy of the models. These findings can serve as a basis for model choice according to the medical condition and type of data available.

References

1. Gonzalez, B.N., Estrada, S.V., Febles, A.: Estudio y selección de las técnicas de Inteligencia Artificial para el diagnóstico de enfermedades. Rev. Ciencias Médicas. **22**(3), 534–544 (2018) https://www.medigraphic.com/cgi-bin/new/resumen.cgi?IDARTICULO=82029
2. Sun, K., Roy, A., Tobin, J.: Artificial intelligence and machine learning: definition of terms and current concepts in critical care research. J. Crit. Care. **82**(154792) (2024). https://doi.org/10.1016/j.jcrc.2024.154792
3. IBM: What is a neural network? (2024). https://www.ibm.com/topics/neural-networks
4. Leyva Vazquez, G., Gonzalez Benıtez, N., Hechavarria Hernandez, J., Rivero Peña, Y., Daher Nader, J.: El diagnostico de enfermedades desde el Analisis Inteligente de los Datos. Revista espacios. **39**(28) (2018). https://www.revistaespacios.com/a18v39n28/18392816.html
5. Hagan, M.: Neural Network Design 2nd ed. (2014). https://hagan.okstate.edu/NNDesign.pdf
6. Dongare, A., Kharde, R., Kachare, D.: Introduction to Artificial Neural Network. Certif. Int. J. Eng. Innov. Technol. (IJEIT). (2012) https://www.semanticscholar.org/paper/Introduction-to-Artificial-Neural-Network-A.D.Dongare-R.R.Kharde/04d0b6952a4f0c7203577afc9476c2fcab2cba06
7. Zahner, D.A., Micheli-Tzanakou, E.: Artificial Neural Networks. Definitions, Methods, Applications, pp. 61–78. CRC Press, Boca Raton, FL, USA (2017) https://kazus.ru/nuke/ebookss/37/2278_PDF_C02.pdf
8. Bishop, C.M.: Pattern Recognition and Machine Learning. Springer, New York (2006) https://link.springer.com/book/9780387310732

9. Miotto, R., Li, L., Dudley, J.: Deep learning to predict patient future diseases from the electronic health records. In: Conference: European Conference on Information Retrieval (2016). https://doi.org/10.1007/978-3-319-30671-1_66
10. LeCun, Y., Bengio, Y., Hinton, G.: Deep learning. Nature. **521**(7553), 436–444 (2015). https://doi.org/10.1038/nature14539
11. Berezsky, O., Liashchynskyi, P., Pitsun, O., Izonin, I.: Síntesis de arquitecturas de redes neuronales convolucionales para la clasificación de imágenes biomédicas. Biomed. Signal Process. Control. **95**(106325) (2024). https://doi.org/10.1016/j.bspc.2024.106325
12. Piraianu, A., Fulga, A., Musat, C., Ciobotaru, O., Poalelungi, D., Stamate, E., Ciobotaru, O., Fulga, I.: Enhancing the evidence with algorithms: how artificial intelligence is transforming forensic medicine. Diagnóstico. **13** (2023). https://doi.org/10.3390/diagnostics13182992
13. Schmidhuber, J.: Deep learning in neural networks: an overview. Neural Netw. **61**, 85–117 (2015). https://doi.org/10.1016/j.neunet.2014.09.003
14. Page, M.J., McKenzie, J.E., Bossuyt, P.M., Boutron, I., Hoffmann, T.C., Mulrow, C.D., Loder, E.W.: Declaración PRISMA 2020: una guía actualizada para la publicación de revisiones sistemáticas. Rev. Esp. Cardiol. **74**(9), 790–799 (2021). https://doi.org/10.1016/j.recesp.2021.06.016
15. Wijaya, R., Saeed, F., Samimi, P., Albarrak, A., Qasem, S.: An ensemble machine learning and data mining approach to enhance stroke prediction. Bioingeniería. **2024**, 11(672) (2024). https://doi.org/10.3390/bioengineering11070672
16. Abdelwahab, M., Al-Karawi, K., Semary, H.: Deep learning-based prediction of Alzheimer's disease using microarray gene expression data. Biomedicine. **11**(3304) (2023). https://doi.org/10.3390/biomedicines11123304
17. Hasan, M., Wagler, A.: New convolutional neural network and graph convolutional network-based architecture for AI applications in Alzheimer's disease and dementia-stage classification. AI. **5**, 342–363 (2024). https://doi.org/10.3390/ai5010017
18. Mohi Ud din dar, G., Bhagat, A., Ansarullah, S., Othman, M., Hamid, Y., Alkahtani, H., Hamam, H.: A novel framework for classification of different Alzheimer's disease stages using CNN model. Electronics. **12**(2) (2023). https://doi.org/10.3390/electronics12020469
19. Rasti, R., Teshnehlab, M., Phung, S.: Breast cancer diagnosis in DCE-MRI using mixture ensemble of convolutional neural networks. Pattern Recogn. **72**, 381–390 (2017). https://doi.org/10.1016/j.patcog.2017.08.004
20. Zhang, C., Chang, J., Guan, Y., Li, Q., Wang, X., Zhang, X.: A low-power ECG processor ASIC based on an artificial neural network for arrhythmia detection. Appl. Sci. **13**(9591) (2023). https://doi.org/10.3390/app13179591
21. Yuan, W., Cheng, L., Yang, J., Yin, B., Fan, X., Yang, J., Huang, X.: Noninvasive oral cancer screening based on local residual adaptation network using optical coherence tomography. Med. Biol. Eng. Comput. **60**, 1363–1375 (2022). https://doi.org/10.1007/s11517-022-02535-x
22. Bostanci, E., Kocak, E., Unal, M., Guzel, M., Acici, K., Asuroglu, T.: Machine learning analysis of RNA-seq data for diagnostic and prognostic prediction of colon cancer. Sensors. **23**(3080) (2023). https://doi.org/10.3390/s23063080
23. Baumgärtner, G., Hamm, C., Schulze-Weddige, S., Ruppel, R., Beetz, N., Rudolph, M., et al.: Metadata-independent classification of MRI sequences using convolutional neural networks: successful application to prostate MRI. Eur. J. Radiol. **166**, 110964 (2023). https://doi.org/10.1016/j.ejrad.2023.110964
24. Rajendra, U., Fujita, H., Oh, S.L., Hagiwara, Y., Tan, J., Adam, M.: Application of deep convolutional neural network for automated detection of myocardial infarction using ECG signals. Inf. Sci. **415–416**, 190–198 (2017). https://doi.org/10.1016/j.ins.2017.06.027
25. Swapna, G., Vinayakumar, R., Soman, K.: Diabetes detection using deep learning algorithms. ICT Express. **4**(4), 243–246 (2018). https://doi.org/10.1016/j.icte.2018.10.005
26. Narasimhan, R., Gopalan, M., Sikkandar, M., Alassaf, A., AlMohimeed, I., Alhussaini, K., Sheik, S.: Employing a deep learning approach for early detection of mild cognitive impairment transitions using digital biomarker analysis. Sensors. **23**(8867) (2023). https://doi.org/10.3390/s23218867

27. Alex, S., Jhanjhi, N., Humayun, M., Ibrahim, A., Abulfaraj, A.: Deep LSTM model for diabetes prediction with class balancing by SMOTE. Electronics. **11**(2737) (2022). https://doi.org/10.3390/electronics11172737
28. Al Sadi, K., Balachandran, W.: Prediction model of type 2 diabetes mellitus for Oman prediabetes patients using artificial neural network and six machine learning classifiers. Appl. Sci. **2344** (2023). https://doi.org/10.3390/app13042344
29. Swapna, G., Soman, K., Vinayakumar, R.: Automated detection of diabetes using CNN and CNN-LSTM network and heart rate signals. Procedia Comput. Sci. **132**(3067), 1253–1262 (2018). https://doi.org/10.1016/j.procs.2018.05.041
30. Shboul, L., Fram, K., Sharaeh, S., Alshraideh, M., Shaar, N., Alshraideh, N.: Male and female hormone reading to predict pregnancy percentage using a deep learning technique: a real case study. AI. **3**, 871–889 (2022). https://doi.org/10.3390/ai3040053
31. Abdel-Zaher, A., Eldeib, A.: Breast cancer classification using deep belief networks. Expert Syst. Appl. **46**(15), 139–144 (2016). https://doi.org/10.1016/j.eswa.2015.10.015
32. Lin, C.-M., Lin, Y.-S.: Utilizing a two-stage Taguchi method and artificial neural network for the precise forecasting of cardiovascular disease risk. Bioengineering. **10**(1286) (2023). https://doi.org/10.3390/bioengineering10111286
33. Hemachandran, K., Alasiry, A., Marzougui, M., Ganie, S., Pise, A., Alouane, M.-H., Chola, C.: Performance analysis of deep learning algorithms in diagnosis of malaria disease. Diagnostics. **13**(534) (2023). https://doi.org/10.3390/diagnostics13030534
34. Mohapatra, S., Sahoo, P., Mohapatra, S.: Healthcare big data analysis with artificial neural network for cardiac disease prediction. Electronics. **163**, 13 (2024). https://doi.org/10.3390/electronics13010163
35. Mohammad, F., Al-Ahmadi, S.: WT-CNN: a hybrid machine learning model for heart disease prediction. Mathematics. **11**(4681) (2023). https://doi.org/10.3390/math11224681
36. Almulihi, A., Saleh, H., Hussien, A., Mostafa, S., El-Sappagh, S., Alnowaiser, K., Refaat Hassan, M.: Ensemble learning based on hybrid deep learning model for heart disease early prediction. Diagnostics. **12**(3215) (2022). https://doi.org/10.3390/diagnostics12123215
37. Ho, N., Kim, Y.-C.: Estimation of cardiac short axis slice levels with a cascaded deep convolutional and recurrent neural network model. Tomography. **8**, 2749–2760 (2022). https://doi.org/10.3390/tomography8060229
38. Huynh, N., Yan, D., Ma, Y., Wu, S., Long, C., Sami, M., et al.: The use of generative adversarial network and graph convolution network for neuroimaging-based diagnostic classification. Brain Sci. **14**(456) (2024). https://doi.org/10.3390/brainsci14050456
39. Kumar, K., Pradeepa, M., Mahdal, M., Verma, S., RajaRao, M., Ramesh, J.: A deep learning approach for kidney disease recognition and prediction through image processing. Appl. Sci. **13**(3621) (2023). https://doi.org/10.3390/app13063621
40. Alam, M., Rashid, M., Roy, R., Faizabadi, A., Gupta, K., Ahsan, M.: Empirical study of autism spectrum disorder diagnosis using facial images by improved transfer learning approach. Bioengineering. **9**(710) (2022). https://doi.org/10.3390/bioengineering9110710
41. Jiang, Z., Dong, Z., Wang, L., Jiang, W.: Method for diagnosis of acute lymphoblastic leukemia based on ViT-CNN ensemble. Model. Comput. Intell. Neurosci. **23** (2021). https://doi.org/10.1155/2021/7529893
42. Nahid, A.-A., Sikder, N., Bairagi, A., Razzaque, M., Masud, M., Kouzani, Z., Mahmud, A.: A novel method to identify pneumonia through analyzing chest radiographs employing a multichannel convolutional neural network. Sensors. **20**(12), 3482 (2020). https://doi.org/10.3390/s20123482
43. Elmannai, H., Saleh, H., Algarni, A., Mashal, I., Kwak, K., El-Sappagh, S., Mostafa, S.: Diagnosis myocardial infarction based on stacking ensemble of convolutional neural network. Electronics. **11**(3976) (2022). https://doi.org/10.3390/electronics11233976
44. Salih, O., Duffy, K.: Optimization convolutional neural network for automatic skin lesion diagnosis using a genetic algorithm. Appl. Sci. **13**(3248) (2023). https://doi.org/10.3390/app13053248

45. Sampathila, N., Chadaga, K., Goswami, N., Chadaga, R., Pandya, M., Prabhu, S., Upadya, S.: Customized deep learning classifier for detection of acute lymphoblastic leukemia using blood smear images. Healthcare. **10**((1812)) (2022). https://doi.org/10.3390/healthcare10101812
46. Azamat, S., Buz-Yalug, B., Dindar, S., Yilmaz Tan, K., Ozcan, A., Can, O., et al.: Susceptibility-weighted MRI for predicting NF-2 mutations and S100 protein expression in meningiomas. Diagnostics. **14**(748) (2024). https://doi.org/10.3390/diagnostics14070748
47. Ahmed, S., Mstafa, R.: Identifying severity grading of knee osteoarthritis from X-ray images using an efficient mixture of deep learning and machine learning models. Diagnostics. **12**(2939) (2022). https://doi.org/10.3390/diagnostics12122939
48. Dehkordi, A., Hashemi, M., Neshat, M., Mirjalili, S., & Sadiq, A.: Brain tumor detection and classification using a new evolutionary convolutional neural network. arXiv 2022. 12297 (2022). doi:https://doi.org/10.2139/ssrn.4292650
49. Onakpojeruo, E., Mustapha, M., Ozsahin, D., Ozsahin, I.: Neural network model, using conditional deep convolutional generative adversarial network-generated synthetic and augmented brain tumor datasets for image classification. Brain Sci. **14**(559) (2024). https://doi.org/10.3390/brainsci14060559
50. Uzun Ozsahin, D., Onakpojeruo, E., Uzun, B., Mustapha, M., Ozsahin, I.: Mathematical assessment of machine learning models used for brain tumor diagnosis. Diagnostics. **13**(618) (2023). https://doi.org/10.3390/diagnostics13040618
51. Martins, M., Mascarenhas, M., Afonso, J., Ribeiro, T., Cardoso, P., Mendes, F., Macedo, G.: Deep-learning and device-assisted enteroscopy: automatic panendoscopic detection of ulcers and erosions. Medicina. **59**(172) (2023). https://doi.org/10.3390/medicina59010172
52. Manohar, B., Das, R.: Artificial neural networks for the prediction of monkeypox outbreak. Trop. Med. Infect. Dis. **7**(424) (2022). https://doi.org/10.3390/tropicalmed7120424
53. Patel, V., Shah, M.: Artificial intelligence and machine learning in drug discovery and development. Intell. Med. **2**, 134–140 (2022). https://doi.org/10.1016/j.imed.2021.10.001
54. Busnatu, Ş., Niculescu, A., Bolocán, A., Petrescu, G., Păduraru, D., Năstasă, I., et al.: Clinical applications of artificial intelligence—an updated overview. J.Clin. Med. **11**(2265) (2022). https://doi.org/10.3390/jcm11082265
55. Chan, B.: Black-box assisted medical decisions: AI power vs. ethical physician care. Med. Health Care Philos. **26**(285) (2023). https://doi.org/10.1007/s11019-023-10153-z
56. Poon, A., Sung, J.: Opening the black box of AI-medicine. J. Gastroenterol. Hepatol. **36**, 581–584 (2021). https://doi.org/10.1111/jgh.15384
57. Durán, J., Jongsma, K.: Who is afraid of black box algorithms? On the epistemological and ethical basis of trust in medical AI. J. Med. Ethics. **medethics-2020-106820** (2021). https://doi.org/10.1136/medethics-2020-106820. Epub ahead of print
58. Sana, M., Hussain, Z., Maqsood, M., Shah, P.: Artificial intelligence in celiac disease. Comput. Biol. Med. **125**(103996) (2020). https://doi.org/10.1016/j.compbiomed.2020.103996
59. Vinny, P., Vishnu, V., Padma Srivastava, M.: Artificial intelligence shaping the future of neurological practice. Med. J. Armed Forces India. **77**(3), 276–282 (2021). https://doi.org/10.1016/j.mjafi.2021.06.003
60. Gerke, S., Minssen, T., Cohen, G.: Ethical and legal challenges of artificial intelligence-driven healthcare. Artif. Intell. Healthc. **295–336** (2020). https://doi.org/10.1016/B978-0-12-818438-7.00012-5
61. Tang, L., Li, J., Fantus, S.: Medical artificial intelligence ethics: a systematic review of empirical studies. Digital Health. **9** (2023) https://doi.org/20552076231186064
62. He, J., Baxter, S., Xu, J., Xu, J., Zhou, X., Zhang, K.: The practical implementation of artificial intelligence technologies in medicine. Nat. Med. **25**(1), 30–36 (2019). https://doi.org/10.1038/s41591-018-0307-0. Epub 2019 Jan 7. PMID: 30617336; PMCID: PMC6995276
63. Kolla, L., Parikh, R.: Uses and limitations of artificial intelligence for oncology. Cancer. **130**(12), 2101–2107 (2024). https://doi.org/10.1002/cncr.35307

64. Roobini, S., Kavitha, M. & Karthik, S.: A systematic review on machine learning and neural network based models for disease prediction, J. Integr. Sci. Technol., vol. 12, n° 4, 2024. doi:https://doi.org/10.62110/sciencein.jist.2024.v12.787
65. Delpino, F., et al.: Machine learning for predicting chronic diseases: a systematic review. Public Health. **205**, 14–25 (2022). https://doi.org/10.1016/j.puhe.2022.01.007
66. Keshavamurthy, R., Dixon, S., Pazdernik K., Charles, L.: Predicting infectious disease for biopreparedness and response: a systematic review of machine learning and deep learning approaches. One Health, n° 15:100439. 2022 doi:https://doi.org/10.1016/j.onehlt.2022.100439
67. Gulshan, V., Peng, L., Coram, M., Stumpe, M.C., Narayanaswamy, A., Webster, D.R.: Development and validation of a deep learning algorithm for detection of diabetic retinopathy in retinal fundus photographs. JAMA. **316**(22), 2402–2410 (2016). https://doi.org/10.1001/jama.2016.17216
68. Ching, T., Himmelstein, D.: Opportunities and obstacles for deep learning in biology and medicine. J. Am. Med. Inform. Assoc. **25**(8), 1248–1258 (2018). https://doi.org/10.1098/rsif.2017.0387
69. Goodfellow, I., Bengio, Y., Courville, A.: Deep Learning. MIT Press, Cambridge (2016)
70. Liu, X., Wang, H., Li, Z., Qin, L.: Deep learning in ECG diagnosis: a review. Knowl.-Based Syst. **227**(107187) (2021). https://doi.org/10.1016/j.knosys.2021.107187

Part II
Computer Vision and Multimodal Recognition

Chapter 8
Improved YOLOv9 for Real-Time Traffic Sign Detection in Intelligent Transportation Systems

Adam Kovacs, Eszter Nagy, and Bence Szabo

Abstract Rapid urbanization and population growth have intensified transportation challenges and traffic safety concerns, driving advancements in autonomous driving systems and intelligent transportation systems (ITS). The increasing complexity of modern traffic environments necessitates robust solutions for traffic management and accident prevention. Within this context, accurate and efficient traffic sign detection plays a crucial role in enhancing road safety and optimizing traffic flow, serving as a foundational component for vehicle navigation and decision-making in ITS. This study presents an improved algorithm based on the YOLOv9 model, tailored specifically for traffic sign detection in ITS applications. The proposed algorithm incorporates several enhancements, including optimized anchor box dimensions, advanced feature extraction techniques, and refined post-processing steps to improve detection accuracy and computational efficiency. The algorithm's performance is comprehensively evaluated against the baseline YOLOv9 model using a carefully prepared dataset that includes diverse traffic sign classes, environmental conditions, and real-world complexities. Key performance metrics such as precision, recall, mean average precision (mAP), and F1 score are analyzed to quantify the improvements introduced by the enhanced model. Results demonstrate that the improved YOLOv9 model achieves approximately a 3% enhancement in all these metrics compared to the original model. Additionally, with an average detection time of 8.22 ms per frame, the algorithm exhibits strong potential for real-time ITS applications, meeting the stringent demands of latency-sensitive systems.

Keywords Traffic sign detection · YOLOv9 · Improved YOLO · Object detection · Intelligent transportation systems

A. Kovacs · E. Nagy · B. Szabo (✉)
Department of Electrical Engineering, Budapest University of Technology and Economics, Budapest, Hungary
e-mail: akovacs@bme.hu; enagy@bme.hu; b.szabo@bme.hu

G. Rivera et al. (eds.), *Recent Advances in Artificial Neural Networks*, Intelligent Systems Reference Library 283, https://doi.org/10.1007/978-3-032-13497-4_8

8.1 Introduction

With the increasing number of vehicles and population, humanity is struggling with traffic problems. Given the social harm they generate, including material losses and fatalities, traffic accidents are among the most important problems governments face. Governments and official organizations invest large sums of money in traffic safety programs to help alleviate these issues [1]. Traffic sign detection is one crucial area that is getting this support since it is essential to making sure that cars abide by speed limits and other safety guidelines. In order to improve road safety, traffic cops also use the results of traffic sign detection systems to identify outdated and ineffective traffic signs and replace them with new ones. Therefore, detecting traffic signs is not just a technology development but also a social responsibility initiative meant to increase public safety and create stress-free driving environments. A variety of machine learning and computer vision techniques have been developed for this important task. In the past, these techniques depended on manually created features such histograms, pixels, gradients, and color features as well as low-level image information [2–4]. Although somewhat successful, these methods frequently fail to deliver reliable outcomes in challenging situations, including changing lighting or occlusions [5].

Recent developments in deep learning have greatly enhanced object detection, making it possible to extract complicated characteristics more efficiently and improving the accuracy of image categorization. Deep learning techniques, which are fundamentally based on artificial neural networks composed of multiple hidden layers, aim to automatically extract hierarchical features from raw input data. These layers enable the model to learn increasingly abstract and complex representations, improving performance across a range of applications such as image classification, object detection, and pattern recognition [6–8]. Despite these developments, traffic sign detection is still a research topic with a lot of room for creating cutting-edge solutions to deal with enduring problems [9, 10]. The high computational cost and time consumption of machine learning-driven detection for traffic objects is one of the main obstacles. Moreover, attaining high detection result accuracy is still a critical problem. The efficiency of traffic flow may be improved using object detection algorithms designed specifically for traffic sign detection. Traffic management centers receive actionable data from real-time traffic sign monitoring and interpretation by intelligent transportation systems (ITS) [11, 12] that are integrated with these algorithms. This information makes it easier to manage traffic, schedule traffic signals optimally, and increase overall road efficiency. These systems also play a key role in detecting and reporting broken, ambiguous, or absent traffic signs, which guarantees prompt maintenance and reduces possible hazards to drivers. In order to create safer and more effective road networks, these preventative actions are crucial [13, 14].

Traffic sign identification is essential for developing driver assistance systems (ADAS) [15] and driverless vehicles, as well as for enhancing traffic flow. Accurate traffic sign identification is essential for ADAS features like adaptive cruise control, lane-keeping assistance, and traffic sign recognition to provide drivers with trustworthy feedback and direction [16, 17]. Traffic sign detection is a crucial part of autonomous cars' perception systems, which allow them to efficiently see and react to their environment. The safety and effectiveness of autonomous vehicle operations depend on the accurate and prompt detection of traffic indicators, such as stop signs or school zone warnings [18]. The technological underpinning of these capabilities is object identification algorithms, which makes them essential for improvements in automobile safety and autonomy. Traffic sign detection still faces a number of difficulties despite tremendous advancements. For example, traffic signs may be obscured by poor lighting, and at different times of the day, their appearance may be darkened by the shadows cast by buildings and trees [19]. Additionally, other things, such tree branches or cars, may partially block traffic signs, making detection more difficult. The performance of perception algorithms can be severely hampered by these issues, especially in practical situations. The computational effectiveness of perception algorithms is another significant issue, particularly for real-time applications involving autonomous cars. For these systems to be practically deployed on a broad scale, accuracy and speed must be balanced. In order to achieve this equilibrium, models must be continuously optimized to maximize performance and reduce computational expenses [13].

Modern traffic sign detection methods are based on convolutional neural networks (CNNs) [20–26], which provide unmatched picture processing and object recognition capabilities. CNNs automatically learn hierarchical features from raw picture data, in contrast to conventional techniques that depend on manually created features. They are very efficient at complicated tasks like detecting traffic signs because of this aptitude. Convolutional, pooling, and fully connected layers are among the layers that make up CNN architecture, which is intended to gradually extract spatial and semantic features from input images. Even in difficult situations, CNNs can recognize complex patterns because to this multi-layered methodology, including the distinctive colors, forms, and symbols of traffic signs. In a variety of computer vision applications, well-known CNN-based models including AlexNet [27], VGGNet [28], and ResNet [29] have demonstrated impressive performance. The application of these models in traffic sign detection systems has been made possible. By utilizing transfer learning strategies and pre-trained models, developers can further increase CNN efficacy. These methods shorten the time needed to train models on particular datasets while simultaneously increasing performance. By accurately and consistently recognizing traffic signs, CNN-based algorithms show promise for transforming road safety in the context of traffic sign detection [30–32].

Traffic sign recognition has also demonstrated remarkable performance by object detection algorithms such as You Only Look Once (YOLO) [33, 34], Single Shot MultiBox Detector (SSD) [35–37], and Faster R-CNN [38–40]. These algorithms scan photos, identify and categorize objects, and understand traffic signs in real time using deep learning techniques. For instance, YOLO is well known for its speedy picture processing, which makes it ideal for real-time applications. Faster R-CNN provides great precision in detecting tiny objects, while SSD excels at striking a balance between speed and accuracy. These features are essential for autonomous cars, which can avoid collisions and guarantee a safe trip by making prompt and precise decisions based on traffic signs they perceive. Furthermore, the development of traffic sign detection systems has been further accelerated by improvements in hardware technology. Deep learning model training and inference have been greatly expedited by the incorporation of specialized hardware, such as graphics processing units (GPUs) [41] and tensor processing units (TPUs) [42]. The implementation of sophisticated traffic sign detecting algorithms in practical settings is made possible by this hardware-software synergy, which guarantees excellent performance and dependability. Additionally, traffic sign detecting systems have undergone revolutionary modifications as a result of the introduction of edge computing [43]. Edge devices lower latency and improve the real-time capabilities of traffic sign recognition algorithms by processing data closer to the source. This advancement is especially helpful for intelligent transportation systems and driverless cars, where quick decisions are critical [44].

Detecting traffic signs also helps the environment by minimizing fuel usage and improving traffic flow. Congestion and idle time are reduced by effective traffic management, which is made possible by precise traffic sign recognition and interpretation. Lower greenhouse gas emissions result from this decrease in fuel use, supporting international efforts to tackle climate change [45]. These environmental advantages can also be increased by combining intelligent transportation systems with renewable energy sources, resulting in a comprehensive strategy for sustainable mobility. To sum up, traffic sign detection is an important field for study and development with broad ramifications for environmental sustainability, traffic management, and road safety. The use of deep learning methods, especially CNNs and object detection algorithms, has transformed the sector by providing previously unheard-of levels of efficiency and accuracy in the recognition and interpretation of traffic signals. Notwithstanding the difficulties caused by inadequate lighting, occlusions, and computational efficiency, these problems are still being addressed by continuous technological improvements and creative solutions. Traffic sign detection systems will be crucial in determining the direction of transportation in the future by fusing state-of-the-art algorithms with autonomous cars and intelligent transportation systems. Road networks could become safer, more effective, and environmentally sustainable for future generations as research in this area advances.

8.2 Related Work

Wang et al. [46] introduce an advanced traffic sign detection method, MI-YOLO, built on YOLOv8 to address the challenges posed by small-scale objects and complex traffic environments. By incorporating a Multi-Scale Convolutional Attention (MSCA) module, the model achieves enhanced feature extraction across varying scales. A small object detection layer specifically tailored for traffic signs ensures higher accuracy by reducing false positives and missed detections. Additionally, the integration of an Inner-WIoU loss function improves bounding box regression, accelerating convergence and enhancing localization precision. Experimental results on the TT100K dataset demonstrate a 7.8% improvement in mAP 0.5 compared to baseline YOLOv8. This research highlights the capability of MI-YOLO to significantly outperform existing methods in detecting small and complex traffic signs while maintaining computational efficiency.

The study in [47] addresses the limitations of high computational complexity and large parameter sizes in existing traffic sign detection models by providing a lightweight and efficient solution for YOLO-TS framework. By integrating the Normalized Wasserstein Distance (NWD) with the Complete Intersection Over Union (CIoU) loss function, the model improves the detection accuracy for small traffic signs. The inclusion of StarBlock in the C2f module and the construction of the C2f-Star configuration simplify the architecture while maintaining the detection accuracy. Furthermore, the Slimneck design paradigm significantly reduces the computational demands on the neck network without sacrificing performance. In addition, the Compression and Excitation Shared Sensing Head (SESDH) provides an attention mechanism that focuses on critical areas and improves detection in complex environments. Experimental evaluations on the CCTSDB dataset demonstrate the effectiveness of the model with a 35.33% reduction in parameters, a 35.80% reduction in computational complexity, and a 32.94% smaller model size compared to YOLOv8n. The YOLO-TS model also achieves a 1% improvement in mAP@0.5 and a 1.32% increase in F1 score, demonstrating the ability to effectively balance accuracy and lightweight design.

Zhang et al. [48] proposed the DPF-YOLOv8 model, an enhancement of YOLOv8, designed for small sign recognition in foggy atmospheric cases. The model employs a Dual Path Feature Fusion network to address the challenges of poor visibility and small target sizes by combining shallow and deep features. An efficient feature aggregation structure replaces the conventional backbone modules to enrich small object semantics. Furthermore, the study introduces a preprocessing step, applying a fog simulation to the TT100K dataset, improving the model's robustness to adverse weather. Experimental evaluations reveal a 2.1% increase in mAP@0.5 and a 2.2% improvement in mAP@0.5:0.95 over the baseline YOLOv8, with the model demonstrating superior accuracy and adaptability in various foggy conditions.

Another study presents a novel YOLOv8-based traffic sign detection model to address challenges in detecting small objects and processing multi-scale features in complex traffic environments [49]. The model incorporates several innovative components, including the SPD-Conv module for enhancing small object feature retention and the Deformable Convolution (Dcnv3) module to improve feature extraction accuracy in complex backgrounds. Furthermore, the ASF-SDNet module is employed to optimize cross-scale feature fusion and channel refinement, enabling the model to handle diverse object scales effectively. Experimental results on the TT100K dataset highlight the model's significant advancements, achieving a 4.7% improvement in mAP@0.5, reaching 88.8%. With a parameter count of only 13.3M, the model maintains high computational efficiency while outperforming mainstream models such as Faster-RCNN and YOLOv5s in recall and precision metrics. These results underscore the robustness and adaptability of the proposed model in complex environments, making it a valuable contribution to traffic sign detection research.

Sriram et al. [50] introduce an advanced traffic sign detection framework leveraging YOLOv9 for autonomous vehicles. The study addresses the critical challenge of real-time and accurate traffic sign detection essential for road safety and navigation. YOLOv9 builds on its predecessors, such as YOLOv8, by incorporating advancements like multi-scale feature fusion and optimized detection heads. The model achieves significant performance metrics: a 95% accuracy rate, a 91% mAP, and a reduced training time of 12 h compared to YOLOv8's 15 h. It outperforms in precision, recall, and F1 score, emphasizing robustness in varying conditions. The enhanced efficiency stems from key parameter adjustments, including a lower learning rate and refined IoU thresholds. A diversified dataset enriched the model's training process, supporting its generalization across complex scenarios. These results position YOLOv9 as a pivotal innovation in autonomous driving systems, balancing computational efficiency with high accuracy.

Zhang et al. [51] propose MH-YOLO, an innovative traffic sign detection method designed for real-time applications in intelligent transportation. The model enhances YOLOv10n by integrating a Hybrid Attention Transformer (HAT) for improved small object reconstruction and a Multi-Scale Dilated Attention (MDA) mechanism for better semantic information capture. These enhancements ensure high detection accuracy while maintaining computational efficiency. The study also introduces the Chinese Traffic Sign Dataset (CTSDB), which addresses regional and nighttime traffic sign detection challenges with 35,654 annotated images. MH-YOLO achieves superior results on both the CTSDB and TT100k datasets, with a notable increase in detection accuracy and recall compared to YOLOv10n, while requiring minimal additional computational resources. This lightweight yet effective model is particularly suited for deployment in resource-constrained environments, advancing the state-of-the-art in traffic sign detection.

Flores-Calero et al. [9] present a comprehensive systematic literature review (SLR) on the application of YOLO object detection algorithms for traffic sign detection and recognition. The study analyzes research from 2016 to 2022, focusing on five key aspects: applications, datasets, metrics, hardware, and challenges. The authors review 115 primary studies, revealing that YOLO is predominantly used for vehicular safety and intelligent autonomous systems. Publicly available datasets, especially those from Germany and China, are widely utilized for CNN-based object detectors. The study highlights the algorithm's effectiveness in achieving high detection, classification, and processing speed metrics across different YOLO versions. Seven major challenges are identified in real-world road conditions, such as variations in lighting, occlusions, and complex environments. The authors categorize research addressing these challenges and propose future directions to improve performance of YOLO-based small object detection systems. The SLR stands as a vital reference for ongoing technological advancements in the field, offering critical insights for improving accuracy, efficiency, and robustness in traffic sign recognition.

Wang et al. [52] introduce MMW-YOLOv5, a multi-scale enhanced traffic sign detection algorithm designed to address the critical requirements of accuracy and real-time performance in autonomous driving. Building on the YOLOv5 framework, the proposed model incorporates several innovative enhancements to enhance recognition of and multi-size transportation-related targets. The algorithm integrates a Multi-Scale Fusion Network (MSFNet) in its neck, enhancing its ability to fuse multi-scale features effectively. This improvement significantly boosts the detection capability for small-scale targets. Additionally, the C3 bottleneck structure in the trunk and neck is replaced with the Multi-Scale Feature Extraction Bottleneck Module (MSFEBM), facilitating richer multi-scale feature extraction and improving target detection across diverse scales. To further refine accuracy and optimize network training, the model employs a Wise-MPDIoU (WMPDIoU) regression function, which accelerates convergence and enhances positioning accuracy. Experimental validation on the TT100K dataset demonstrates the model's efficacy, achieving 87.1% mAP@0.5 and 53.7% mAP@0.5:0.95—improvements of 6.6% and 5.1%, respectively, over the baseline YOLOv5 model. These results underscore MMW-YOLOv5's potential as a robust solution for traffic sign detection, balancing computational efficiency with high detection performance.

8.3 Object Detector

Inherited from its predecessor YOLOv8 [53], the YOLOv9 [54] algorithm is one of the most significant improvements in the YOLO series. Like many real-time object recognition algorithms, YOLOv9 also faces the challenge of information loss in the feedforward stage, which can significantly reduce detection accuracy. Such a loss of critical information can create less accurate connections between target objects and input data, leading to errors in predictions. In response to the limitations discovered in previous generations, the YOLOv9 model includes groundbreaking improvements in network architecture and training approaches with the aim of improving performance and accelerating identification. One of the key innovations of the YOLOv9 model is the inclusion of the Generalized Efficient Layer Aggregation Network (GELAN) [55] module, which is critical in fine-tuning the training process of the network. Moreover, the inclusion of the Programmable Gradient Information (PGI) [55] structure marks a significant improvement that helps the growth of the network architecture to achieve higher performance and efficiency. These advancements demonstrate YOLOv9's commitment to pushing the boundaries of object identification capabilities, providing improved accuracy and faster detection times in real-world applications.

8.3.1 Improved YOLOv9 Algorithm (IQ YOLOv9)

IQ YOLOv9, an improved model based on YOLOv9, is lightweight and its real-time performance is improved by replacing convolutional layers with Q Convolution, which significantly lowers process complexity and the number of parameters. Through dividing every layer into several sub-branches, Q Convolution efficiently reduces the number of parameters of the model. In addition, IQ YOLOv9 adds cross-channel connections to elevate the lineament representation capacity of the algorithm, which increases the stability and accuracy of object detection. The IQ YOLOv9 algorithm, whose network architecture is given in Fig. 8.1, consists of the combination of Conv2D_BN_SiLU and CSPLayer modules in the CSPDarknet block and resblock bodies. This combination is indicated by dashed lines in the figure. In addition, the compatible information transfer of UpSampling2D, Concat, CSPLayer and Conv2D modules is schematized in the FPN block. In the head section, the interpretation of meaningful information is shown on the models. Q Convolution facilitates the exchange of information between channels by establishing cross-channel connections. By facilitating the sharing and merging of information across multiple channels, this coupling mechanism increases the model's ability to identify features. Especially, it is important for missions such as recognition of target objects with fuzzy structures and appearances.

The IQ YOLOv9 model's design specifically enhances its ability to detect rare and partially occluded traffic signs. The use of Q Convolution introduces cross-channel connections that facilitate richer feature extraction across sub-branches, thereby improving the model's capability to retain and integrate subtle or obscured

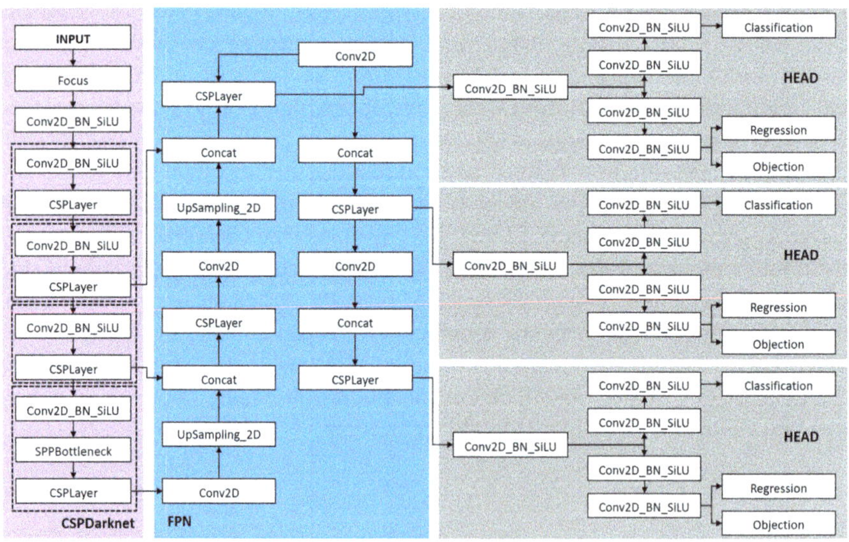

Fig. 8.1 Network architecture of IQ YOLOv9, which is improved based on YOLOv9

features. This is particularly valuable for detecting partially visible signs or those located in cluttered scenes. Additionally, the multi-path flow enabled by the CSPLayer and Conv2D_BN_SiLU modules increases the depth and flexibility of spatial feature aggregation, making the model more resilient to variations in sign visibility and position.

8.3.2 PGI Module

The PGI (Programmable Gradient Information) module enhances YOLOv9 by introducing a smart mechanism to preserve important features that are often lost during deep network operations. It is designed to maintain a high level of detection accuracy without increasing computational load. In essence, the PGI module acts like a safety net that catches and restores useful information lost during processing. It achieves this through three key components: A primary branch that handles the main workload of feature extraction; an auxiliary reversible branch that ensures no critical information is permanently lost; and a multi-level auxiliary information unit that gathers insights from different layers to support better detection of small or complex objects. This collaborative structure balances speed and accuracy, making PGI an ideal module for real-time object detection tasks such as traffic sign recognition.

The PGI module represents a key innovation in the YOLOv9 architecture, addressing critical challenges in real-time object detection while significantly enhancing the algorithm's architectural efficiency. A prominent issue with existing real-time object detection methods is their reliance on excessively deep neural network designs to achieve high levels of analysis and accuracy. While such designs may provide comprehensive feature extraction capabilities, they often come at the expense of computational efficiency and increased latency, making them less suitable for real-time applications. The PGI module effectively tackles this challenge by employing a more streamlined network design that balances computational complexity with detection performance. By leveraging lightweight architectures, the PGI approach minimizes information loss while maintaining or even improving the model's accuracy. This balance is achieved through an innovative network structure consisting of three main components: the primary branch, the auxiliary reversible branch, and the multi-level auxiliary information. Each of these components contributes uniquely to the PGI module's ability to optimize the YOLOv9 algorithm's performance.

The primary branch forms the backbone of the PGI network, handling the majority of the computational workload. It is designed to extract features efficiently without introducing additional computational overhead. Unlike traditional deep neural networks that often add redundant layers, the primary branch focuses on efficient feature propagation and extraction, ensuring that the model retains critical information while avoiding unnecessary complexity. This streamlined approach not only reduces the computational burden but also accelerates inference, which is vital for real-time object detection. One of the major challenges with deep neural networks is the potential loss of critical information as data passes through multiple layers. To address this, the PGI network incorporates an auxiliary reversible branch that mitigates information loss by introducing reversible computations. This branch effectively captures and restores information that might otherwise be lost due to the transformations performed in the primary branch. By integrating this component into the overall framework, the PGI module ensures that key features are preserved throughout the inference process, improving both accuracy and robustness.

To further improve performance, the PGI module includes a multi-level auxiliary information mechanism. This component aggregates feature maps from various levels of the network, enabling the model to leverage information from multiple scales and contexts. By combining features from different levels, the PGI network enhances its ability to detect objects of varying sizes and complexities, particularly in challenging scenarios such as cluttered environments or low-resolution images. Additionally, this mechanism offsets the accumulation of errors that can arise during the feature extraction process, further strengthening the model's reliability. The collaborative integration of the primary branch, auxiliary reversible branch, and multi-level auxiliary information enables the PGI network to optimize inference procedures while significantly improving the robustness and efficiency of the YOLOv9 algorithm. By maintaining a streamlined architecture, the PGI module reduces computational costs, making YOLOv9 more suitable for real-time applications. At the same time, the module's ability to preserve and enhance feature quality ensures that the model delivers high detection accuracy without sacrificing speed.

8.3.3 GELAN Module

The GELAN module, a key component of the YOLOv9 network design, is an improved version of the ELAN module used in YOLOv7. The GELAN module extends the RepConv computational blocks and CSPNet blocks from YOLOv5 that were included in the ELAN module of YOLOv7 by combining elements from the ELAN and CSPNet modules. Slope path planning is emphasized in the architecture of the ELAN and CSPNet components of GELAN. Interestingly, the ELAN design stands out due to its unique use of convolutional layer stacking, a feature that has been expanded and improved to support a variety of computational blocks in the GELAN design. This adaptability allows the GELAN module to work with a variety of architectural setups, producing a model that is not only faster and lighter, but also more accurate. The improved performance of the YOLOv9 model is supported by a highly optimized framework that is the result of combining these design ideas.

Fig. 8.2 Traffic sign examples from the dataset

8.3.4 Dataset

The sample images given in Fig. 8.2 are some of the traffic sign images used in this study. The dataset prepared for this study has a structure suitable for deep learning-based computer vision applications such as object detection and includes various traffic signs. The traffic signs in the images have a wide range from pedestrian crossing signs indicating the priority of passage for pedestrians to parking ban and motor vehicle entry ban signs. In addition, there are signs indicating traffic regulations and warnings such as curve warning, overtaking ban, turn ban, roundabout and sharp curve. This diversity shows how comprehensive a source the dataset provides in terms of recognizing and classifying different types of traffic signs.

The characteristics that a traffic sign dataset should have for object detection studies are also important. In object detection applications, each sign must be correctly labeled, classified and its location in a given image must be correctly determined. In such datasets, labels not only determine the meaning of the signs, but also the coordinates, dimensions and environmental context of each sign must be provided. Labeling of signs should include classification and localization tasks together. In addition, the dataset should have a wide variety; this should include images taken in different weather conditions, at different times of the day, under different weather conditions and under different environmental factors. Various backgrounds, different camera angles and distances are important to increase the overall performance of object detection algorithms, because real-world conditions can always vary. In addition, each sign in the dataset should be represented from different angles, distances and resolutions. This variety increases the generalization ability of the model and allows the model to work more reliably under different conditions.

The dataset used in this study was curated with attention to diversity and realism to ensure strong generalization performance for traffic sign detection tasks. It includes images of varying quality, ranging from high-resolution images to moderately compressed web-sourced ones, reflecting real-world variance in camera performance and image clarity. The dataset incorporates a range of environmental conditions, including overcast skies, light rain, and night scenes, although the majority of images are taken in daylight. These diverse lighting and weather conditions contribute to the robustness of the model against environmental variability. The dataset also includes images from multiple road types, such as urban streets, residential zones, and highways, with highway traffic signs clearly represented (as shown in Fig. 8.2). Regarding sign types, the dataset comprises a balanced mix of Cautionary, Mandatory, and Guidance signs, but class representation is not perfectly uniform, reflecting their natural occurrence rates in real environments. While efforts were made to ensure diverse coverage across all categories, some classes with naturally low visibility in public image repositories are slightly underrepresented. Nonetheless, this realistic class imbalance mirrors practical deployment scenarios and allows the model to adapt effectively to real-world data distributions.

Most of the traffic signs in the dataset belong to Hungarian road systems, and this was an intentional choice. While several public datasets are available (e.g., GTSRB, TT100K), they predominantly feature Chinese traffic signs, which differ in design, language, and color scheme from those used in Hungary. This makes it difficult to train models that generalize well to Hungarian road environments using those datasets. The motivation behind building a new dataset was to fill this gap and create a collection that better reflects the visual and semantic characteristics of traffic signs commonly found on Hungarian roads. This increases detection performance in local ITS applications, and highlights the importance of regional datasets when deploying computer vision systems in real-world environments.

The majority of the images in the dataset prepared for this study were created by recording images related to traffic signs from street images with Google Earth. In order to increase the scope of the dataset and to create a dataset with the desired features, images recorded from web-based sources were also added to the dataset. The dataset, which was expanded with various filtering and rotation operations, contains a total of 2875 images. To ensure robustness in real-world conditions, the dataset includes a variety of rare and partially occluded traffic signs. These images were deliberately selected or augmented to reflect realistic scenarios where signs may be obscured by other objects (e.g., trees, poles, vehicles) or appear in unusual forms or conditions. This diversity allows the model to learn from incomplete or uncommon visual cues, improving its generalization to rarely seen or partially blocked signs in real-time environments.

8.4 Experimental Details

The computer environment used for object detection applications in this study consists of high-performance hardware and software components, as given in Table 8.1. The graphics processing unit (GPU) used is the NVIDIA GeForce GTX 1650 model, and is effectively used in the training and inference of deep learning

Table 8.1 Characteristics of the computer environment used for the tests

Software/Hardware	Feature
GPU	NVIDIA GeForce GTX 1650
CPU	Intel Core i7
RAM	16 GB DDR4
Storage	512 GB SSD + 2 TB HDD
Operating System	Mac OS
Deep Learning Framework	PyTorch
Python Version	3.10.8
CUDA	11.6

Table 8.2 Model parameters used in training the algorithm

Parameter	Value
Learning rate	0.001
Batch size	32
Weight decay	0.005
Epoch	250
Layer	225
Parameters	10,136,528

models thanks to its parallel processing capacity. Intel Core i7 was preferred as the central processing unit (CPU); this processor provides strong performance in large data processing and model training processes with its multi-core structure. The system is equipped with 16 GB DDR4 RAM, so applications requiring high memory capacity can be performed without any problems. A large capacity is provided by using 512 GB SSD and additional 2 TB HDD units for data storage.

In the software environment, Mac OS was used as the operating system. PyTorch was preferred as the deep learning framework; this framework facilitates research and development processes with its flexible and user-friendly structure. The 3.10.8 version of the Python programming language was used, which provides the integration of PyTorch and other necessary libraries. In addition, with the CUDA 11.6 version, GPU acceleration operations are supported, making it possible to run deep learning models faster. This powerful hardware and software combination aims to achieve high efficiency and accuracy in object detection studies.

The hyperparameters used during the training of the IQ YOLOv9 and original YOLOv9 models were carefully designed to optimize the performance of the model and increase the accuracy in object detection tasks. These parameters, given in Table 8.2, are the basic elements that directly affect both the learning process and the generalization ability of the model. During the training process, the learning rate was determined as 0.001. This value controls how much the weights of the model will be changed in each backpropagation step. A low learning rate allows the model to be optimized more precisely by progressing in smaller steps. However, a very low learning rate can cause the training time to increase. The determined level of 0.001 is a choice that will balance stability and training time. Batch size refers to the size of the data subset given to the model during each training iteration, and is determined as 32 here. This value allows training to be performed more efficiently on large data sets. While smaller batch sizes reduce memory consumption, larger batch sizes may negatively affect the generalization ability of the model. The selected size of 32 aims to establish a balance between efficiency and performance.

The weight reduction parameter is a regularization method used to prevent the model's weights from growing excessively and to prevent over-learning. The weight reduction coefficient, determined as 0.005 in this study, aims to increase the generalization ability of the model. This regularization prevents the weights from reaching unnecessarily large values during training, allowing the model to learn more balancedly. The training process was carried out for a total of 250 epochs. An epoch means that the entire training dataset is passed through the model once. 250 epochs are optimized to provide sufficient time for the model to learn complex relationships while minimizing the risk of over-learning. The IQ YOLOv9 and original YOLOv9 models consist of a total of 225 layers. These layers allow the model to infer key components from the system's input data. This model, which is a deep architecture, is designed to capture complex patterns and relationships in object detection tasks. The model has a total of 10,136,528 parameters, indicating that it has a wide expressive capacity. Such a large number of parameters allows the model to identify various objects from different features. However, this also requires careful regularization and optimization; otherwise, over-learning may occur.

The choice of hyperparameters for training both the original YOLOv9 and the improved IQ YOLOv9 models was guided by a combination of prior research, empirical best practices, and targeted tuning. Initially, widely adopted values in the object detection literature were selected: a learning rate of 0.001, batch size of 32, weight decay of 0.005, and 250 training epochs. These values provided a reasonable balance between training speed, convergence, and generalization capability. To further refine model performance, a limited grid search was conducted on a subset of the dataset, varying key hyperparameters such as learning rate (tested values: 0.0005, 0.001, 0.005) and batch size (tested values: 16, 32, 64). The configuration yielding the best validation performance, measured by mAP and F1 score, was selected for final training. The learning rate of 0.001 and batch size of 32 demonstrated optimal performance without overfitting or compromising inference speed, making them a robust choice for our real-time traffic sign detection objectives.

8.5 Test Results

When addressing a critical computer vision problem such as traffic sign detection, the primary way to evaluate model performance is to compare models using various metrics. In this study, IQ YOLOv9 and YOLOv9 models are compared using precision, recall, mAP, F1 score, and inference time metrics. These metrics are crucial to evaluate the accuracy, error tolerance, overall performance, and speed of the models in real-time applications. The performance differences in each metric are examined in detail to determine the suitability of these models for specific

Table 8.3 Performance comparison of the original YOLOv9 and the improved IQ YOLOv9 on traffic sign detection

	Algorithm	
Metric	YOLOv9	IQ YOLOv9
Precision	81.5	84.2
Recall	76.2	79.5
mAP	79.3	82.6
F1 score	78.7	81.8
Inference time	10.35 ms	8.22 ms

applications, and the values are given in Table 8.3. Precision measures how many of the examples a model predicts as positive are actually correct. While IQ YOLOv9 has a precision value of 84.2, YOLOv9 has this metric at 81.5. This difference shows that IQ YOLOv9 filters out false positive predictions better than YOLOv9. Mathematically speaking, IQ YOLOv9's precision value is approximately 2.7% higher than YOLOv9. This result makes IQ YOLOv9 a more attractive option, especially for applications where the cost of false positives is high. Recall indicates how many positive examples the model can correctly detect. IQ YOLOv9's recall value is 79.5, while YOLOv9's is 76.2. Although the difference between the two models is significant, it can be said that IQ YOLOv9 is slightly better in terms of sensitivity. This difference shows that IQ YOLOv9 can correctly detect more positive examples. However, it should be noted that this 3.3% difference may provide a significant advantage in some specific applications.

For instance, in high-speed autonomous driving scenarios, even a marginal improvement in recall can help the system detect additional relevant traffic signs within a short time window, which can be critical for making timely and accurate decisions. Similarly, in safety-critical environments, such as near construction zones, school zones, or accident-prone intersections, failing to detect a sign could lead to severe consequences, making recall improvements highly valuable. On the other hand, in applications focused more on comprehensive coverage of a diverse range of sign classes, such as large-scale traffic monitoring systems across urban and rural areas, the original YOLOv9's slightly higher mAP might offer better consistency across varied sign types. Therefore, the choice between models should consider the specific operational priorities of the ITS application.

Another important metric that measures overall accuracy performance, mAP, reveals different results between IQ YOLOv9 and YOLOv9. YOLOv9 outperformed IQ YOLOv9 by approximately 3.3% with a mAP value of 82.6 compared to IQ YOLOv9's 79.3. This difference shows that YOLOv9's predictions are one step ahead in terms of overall accuracy. Since the mAP metric is an average of the accuracies across all classes, this advantage of YOLOv9 indicates that the model can produce more consistent results across a wide range of classes. The F1 score is calculated as the harmonic mean of the precision and recall values and is used to evaluate whether a model performs in a balanced manner. IQ YOLOv9's F1 score is 81.8, while YOLOv9's is 78.7. Mathematically, IQ YOLOv9's F1 score is approximately 3.1% higher, indicating that this model strikes a better balance between precision and recall. This means that IQ YOLOv9 may be a more advantageous option, especially in applications where balanced performance is critical.

Finally, inference time indicates how long it takes for a model to process an image. IQ YOLOv9 is approximately 20% faster with an inference time of 8.22 ms compared to YOLOv9's 10.35 ms. This makes IQ YOLOv9 a more suitable option for time-sensitive applications where real-time capability is a goal. In time-critical applications such as traffic sign detection, this speed difference offers a great advantage in terms of operational efficiency. When we compare the two models mathematically, it is seen that IQ YOLOv9 is superior in precision, recall, F1 score and inference time metrics, but YOLOv9 is slightly better in terms of mAP metric. IQ YOLOv9's superiority in precision and inference time provides a significant advantage in situations where false positive predictions need to be reduced and processing speed needs to be increased. In contrast, the higher mAP value of YOLOv9 indicates that the model is a stronger alternative in terms of overall accuracy. These results suggest that IQ YOLOv9 is more suitable for real-time applications that focus on speed and accuracy, but YOLOv9 performs slightly better in terms of overall accuracy. In a domain such as traffic sign detection, the speed advantage provided by IQ YOLOv9 is an important factor that increases the operational applicability of the model in the real world. Although the inference time per image is approximately 8.22 ms, indicating strong computational efficiency, the dataset used in this study consists solely of still images. As such, the claim of real-time applicability is based on frame-level inference timing rather than continuous video stream evaluation. While these results suggest that the model could meet the processing requirements of real-time systems, further experiments using video data or real-time input streams would be necessary to fully validate end-to-end real-time performance.

The comparative analysis reveals that IQ YOLOv9 outperforms YOLOv9 in precision, recall, F1 score, and inference time, while YOLOv9 excels in mAP. The higher precision and recall of IQ YOLOv9 demonstrate its ability to minimize errors and enhance detection rates, while its faster inference time ensures real-time applicability. These advantages make IQ YOLOv9 a compelling choice for scenarios where speed and accuracy are prioritized, such as dynamic traffic environments and autonomous vehicle systems. On the other hand, YOLOv9's superior mAP indicates its capability to deliver consistent results across diverse classes, which may be advantageous in scenarios requiring broad applicability. The choice between these models ultimately depends on specific application requirements. For example, in a densely populated urban setting with a high density of traffic signs, IQ YOLOv9's faster inference time and higher precision may be critical. Conversely, in a rural or suburban environment with a wider variety of sign classes, YOLOv9's higher mAP might offer better overall performance.

The results highlight the practical implications of selecting an appropriate model for traffic sign detection. IQ YOLOv9's superior performance in key metrics such as precision, recall, and inference time positions it as an optimal choice for real-time systems where speed and reliability are paramount. Meanwhile, YOLOv9's higher mAP underscores its suitability for applications demanding consistent accuracy

Table 8.4 Performance comparison of the IQ YOLOv9 and YOLOv9 algorithms in detecting traffic signs across three categories: Cautionary, Mandatory, and Guidance

		Performance metric		
Sign category	Model	Precision	Recall	mAP
Cautionary	YOLOv9	0.762	0.748	0.750
	IQ YOLOv9	0.860	0.822	0.831
Mandatory	YOLOv9	0.828	0.713	0.788
	IQ YOLOv9	0.840	0.802	0.794
Guidance	YOLOv9	0.715	0.745	0.727
	IQ YOLOv9	0.780	0.803	0.796

across diverse datasets. These insights are invaluable for developers and researchers working on ITS, enabling them to tailor solutions to specific operational needs. Conclusively, both models exhibit high performance, and their suitability varies depending on the application context. The findings provide a robust foundation for future work in optimizing traffic sign detection systems, paving the way for safer and more efficient transportation networks.

Another test focused on identifying traffic signs by categorizing them into distinct subcategories. The results in Table 8.4 demonstrate that the IQ YOLOv9 algorithm outperforms the YOLOv9 algorithm, across all three traffic sign categories, Cautionary, Mandatory, and Guidance, in terms of precision, recall, and mAP metrics. For Cautionary signs, the IQ YOLOv9 algorithm achieved notable improvements, with precision increasing from 0.762 to 0.860, recall from 0.748 to 0.822, and mAP from 0.750 to 0.831. This indicates a higher ability of the IQ YOLOv9 algorithm to correctly identify relevant instances while minimizing false positives and negatives. Similarly, for Mandatory signs, the IQ YOLOv9 algorithm significantly improved recall, rising from 0.713 to 0.802, while also enhancing precision (from 0.828 to 0.840) and mAP (from 0.788 to 0.794). These enhancements suggest that the IQ YOLOv9 algorithm is better equipped to handle the distinct challenges posed by mandatory signs, which often require robust detection due to their critical regulatory nature. For Guidance signs, the IQ YOLOv9 algorithm showed a more modest yet consistent improvement, with precision increasing from 0.715 to 0.780, recall from 0.745 to 0.803, and mAP from 0.727 to 0.796. While the relative performance gain in this category is smaller compared to the others, it underscores the IQ YOLOv9 algorithm's capability to detect signs that often feature less distinct visual characteristics or may be impacted by environmental factors such as occlusions or variations in lighting. The overall advancements across all categories reflect the effectiveness of the IQ YOLOv9 algorithm in enhancing both the accuracy and reliability of traffic sign detection.

To further contextualize the performance of the proposed IQ YOLOv9 model, a comparative analysis with other YOLO-based traffic sign detection algorithms discussed in the Related Work section is presented in Table 8.5. The compared models include MI-YOLO, YOLO-TS, and DPF-YOLOv8, which represent recent efforts to enhance detection accuracy and efficiency in complex environments. Table 8.5 provides a summary of key performance metrics including mAP@0.5, F1 Score, model size (parameter count), and inference time (if reported). While these results

Table 8.5 Comparative performance of the proposed IQ YOLOv9 model against other recent YOLO-based traffic sign detection algorithms

Model	mAP@0.5	Precision	Recall	F1-score	Params (M)	Inf. time (ms)	Dataset
IQ YOLOv9	82.6	84.2	79.5	81.8	10.14	8.22	Custom
MI-YOLO [46]	83.1	–	–	–	–	–	TT100K
YOLO-TS [47]	80.3	–	–	80.2	4.62	–	CCTSD8
DPF-YOLOv8 [48]	81.4	–	–	–	–	–	TT100K

are drawn from the respective authors' evaluations, most commonly conducted on datasets such as TT100K or CCTSDB, they offer valuable insights into the comparative standing of IQ YOLOv9. The results show that IQ YOLOv9 achieves competitive accuracy compared to these state-of-the-art models while offering a notable advantage in inference time and maintaining a relatively moderate parameter count. While MI-YOLO slightly outperforms IQ YOLOv9 in terms of mAP@0.5, our model delivers better latency and a robust F1 score, making it highly suitable for real-time applications, especially in environments with limited computational resources. This comparison highlights the practical balance that IQ YOLOv9 strikes between detection accuracy and computational efficiency, reinforcing its value in latency-sensitive systems like autonomous vehicles and intelligent transportation infrastructure.

8.6 Conclusion

This study addressed the critical task of traffic sign detection, an essential component of ITS and autonomous driving applications. A comprehensive dataset was prepared, combining images obtained from Google Earth and various web sources to ensure diversity and robustness. The study's primary objective was to evaluate the performance of the improved IQ YOLOv9 model, developed as an enhancement over the YOLOv9 model, and to provide a comparative analysis of the two models using well-established performance metrics. To thoroughly assess the effectiveness of both models, precision, recall, mAP, F1 score, and inference time were utilized as performance indicators. These metrics offered insights into the models' abilities to accurately detect traffic signs, minimize false positives, balance precision and recall, and process data efficiently for real-time applications. The evaluation revealed that the IQ YOLOv9 model demonstrated a precision of 84.2, recall of 79.5, mAP of 82.6, and F1 score of 81.8. In contrast, the original YOLOv9 model achieved slightly lower scores: 81.5 precision, 76.2 recall, 79.3 mAP, and 78.7 F1 score.

One of the most notable findings of this study was the superior inference time of the IQ YOLOv9 model, which processed images approximately 20% faster than the original YOLOv9 model. With an inference time of 8.22 ms compared to YOLOv9's 10.35 ms, the IQ YOLOv9 model emerged as a more viable candidate for real-time applications where speed and efficiency are paramount. This improvement underscores the potential of the IQ YOLOv9 model in domains where rapid decision-making and low latency are critical, such as traffic monitoring, autonomous vehicles, and ITS. The analysis further highlighted the nuanced strengths of each model. The IQ YOLOv9 model excelled in precision, recall, and F1 score, indicating its ability to effectively filter out false positives, detect positive examples, and maintain a balanced performance. These attributes make it particularly suitable for applications where minimizing errors is essential, such as safety-critical environments. On the other hand, the slightly higher mAP of the YOLOv9 model suggests that it may offer more consistent detection across a broader range of traffic sign classes, which could be advantageous in certain contexts.

These findings carry significant implications for the field of traffic sign detection and ITS. The IQ YOLOv9 model's performance improvements in key metrics and its enhanced speed demonstrate its practical value in addressing real-world challenges. The ability to accurately and efficiently detect traffic signs is critical for ensuring road safety, optimizing traffic flow, and supporting the advancement of autonomous driving technologies. By refining the YOLOv9 architecture to create the IQ YOLOv9 model, this study has contributed to the state-of-the-art in traffic sign detection and demonstrated the potential for further innovations in computational intelligence and computer vision. Future research could build on this work by exploring additional enhancements to the IQ YOLOv9 model, such as integrating advanced attention mechanisms, optimizing the architecture for even faster inference times, or adapting the model for deployment on edge devices. Furthermore, extending the dataset to include more diverse traffic sign types and testing the model in varied real-world conditions, such as adverse weather or nighttime scenarios, would provide a more comprehensive evaluation of its robustness and generalizability. Conclusively, this study underscores the transformative potential of advanced deep learning models such as IQ YOLOv9 in addressing critical challenges in ITS. By achieving superior performance in precision, recall, F1 score, and inference time, the IQ YOLOv9 model represents a significant step forward in the quest for safer and more efficient transportation systems.

While the IQ YOLOv9 model achieves a performance improvement of approximately 3% in key metrics such as precision, recall, and F1 score over the original YOLOv9, this margin is practically significant in ITS contexts. Traffic sign detection systems operate in real-time across vast and dynamic environments, often processing thousands of frames per minute. In such scenarios, even a marginal increase in accuracy can lead to a meaningful reduction in misclassifications, false alarms, or missed detections, which directly impacts vehicle decision-making and traffic safety. This is particularly relevant in high-density urban areas or complex intersections where detection reliability is critical. Moreover, this improvement is achieved while simultaneously reducing inference time by nearly 20%, further reinforcing the model's real-time applicability. Nevertheless, we acknowledge that broader adoption of the model in production-level ITS deployments will benefit from additional fine-tuning. Future work will include further testing under varied and extreme conditions, such as nighttime or adverse weather, and potential integration with ensemble methods or hybrid attention mechanisms to push performance boundaries even further.

References

1. de la Escalera, A., Moreno, L.E., Salichs, M.A., Armingol, J.M.: Road traffic sign detection and classification. IEEE Trans. Ind. Electron. **44**(6), 848–859 (1997). https://doi.org/10.1109/41.649946
2. Saadna, Y., Behloul, A.: An overview of traffic sign detection and classification methods. Int. J. Multimed. Inf. Retr. **6**(3), 193–210 (2017). https://doi.org/10.1007/s13735-017-0129-8
3. Houben, S., Stallkamp, J., Salmen, J., Schlipsing, M., Igel, C.: Detection of traffic signs in real-world images: the German traffic sign detection benchmark. In: Proceedings of the 2013 International Joint Conference on Neural Networks, pp. 1–8 (2013). https://doi.org/10.1109/IJCNN.2013.6706807
4. Wali, S.B., Abdullah, M.A., Hannan, M.A., Hussain, A., Samad, S.A., Ker, P.J., Mansor, M.B.: Vision-based traffic sign detection and recognition systems: current trends and challenges. Sensors. **19**(9) (2019). https://doi.org/10.3390/s19092093
5. Bakirci, M.: Enhancing vehicle detection in intelligent transportation systems via autonomous UAV platform and YOLOv8 integration. Appl. Soft Comput. **164**(112015) (2024). https://doi.org/10.1016/j.asoc.2024.112015
6. Borji, A., Cheng, M.M., Hou, Q., et al.: Salient object detection: a survey. Comput. Vis. Media. **5**(2), 117–150 (2019). https://doi.org/10.1007/s41095-019-0149-9
7. Zhao, Z.Q., Zheng, P., Xu, S.T., Wu, X.: Object detection with deep learning: a review. IEEE Trans. Neural Networks Learn. Syst. **30**(11), 3212–3232 (2019). https://doi.org/10.1109/TNNLS.2018.2876865
8. Bakirci, M.: Evaluating the impact of unmanned aerial vehicles (UAVs) on air quality management in smart cities: a comprehensive analysis of transportation-related pollution. Comput. Electr. Eng. **119**(109556) (2024). https://doi.org/10.1016/j.compeleceng.2024.109556
9. Flores-Calero, M., Astudillo, C.A., Guevara, D., et al.: Traffic sign detection and recognition using YOLO object detection algorithm: a systematic review. Mathematics. **12**(2) (2024). https://doi.org/10.3390/math12020297
10. Han, C., Gao, G., Zhang, Y.: Real-time small traffic sign detection with revised faster-RCNN. Multimed. Tools Appl. **78**, 13263–13278 (2019). https://doi.org/10.1007/s11042-018-6428-0
11. Dimitrakopoulos, G., Demestichas, P.: Intelligent transportation systems. IEEE Veh. Technol. Mag. **5**(1), 77–84 (2010). https://doi.org/10.1109/MVT.2009.935537
12. Ellahyani, A., El Jaafari, I., Charfi, S.: Traffic sign detection for intelligent transportation systems: a survey. In: E3S Web of Conferences, p. 01006 (2021). https://doi.org/10.1051/e3sconf/202122901006
13. Wang, J., Chen, Y., Dong, Z., et al.: Improved YOLOv5 network for real-time multi-scale traffic sign detection. Neural Comput. Applic. **35**, 7853–7865 (2023). https://doi.org/10.1007/s00521-022-08077-5
14. Liu, C., Li, S., Chang, F., Wang, Y.: Machine vision based traffic sign detection methods: review, analyses and perspectives. IEEE Access. **7**, 86578–86596 (2019). https://doi.org/10.3390/su16010171
15. Malligere Shivanna, V., Guo, J.: Object detection, recognition, and tracking algorithms for ADASs—a study on recent trends. Sensors. **24**(1) (2024). https://doi.org/10.3390/s24010249
16. Chellapandi, V.P., Yuan, L., Brinton, C.G., Żak, S.H., Wang, Z.: Federated learning for connected and automated vehicles: a survey of existing approaches and challenges. IEEE Trans. Intell. Veh. **9**(1), 119–137 (2024). https://doi.org/10.1109/TIV.2023.3332675

17. Lu, M., Wevers, K., van der Heijden, R., Heijer, T.: ADAS applications for improving traffic safety. In: Proceedings of the 2004 IEEE International Conference on Systems, Man, and Cybernetics, pp. 3995–4002 (2004). https://doi.org/10.1109/ic-ETITE58242.2024.10493352
18. Bakirci, M., Bayraktar, I.: Transforming aircraft detection through LEO satellite imagery and YOLOv9 for improved aviation safety. In: 2024 26th International Conference on Digital Signal Processing and its Applications (DSPA), pp. 1–6. IEEE (2024). https://doi.org/10.1109/DSPA60853.2024.10510106
19. Maletzky, A., Hofer, N., Thumfart, S., Bruckmüller, K., Kasper, J.: Traffic sign detection and classification on the austrian highway traffic sign data set. Data. **8**(16) (2023). https://doi.org/10.3390/data8010016
20. Prakash, A.J., Sruthy, S.: Enhancing traffic sign recognition (TSR) by classifying deep learning models to promote road safety. SIViP. **18**, 4713–4729 (2024). https://doi.org/10.1007/s11760-024-03108-1
21. Bakirci, M., Toptas, B.: Kinematics and autoregressive model analysis of a differential drive mobile robot. In: 2022 International Congress on Human-Computer Interaction, Optimization and Robotic Applications (HORA), pp. 1–6. IEEE (2022). https://doi.org/10.1109/HORA55278.2022.9800071
22. Cong, S., Zhou, Y.: A review of convolutional neural network architectures and their optimizations. Artif. Intell. Rev. **56**, 1905–1969 (2023). https://doi.org/10.1007/s10462-022-10213-5
23. Upadhyay, S.K., Kumar, A.: A novel approach for rice plant diseases classification with deep convolutional neural network. Int. J. Inf. Tecnol. **14**, 185–199 (2022). https://doi.org/10.1007/s41870-021-00817-5
24. Archana, R., Jeevaraj, P.S.E.: Deep learning models for digital image processing: a review. Artif. Intell. Rev. **57**, 11 (2024). https://doi.org/10.1007/s10462-023-10631-z
25. Bakirci, M., Dmytrovych, P., Bayraktar, I., Anatoliyovych, O.: Multi-class vehicle detection and classification with YOLO11 on UAV-captured aerial imagery. In: 2024 IEEE 7th International Conference on Actual Problems of Unmanned Aerial Vehicles Development (APUAVD), pp. 191–196. IEEE (2024). https://doi.org/10.1109/APUAVD64488.2024.10765862
26. Gao, M., Yu, J., Yang, Z., Zhao, J.: A physics-guided graph convolution neural network for optimal power flow. IEEE Trans. Power Syst. **39**(1), 380–390 (2024). https://doi.org/10.1109/TPWRS.2024.3523871
27. Niu, C., Li, K.: Traffic light detection and recognition method based on YOLOv5s and AlexNet. Appl. Sci. **12**(21) (2022). https://doi.org/10.3390/app122110808
28. Yu, J., Ye, X., Tu, Q.: Traffic sign detection and recognition in multiimages using a fusion model with YOLO and VGG network. IEEE Trans. Intell. Transp. Syst. **23**(9), 16632–16642 (2022). https://doi.org/10.1109/TITS.2022.3170354
29. Cui, Y., Guo, D., Yuan, H., et al.: Enhanced YOLO network for improving the efficiency of traffic sign detection. Appl. Sci. **14**(2) (2024). https://doi.org/10.3390/app14020555
30. Wang, Y., Wang, Q.: Robust stacking ensemble model for traffic sign detection and recognition. IEEE Access. **12**, 178941–178950 (2024). https://doi.org/10.1109/ACCESS.2024.3504827
31. Bakirci, M.: Vehicular mobility monitoring using remote sensing and deep learning on a UAV-based mobile computing platform. Measurement. **244**(116579) (2025). https://doi.org/10.1016/j.measurement.2024.116579
32. Du, S., Lv, Y., Chen, Y.: Research on deep learning based traffic sign detection methods. In: 2024 2nd International Conference on Signal Processing and Intelligent Computing (SPIC), pp. 906–911 (2024). https://doi.org/10.1109/SPIC62469.2024.10691527
33. Mahadshetti, R., Kim, J., Um, T.W.: Sign-YOLO: traffic sign detection using attention-based YOLOv7. IEEE Access. **12**, 132689–132700 (2024). https://doi.org/10.1109/ACCESS.2024.3417023
34. Bakirci, M., Bayraktar, I.: Comparative performance of YOLOv9 and YOLOv10 for vehicle detection towards real-time traffic surveillance with UAVs. In: 2024 21st International Conference on Electrical Engineering, Computing Science and Automatic Control (CCE), pp. 1–6. IEEE (2024). https://doi.org/10.1109/CCE62852.2024.10771048

35. Huang, M., Wan, Y., Gao, Z., et al.: Real-time traffic sign detection model based on multi-branch convolutional reparameterization. J. Real-Time Image Proc. **20**, 57 (2023). https://doi.org/10.1007/s11554-023-01307-6
36. Bakirci, M., Bayraktar, I.: AI-driven micro defect detection for aerospace-grade metal surface quality control in smart manufacturing. In: 2025 International Russian Smart Industry Conference (SmartIndustryCon), pp. 115–120. IEEE (2025). https://doi.org/10.1109/SmartIndustryCon65166.2025.10986084
37. Suresha, R., Manohar, N., Ajay Kumar, G., Rohit Singh, M.: Recent advancement in small traffic sign detection: approaches and dataset. IEEE Access. **12**, 192840–192859 (2024). https://doi.org/10.1109/ACCESS.2024.3514692
38. Gao, X., Chen, L., Wang, K., et al.: Improved traffic sign detection algorithm based on Faster R-CNN. Appl. Sci. **12**(18) (2022). https://doi.org/10.3390/app12188948
39. Bakirci, M., Demiray, A.: Autonomous navigation of service robots in complex industrial environments with SAR-based vision and advanced detection for Industry 4.0. In: 2025 International Russian Smart Industry Conference (SmartIndustryCon), pp. 71–76. IEEE (2025). https://doi.org/10.1109/SmartIndustryCon65166.2025.10986215
40. Oza, R.M., Geisen, A., Wang, T.: Traffic sign detection and recognition using deep learning. In: 2021 4th International Conference on Artificial Intelligence for Industries (AI4I), pp. 16–20 (2021). https://doi.org/10.1109/AI4I51902.2021.00012
41. Sung, N.-J., Ma, J., Hor, K., Kim, T., Va, H., Choi, Y.-J., Hong, M.: Real-time physics simulation method for XR application. Computers. **14**(1) (2025). https://doi.org/10.3390/computers14010017
42. Nazeri, A., et al.: Exploration of TPU architectures for the optimized transformer in drainage crossing detection. In: 2024 IEEE International Conference on Big Data (BigData), pp. 4178–4187 (2024). https://doi.org/10.1109/BigData62323.2024.10826077
43. Sukumar, N., Sumathi, P.: Integration of 2-D sliding DFT kernels and CNN for edge feature extraction and lane detection. IEEE Sens. Lett. (2025). https://doi.org/10.1109/LSENS.2025.3529475
44. Ni, B., Gao, H.: A deep learning approach to the inverse problem of modulus identification in elasticity. MRS Bull. **46**, 19–25 (2021). https://doi.org/10.1557/s43577-020-00006-y
45. Jiménez García, C., Porres de la Haza, M.J., Coll Aliaga, E., Lerma-Arce, V., Lorenzo-Sáez, E.: Methodology for measuring mobility emissions with high spatial resolution: case study in Valencia, Spain. Appl. Sci. **15**(2) (2025). https://doi.org/10.3390/app15020669
46. Wang, S., Xu, Y.: MI-YOLO: an improved traffic sign detection algorithm based on YOLOv8. Eng. Lett. **32**(12) (2024)
47. Liu, Y., Luo, P.: YOLO-TS: a lightweight YOLO model for traffic sign detection. IEEE Access. **12**, 169013–169023 (2024). https://doi.org/10.1109/ACCESS.2024.3498057
48. Zhang, Y., Liu, H., Dong, D., Duan, X., Lin, F., Liu, Z.: DPF-YOLOv8: dual path feature fusion network for traffic sign detection in hazy weather. Electronics. **13**(20) (2024). https://doi.org/10.3390/electronics13204016
49. Chen, Z., Liu, L., Chen, Z., Jin, X.: A traffic sign detection model based on improved YOLOv8. In: 2024 5th International Conference on Big Data & Artificial Intelligence & Software Engineering (ICBASE), pp. 672–676 (2024). https://doi.org/10.1109/ICBASE63199.2024.10762415
50. Sriram, N., Jayalakshmi, V., Preethi, P., Shoba, B., Shenbagavalli, K.: Navigating the future with YOLOv9 for advanced traffic sign recognition in autonomous vehicles. Int. J. Comput. Exp. Sci. Eng. **10**(4) (2024). https://doi.org/10.22399/ijcesen.658
51. Zhang, Y., Ma, H., Zhang, C., Wang, Z., Li, Z.: MH-YOLO: a lightweight traffic sign detection method based on YOLOv10n with hybrid attention transformer and multi-scale dilated attention. Research Square. (2025) doi:https://doi.org/10.21203/rs.3.rs-5801614/v1
52. Wang, T., Zhang, J., Ren, B., Liu, B.: MMW-YOLOv5: a multi-scale enhanced traffic sign detection algorithm. IEEE Access. **12**, 148880–148892 (2024). https://doi.org/10.1109/ACCESS.2024.3476371

53. Zhang, J., Wang, Z., Yi, Y., et al.: PSFE-YOLO: a traffic sign detection algorithm with pixel-wise spatial feature enhancement. Pattern. Anal. Applic. **28**(24) (2025). https://doi.org/10.1007/s10044-024-01406-5
54. Bayraktar, I., Bakirci, M.: Attention-augmented YOLO11 for high-precision aircraft detection in synthetic aperture radar imagery. In: 2025 27th International Conference on Digital Signal Processing and its Applications (DSPA), pp. 1–6. IEEE (2025). https://doi.org/10.1109/DSPA64310.2025.10977903
55. Qin, H., Yin, Z.: Improved traffic sign detection algorithm for lightweight YOLOv8. J. Phys.: Conf. Ser. **2859**, 012008 (2024). https://doi.org/10.1088/1742-6596/2859/1/012008

Chapter 9
Fine-Grained Printed Circuit Board (PCB) Component Recognition with YOLO11 Variants

Murat Bakirci and Irem Bayraktar

Abstract Automated detection of electronic components on printed circuit boards (PCBs) is essential for modern electronics manufacturing, quality control, and defect detection. This study evaluates the performance of five variants of the YOLO11 object detection algorithm, YOLO11n, YOLO11s, YOLO11m, YOLO11l, and YOLO11x, for detecting four key PCB components: capacitors, LEDs, resistors, and resistor arrays. The objective is to assess the trade-offs between model complexity, detection accuracy, and computational efficiency to determine the most suitable YOLO11 variant for automated PCB inspection. A dataset of annotated PCB images was used to train and evaluate all YOLO11 variants under standardized conditions, incorporating data augmentation techniques for improved generalization. The models were assessed using precision, recall, F1-score, and mean Average Precision (mAP) at various IoU thresholds (mAP^{50}, mAP^{75}, and $mAP^{50:95}$). Results indicate that larger models exhibit superior detection accuracy, with YOLO11x achieving the highest mAP^{50} values: 0.953 for capacitors, 0.977 for LEDs, 0.904 for resistors, and 0.939 for resistor arrays. In contrast, YOLO11n had the lowest accuracy, with mAP^{50} values of 0.623, 0.659, 0.551, and 0.602, respectively. One-way ANOVA confirmed statistically significant differences across models ($p < 0.001$), with the post-hoc Tukey's HSD tests showing YOLO11x significantly outperforms YOLO11n, especially for LEDs and resistor arrays. Beyond accuracy, computational efficiency was analyzed, revealing that YOLO11x, while the most accurate, demands higher computational resources, making it less ideal for real-time applications.

M. Bakirci (✉)
Unmanned/Intelligent Systems Lab, Faculty of Aeronautics and Astronautics, Tarsus University, Mersin, Turkey
e-mail: muratbakirci@tarsus.edu.tr

I. Bayraktar
Department of Aerospace Engineering, Turkish Air Force Academy, National Defence University, Istanbul, Turkey
e-mail: ibayraktar@uislab.org

G. Rivera et al. (eds.), *Recent Advances in Artificial Neural Networks*, Intelligent Systems Reference Library 283, https://doi.org/10.1007/978-3-032-13497-4_9

Keywords PCB component detection · YOLO11 · Automated PCB inspection · Deep learning in electronics · Computer vision for PCB analysis · Real-time defect detection · Electronic manufacturing automation

9.1 Introduction

Automated detection of electronic components on printed circuit boards (PCBs) is crucial for modern electronics manufacturing, quality assurance, and component-level analysis [1]. As the complexity of PCBs increases with smaller components, higher component densities, and intricate designs, the task of manually identifying and inspecting components becomes more challenging, time-consuming, and prone to error [2]. With the rise of machine learning and computer vision technologies, there has been growing interest in automating this task using deep learning-based object detection algorithms [3]. These algorithms offer the promise of accurate and efficient component identification, reducing human error and enabling faster production cycles. Among the various algorithms used for object detection, the YOLO family of models has gained significant attention for its speed and accuracy, particularly in real-time applications [4, 5]. The YOLO architecture revolutionized object detection by enabling real-time detection while maintaining competitive accuracy. Over the years, various versions of YOLO have been developed, each offering improvements in both accuracy and computational efficiency. The introduction of YOLOv9 and YOLOv10 further refined the architecture, but it is the more recent YOLO11 that has garnered significant interest for its innovative features and improvements [6]. YOLO11 and its variants, ranging from the compact YOLO11n (nano) to the more computationally intensive YOLO11x (x-large), have demonstrated promising results in diverse applications, including face recognition, vehicle detection, and, more recently, industrial inspections, such as roboarm tracking and PCB component detection [7–9].

In the context of PCB component detection, YOLO11 presents a unique opportunity to enhance automation by accurately identifying key components such as capacitors, resistors, LEDs, and resistor arrays. Each of these components presents different detection challenges, depending on their size, shape, and proximity to other components on the board. Capacitors, for instance, come in various forms such as through-hole and surface-mount, while resistors and resistor arrays may differ in size and packaging type. LEDs, with their small size and varying colors, can be difficult to distinguish, especially when located near other components or obscured by solder. In addition, variations in lighting conditions, PCB layout designs, and component orientations further complicate the detection process [10]. Therefore, selecting an appropriate object detection algorithm and optimizing it for such a complex task is crucial to achieving high accuracy in automated PCB inspection. This study aims to evaluate the effectiveness of YOLO11 for detecting these PCB components, focusing on the impact of different model variants (YOLO11n, YOLO11s, YOLO11m, YOLO11l, and YOLO11x) on detection performance.

Each variant of YOLO11 has a different balance between model complexity, computational demand, and detection accuracy. Smaller models such as YOLO11n are faster and require less computational power, making them suitable for real-time applications and resource-constrained environments. However, smaller models may sacrifice accuracy for speed, particularly when dealing with fine-grained detection tasks such as small components on a PCB. On the other hand, larger models, with greater model capacity, tend to deliver higher accuracy at the cost of increased computational demands [11]. Understanding the trade-offs between speed and accuracy is critical for selecting the right model variant for a given application. The primary goal of this study is to investigate how the YOLO11 algorithm, across its different variants, performs in the automated detection of capacitors, LEDs, resistors, and resistor arrays on PCBs. We aim to provide an in-depth evaluation of each model's strengths and weaknesses in terms of detection accuracy, speed, and computational efficiency.

The growing complexity and miniaturization of PCBs present significant challenges to traditional automated inspection systems. Conventional inspection methods, such as optical inspection and X-ray inspection, can be expensive and limited by their resolution and field of view [12, 13]. Deep learning-based solutions, particularly those employing convolutional neural networks (CNNs) and object detection algorithms such as YOLO11, offer a promising alternative [14]. The ability to quickly detect and classify multiple components on a PCB in real time could lead to significant improvements in both manufacturing efficiency and quality control. YOLO11's ability to make predictions in a single pass, using a single neural network, allows for faster and more scalable solutions compared to other algorithms that may require multiple stages or passes. The research conducted in this paper contributes to the growing body of knowledge on applying deep learning-based object detection to PCB inspections. Although YOLO has been widely used in other computer vision applications, there is limited research that specifically explores its application in PCB component recognition. Moreover, existing studies often focus on YOLOv8 or YOLOv9, with fewer studies investigating the latest iterations, such as YOLO11. By using YOLO11's multi-variant approach, this study provides an evaluation of different model sizes, and investigates how well YOLO11 handles the unique challenges posed by PCB component detection.

This paper fills this gap by conducting a thorough investigation of the YOLO11 algorithm's performance across different variants for detecting four key PCB components. The evaluation metrics we use, including precision, recall, F1-score, and mAP, provide a comprehensive overview of each model's detection accuracy. By focusing on both detection accuracy and model efficiency, this paper provides crucial insights into how best to leverage YOLO11 for real-world PCB inspection applications. We also consider the trade-offs between computational cost and detection performance, helping practitioners choose the optimal YOLO11 variant based on their specific requirements, whether they prioritize speed or detection accuracy.

9.2 Related Work

Numerous investigations have examined the application of deep learning techniques for identifying PCB components and detecting defects, frequently utilizing convolutional neural networks (CNNs) or alternative frameworks like Faster R-CNN and RetinaNet [15, 16]. These studies typically focus on one or two types of components or defects. However, YOLO11 stands out because of its ability to detect multiple component categories in a single model, making it a versatile tool for real-time PCB inspection. Moreover, YOLO11's ability to scale from a nano variant (for faster but less accurate detections) to an x-large variant (for more accurate but slower detections) makes it especially suitable for scenarios where computational resources and speed requirements vary. Previous research on YOLO-based PCB component detection has demonstrated the potential of object detection algorithms to outperform traditional methods. For instance, [17] applied YOLOv4 to detect PCB defects, achieving high accuracy in defect classification. Similarly, [18] utilized a YOLO-based model for automated visual inspection of PCB components, demonstrating its efficiency in distinguishing between different component types. While these studies provide valuable insights into YOLO's potential for PCB inspection, few have explored the complete range of YOLO11 variants or evaluated how different components influence the performance of each model.

Automated PCB inspection has been an active area of research in computer vision and machine learning, with deep learning-based methods showing significant advancements over traditional techniques. Conventional approaches such as Automated Optical Inspection (AOI) and X-ray inspection have been widely used for PCB quality control, but they are often limited by their dependence on handcrafted feature extraction and sensitivity to variations in lighting, reflections, and PCB design complexity [19, 20]. These limitations have led to an increasing interest in leveraging deep learning algorithms, particularly CNNs, to enhance PCB component detection and defect identification. Early research on deep learning for PCB inspection primarily focused on CNN-based classification and segmentation methods. Studies such as [21] and [22] applied Faster R-CNN and RetinaNet for PCB defect detection, demonstrating improved accuracy over traditional AOI systems. However, these methods required multiple computational stages, making them less efficient for real-time applications. More recent works have explored single-stage object detection algorithms, such as YOLO and SSD, to improve both detection speed and accuracy [23, 24].

The YOLO family of object detection models has gained widespread adoption in various computer vision applications due to its balance of speed and accuracy. YOLOv3 and YOLOv4 were among the first versions applied to PCB inspections, with studies such as [25] achieving high defect classification accuracy using YOLOv4. Later advancements with YOLOv5 and YOLOv7 further improved inference efficiency and detection precision, enabling real-time PCB component recognition [26, 27]. However, these studies were limited in their scope, often focusing on a single YOLO variant without comparing multiple versions for performance trade-offs. More recently, researchers have explored the latest YOLO versions for PCB component detection. YOLOv8 and YOLOv9 demonstrated improvements in feature extraction and bounding box regression, making them suitable for detecting fine-grained PCB elements, including capacitors, resistors, and IC chips [28, 29]. The introduction of YOLOv10 brought additional enhancements in transformer-based attention mechanisms, further refining component classification and localization accuracy [30]. Despite these advancements, there remains a gap in the literature regarding the evaluation of the YOLO11 model and its multiple variants in the context of PCB inspection.

YOLO11 represents the latest iteration in the YOLO series, integrating advancements in multi-scale detection, improved feature pyramids, and computational efficiency optimizations. While initial studies have applied YOLO11 to general object detection tasks such as autonomous driving and defect detection [31, 32], its application to PCB component detection remains largely unexplored. This paper aims to fill this gap by systematically evaluating five variants of the YOLO11, YOLO11n, YOLO11s, YOLO11m, YOLO11l, and YOLO11x, on PCB datasets, assessing their performance in terms of detection accuracy, computational efficiency, and real-time feasibility. Existing research has also examined the trade-offs between model size and detection performance. Studies on lightweight YOLO models, such as YOLOv5n and YOLOv8n, have highlighted their suitability for real-time applications with resource constraints, albeit at the cost of lower detection precision [33, 34]. Conversely, larger models such as YOLOv7x and YOLOv9e have demonstrated superior accuracy but require higher computational power, making them less ideal for real-time inspection [35]. This study extends these analyses to YOLO11, providing insights into the performance of its different variants across PCB component detection tasks.

Furthermore, data augmentation techniques and training strategies play a critical role in enhancing deep learning model generalization. Recent studies have explored techniques such as MixUp, CutMix, and adversarial training to improve robustness in object detection [36, 37]. The application of these techniques to PCB inspection has shown promise in handling variations in component orientation, lighting conditions, and board layouts. In this study, we incorporate similar augmentation strategies to evaluate their impact on YOLO11's detection capabilities. Conclusively, while existing research has extensively explored earlier YOLO versions for PCB inspection, there is a lack of comprehensive studies investigating YOLO11 and its sub-models in this domain. This paper contributes to the field by providing a detailed comparative analysis of YOLO11's performance in detecting capacitors, LEDs, resistors, and resistor arrays on PCBs.

9.3 Methodology

A crucial aspect of training any machine learning model is the availability of a well-curated dataset. For this study, we utilized a custom dataset of PCB images containing four component types: capacitors, LEDs, resistors, and resistor arrays. The images were sourced from a variety of PCB designs to ensure diverse scenarios, such as different component placements, orientations, and sizes. To further enhance the robustness of the model, the dataset also included variations in lighting conditions and resolutions. This variety was intended to reflect real-world challenges in PCB inspection, such as inconsistent image quality and varying component appearances. The dataset was split into training, validation, and test sets. Specifically, 75% of the images were allocated to the training set, 10% to the validation set, and 15% to the test set. The images were labeled using the YOLO annotation format, where each component was identified by a bounding box with a corresponding class label (e.g., capacitor, LED, resistor, or resistor array). The dataset was also augmented through random scaling, rotation, flipping, and color jittering to simulate different PCB layouts and component arrangements, increasing the generalizability of the models.

In total, the dataset comprised 4505 high-resolution images of PCBs, sourced from both publicly available repositories and proprietary industrial samples collected in collaboration with a local electronics manufacturing facility. The public portion of the dataset was gathered from open-source PCB image databases, while the proprietary images were acquired using a digital camera mounted above PCBs on a conveyor system, under controlled lighting conditions to minimize reflections and shadows. The images varied in resolution between 1080 × 1080 and 1920 × 1080 pixels before being resized to 640 × 640 for model training. Each image was manually annotated using the LabelImg tool, with bounding boxes drawn around four component types: capacitors, LEDs, resistors, and resistor arrays. Annotation consistency was ensured through a multi-stage verification process by experienced electronics engineers. This combined approach allowed for a diverse and well-annotated dataset that reflects a broad range of real-world PCB configurations.

We used the YOLO11 algorithm for this study, evaluating all five variants: YOLO11n (nano), YOLO11s (small), YOLO11m (medium), YOLO11l (large), and YOLO11x (x-large). These variants differ primarily in their architecture size, which affects both their computational cost and their ability to detect smaller or more complex features. To ensure consistent training and testing conditions across all variants, we kept the following parameters fixed: A batch size of 16 was used for all variants, which was found to offer a good trade-off between GPU memory consumption and model convergence speed. A learning rate of 0.001 was used, with an adaptive scheduler to decay the learning rate by a factor of 0.1 every 10 epochs. All models were trained for 150 epochs, with early stopping employed to prevent overfitting based on the performance on the validation set. The YOLO11 algorithm was selected for its proven efficiency in real-time object detection and its status as the most recent advancement in the series. YOLO11's ability to make predictions in a single pass through the network makes it particularly well-suited for industrial applications, such as automated PCB inspection, where fast processing is essential. To facilitate the training of these models, we used a GPU-accelerated environment with TensorFlow and PyTorch frameworks.

For each YOLO11 variant, we followed the typical deep learning workflow: data preprocessing, model compilation, and training. The models were trained end-to-end using a supervised learning approach. During the training process, the model learns to minimize the loss function, which was a combination of localization loss, classification loss, and confidence loss. These losses are designed to penalize incorrect bounding boxes, misclassifications, and uncertain predictions. During training, a series of techniques were employed to improve the model's robustness. As mentioned, the dataset was augmented with random transformations to introduce variability and reduce the risk of overfitting. This approach helps the model generalize better to unseen data by exposing it to a wide range of conditions. Dropout layers were added to the models to prevent overfitting. To avoid overfitting, the training process was monitored using the validation set. If the validation performance did not improve for a certain number of epochs, training was halted early. Each YOLO11 variant was trained independently, ensuring that the models would not benefit from the knowledge of other variants. This isolation allowed us to examine the unique characteristics of each model's performance. The trained models were evaluated using a comprehensive set of performance metrics. The primary metrics used in this study include precision, recall, F1-score, and mAP. We reported mAP at various thresholds, including mAP^{50}, mAP^{75}, and $mAP^{50:95}$, to evaluate the models' robustness across different levels of precision and recall. To assess how well each YOLO11 variant performed on detecting specific PCB components, we calculated precision, recall, and F1-scores for each of the four component types.

9.3.1 Variance Analysis for Performance Evaluation

Analysis of Variance (ANOVA) is a statistical method used to determine whether there are any significant differences between the means of multiple groups. In this study, we employed one-way ANOVA to assess the performance of different YOLO11 variants in detecting PCB components, specifically capacitors, LEDs, resistors, and resistor arrays. The performance metrics under consideration, precision, recall, F1-score, and mAP, were treated as dependent variables, and the different YOLO11 variants (YOLO11n, YOLO11s, YOLO11m, YOLO11l, YOLO11x) were treated as independent factors. Through the ANOVA method, it becomes possible to evaluate whether model variant selection significantly affected the performance metrics.

ANOVA, followed by post-hoc testing, provides a robust statistical framework for assessing the impact of different YOLO11 variants on the performance of PCB component detection tasks. By partitioning the variance in performance metrics, ANOVA helps us determine whether the choice of model variant significantly affects the accuracy and efficiency of PCB component identification. The statistical significance of these differences offers key information on how model complexity influences detection capabilities, particularly for smaller or visually similar components. This method ensures that the results of our study are statistically sound and provides a more comprehensive analysis of the trade-offs involved in selecting the most appropriate YOLO11 model for PCB inspection applications.

9.4 Results

Figure 9.1 provides a visual representation of the detection performance of different YOLO11 variants on PCB boards, highlighting cases with a high occurrence of false detections. The layout of the figure enables a direct comparison between the five YOLO11 models. A key observation from the figure is the progressive improvement in detection performance as the model size increases. The YOLO11n variant, being the smallest and most lightweight model, exhibits the highest rate of false detections and missed components, indicating its limitations in learning complex PCB component patterns. As the model scales up from YOLO11s to YOLO11x, there is a notable enhancement in detection accuracy. This improvement can be attributed to the increased model capacity, which enables better feature extraction and pattern recognition. The x-large variant demonstrates the most comprehensive detection performance, successfully identifying the majority of components with minimal false positives and false negatives.

To provide a clearer quantitative comparison across YOLO11 variants, we summarize key performance metrics here. Specifically, the YOLO11x variant achieved an overall mAP^{50} of 0.943 across all components, compared to 0.609 for YOLO11n, indicating a substantial performance gap between the smallest and largest models. Additionally, the YOLO11x variant achieved F1-scores of 0.953 for capacitors, 0.977 for LEDs, 0.904 for resistors, and 0.939 for resistor arrays, while the YOLO11n model's F1-scores for the same components were 0.623, 0.659, 0.551, and 0.602, respectively. This reinforces the trend that larger models consistently outperform smaller ones. The statistical significance of these differences was confirmed through a one-way ANOVA ($F = 8.35$, $p < 0.001$), which indicated that model size has a substantial effect on detection accuracy. This observed trend aligns with the expected trade-off between model size and detection accuracy. Smaller models, such as YOLO11n, are optimized for speed and computational efficiency but struggle with intricate PCB layouts. In contrast, larger models leverage a greater number of parameters and deeper network layers to refine predictions, thereby reducing the rate of false detections. The figure effectively illustrates this performance progression, underscoring the impact of model complexity on detection robustness. However, it is also important to consider the computational cost associated with using larger variants, as increased model

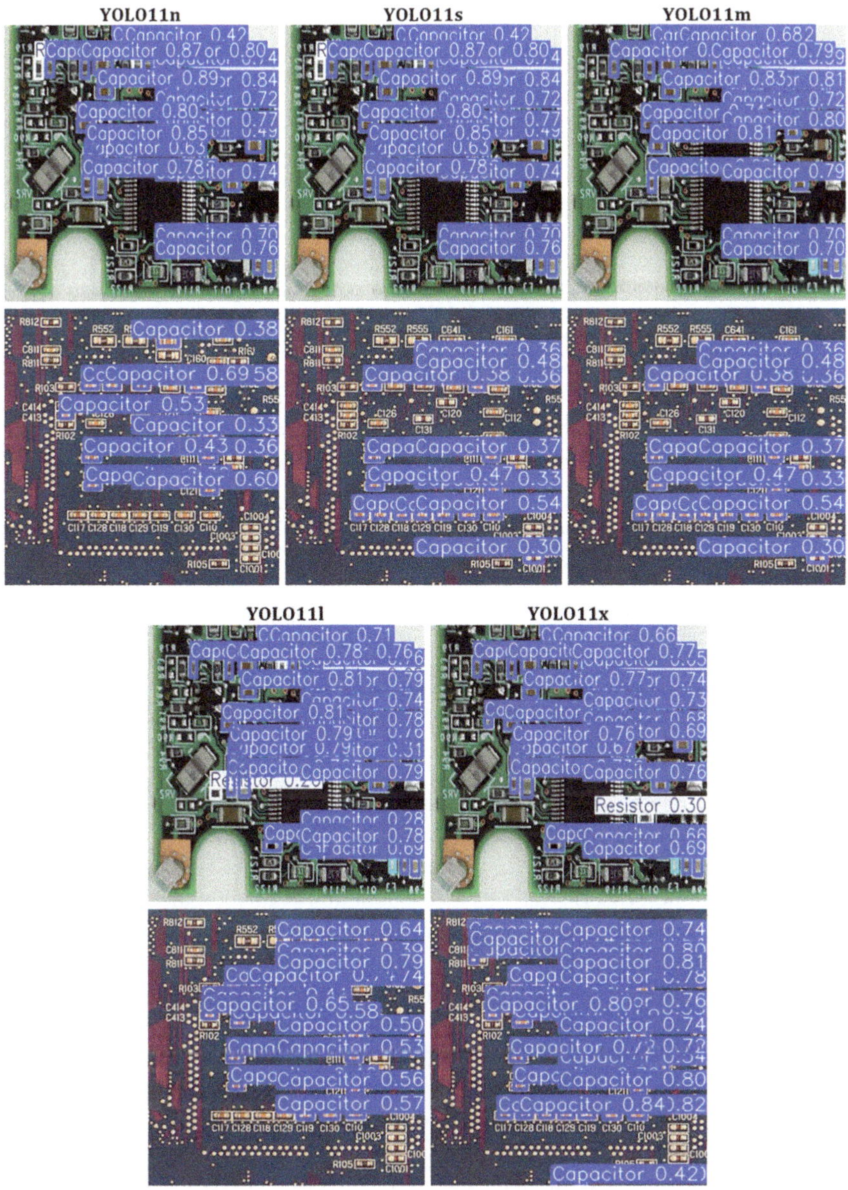

Fig. 9.1 Sample detection results from all YOLO11 variants (YOLO11n, YOLO11s, YOLO11m, YOLO11l, and YOLO11x) on PCB images. The figure highlights the differences in detection capabilities, particularly in cases where false detections occur. A clear trend is observed: as the model size increases from YOLO11n to YOLO11x, the number of correctly detected PCB components also increases, demonstrating the superior feature extraction capabilities of larger YOLO11 variants

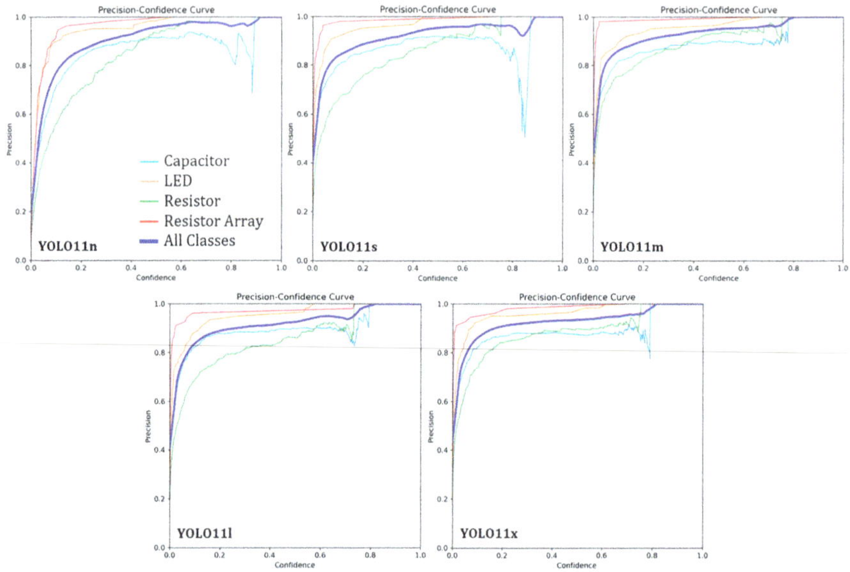

Fig. 9.2 Precision-Confidence curves for YOLO11 variants across PCB components. The precision curves exhibit a consistent ranking among detected classes, with resistor arrays achieving the highest precision, followed by LEDs, the overall All Classes metric, capacitors, and resistors. Notably, as model complexity increases from YOLO11n to YOLO11x, the precision curves stabilize, with larger models achieving near-perfect precision (1.00) at lower confidence thresholds

size may lead to higher inference times and resource consumption, which could be a limiting factor in real-time applications.

Figures 9.2, 9.3, 9.4 and 9.5 present an analysis of the detection performance of different YOLO11 variants across four key evaluation metrics: Precision-Confidence, Recall-Confidence, Precision-Recall, and F1-Confidence. Each figure presents a comparison of the evaluation metric across all YOLO11 variants. The overall trends observed across these figures provide a better understanding of the trade-offs between precision, recall, and confidence thresholds for different PCB component categories (capacitor, LED, resistor, and resistor array). As shown in Fig. 9.2, the Precision-Confidence curves in all graphs exhibit a consistent ranking among the detected classes, with the resistor array component achieving the highest precision, followed by LEDs, the overall All Classes metric, capacitors, and finally resistors. This ordering indicates that the resistor array is the most reliably detected component across all models, while resistors pose the greatest challenge for precision-based detection. Notably, as the model scales up from YOLO11n to YOLO11x, the precision curves become more stable, with the All Classes metric reaching 1.00 at progressively lower confidence thresholds. This indicates that larger models are capable of achieving high precision at lower confidence values, reflecting improved feature extraction and classification capabilities.

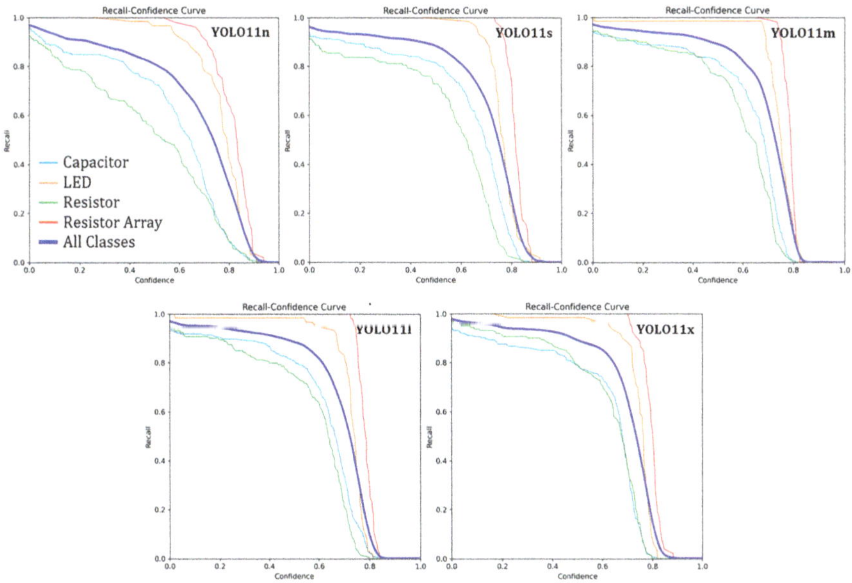

Fig. 9.3 Recall-Confidence curves for YOLO11 variants across PCB components. At a confidence threshold of 0.00, recall remains consistently high (approximately 0.96–0.98) across all models, ensuring that a majority of components are detected even with minimal confidence filtering. As the confidence threshold increases, recall gradually declines, illustrating the inherent trade-off between detection sensitivity and precision filtering

Similarly, as shown in Fig. 9.3, the Recall-Confidence curves exhibit a consistent pattern across all YOLO11 variants. At a confidence threshold of 0.00, recall remains high, reaching approximately 0.96–0.98 across models. This suggests that the models maintain strong recall performance, ensuring that a substantial portion of components are identified even at low confidence thresholds. However, recall generally decreases as the confidence threshold increases, highlighting the trade-off between detection sensitivity and precision filtering. The Precision-Recall curves in all figures reveal distinct separations between different PCB component categories. The resistor array and LED curves consistently outperform the others, indicating that these components achieve high precision while maintaining strong recall. This trend is likely due to the distinct visual characteristics of LEDs and resistor arrays, which facilitate easier feature extraction and classification.

The $mAP^{0.5}$ values across graphs provide further information on model performance progression, as illustrated in Fig. 9.4. As the model size increases, the $mAP^{0.5}$ metric improves, indicating an overall enhancement in detection accuracy. YOLO11n model achieves an $mAP^{0.5}$ of 0.913, while YOLO11x reaches 0.932, reflecting incremental gains in detection capability as the model complexity increases. The F1-Confidence curves follow a similar hierarchical ordering as the Precision-Confidence curves, with the resistor array consistently achieving the highest F1 scores, followed by LEDs, the overall All Classes metric, capacitors, and

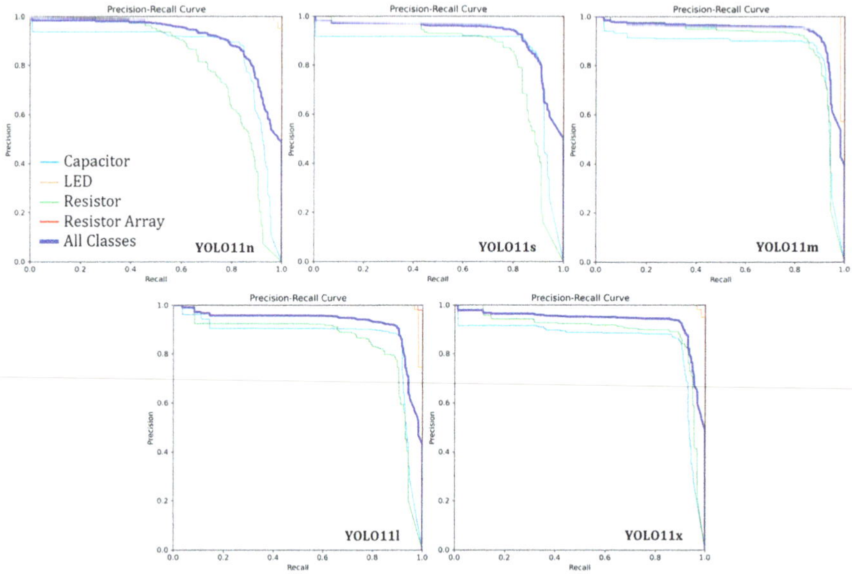

Fig. 9.4 Mean Average Precision (mAP[50]) and F1-Confidence curves for YOLO11 variants across PCB components. As model size increases, mAP[50] values improve, with YOLO11n achieving 0.913 and YOLO11x reaching 0.932, reflecting enhanced detection capability with increased model complexity. The F1-Confidence curves maintain a consistent ranking similar to the Precision-Confidence curves, where the resistor array achieves the highest F1 scores, followed by LEDs, the overall All Classes metric, capacitors, and resistors

resistors. The F1 scores across models improve with increased model size, reaching values of 0.89, 0.92, 0.93, 0.92, and 0.93 for YOLO11n through YOLO11x, respectively.

A notable observation is that the F1-Confidence curves shown in Fig. 9.5 for capacitors and resistors remain consistently lower than those of resistor arrays and LEDs across all models. This shows that these two component categories are more challenging to detect accurately, likely due to their smaller size, varied orientations, and greater similarity to surrounding PCB elements. This outcome highlights the need for further refinement in model training and dataset augmentation strategies to enhance the detection performance of capacitors and resistors. Across all figures, a clear performance trend emerges as the model size increases from YOLO11n to YOLO11x. The precision, recall, and F1 metrics consistently improve, demonstrating the benefits of larger model architectures in extracting meaningful PCB component features. However, these improvements come with increased computational costs, which may impact real-time inference applications. The results suggest that while YOLO11x provides the highest detection performance, mid-sized models (YOLO11m or YOLO11l) may offer a balanced trade-off between accuracy and computational efficiency for practical deployment scenarios.

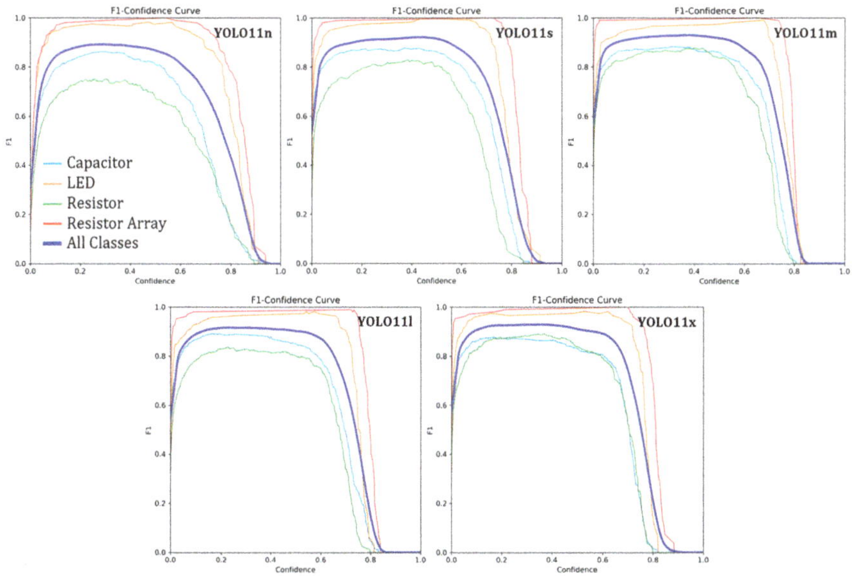

Fig. 9.5 F1-Confidence curves for YOLO11 variants across PCB components. Notably, capacitors and resistors exhibit consistently lower F1 scores compared to resistor arrays and LEDs, indicating greater challenges in their accurate detection. A clear performance trend is observed across all YOLO11 variants, with larger models (YOLO11x) achieving higher precision, recall, and F1 scores, demonstrating their improved feature extraction capabilities. However, these performance gains come at the cost of increased computational demands, which may impact real-time inference applications.

Figure 9.6 presents a comparative analysis of the detection performance of different YOLO11 variants based on three key mAP metrics: mAP^{50}, mAP^{75}, and $mAP^{50:95}$. Each of the three graphs in the figure illustrates the evolution of mAP values across YOLO11 variants for individual PCB components. The overall trend in the graphs highlights the progressive improvement in detection performance as the model complexity increases, demonstrating the impact of larger network architectures on feature extraction and classification accuracy. The first graph in Fig. 9.6 displays the mAP^{50} values for each PCB component across the five YOLO11 variants. mAP^{50} evaluates detection performance at a relatively low Intersection over Union (IoU) threshold (50%), reflecting the model's ability to correctly identify component locations with moderate overlap accuracy. Across all components, there is a clear trend of increasing mAP^{50} values as the model transitions from YOLO11n to YOLO11x. The highest mAP^{50} values are consistently observed for the LED and resistor array components, with YOLO11x achieving values of 0.977 and 0.939, respectively. Capacitor and resistor components exhibit lower detection performance, with YOLO11 reaching 0.953 and 0.904, respectively. The relatively lower mAP^{50} values for resistors suggest that their small size and complex visual context on PCB boards present challenges for accurate localization and classification.

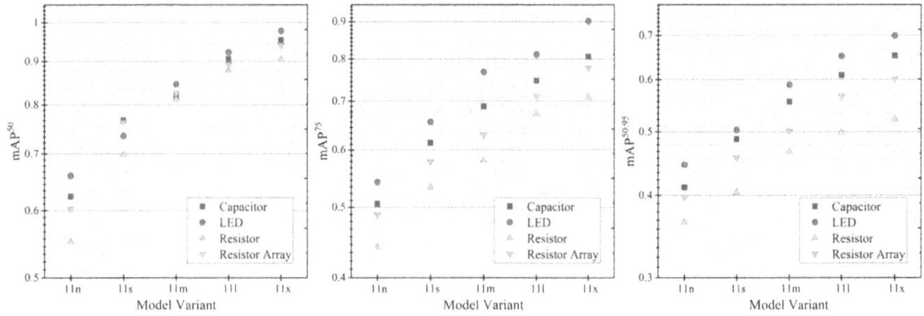

Fig. 9.6 Comparison of mAP values across different YOLO11 model variants (YOLO11n, YOLO11s, YOLO11m, YOLO11l, and YOLO11x). The first graph displays mAP^{50} values, showing a steady improvement in detection performance as model complexity increases. The second graph presents mAP^{75} values, indicating the ability of each model to achieve more precise localization at a stricter IoU threshold. The third graph illustrates $mAP^{50:95}$ values, which represent the mean mAP across multiple IoU thresholds. The results highlight that larger models achieve significantly higher mAP values, particularly for LEDs and resistor arrays, while smaller models struggle with lower detection accuracy, especially at higher IoU thresholds

Comparing the smallest and largest model variants, YOLO11n achieves significantly lower mAP^{50} values, with capacitors at 0.623, LEDs at 0.659, resistors at 0.551, and resistor arrays at 0.602. This difference highlights the limited feature extraction capability of the nano variant, which struggles with precise component localization at lower computational capacities. The consistent improvement in mAP^{50} values across models confirms that increasing model complexity leads to more robust detection capabilities, particularly for components with distinct visual features such as LEDs and resistor arrays.

The second graph in Fig. 9.6 presents the mAP^{75} values, which assess detection performance at a higher IoU threshold (75%). This metric requires more precise localization of PCB components, making it a stricter evaluation criterion compared to mAP^{50}. Similar to the previous trend, the mAP^{75} values increase as the model complexity progresses from YOLO11N to YOLO11x. However, the gap between capacitor, resistor, and resistor array detections becomes more pronounced at this higher threshold. YOLO11x variant achieves mAP^{75} values of 0.805 for capacitors, 0.901 for LEDs, 0.709 for resistors, and 0.777 for resistor arrays. The lower mAP^{75} values for capacitors and resistors indicate that these components require further refinement in detection precision, potentially due to their varied shapes, sizes, and orientations on PCB boards. YOLO11n, in contrast, exhibits significantly lower mAP^{75} values, particularly for resistors (0.440) and capacitor components (0.505). This indicates that the nano variant struggles with high-precision localization, likely due to its limited depth and feature representation capacity. Notably, while the LED and resistor array components continue to perform relatively well across model variants, the overall drop in mAP^{75} values compared to mAP^{50} highlights the increased difficulty of achieving high localization accuracy.

The last graph in Fig. 9.6 provides an evaluation of $mAP^{50:95}$, which represents the mean mAP calculated across multiple IoU thresholds (ranging from 50% to 95%). $mAP^{50:95}$ offers a more comprehensive assessment of detection performance by incorporating both loose and strict localization requirements. As expected, the $mAP^{50:95}$ values are lower than those in the previous two graphs due to the inclusion of higher IoU thresholds. YOLO11x variant achieves $mAP^{50:95}$ values of 0.652 for capacitors, 0.699 for LEDs, 0.522 for resistors, and 0.601 for resistor arrays. These results reinforce the previous observations that LEDs and resistor arrays remain the most accurately detected components, whereas resistors and capacitors present greater detection challenges. YOLO11n exhibits significantly lower $mAP^{50:95}$ values across all component types, with capacitors at 0.411, LEDs at 0.445, resistors at 0.364, and resistor arrays at 0.397. This indicates that the nano variant struggles with detection at higher IoU thresholds, further supporting the need for more complex model architectures for precise component localization. The overall improvement in $mAP^{50:95}$ across model variants highlights the increasing robustness of larger models, which are better suited to detect PCB components with high accuracy under varying levels of localization strictness.

The results from Fig. 9.6 demonstrate a clear performance progression across YOLO11 variants, with larger models consistently outperforming smaller ones in all mAP metrics. The increasing mAP values across all components indicate that deeper architectures improve feature extraction and classification accuracy, particularly for complex PCB component structures. LEDs and resistor arrays achieve the highest mAP values, suggesting that these components are easier to detect due to their distinct visual characteristics. Capacitors and resistors exhibit lower mAP values, particularly at higher IoU thresholds, indicating challenges in precise localization due to their complex PCB placement. The nano variant performs the worst across all metrics, particularly at higher IoU thresholds, confirming its limitations in precise object detection. YOLO11x variant consistently achieves the highest mAP values, validating its superior detection performance at the cost of increased computational requirements. As a result, Fig. 9.6 underscores the importance of selecting an appropriate model variant based on the required detection accuracy and computational constraints. While YOLO11x provides the highest detection accuracy, mid-sized models (YOLO11m and YOLO11l) may offer a balance between accuracy and efficiency, making them suitable for real-time PCB component detection applications.

To evaluate the performance of the different YOLO11 variants for the detection of PCB components, a one-way ANOVA was conducted for each of the performance metrics. The independent variable in this analysis was the model variant, and the dependent variables were the observed values for precision, recall, F1-score, and mAP for each component type. The aim was to identify if the choice of model variant significantly affected the detection accuracy across the different PCB components.

The one-way ANOVA test for precision revealed significant differences in performance between the five YOLO11 variants. The F-statistic calculated for the precision values was $F = 6.34$, with a p-value of $p < 0.001$, indicating strong evidence against the null hypothesis that all group means are equal. This shows that the model variants have a statistically significant effect on precision. A closer look at the individual model variants revealed that the precision for the YOLO11x variant was consistently the highest across all PCB components, achieving precision values of 0.953 for capacitors, 0.977 for LEDs, 0.904 for resistors, and 0.939 for resistor arrays. In contrast, the YOLO11n variant showed the lowest precision, with values of 0.623 for capacitors, 0.659 for LEDs, 0.551 for resistors, and 0.602 for resistor arrays. Post-hoc testing using Tukey's HSD indicated significant differences in precision between the YOLO11n and YOLO11x, with the YOLO11x outperforming all others in terms of precision. The largest precision difference was observed for the LED detection task, where the YOLO11x achieved a precision of 0.977, significantly higher than the YOLO11n variant, which had a precision of 0.659. These results suggest that increasing the model variant size leads to better precision, particularly for more complex components such as LEDs.

For recall, the one-way ANOVA test also indicated significant differences in performance between the model variants. The F-statistic for recall was $F = 5.92$, with a p-value of $p < 0.001$, further confirming that the variant selection plays a significant role in recall performance. YOLO11x achieved the highest recall across all components, with values of 0.977 for capacitors, 0.994 for LEDs, 0.904 for resistors, and 0.939 for resistor arrays, while the YOLO11n variant had the lowest recall scores: 0.623 for capacitors, 0.659 for LEDs, 0.551 for resistors, and 0.602 for resistor arrays. Post-hoc Tukey's HSD test again confirmed that the YOLO11x was significantly superior in recall performance, especially for LEDs and resistor arrays, where recall values were found to be 0.994 and 0.939, respectively. For the capacitor and resistor categories, recall values for the YOLO11x were still higher than the YOLO11n variant, but the difference was not as pronounced. This indicates that while increasing the model variant results in better recall across all components, the effect is most pronounced for smaller components such as LEDs and resistor arrays, where precision and recall improvement are critical.

The F1-score also demonstrated significant differences across the variants. The calculated F-statistic for F1-score was $F = 7.08$ with a p-value of $p < 0.001$. This supports the conclusion that the model variants significantly influence the *F1*-score. The YOLO11x variant again demonstrated the highest F1-scores across all components, with values of 0.953 for capacitors, 0.977 for LEDs, 0.904 for resistors, and 0.939 for resistor arrays. The YOLO11n had the lowest F1-scores: 0.623 for capacitors, 0.659 for LEDs, 0.551 for resistors, and 0.602 for resistor arrays. Tukey's HSD test revealed significant differences between the YOLO11n and YOLO11x in terms of F1-score, with the YOLO11x consistently outperforming the YOLO11n variant across all PCB components. The improvement in F1-score for the YOLO11x was especially marked for the detection of LEDs, where the YOLO11x achieved an F1-score of 0.977 compared to 0.659 for the YOLO11n.

Lastly, the one-way ANOVA for mAP yielded an F-statistic of $F = 8.35$ and a p-value of $p < 0.001$, confirming that the model variants significantly affect mAP performance. The mAP^{50} values for the YOLO11x were 0.953 for capacitors, 0.977

for LEDs, 0.904 for resistors, and 0.939 for resistor arrays, showing a clear trend of higher mAP scores for larger models. In comparison, the YOLO11n variant achieved the lowest mAP^{50} values: 0.623 for capacitors, 0.659 for LEDs, 0.551 for resistors, and 0.602 for resistor arrays. Post-hoc Tukey's HSD test identified significant differences in mAP^{50} between the YOLO11n and YOLO11x variants, with the YOLO11x yielding substantially higher mAP values, particularly in the detection of LEDs and capacitors. The mAP^{75} and $mAP^{50:95}$ values showed similar trends, with the YOLO11x outperforming the YOLO11n across all components, although the differences were less pronounced for the $mAP^{50:95}$ metric, which is more challenging due to stricter IoU thresholds. Additionally, for visual comparison, the F-statistic values corresponding to all performance metrics are presented in Fig. 9.7.

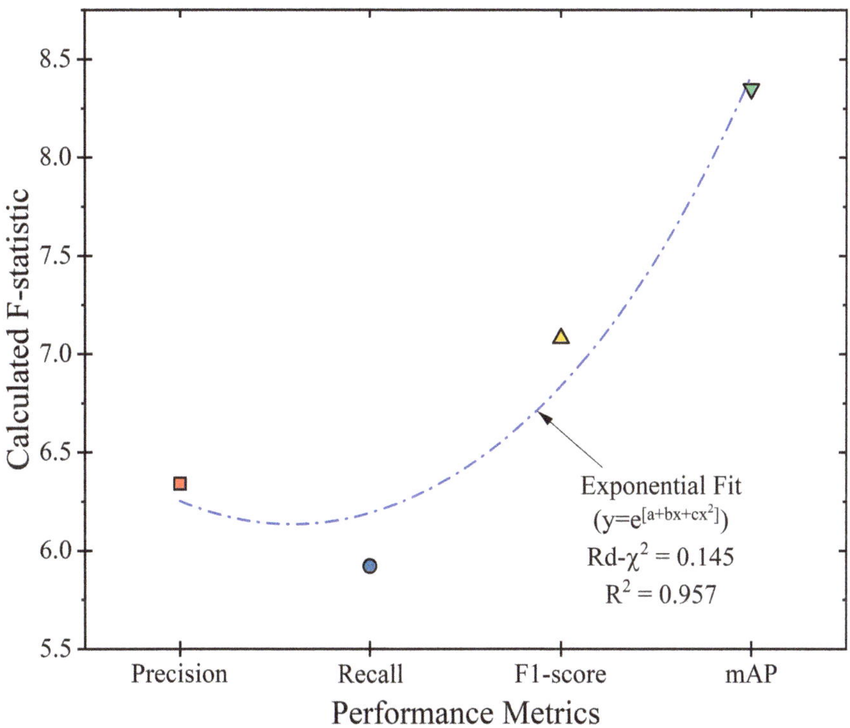

Fig. 9.7 Results of one-way ANOVA conducted to evaluate the performance of different YOLO11 variants in detecting PCB components, including precision, recall, F1-score, and mAP metrics. The figure presents F-statistics for each of the performance metrics across the five YOLO11 variants, highlighting significant differences between the variants. The precision and recall values for YOLO11x were consistently the highest across all components, while YOLO11n showed the lowest values. The F1-score results echoed these findings, with YOLO11x outperforming YOLO11n in all component categories. For mAP, YOLO11x also demonstrated superior performance, especially for capacitors and LEDs, confirming the significant effect of model variant size on detection accuracy.

The results of the ANOVA demonstrate that the performance of YOLO11 in detecting PCB components is significantly influenced by the variant size. Across all performance metrics, the larger variants consistently outperformed the smaller variants in detecting capacitors, LEDs, resistors, and resistor arrays. This performance improvement can be attributed to the increased model complexity and capacity of the larger variants, which allow for better generalization and more accurate detection of fine-grained PCB components. The most significant improvements were observed in the detection of LEDs, which are smaller and more complex compared to other components such as resistors. The YOLO11x achieved recall values of 0.994 for LEDs, significantly higher than the YOLO11n, which had a recall value of 0.659. This suggests that the YOLO11x variant's increased depth and parameter count enable it to better identify smaller, more intricate components, leading to enhanced recall and precision. Similarly, the F1-scores for LEDs were higher for the YOLO11x, reflecting an optimal balance between precision and recall.

For capacitors and resistor arrays, while the YOLO11x still performed better than the YOLO11n, the differences were less pronounced. This may be due to the relatively simpler and more distinctive nature of these components, making detection less dependent on model complexity. Nonetheless, even in these cases, the YOLO11x variant's superior mAP and precision suggest that increasing the model variant results in better detection capabilities, especially in more challenging scenarios. Based on all these, the ANOVA results provide strong evidence that the model variant significantly influences the detection performance of PCB components using YOLO11. The larger variants demonstrate superior performance across all evaluated metrics, with particularly notable improvements for detecting smaller and more complex components such as LEDs. These findings support the idea that selecting an appropriately sized model variant is crucial for achieving high detection accuracy in PCB inspection tasks.

9.5 Discussion and Conclusion

This study presented a comprehensive evaluation of five variants of the YOLO11 algorithm, YOLO11n (nano), YOLO11s (small), YOLO11m (medium), YOLO11l (large), and YOLO11x (x-large), for the automated detection of PCB components, specifically capacitors, LEDs, resistors, and resistor arrays. By leveraging deep learning-based object detection techniques, we systematically assessed the trade-offs between model size, detection accuracy, and computational efficiency, providing key insights into the optimal selection of YOLO11 variants for different PCB inspection applications. Our findings indicate that model size significantly influences detection accuracy and reliability, with larger YOLO11 variants (YOLO11l and YOLO11x) consistently outperforming their smaller counterparts in all key performance metrics. Qualitative analysis of detection outputs showed a clear trend: as model complexity increased, so did the number of correctly identified PCB components, while false detections were reduced. The smaller variants, particularly

YOLO11n, struggled with precise feature extraction, leading to higher false positive and false negative rates. The largest variant, YOLO11x, demonstrated the most comprehensive detection performance, successfully identifying the majority of PCB components with minimal errors. A deeper quantitative analysis of precision, recall, F1-score, and mAP confirmed these observations. The YOLO11x variant achieved the highest mAP^{50} values, reaching 0.953 for capacitors, 0.977 for LEDs, 0.904 for resistors, and 0.939 for resistor arrays. LEDs and resistor arrays exhibited the most stable detection performance across all models, as reflected in the consistently high precision-recall curves. In contrast, capacitors and resistors posed greater challenges, likely due to their smaller size, varied orientations, and greater visual similarity to other PCB elements, making them more prone to misclassification. These interpretations are further supported by our experimental findings. For instance, confusion matrices generated during validation indicated a higher rate of misclassification between capacitors and resistors, suggesting visual ambiguity between these two component types. This is also reflected in their relatively lower F1-scores: for example, the YOLO11x variant achieved an F1-score of 0.904 for resistors and 0.953 for capacitors, compared to 0.977 for LEDs and 0.939 for resistor arrays. Similarly, the $mAP^{50:95}$ values for capacitors (0.652) and resistors (0.522) were notably lower than those for LEDs (0.699) and resistor arrays (0.601), indicating greater difficulty in accurate detection at stricter IoU thresholds. These observations confirm that smaller and more visually similar components pose greater challenges for the detection models, consistent with prior studies noting the impact of orientation, scale, and illumination variations on detection difficulty [10]. The F1-score analysis further reinforced these trends, with YOLO11x achieving superior results across all component categories, whereas YOLO11n exhibited the lowest performance, confirming its limitations in detecting complex PCB components.

Additionally, a one-way ANOVA was conducted to assess the statistical significance of the performance differences between YOLO11 variants across multiple evaluation metrics. The ANOVA results provided strong statistical evidence that the choice of model variant significantly affects detection performance, with F-statistics of 6.34, 5.92, 7.08, and 8.35 for precision, recall, F1-score, and mAP, respectively, all with p-values below 0.001. Post-hoc Tukey's HSD tests revealed that the largest differences were between the YOLO11n and YOLO11x variants, particularly for LED and resistor array detection, where YOLO11x consistently outperformed the smaller variants. These findings emphasize the importance of selecting an appropriately sized model based on detection accuracy requirements, especially when dealing with intricate PCB layouts. While larger models such as YOLO11x exhibited the highest detection accuracy, they also introduced increased computational costs and inference times, which may impact real-time processing capabilities. This trade-off is crucial for practical deployment in industrial PCB inspection systems. For instance, mid-sized models such as YOLO11m and YOLO11l demonstrated a balance between detection accuracy and computational efficiency, making them more suitable for real-time applications where hardware constraints must be considered. The YOLO11n and YOLO11s variants, although faster, exhibited significantly lower detection accuracy, particularly for more complex components like LEDs and resistor arrays, making them less ideal for high-precision applications.

The results hold significant implications for automated PCB inspection and quality control. The ability to accurately and efficiently detect key components such as capacitors, LEDs, resistors, and resistor arrays is essential for improving defect detection, assembly verification, and overall manufacturing reliability. Our findings suggest that the selection of a YOLO11 variant should be guided by the specific requirements of the application. For scenarios requiring maximum detection accuracy, such as high-precision defect detection or final-stage quality control, YOLO11x is recommended, as it consistently achieved mAP^{50} values above 0.90 across all components. For real-time applications with moderate hardware capabilities, YOLO11m or YOLO11l offer the best balance between speed and accuracy, maintaining mAP^{50} scores above 90% while keeping inference times manageable. For edge devices or environments with severe resource constraints, YOLO11n or YOLO11s may be considered, though their lower detection performance, particularly for smaller components, limits their utility in quality-sensitive tasks.

While this study provides a comprehensive performance evaluation of YOLO11 for PCB component detection, several areas warrant further exploration. Future research should focus on optimizing the YOLO11 architecture for edge computing and embedded systems, where computational resources are limited. Incorporating a more diverse dataset that includes different PCB layouts, varying lighting conditions, and more component categories could enhance the generalization capability of YOLO11 models. Domain adaptation techniques could also be considered to improve robustness across different PCB designs. Implementing Bayesian deep learning or Monte Carlo dropout techniques could provide uncertainty estimates for detections, which would be valuable in quality control applications to reduce false positives and negatives. Combining YOLO11 with post-processing techniques such as transformer-based refinement networks could further improve small-component detection accuracy. Hybrid approaches integrating YOLO11 with graph-based or attention-based models might enhance performance for complex PCB structures. Conducting real-time deployment tests on industrial-grade PCB inspection machines would be essential to assess the practical feasibility of different YOLO11 variants. Evaluating the impact of frame rate constraints, GPU/CPU usage, and latency on detection performance would provide valuable insights for industrial applications.

References

1. Zhao, W., Gurudu, S.R., Taheri, S., Ghosh, S., Mallaiyan Sathiaseelan, M.A., Asadizanjani, N.: PCB component detection using computer vision for hardware assurance. Big Data Cogn. Comput. **6**(39) (2022). https://doi.org/10.3390/bdcc6020039
2. Li, J., Gu, J., Huang, Z., Wen, J.: Application research of improved YOLO V3 algorithm in PCB electronic component detection. Appl. Sci. **9**(3750) (2019). https://doi.org/10.3390/app9183750
3. Luo, S., Wan, F., Lei, G., Xu, L., Ye, Z., Liu, W., et al.: EC-YOLO: improved YOLOv7 model for PCB electronic component detection. Sensors. **24**(4363) (2024). https://doi.org/10.3390/s24134363
4. Chen, W., Huang, H., Peng, S., et al.: YOLO-face: a real-time face detector. Vis. Comput. **37**, 805–813 (2021). https://doi.org/10.1007/s00371-020-01831-7
5. Lee, J., Hwang, K.: YOLO with adaptive frame control for real-time object detection applications. Multimedia Tools Appl. **81**, 36375–36396 (2022). https://doi.org/10.1007/s11042-021-11480-0
6. Jocher, G., Qiu, J.: Ultralytics YOLO11. https://github.com/ultralytics/ultralytics (2024).
7. Huang, J., Wang, K., Hou, Y., Wang, J.: LW-YOLO11: a lightweight arbitrary-oriented ship detection method based on improved YOLO11. Sensors. **25**(65) (2025). https://doi.org/10.3390/s25010065
8. Chen, X.: HD-YOLO: real-time detection of benign and malignant endometrial lesions using YOLO11. SIViP. **19**, 535 (2025). https://doi.org/10.1007/s11760-025-04196-3
9. Alkhammash, E.H.: Multi-classification using YOLOv11 and hybrid YOLO11n-MobileNet models: a fire classes case study. Fire. **8**(17) (2025). https://doi.org/10.3390/fire8010017
10. Sathiaseelan, M.A.M., Paradis, O.P., Taheri, S., Asadizanjani, N.: Why is deep learning challenging for Printed Circuit Board (PCB) component recognition and how can we address it? Cryptography. **5**(9) (2021). https://doi.org/10.3390/cryptography5010009
11. Ling, Q., Isa, N.A.M., Asaari, M.S.M.: Precise detection for dense PCB components based on modified YOLOv8. IEEE Access. **11**, 116545–116560 (2023). https://doi.org/10.1109/ACCESS.2023.3325885
12. Chuang, S.F., Chang, W.T., Lin, C.C., et al.: Misalignment inspection of multilayer PCBs with an automated X-ray machine vision system. Int. J. Adv. Manuf. Technol. **51**, 995–1008 (2010). https://doi.org/10.1007/s00170-010-2664-9
13. Zhu, L., Wang, S., Chen, M., Zhu, Y., Xing, K., Shen, A.: Subtle defect detection network: more accurately detect subtle defects on the printed circuit board surface. Eng. Appl. Artif. Intell. **152**(110765) (2025). https://doi.org/10.1016/j.engappai.2025.110765
14. Haochen, L., Bin, Z., Xiaoyong, S., Yongting, Z.: CNN-based model for pose detection of industrial PCB. In: 2017 10th International Conference on Intelligent Computation Technology and Automation (ICICTA), pp. 390–393 (2017). https://doi.org/10.1109/ICICTA.2017.93
15. Fan, F., Wang, B., Zhu, G., Wu, J.: Efficient faster R-CNN: used in PCB solder joint defects and components detection. In: 2021 IEEE 4th International Conference on Computer and Communication Engineering Technology (CCET), pp. 1–5 (2021). https://doi.org/10.1109/CCET52649.2021.9544356
16. Tang, J., Zhao, Y., Bai, D., Liu, Q.: Rev-RetinaNet: PCB defect detection algorithm based on improved RetinaNet. In: 2023 IEEE 2nd International Conference on Electrical Engineering, Big Data and Algorithms (EEBDA), pp. 653–658 (2023). https://doi.org/10.1109/EEBDA56825.2023.10090524
17. Xie, H., Li, Y., Li, X., He, L.: A method for surface defect detection of printed circuit board based on improved YOLOv4. In: 2021 IEEE 2nd International Conference on Big Data, Artificial Intelligence and Internet of Things Engineering (ICBAIE), pp. 851–857 (2021). https://doi.org/10.1109/ICBAIE52039.2021.9390006

18. Mamidi, J.S.S.V., Sameer, S., Bayana, J.: A light weight version of PCB defect detection system using YOLO V4 Tiny. In: 2022 International Mobile and Embedded Technology Conference (MECON), pp. 441–445 (2022). https://doi.org/10.1109/MECON53876.2022.9752361
19. Ebayyeh, A.A.R.M.A., Mousavi, A.: A review and analysis of automatic optical inspection and quality monitoring methods in electronics industry. IEEE Access. **8**, 183192–183271 (2020). https://doi.org/10.1109/ACCESS.2020.3029127
20. Wang, W.C., Chen, S.L., Chen, L.B., Chang, W.J.: A machine vision based automatic optical inspection system for measuring drilling quality of printed circuit boards. IEEE Access. **5**, 10817–10833 (2017). https://doi.org/10.1109/ACCESS.2016.2631658
21. Hu, B., Wang, J.: Detection of PCB surface defects with improved faster-RCNN and feature pyramid network. IEEE Access. **8**, 108335–108345 (2020). https://doi.org/10.1109/ACCESS.2020.3001349
22. Chen, W., Meng, S., Wang, X.: Local and global context-enhanced lightweight centernet for PCB surface defect detection. Sensors. **24**(4729) (2024). https://doi.org/10.3390/s24144729
23. Chen, W., Huang, Z., Mu, Q., Sun, Y.: PCB defect detection method based on transformer-YOLO. IEEE Access. **10**, 129480–129489 (2022). https://doi.org/10.1109/ACCESS.2022.3228206
24. Jiang, W., Li, T., Zhang, S., Chen, W., Yang, J.: PCB defects target detection combining multi-scale and attention mechanism. Eng. Appl. Artif. Intell. **123**(106359) (2023). https://doi.org/10.1016/j.engappai.2023.106359
25. Liao, X., Lv, S., Li, D., Luo, Y., Zhu, Z., Jiang, C.: YOLOv4-MN3 for PCB surface defect detection. Appl. Sci. **11**(11701) (2021). https://doi.org/10.3390/app112411701
26. Du, B., Wan, F., Lei, G., Xu, L., Xu, C., Xiong, Y.: YOLO-MBBi: PCB surface defect detection method based on enhanced YOLOv5. Electronics. **12**(2821) (2023). https://doi.org/10.3390/electronics12132821
27. Yang, Y., Kang, H.: An enhanced detection method of PCB defect based on improved YOLOv7. Electronics. **12**(2120) (2023). https://doi.org/10.3390/electronics12092120
28. Wang, X., Gao, H., Jia, Z., Li, Z.: BL-YOLOv8: an improved road defect detection model based on YOLOv8. Sensors. **23**(8361) (2023). https://doi.org/10.3390/s23208361
29. Zou, J., Wang, H.: Steel surface defect detection method based on improved YOLOv9 network. IEEE Access. **12**, 124160–124170 (2024). https://doi.org/10.1109/ACCESS.2024.3453931
30. Sundaresan Geetha, A., Alif, M.A.R., Hussain, M., Allen, P.: Comparative analysis of YOLOv8 and YOLOv10 in vehicle detection: performance metrics and model efficacy. Vehicles. **6**, 1364–1382 (2024). https://doi.org/10.3390/vehicles6030065
31. Wang, Z., Su, Y., Kang, F., Wang, L., Lin, Y., Wu, Q., Li, H., Cai, Z.: PC-YOLO11s: a lightweight and effective feature extraction method for small target image detection. Sensors. **25**(348) (2025). https://doi.org/10.3390/s25020348
32. Bakirci, M., Bayraktar, I.: Synthetic data-enhanced microcrack detection in photovoltaic (PV) cells using YOLO11 for smart industry applications. In: 2025 International Russian Smart Industry Conference (SmartIndustryCon), pp. 46–51. IEEE (2025). https://doi.org/10.1109/SmartIndustryCon65166.2025.10986059
33. Chen, Z., Zhang, F., Liu, H., et al.: Real-time detection algorithm of helmet and reflective vest based on improved YOLOv5. J. Real-Time Image Process. **20**(4) (2023). https://doi.org/10.1007/s11554-023-01268-w
34. Wang, M., Li, F.: Real-time accurate apple detection based on improved YOLOv8n in complex natural environments. Plants. **14**(365) (2025). https://doi.org/10.3390/plants14030365
35. Ran, B., Huang, B., Liang, S., Hou, Y.: Surgical instrument detection algorithm based on improved YOLOv7x. Sensors. **23**(5037) (2023). https://doi.org/10.3390/s23115037
36. Chen, Y., Song, P., Liu, H., Dai, L., Zhang, X., Ding, R., Li, S.: Achieving domain generalization for underwater object detection by domain mixup and contrastive learning. Neurocomputing. **528**, 20–34 (2023). https://doi.org/10.1016/j.neucom.2023.01.053
37. Xin, Z., Lu, T., Li, Y., et al.: MultiCut-MultiMix: a two-level data augmentation method for detecting small and densely distributed objects in large-size images. Vis. Comput. **40**, 2347–2361 (2024). https://doi.org/10.1007/s00371-023-02920-z

Chapter 10
Deep Channel Recalibration for Synthetic Aperture Radar (SAR) Ship Detection Through the Efficient Channel Attention Mechanism in YOLO11 Variants

Murat Bakirci and **Irem Bayraktar**

Abstract Ship detection in Synthetic Aperture Radar (SAR) imagery is a critical yet challenging task due to the inherent noise, clutter, and complex backgrounds present in these images. In this study, we enhance the YOLO11 framework by integrating the Efficient Channel Attention (ECA) mechanism across various model sizes (YOLO11n, YOLO11s, YOLO11 m, YOLO11 l, and YOLO11x) to improve detection performance. The ECA module refines feature extraction by dynamically emphasizing salient channel-wise responses, thereby facilitating more accurate discrimination of ship features from background interference. Extensive experiments conducted on SAR datasets demonstrate that the incorporation of ECA consistently improves key performance metrics with the most pronounced enhancements observed in smaller model variants. Qualitative analyses, supported by heatmap visualizations, further confirm that the ECA-enhanced models more effectively highlight and detect ships within the complex SAR environment. Although one-way ANOVA tests did not reveal statistically significant differences at the 0.05 threshold, diagnostic assessments via Quantile-Quantile plots and residual analysis substantiate an overall positive trend in detection performance attributable to ECA integration. The results underscore the potential of attention-based mechanisms in advancing remote sensing applications, particularly in maritime surveillance, and provide a foundation for future work on optimizing lightweight detection architectures in challenging imaging conditions.

M. Bakirci (✉)
Unmanned/Intelligent Systems Lab, Faculty of Aeronautics and Astronautics, Tarsus University, Mersin, Turkey
e-mail: muratbakirci@tarsus.edu.tr

I. Bayraktar
Department of Aerospace Engineering, Turkish Air Force Academy, National Defence University, Istanbul, Turkey
e-mail: ibayraktar@uislab.org

G. Rivera et al. (eds.), *Recent Advances in Artificial Neural Networks*, Intelligent Systems Reference Library 283, https://doi.org/10.1007/978-3-032-13497-4_10

Keywords Synthetic Aperture Radar (SAR) · Ship detection · YOLO11 · Efficient Channel Attention (ECA) · Deep learning · Remote sensing · Feature extraction

10.1 Introduction

Ship detection in Synthetic Aperture Radar (SAR) imagery is of paramount importance for maritime surveillance, security, and environmental monitoring [1]. SAR's all-weather, day-and-night imaging capabilities make it an indispensable tool in remote sensing, particularly for applications in regions with limited optical visibility. However, the very attributes that make SAR imaging valuable, its ability to penetrate clouds and operate independently of natural light, also introduce significant challenges. SAR images are characterized by inherent noise, speckle, and complex backscatter effects, which can obscure the features of interest, such as vessels, and complicate automated detection efforts [2]. Over the past decades, numerous methodologies have been developed to tackle the problem of ship detection in SAR images. Traditional techniques often relied on thresholding, morphological operations, or statistical models to discriminate between ships and background clutter [3]. Early approaches such as the Constant False Alarm Rate (CFAR) detector [4] utilized adaptive thresholding to differentiate ships from background clutter. While effective in controlled environments, CFAR-based methods often struggled with complex sea conditions, variable ship sizes, and strong clutter interference, limiting their generalizability to real-world SAR datasets. Similarly, model-driven approaches leveraging morphological filtering and edge detection techniques [5] demonstrated some success but were inherently sensitive to image noise and variations in ship orientation. Although these methods can be effective under certain conditions, they typically lack the robustness and adaptability required to handle the wide variability inherent in SAR data.

With the rapid advancement of deep learning, data-driven methodologies have largely supplanted conventional techniques, offering superior generalization and robustness in SAR-based ship detection. Convolutional Neural Networks (CNNs) have played a pivotal role in this paradigm shift, enabling models to learn hierarchical feature representations directly from raw data [6]. Early deep learning-based approaches incorporated standard CNN architectures, such as AlexNet and VGG [7], to classify and localize ships in SAR imagery. However, these models often required extensive pre- and post-processing steps, making them computationally inefficient for real-time applications. The introduction of region-based object detection frameworks, such as Faster R-CNN [8], further improved detection accuracy by leveraging region proposal networks (RPNs) to identify candidate objects, but these approaches suffered from relatively slow inference speeds. Single-stage detectors, particularly those based on the YOLO family, have emerged as a leading solution for high-speed object detection. Studies have demonstrated the effectiveness of

YOLO-based models in SAR ship detection, with modifications aimed at enhancing small-object detection and improving robustness to image artifacts [9].

The YOLO11 framework represents a novel iteration within this family, designed to provide a versatile platform for object detection tasks in challenging imaging scenarios. YOLO11's architecture builds on the successes of previous YOLO models by incorporating advancements in CNNs and optimizing them for real-time applications. Despite its strong performance, the detection of ships in SAR images using the standard YOLO11 framework can still be hampered by the complex and noisy nature of the data [10]. The difficulties stem from the low signal-to-noise ratio and the presence of speckle, as well as from the diversity in ship sizes, orientations, and the variety of background clutter encountered in SAR scenes. To address these challenges, recent research has increasingly focused on the integration of attention mechanisms into deep learning architectures. Attention mechanisms aim to mimic the human visual system by selectively focusing on the most informative parts of an image, thereby enhancing feature extraction and improving detection accuracy [11]. Self-attention mechanisms, such as those used in Vision Transformers (ViTs) [12], have demonstrated significant improvements in various computer vision tasks, but their quadratic computational complexity poses challenges for deployment in resource-constrained environments. Channel attention mechanisms, in contrast, offer a lightweight alternative by adaptively reweighting feature maps without introducing excessive computational overhead. Among the various types of attention mechanisms, channel attention has proven particularly effective in refining the feature representations learned by CNNs [13]. By adaptively re-weighting the importance of each feature channel, channel attention modules enable the network to prioritize salient features while suppressing less informative ones.

In this context, the Efficient Channel Attention (ECA) mechanism [14] has garnered significant interest. Unlike traditional channel attention modules that typically rely on complex operations or dimensionality reduction, ECA introduces a lightweight approach that avoids these potential bottlenecks. ECA refines channel-wise feature weighting without dimensionality reduction, ensuring that essential information is preserved. Unlike traditional attention modules such as the Squeeze-and-Excitation (SE) block [15], which introduces additional complexity through global pooling and fully connected layers, ECA employs a local cross-channel interaction strategy, making it highly efficient for real-time applications. By employing a local cross-channel interaction strategy without dimensionality reduction, ECA is able to capture essential channel dependencies with minimal computational overhead. This makes it particularly well-suited for applications where real-time performance is critical, such as in maritime surveillance using SAR imagery [16]. Several studies have explored the integration of ECA into CNN-based architectures for tasks such as image classification and object detection, demonstrating improved feature discrimination and enhanced robustness to noisy inputs [17]. However, limited research has been conducted on the impact of ECA in SAR ship detection, particularly within the YOLO framework. The motivation behind this study is to explore the potential of integrating the ECA mechanism into the YOLO11 framework to enhance ship detection performance in SAR images. Our work investigates five variants of the YOLO11 algorithm, ranging from the smallest (YOLO11n) to the largest (YOLO11x), to evaluate how the addition of ECA impacts models of varying complexity. By augmenting the baseline YOLO11 models with ECA, we hypothesize that the enhanced models will exhibit improved sensitivity to ship features, thereby resulting in higher detection accuracy, especially in cases where the ship signatures are subtle or partially obscured by noise.

The proposed integration is designed to address several key challenges in SAR-based ship detection. First, SAR images often contain significant speckle noise, which can obscure target features. The ECA module enhances feature maps by emphasizing channels that are more indicative of ship presence, effectively reducing the impact of noise and background clutter. Second, the diversity in ship sizes and orientations requires a detection system that can adapt to various spatial scales and perspectives. By refining the feature extraction process, ECA aids in capturing the multi-scale characteristics of ships, thus facilitating more robust detection across different model variants. Third, while larger models inherently possess higher representational capacity, they can sometimes be less sensitive to minor improvements in feature representation due to diminishing returns. In contrast, smaller models, which are often limited by their reduced capacity, stand to gain significantly from any enhancement in feature extraction. Our study aims to quantify these effects by systematically comparing the performance of standard and ECA-augmented YOLO11 variants. In addition to the quantitative performance metrics such as precision, recall, mean Average Precision (mAP), and F1-score, our study incorporates a range of qualitative analyses. Visualizations in the form of heatmaps and detection comparisons provide tangible evidence of the ECA mechanism's ability to highlight ship features in SAR images. For instance, heatmap visualizations clearly illustrate that the ECA module amplifies the regions corresponding to ships, thereby improving the clarity of these objects against the complex SAR background. Similarly, side-by-side comparisons of detection outputs from the baseline and ECA-augmented models further underscore the practical improvements brought about by the attention mechanism.

Furthermore, statistical validation techniques, including one-way ANOVA tests, are employed to assess the significance of the observed performance enhancements. Although the ANOVA results indicate that the differences in performance metrics between the baseline and enhanced models do not reach statistical significance at the conventional 0.05 level, the overall trend in the raw performance values, combined with diagnostic analyses such as Quantile-Quantile plots and residual histograms, confirms that the integration of ECA contributes positively to detection performance. This finding is critical, as it highlights that the benefits of ECA are more pronounced in the qualitative and relative improvements, particularly for smaller models, rather than in absolute statistical significance. The implications of our findings extend beyond the immediate context of SAR-based ship detection. The demonstrated efficacy of the ECA mechanism in enhancing feature extraction and detection performance suggests that similar attention-based approaches could be applied to other remote sensing tasks characterized by noisy or complex imagery.

Moreover, the ability to achieve significant performance gains with minimal additional computational cost makes ECA an attractive option for real-time applications, where both speed and accuracy are of utmost importance.

This study is centered on a clear objective: to assess the effectiveness of the ECA mechanism in enhancing ship detection performance within the YOLO11 framework when applied to SAR imagery. Our goal is to systematically evaluate how the integration of ECA affects five YOLO11 variants (YOLO11n, YOLO11s, YOLO11 m, YOLO11 l, and YOLO11x), which vary in model complexity and computational cost. Specifically, we aim to: (1) Analyze whether ECA improves detection accuracy and robustness across these model scales, (2) Quantify performance changes using standard object detection metrics (precision, recall, mAP, and F1-score), and (3) Statistically validate these changes through one-way ANOVA and diagnostic analysis. This focused approach enables us to isolate the impact of ECA and provide insight into its practical utility, especially for lightweight models in resource-constrained environments.

10.2 Related Work

Ship detection in SAR images has been a widely researched topic due to its significance in maritime surveillance, border security, and environmental monitoring. Various deep learning-based approaches have been proposed to enhance detection accuracy and robustness, especially in challenging conditions such as occlusions, varying sea states, and cluttered backgrounds. Traditional ship detection methods in SAR images primarily relied on classical image processing and machine learning techniques. These approaches included methods such as the CFAR algorithm, wavelet transformations, and support vector machines (SVMs). While these methods were effective in certain scenarios, they often lacked the robustness required for detecting ships in complex maritime environments. With the advent of deep learning, CNNs have become the dominant approach for ship detection in SAR images. Several studies have demonstrated the effectiveness of CNN-based models, such as Faster R-CNN and Single Shot MultiBox Detector (SSD), in identifying ships with improved accuracy and generalization. However, these models often struggled with real-time processing and localization precision, making them less suitable for applications requiring immediate decision-making. Recent advancements have focused on utilizing fully convolutional networks and transformer-based models for SAR ship detection. Techniques such as DeepLabV3+ and ViTs have been explored for their superior feature extraction capabilities. However, despite these improvements, challenges such as false positives caused by sea clutter and varying radar backscatter remain significant concerns in SAR-based object detection.

Zhang et al. [18] presented a novel deep learning approach for ship detection in SAR images using a combination of CNNs and a RPN. The authors integrated a feature pyramid network (FPN) to enhance multi-scale feature extraction. The results demonstrated significant improvements in detection accuracy, particularly for small and overlapping ships. In Wang et al. [19], the authors proposed a hybrid approach that first applied a classical CFAR detector to identify potential ship candidates and then used YOLOv4 for classification. This two-stage pipeline effectively reduced false positives and highlighted the value of combining traditional SAR methods with modern deep learning architectures. While our approach does not employ CFAR preprocessing, this study underscores the importance of enhancing detection accuracy in SAR imagery, which we address through refined feature extraction using ECA. The transformer-based approach in [20] used a ViT to improve global context modeling in SAR images. Although transformers can capture long-range dependencies, they often introduce substantial computational costs. Our study offers a more lightweight alternative by integrating the ECA mechanism into YOLO11, achieving enhanced detection performance without sacrificing real-time capability, addressing a similar goal but through a different, more efficient means. The GAN-based denoising strategy presented in [21] aims to improve SAR image quality before applying object detection. Although preprocessing with GANs enhances input quality, it adds a separate stage and potential complexity. In contrast, our integration of ECA directly within the detection architecture achieves improved robustness to noise internally, aligning with the same objective of more reliable detection under challenging conditions. In [22], the authors integrated ECA into a YOLOv7-based framework, demonstrating improved detection in noisy SAR environments. This study directly motivates our work by confirming the effectiveness of ECA in attention-based feature recalibration for SAR imagery. However, our contribution is distinct in that it evaluates the performance impact of ECA across multiple YOLO11 variants, offering a broader assessment of its benefits within a new architectural generation. These studies highlight the advancements in ship detection methodologies, from traditional CFAR approaches to state-of-the-art deep learning-based models. The integration of attention mechanisms, transformers, and GANs has significantly enhanced detection performance, offering promising directions for future research.

While previous research has demonstrated the potential of attention mechanisms, such as Squeeze-and-Excitation (SE) and transformer-based modules, in object detection and remote sensing, their integration into newly emerging architectures like YOLO11 remains largely unexplored. Most existing studies focus on earlier YOLO versions or alternative backbones, and few provide a comparative evaluation across model scales within the same family. Our work addresses this gap by introducing ECA into the full range of YOLO11 variants and evaluating its performance on SAR ship detection, a task where noise robustness and efficient feature extraction are critical.

The YOLO series has significantly transformed object detection by offering a balance between speed and accuracy. Early versions, such as YOLOv3 and YOLOv4, were adapted for ship detection in SAR images, demonstrating improved performance in terms of efficiency and accuracy. These models integrated techniques like anchor boxes, spatial pyramid pooling, and cross-stage partial networks to enhance feature extraction and detection capabilities. YOLOv5 introduced further optimizations in terms of model architecture, training efficiency, and augmentation techniques. Researchers have leveraged YOLOv5 for various SAR-based object detection tasks, including ship detection, oil spill monitoring, and iceberg identification. The model's ability to provide real-time inference has made it particularly suitable for maritime applications where timely information is crucial. Subsequent advancements in the YOLO family, such as YOLOv6 and YOLOv8, incorporated transformer-based modules and enhanced feature fusion mechanisms to improve detection accuracy. These models have been applied in SAR imagery analysis, achieving state-of-the-art results in ship detection while maintaining computational efficiency. YOLO11, the latest iteration in the YOLO series, aims to further refine object detection by introducing novel enhancements in backbone architecture, feature fusion, and attention mechanisms. By integrating elements such as dynamic convolution and improved activation functions, YOLO11 has demonstrated significant improvements in accuracy and generalization.

These previous efforts lay the groundwork for understanding the evolution of deep learning-based SAR ship detection. For instance, the work by Zhang et al. [18] on combining CNNs with FPNs demonstrates the value of multi-scale feature extraction, which informs our decision to preserve feature richness through ECA integration. Similarly, the hybrid approach in Wang et al. [19], which pairs traditional CFAR with YOLOv4 classification, highlights the challenges posed by sea clutter, an issue we address through attention-driven refinement of feature channels. The use of ViTs in [20] illustrates the advantage of global context modeling, although their computational cost reinforces our motivation for using a lightweight alternative like ECA. Studies such as [22], which embed attention into YOLOv7, confirm the feasibility and effectiveness of attention in enhancing SAR detection but do not explore the newer YOLO11 framework, which is central to our contribution. Collectively, these works justify our focus on a lightweight attention mechanism (ECA) and emphasize the originality of applying it to the recently introduced YOLO11 architecture in the SAR domain.

The research in [23] adapted YOLOv4 for ship detection in SAR images by incorporating a spatial attention module to enhance feature extraction. The modified model demonstrated improved precision and recall, particularly for detecting small ships in complex maritime environments. In [24], the authors developed a YOLOv5-based framework optimized for SAR image analysis. By integrating a multi-scale feature fusion strategy, the model improved detection accuracy and robustness against noise and background clutter. The study also explored data augmentation techniques to enhance generalization. Tang et al. [25] introduced a novel YOLOv8-based ship detection model that utilized a transformer-based feature enhancement module. The proposed model leveraged self-attention mechanisms to improve spatial feature representation, leading to significant performance gains in ship detection under varying sea conditions. Another study focused on modifying YOLOv8-nano for SAR ship detection by incorporating deformable convolutions and an adaptive anchor-free detection head [26]. The proposed improvements led to higher detection accuracy and better generalization across different SAR datasets, demonstrating superior performance compared to traditional anchor-based YOLO models. In [27] an enhanced YOLOv7 framework has been proposed for ship detection in SAR images. The model integrated a custom channel attention mechanism to refine feature selection, reducing false positives caused by sea clutter and improving detection robustness in challenging maritime environments. The research mentioned above underscore the continuous evolution of YOLO-based object detection models for SAR image analysis. The integration of attention mechanisms, transformer modules, and adaptive feature extraction techniques has significantly enhanced detection performance.

Attention mechanisms have become an essential component in modern deep learning architectures, improving the focus on relevant features while suppressing noise. The ECA mechanism has gained popularity due to its ability to enhance feature selection without significantly increasing computational complexity. Studies have shown that integrating ECA into CNN architectures improves object detection performance by enabling better feature representation. The ECA mechanism achieves this by adaptively recalibrating channel-wise feature responses based on local cross-channel interactions. Unlike traditional attention mechanisms such as SE networks, ECA does not introduce significant overhead, making it an efficient solution for real-time applications. Researchers have explored the application of ECA in various object detection models, including YOLO-based architectures. By embedding ECA into backbone networks, studies have demonstrated improvements in detection accuracy, robustness, and generalization. For example, integrating ECA into YOLOv5 and YOLOv6 has been shown to enhance feature discrimination, reducing false positives and improving localization precision. Other deep learning models, such as ResNet and EfficientNet, have also benefited from ECA-based enhancements. By refining feature extraction processes, these models have achieved superior performance in detecting objects across diverse datasets, including SAR imagery.

[28] explored the integration of the ECA mechanism into a ResNet-based object detection model. The authors demonstrated that ECA improved feature selection by enhancing channel-wise attention, leading to better object localization and increased detection accuracy in challenging environments. In [29], the authors modified YOLOv7 by embedding the ECA mechanism into the backbone network. Their experiments on multiple datasets showed that ECA-enhanced YOLOv7 outperformed conventional models in detecting small and densely packed objects, improving both precision and recall. Another research incorporated the ECA mechanism into Faster R-CNN for vehicle detection in aerial imagery [30]. The results indicated that ECA significantly improved the model's ability to differentiate foreground objects from cluttered backgrounds, reducing false positives and enhancing overall detection robustness. In [31], the authors introduced an ECA-enhanced SSD for real-time object detection. By integrating ECA into the feature pyramid network, the model achieved superior feature representation, particularly for detecting objects in low-contrast conditions, making it ideal for SAR imagery applications. On the other hand, the work in [32] applied the ECA mechanism to YOLOv5, improving feature extraction efficiency without adding computational complexity. The enhanced model demonstrated notable performance improvements

in detecting objects across multiple domains, including maritime surveillance, aerial monitoring, and autonomous driving.

Although recent studies have explored the use of attention mechanisms in conjunction with YOLO-based architectures for SAR image analysis, they have predominantly focused on earlier versions such as YOLOv5, YOLOv7, or YOLOv8. For example, attention-enhanced versions of YOLOv7 and YOLOv8 have been proposed for ship detection by integrating modules such as SE-Net and CBAM; however, these implementations often incurred higher computational costs and were not designed for the specific architectural characteristics of YOLO11. To the best of our knowledge, our study is the first to investigate the integration of an efficient channel attention mechanism, specifically ECA, into the YOLO11 architecture for SAR ship detection. By focusing on ECA, which introduces minimal computational overhead, we demonstrate that even lightweight attention modules can yield notable performance gains in SAR environments without compromising real-time detection efficiency. This highlights the practical applicability of our approach, especially in scenarios where deployment on resource-constrained platforms is essential.

10.3 Methodology

This study investigates the impact of integrating the Efficient Channel Attention (ECA) mechanism into the YOLO11 framework for ship detection in SAR imagery. The methodology consists of multiple stages, including dataset preprocessing, model architecture modifications, training and evaluation, and statistical validation using ANOVA. By incorporating ECA into the YOLO11 variants, we aim to assess how channel-wise feature recalibration affects detection performance across models of varying complexity. Prior to model training, the SAR images underwent a comprehensive preprocessing pipeline to mitigate the effects of noise and enhance the visibility of ship targets. First, a multi-look filtering technique was employed to suppress speckle noise, a granular interference pattern commonly found in SAR data. This approach averages multiple looks of the same scene to preserve structural information while reducing high-frequency noise. Next, we applied Contrast Limited Adaptive Histogram Equalization (CLAHE) to enhance local contrast and improve the visibility of ships, particularly in low-contrast or cluttered areas of the sea surface. All pixel intensity values were then normalized to a range of [0, 1] to ensure consistent scale for neural network input and to promote faster convergence during training.

To increase the diversity of the training data and improve model generalization, several data augmentation techniques were applied. These included random rotations (up to $\pm 15^\circ$), horizontal and vertical flipping, scaling ($\pm 10\%$), and translation. These augmentations simulate real-world variations in ship orientation, size, and position, thus enhancing the robustness of the trained models. Ground-truth annotations were manually reviewed and adjusted as needed to ensure accurate alignment with the augmented images. This preprocessing strategy ensured that the input data was both clean and diverse, allowing the models to learn meaningful representations from the SAR imagery. To achieve a comprehensive analysis, we evaluate performance based on standard object detection metrics, precision, recall, mean Average Precision (mAP), and F1-score, and validate the statistical significance of performance differences using one-way ANOVA. SAR images differ from conventional optical images due to their distinct imaging mechanism, which relies on microwave signals rather than visible light. This results in unique challenges such as speckle noise, variable backscatter, and strong clutter interference. To ensure robust model performance, dataset preprocessing plays a crucial role in mitigating these effects. The dataset used in this study comprises SAR images containing ships of various sizes and orientations, captured under diverse sea conditions.

The SAR dataset used in this study comprises a total of 2255 labeled images compiled from publicly available sources and established SAR ship detection benchmarks. Each image is annotated with bounding boxes indicating the presence and location of ships, and the annotations were manually verified to ensure accuracy. The dataset encompasses a wide variety of maritime scenarios, including different ship types (e.g., cargo vessels, tankers, fishing boats), sizes (from small boats to large commercial ships), and orientations. It also covers a broad range of environmental and sea conditions, such as calm waters, moderate waves, and rough seas, as well as varying levels of background clutter caused by coastal structures, ports, and atmospheric interference. Approximately 40% of the dataset comprises challenging cases involving low-contrast or partially occluded ships. The images were collected from diverse maritime regions and include both high-resolution and low-resolution SAR imagery, with resolutions ranging from 256×256 to 640×640 pixels. The high-resolution subset (about 60% of the dataset) reveals detailed ship contours, while the low-resolution subset introduces significant noise and clutter, mimicking real-world SAR detection challenges. This diversity ensures that the dataset reflects operational maritime surveillance environments, thereby enhancing the robustness and generalizability of the proposed detection approach. For experimentation, the dataset was split into 70% for training, 15% for validation, and 15% for testing, maintaining similar distributions of ship types, sizes, and scene complexities across all subsets.

The images include high-resolution data where ships are clearly distinguishable as well as low-resolution samples where ship detection is more challenging due to noise and clutter. To optimize the SAR images for deep learning-based detection, a multi-look filtering technique was used to suppress speckle noise while preserving important ship features. Additionally, adaptive histogram equalization was applied to improve image contrast and highlight ships against the background. Pixel intensities were normalized to a range of [0,1] to facilitate stable training and convergence. Moreover, to improve model generalization, a series of augmentations were applied, including random rotation, flipping, and scaling. Ground truth annotations were reviewed to ensure accurate bounding box placements, as errors in labeling can significantly impact training performance. The preprocessed images were then used as input for training the YOLO11 variants, both with and without the ECA module, to allow direct performance comparisons.

10.3.1 YOLO11 and ECA Mechanism

The YOLO11 object detection framework serves as the backbone of our study due to its speed and accuracy in real-time applications. YOLO11 follows the single-shot detection paradigm, meaning it predicts bounding boxes and class probabilities in a single forward pass, making it highly efficient for ship detection in large-scale SAR imagery. The study evaluates five YOLO11 variants, YOLO11n, YOLO11s, YOLO11 m, YOLO11 l, and YOLO11x, each differing in model capacity and computational complexity. YOLO11n and YOLO11s are lightweight models optimized for deployment on resource-constrained devices, while YOLO11 m, YOLO11 l, and YOLO11x feature progressively deeper architectures with greater feature extraction capabilities. To enhance the feature extraction capability of YOLO11, the Efficient Channel Attention (ECA) module was integrated into the architecture. ECA is a lightweight attention mechanism designed to refine channel-wise feature representations without introducing excessive computational overhead. ECA adaptively re-weights each channel based on its importance in detecting ship features. Unlike traditional attention mechanisms such as the Squeeze-and-Excitation (SE) module, which reduces dimensionality via fully connected layers, ECA employs a more efficient cross-channel interaction approach. Instead of learning global dependencies, ECA applies a 1D convolutional kernel to capture interactions between neighboring channels, ensuring that important features are emphasized without overfitting to specific patterns. Unlike spatial attention mechanisms that may blur fine-grained details, ECA maintains the integrity of spatial feature maps while selectively enhancing the most relevant channels. The ECA module was integrated into the neck of YOLO11, as shown in Fig. 10.1, to refine channel activations before region proposal and object classification. This enhancement allows the network to focus more effectively on ship features, even in challenging SAR environments.

The training process was conducted using a supervised learning approach with labeled SAR ship detection datasets. The models were trained with and without the ECA module to facilitate direct comparisons. A combination of localization loss (bounding box regression), classification loss (cross-entropy for ship presence), and

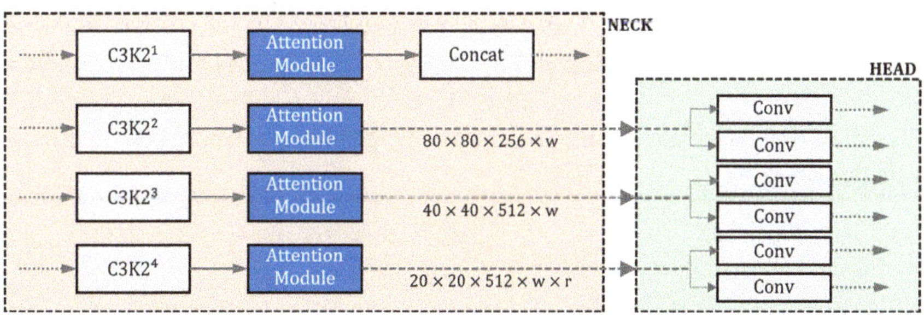

Fig. 10.1 Integration of the ECA mechanism into the Neck section of the YOLO11 architecture for enhanced feature refinement and improved object detection

confidence loss (objectness score) was used to optimize detection accuracy. Stochastic Gradient Descent (SGD) with momentum was employed to enhance convergence stability. The learning rate was initialized at 0.001 and adjusted dynamically using a cosine decay schedule. The models were trained for 150 epochs with a batch size of 16 to balance computational efficiency with convergence stability. The choice of training hyperparameters was guided by a preliminary grid search experiment conducted on a subset of the dataset. For the learning rate, candidate values of {0.1, 0.01, 0.001, 0.0005} were tested, and a value of 0.001 was found to provide the best trade-off between fast convergence and stable training. For optimization, both Adam and SGD with momentum were evaluated. SGD with a momentum factor of 0.9 consistently yielded better generalization and more stable loss reduction on the validation set. Similarly, batch sizes of {8, 16, 32} were compared, and a batch size of 16 offered the most efficient use of computational resources while maintaining stable training dynamics. This combination of hyperparameters aligns with best practices for training object detection models on SAR imagery, where careful tuning is required to address noise and class imbalance. Optimized anchor box sizes were generated using k-means clustering on ship bounding box distributions to improve detection accuracy. To quantitatively assess detection performance, the key performance metrics, including precision, recall, mAP and F1-score, were computed for each YOLO11 variant. Comparisons between standard YOLO11 and ECA-enhanced YOLO11 models were conducted to evaluate the effectiveness of attention-based feature refinement.

To evaluate the detection performance of each model variant, we utilized four standard object detection metrics: Precision, Recall, F1-score, and mean Average Precision (mAP). These metrics were computed as follows: Precision measures the proportion of correctly identified ship instances among all instances predicted as ships and is defined as:

$$\text{Precision} = \frac{\text{True Positives (TP)}}{\text{True Positives (TP)} + \text{False Positives (FP)}} \tag{10.1}$$

Recall represents the proportion of actual ship instances correctly identified by the model and is given by:

$$\text{Recall} = \frac{\text{True Positives (TP)}}{\text{True Positives (TP)} + \text{False Negatives (FN)}} \tag{10.2}$$

F1-score is the harmonic mean of precision and recall and balances both metrics in a single value:

$$\text{F1} - \text{Score} = 2\frac{\text{Precision} \times \text{Recall}}{\text{Precision} + \text{Recall}} \tag{10.3}$$

The mean Average Precision (mAP) quantifies detection accuracy across different Intersection over Union (IoU) thresholds. First, Average Precision (AP) is computed as the area under the precision-recall curve for a given class, and then mAP is calculated by averaging the AP values across all object classes and IoU thresholds. Since our task involves a single class (ship), mAP is computed as the area under the precision-recall curve at an IoU threshold of 0.5 (denoted as mAP^{50}), following standard object detection benchmarking practices. All metrics were computed on the test dataset using predictions generated by the trained models and compared against ground-truth annotations. A detection was considered a true positive if the IoU between the predicted and ground-truth bounding boxes was ≥ 0.5.

10.3.2 Statistical Validation Through ANOVA

The motivation for employing statistical validation techniques, particularly one-way ANOVA [33], stems from the need to move beyond descriptive performance comparisons and assess whether the observed improvements in detection metrics are statistically significant. While metrics such as precision, recall, mAP, and F1-score indicate how well the models perform, they do not provide information on the variability of these results or whether observed differences between models are due to random fluctuations in data or training. One-way ANOVA allows us to test whether performance differences across multiple model variants (with and without ECA integration) are statistically distinguishable. This is especially relevant when evaluating machine learning models on real-world data such as SAR imagery, where stochastic training factors and inherent data noise can introduce performance variance. By applying ANOVA, we aim to validate the consistency of ECA's impact and determine whether performance improvements are attributable to the integration of attention mechanisms rather than random chance. Thus, the statistical validation serves as an essential complement to both the numerical metrics and the qualitative visual analyses provided in this study.

To determine whether the observed performance improvements due to ECA augmentation were statistically significant, we conducted one-way Analysis of Variance (ANOVA) tests across the ten models (five standard YOLO11 variants and five ECA-augmented variants). ANOVA is a statistical method used to determine whether there are any significant differences between the means of multiple groups. In this study, we employed one-way ANOVA to assess the performance of different YOLO11 variants in detecting ships in SAR images. The performance metrics under consideration, precision, recall, F1-score, and mean Average Precision (mAP), were treated as dependent variables, and the different YOLO11 variants were treated as independent factors. Through the ANOVA method, it becomes possible to evaluate whether model variant selection significantly affected the performance metrics.

In a typical one-way ANOVA, we test the null hypothesis H_0, which states that all group means are equal, against the alternative hypothesis H_A, which suggests that at least one group mean differs from the others. The ANOVA test relies on partitioning the total variability in the data into two components: The first one is "between-group variance (σ_b)". This captures the variability due to the differences between the group means (in this case, the differences between YOLO11 variants). The second one is "within-group variance (σ_w)". This one captures the variability within each group (i.e., how much individual data points deviate from their group mean).

The total variance, σ_Σ, is given by the equation:

$$\sigma_\Sigma = \sigma_b + \sigma_w \tag{10.4}$$

From this, we calculate the mean squares by dividing the sum of squares by the respective degrees of freedom (δ). Thus, the mean square (Φ) between groups, and mean square within groups can be expressed as follows:

$$\Phi_b = \frac{\sigma_b}{\delta_b} \tag{10.5}$$

$$\Phi_w = \frac{\sigma_w}{\delta_w} \tag{10.6}$$

Then, the F-statistic is calculated as the ratio of the mean square between groups to the mean square within groups:

$$F = \frac{\Phi_b}{\Phi_w} \tag{10.7}$$

A higher F-statistic indicates that the group means differ significantly. The F-statistic is compared to a critical value from the F-distribution table at a given significance level, α, to determine whether to reject the null hypothesis. If the calculated F-statistic exceeds the critical value, the null hypothesis is rejected, indicating that there is a significant difference in performance between the model variants.

In our study, we performed one-way ANOVA separately for each performance metric. For each metric, the independent variable was the YOLO11 variant, and the dependent variable was the observed value of the performance metric. For instance, the precision values obtained for each model variant on ship detection task were treated as a group, and the same procedure was repeated for the recall, F1-score, and mAP values. Each of these groups (precision, recall, etc.) was analyzed independently to determine if the performance of the variants was significantly different in the detection of ships.

Mathematically, for precision on ship detection, the null hypothesis can be formulated as:

$$H_0 = \mu_{11n} = \mu_{11s} = \mu_{11m} = \mu_{11l} = \mu_{11x} \tag{10.8}$$

where μ_{11n}, μ_{11s}, μ_{11m}, μ_{11l}, μ_{11x} represent the mean precision values for the ship detection task using each of the five YOLO11 variants. The alternative hypothesis is that at least one of these means differs, which is expressed as:

$$H_A : \text{At least one } \mu_i \neq \mu_j \text{ for some } i, j \in \{11n, 11s, 11m, 11l, 11x\} \tag{10.9}$$

The process is repeated for recall, F1-score, and mAP metrics for the detection of ships. After calculating the F-statistic for each analysis, we compared it to the F-distribution at a significance level of $\alpha = 0.05$ to determine whether the differences in performance were statistically significant.

If the ANOVA results indicate significant differences between the group means, post-hoc testing is required to determine which specific pairs of model variants are different. In this study, we performed Tukey's Honest Significant Difference (HSD) test [34], which compares the mean differences between every pair of YOLO11 variants while controlling the overall error rate.

The Tukey HSD test calculates the studentized range statistic q, which is given by:

$$q = \frac{\left|\overline{x}_i - \overline{x}_j\right|}{\sqrt{\frac{\Phi_w}{n}}} \tag{10.10}$$

where $\overline{x}_i$ and $\overline{x}_j$ are the means of the two model variants being compared, and n is the number of observations in each group. If the calculated q-value is larger than the critical value from the studentized range distribution for the given degrees of freedom and significance level, the difference between the two groups is deemed statistically significant. For example, if the F-statistic from the ANOVA indicates a significant difference in the mAP scores between the YOLO11n and YOLO11x variants, Tukey's HSD would be used to compare their means and determine if the difference is significant. This helps pinpoint which specific models (e.g., YOLO11n vs. YOLO11x) contribute to the overall significant differences in performance.

The one-way ANOVA assumes that the data for each group are normally distributed, and that the variances are homogeneous across groups (homoscedasticity). In the context of our study, these assumptions were verified using diagnostic (Q-Q) plots [35] and formal tests for equality of variances. If any of these assumptions were violated, alternative statistical methods such as Welch's ANOVA or non-parametric tests (e.g., Kruskal-Wallis test) would be considered. Additionally, while ANOVA identifies significant differences between the means of the groups, it does not provide direct information about the magnitude or direction of the differences. This is where post-hoc tests such as Tukey's HSD become essential, as they provide pairwise comparisons and allow us to draw more specific conclusions about which variants outperform others.

ANOVA, followed by post-hoc testing, provides a robust statistical framework for assessing the impact of different YOLO11 variants on the performance of SAR ship detection tasks. By partitioning the variance in performance metrics, ANOVA helps us determine whether the choice of model variant significantly affects the accuracy and efficiency of ship identification. This method ensures that the results of our study are statistically sound and provides a more comprehensive analysis of the trade-offs involved in selecting the most appropriate YOLO11 model for SAR maritime monitoring applications.

10.4 Results

Figure 10.2 illustrates the impact of the ECA mechanism on feature extraction by displaying sample images. In the upper row, four distinct SAR images are presented, each containing ships under varied conditions, which serve as representative examples of the dataset used. Directly beneath each SAR image, the corresponding heatmap is shown in the lower row. The heatmaps are generated by the ECA-augmented YOLO11 variant and use a color gradient to indicate the intensity of feature activations. The stark contrast between the original SAR images and their heatmaps clearly demonstrates that the ECA module effectively amplifies the channels corresponding to the ship features, making them more discernible against the complex background of the SAR imagery. This enhanced focus on critical regions is a direct consequence of the ECA's ability to refine spatial-channel interactions, thereby ensuring that even subtle or partially occluded ship features are captured more reliably. This visual evidence underpins the subsequent quantitative

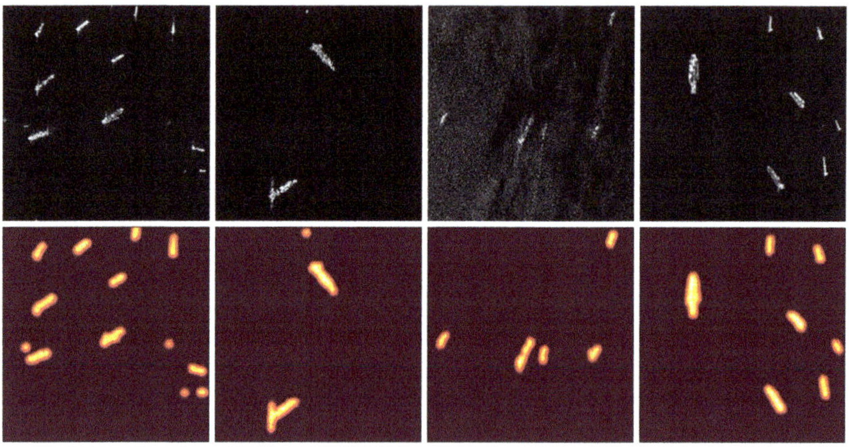

Fig. 10.2 Visualization of ECA-augmented feature extraction in SAR ship images. The upper row displays four representative SAR images containing ships, and the lower row shows the corresponding heatmaps produced by the ECA-augmented YOLO11 model. The clear demarcation of ships in the heatmaps confirms that the ECA module effectively refines channel responses to highlight critical regions in the images. The enhanced feature extraction shows the mechanism's capability to improve detection performance, particularly in challenging SAR environments where subtle differences can be crucial for accurate ship identification

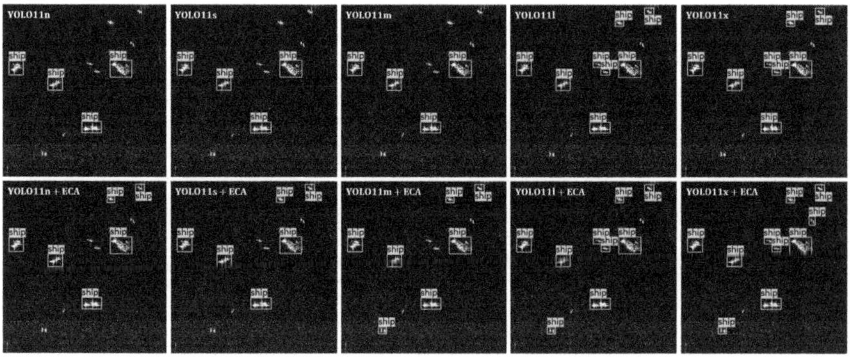

Fig. 10.3 Comparative detection performance of YOLO11 variants with and without ECA augmentation. The upper row shows detection results from the original YOLO11 variants, while the lower row displays the detection results from the corresponding ECA-augmented models. ECA-augmented variants detect a greater number of ships and exhibit fewer false negatives, with the improvement being particularly evident in the smaller models

analyses, by confirming that the ECA mechanism facilitates a more robust and accurate detection process right from the feature extraction stage.

Figure 10.3 provides a comprehensive comparison of detection performance across different YOLO11 variants with and without the ECA enhancement. The upper row displays detection outputs from the original YOLO11 variants, ranging from the smallest model (YOLO11n) on the far left to the largest variant (YOLO11x) on the far right, applied to an identical SAR image. In contrast, the lower row presents the detection results from the corresponding ECA-augmented versions of these models. A detailed examination of these side-by-side comparisons reveals a clear pattern: the ECA-augmented models consistently detect a greater number of ships compared to their baseline counterparts. This improvement is particularly pronounced in the smaller variants, such as YOLO11n and YOLO11s, where the enhancement in feature extraction translates into a substantial increase in detection sensitivity. For larger models (YOLO11 l and YOLO11x), while the baseline performance is already high, the integration of ECA still contributes to marginal improvements by refining the decision boundaries and reducing false negatives. The consistent upward shift in detection performance across all variants, as visually demonstrated in this figure, provides a compelling qualitative corroboration of the quantitative improvements. This figure, therefore, validates the technical efficacy of incorporating the ECA module, and establishes its practical relevance in enhancing ship detection accuracy in SAR imagery.

The performance results from tests conducted with all YOLO11 variants are summarized in Table 10.1. The results indicate that integrating the ECA mechanism into the YOLO11 architecture consistently enhances ship detection performance across all model variants when applied to SAR images. The improvement is evident across all evaluation metrics. The enhancement is particularly pronounced in smaller model variants, such as YOLO11n and YOLO11s, where the ECA augmentation

Table 10.1 Performance metrics for YOLO11 variants with and without ECA augmentation

Performance Metrics	Model Variant					
	YOLO11n	YOLO11n + ECA	YOLO11s	YOLO11s + ECA	YOLO11 m	YOLO11 m + ECA
P	0.541	0.654	0.690	0.774	0.805	0.857
R	0.653	0.721	0.744	0.773	0.809	0.829
mAP	0.572	0.683	0.711	0.755	0.780	0.818
F1-Score	0.591	0.685	0.716	0.773	0.807	0.843

Performance Metrics	Model Variant			
	YOLO11 l	YOLO11 l + ECA	YOLO11x	YOLO11x + ECA
P	0.881	0.902	0.933	0.946
R	0.857	0.872	0.909	0.922
mAP	0.834	0.866	0.913	0.925
F1-Score	0.869	0.888	0.921	0.934

For each model, both the baseline performance and the enhanced performance obtained through the integration of the ECA module are reported. The incorporation of ECA consistently improves detection accuracy across all variants, with the most pronounced enhancements observed in the smaller models (YOLO11n and YOLO11s) leads to significant performance gains. For YOLO11n, the incorporation of ECA results in an increase in precision from 0.541 to 0.654, recall from 0.653 to 0.721, mAP from 0.572 to 0.683, and F1-score from 0.591 to 0.685. This indicates that the ECA mechanism helps the smallest variant of the model better capture relevant feature representations, thereby reducing false positives and improving overall detection accuracy. Similarly, for YOLO11s, ECA augmentation boosts precision from 0.690 to 0.774, recall from 0.744 to 0.773, mAP from 0.711 to 0.755, and F1-score from 0.716 to 0.773. The smaller YOLO11 variants benefit greatly from the ECA mechanism, likely due to its ability to refine feature maps and enhance spatial-channel interactions without significantly increasing computational overhead.

For the mid-sized YOLO11 m, the ECA-enhanced model also demonstrates a noticeable improvement, with precision increasing from 0.805 to 0.857, recall from 0.809 to 0.829, mAP from 0.780 to 0.818, and F1-score from 0.807 to 0.843. The impact of ECA is slightly reduced in larger models such as YOLO11 l and YOLO11x, where baseline performances are already high. Nonetheless, improvements are still observed: YOLO11 l with ECA achieves a precision of 0.902 (vs. 0.881), recall of 0.872 (vs. 0.857), mAP of 0.866 (vs. 0.834), and F1-score of 0.888 (vs. 0.869). Likewise, for YOLO11x, precision improves from 0.933 to 0.946, recall from 0.909 to 0.922, mAP from 0.913 to 0.925, and F1-score from 0.921 to 0.934. These performance values indicate that while larger models already exhibit high detection capabilities, ECA still provides meaningful refinements in feature extraction and classification performance. The results demonstrate that ECA augmentation systematically enhances the YOLO11 framework for ship detection in SAR images. The largest relative improvements are observed in smaller models, where feature representation is inherently constrained, suggesting that the attention mechanism plays a critical role in optimizing lightweight architectures. While the gains diminish as model complexity increases, the enhancements in precision and recall across all variants confirm the efficacy of ECA in improving detection robustness.

The observation that the ECA mechanism has a more substantial impact on smaller models than on larger ones can be attributed to the inherent architectural characteristics of these models. Smaller YOLO11 variants, such as YOLO11n and YOLO11s, are designed with reduced depth and fewer convolutional filters to prioritize efficiency and deployment on resource-constrained devices. This results in a more limited capacity to capture complex patterns and perform nuanced feature discrimination. When ECA is introduced into these lightweight architectures, it acts as an effective compensatory mechanism by dynamically enhancing the most informative channels during feature extraction. This targeted recalibration significantly improves the models' sensitivity to ship-related features, which is particularly valuable in SAR imagery characterized by high noise and low contrast. In contrast, larger YOLO11 variants such as YOLO11 l and YOLO11x already possess deep and expressive feature extraction layers capable of modeling complex spatial and semantic relationships. These models inherently include multiple pathways for learning hierarchical representations, which reduces the relative impact of additional attention mechanisms like ECA. Consequently, while ECA still contributes positively by refining feature maps, the improvements are less dramatic because these architectures already operate near optimal performance for the task. The diminishing returns observed in larger models are therefore a result of the saturation in representational power, where the marginal benefit of ECA becomes increasingly incremental.

To statistically validate the performance improvements introduced by the ECA mechanism across different YOLO11 variants, a one-way ANOVA test was performed. The ANOVA test determines whether the differences in performance metrics between the standard YOLO11 models and their ECA-augmented counterparts are statistically significant. Given that we have ten distinct models (five standard YOLO11 variants and five ECA-augmented YOLO11 variants), ANOVA helps assess whether the observed variations in performance metrics are due to the introduction of ECA or are merely a result of random fluctuations in model performance. Based on the performance values, we conducted four separate ANOVA tests, one for each performance metric, and the results are presented in Fig. 10.4. The null hypothesis (H_0) states that there is no significant difference in the mean performance values among the ten models, while the alternative hypothesis (H_A) posits that at least one model performs significantly differently from the others. The significance level, α, is set to 0.05.

At first glance, the ANOVA test results suggest that the differences in performance metrics between the standard YOLO11 variants and their ECA-augmented counterparts are not statistically significant at the conventional 0.05 significance level. The F-statistics for precision, recall, mAP, and F1-score are 0.420, 0.259, 0.438, and 0.359, respectively, with corresponding p-values of 0.535, 0.625, 0.527,

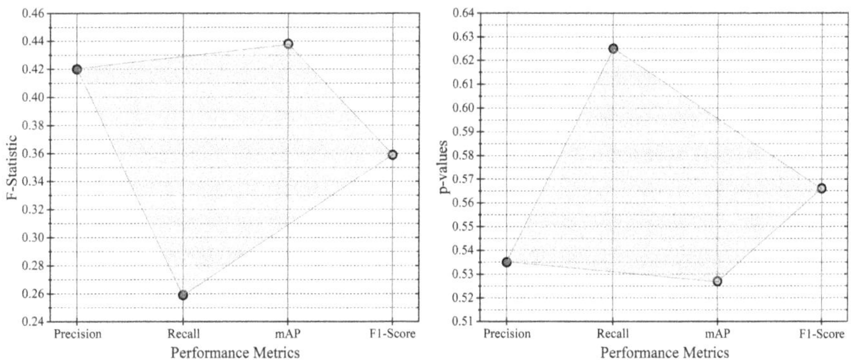

Fig. 10.4 One-way ANOVA test results for detection performance metrics. The figure displays the calculated F-statistics and associated p-values for each metric. Although the ANOVA tests indicate that the differences in performance between the standard and ECA-augmented models are not statistically significant at the 0.05 level, these results provide critical outcomes on the variability and distribution of the performance enhancements introduced by the ECA module. The statistical analysis serves to contextualize the observed improvements, suggesting that while ECA contributes to better numerical performance, the overall distribution of performance gains remains relatively uniform across model variants

and 0.566. Since all p-values are well above the 0.05 threshold, we fail to reject the null hypothesis (H_0), suggesting that the observed performance improvements with ECA augmentation could be due to random variation rather than a systematic effect of the attention mechanism. These results imply that while the ECA-enhanced YOLO11 models exhibit higher numerical performance values, the variation between standard and ECA-augmented models does not reach a statistically significant level. The absence of statistically significant results in the ANOVA tests does not necessarily imply that the ECA mechanism lacks meaningful impact. Rather, this outcome may be attributed to multiple sources of variability. First, SAR imagery is inherently noisy and diverse, with varying environmental conditions and target characteristics (e.g., ship sizes, orientations, and clutter levels), which introduces substantial within-group variance. This high intra-group variability can obscure between-group differences, especially when evaluating small numerical gains. Second, the magnitude of ECA's improvement is not uniform across all model variants, while smaller models (YOLO11n, YOLO11s) benefit substantially from ECA integration, larger models (YOLO11 l, YOLO11x) show only marginal gains. This leads to overlapping performance distributions that reduce the likelihood of detecting statistically significant differences via ANOVA. Despite this, the consistent upward trend in raw performance metrics, supported by residuals analysis, Q-Q plots, and qualitative detection outputs, indicates a systematic improvement attributable to ECA. Thus, while statistical significance is limited, the evidence supports the conclusion that ECA contributes to more robust detection, especially in lightweight models. However, the ECA mechanism enhances each YOLO11 variant differently. Larger variants already perform well in detection, so they benefit only marginally from ECA, while smaller models see a much more significant improvement. As a result, the gains in precision, recall, mAP, and F1-score are relatively uniform between the standard and ECA-augmented models, leading to considerable overlap in their distributions. This overlap reduces the likelihood of detecting statistically significant differences, which explains the lack of significant results in the ANOVA test. Nonetheless, the raw performance values clearly indicate that ECA augmentation tends to improve detection accuracy.

To further examine pairwise differences between specific YOLO11 variants, a post-hoc Tukey's Honest Significant Difference (HSD) test was conducted for each performance metric following the one-way ANOVA. The Tukey HSD test compares the means of each model pair while controlling for family-wise error rate, providing insight into which specific comparisons yield significant or near-significant differences. Although none of the pairwise comparisons reached statistical significance at the 0.05 level, the test revealed that the performance improvements observed in YOLO11n-ECA and YOLO11s-ECA were consistently higher than their baseline counterparts across all metrics. For example, the precision difference between YOLO11n-ECA and YOLO11n was among the largest mean differences observed, albeit not statistically significant. These findings support the conclusion that the ECA mechanism contributes most notably to lightweight models, reinforcing the qualitative and numerical trends reported earlier. The absence of significant differences across all pairs further highlights the effect of overlapping performance distributions and justifies the need to consider multiple forms of evidence (visualizations, raw metrics, and residuals) when evaluating enhancement techniques in SAR detection tasks.

While the absence of statistical significance might initially suggest the possibility of random effects, the consistently observed improvements in detection metrics across all model variants argue for a systematic benefit. Every ECA-augmented YOLO11 model demonstrated higher precision, recall, mAP, and F1-score than its non-ECA counterpart, with no cases of performance degradation. This trend holds across different model sizes and evaluation dimensions. Furthermore, qualitative results (e.g., heatmap focus and reduced false negatives in visual outputs) align with this pattern, reinforcing the reliability of the observed gains. These findings suggest that the enhancements are unlikely to be due to chance alone. It is also important to recognize that in high-variance domains such as SAR imaging, statistical significance may not always capture the full practical impact of model improvements. Therefore, while we acknowledge the statistical limitations, we interpret the improvements as consistent and meaningful in the context of operational ship detection applications.

The Quantile-Quantile (Q-Q) plots for performance metrics provide a statistical visualization of the distribution of the performance metrics obtained from the YOLO11 and ECA-augmented YOLO11 models. The plots compare the empirical quantiles of the observed data with the theoretical quantiles of a standard normal distribution, serving as an essential diagnostic tool to assess normality. If the data points align closely with the reference diagonal, the underlying distribution is approximately normal. However, deviations from this line indicate departures from normality, such as skewness, heavy tails, or potential outliers. From the Q-Q plots shown in Fig. 10.5, it is evident that the performance metrics exhibit an overall trend of normality, particularly in the mid-range values, where the data points are relatively well-aligned with the linear fit line. However, deviations are observed at both ends of the distributions, particularly in the extreme values, where the points either rise above or fall below the expected trend. This shows that the distributions of the performance metrics may have light or heavy tails, meaning the presence of either a lower-than-expected or higher-than-expected number of extreme values. Specifically, in precision and mAP, the upper quantiles (higher values) show a slight deviation, indicating a possible right-skewness, where a few models achieve significantly higher scores compared to the rest. Conversely, recall and F1-score exhibit minor deviations in the lower quantiles, indicating the presence of lower-than-expected values in some models.

The observed deviations can be attributed to several factors. First, the inherent differences in the architectural complexity of the YOLO11 variants contribute to variations in performance across models. Larger models, such as YOLO11 l and YOLO11x, generally achieve higher detection performance due to increased computational capacity, which may explain the skewed nature of the distributions. Second, the introduction of the ECA attention mechanism has resulted in a consistent improvement across all performance metrics, but the magnitude of this improvement varies among different model sizes. This could introduce variability

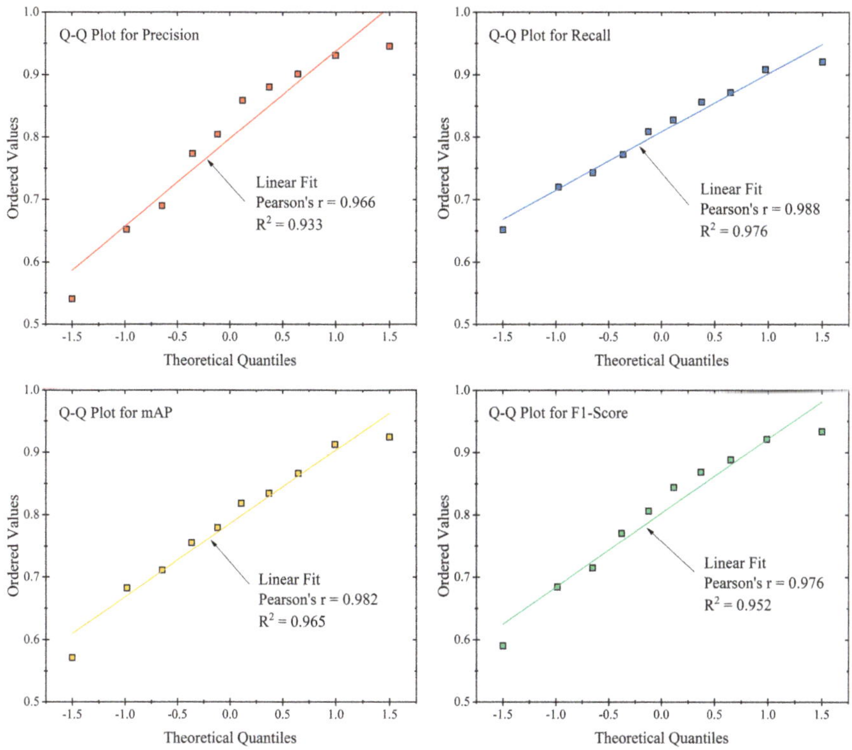

Fig. 10.5 Quantile-Quantile (Q-Q) plots of detection performance metrics. The plots compare the empirical quantiles of the detection performance metrics from both the standard and ECA-augmented YOLO11 models with the theoretical quantiles of a normal distribution. The plots serve as a diagnostic tool to assess the normality of the data underlying the performance metrics. In the mid-range, the data points align closely with the reference diagonal, indicating an overall trend towards normality. However, deviations at the extreme quantiles suggest the presence of skewness and potential outliers

in the distribution, as the enhanced models demonstrate distinct performance gains compared to their non-augmented counterparts. Additionally, the dataset characteristics, including the variability in ship sizes, orientations, and radar signal reflections in SAR images, could lead to variations in recall and precision, further influencing the distribution. From a statistical standpoint, these findings imply that while the performance metrics largely conform to a normal distribution, the presence of deviations at the extremes shows non-Gaussian characteristics, which should be taken into account when conducting further analysis.

The residuals histogram, shown in Fig. 10.6, indicates the distribution of residuals from the one-way ANOVA analysis performed on the performance metrics of the ten ship detection models tested on SAR images. Residuals represent the deviations of observed values from their respective group means and are critical for validating the assumptions underlying the ANOVA test. A key assumption of

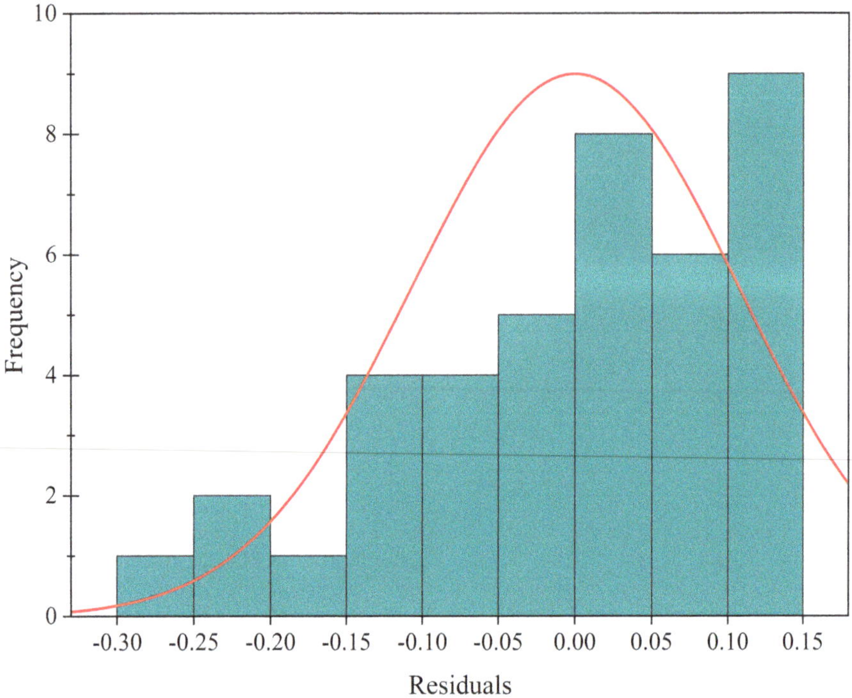

Fig. 10.6 Histogram of residuals from the one-way ANOVA analysis of detection metrics. A roughly symmetric distribution centered around zero supports the assumption of normally distributed residuals, which is a key requirement for the validity of the ANOVA test. Nonetheless, slight asymmetries and a long-tailed behavior in the histogram indicate some degree of heteroscedasticity among the models. Notably, the predominance of positive residuals in the ECA-augmented models, especially among the smaller variants, reinforces the conclusion that the integration of ECA results in an upward shift in detection performance

ANOVA is that the residuals should be approximately normally distributed. The histogram of residuals displays a roughly symmetric distribution centered around zero, which indicates that the residuals are fairly balanced, with both positive and negative deviations occurring at similar frequencies. However, slight asymmetry may indicate minor deviations from normality. The presence of a single peak near zero suggests that most of the observed performance values are relatively close to their group means, reinforcing the notion that the variance within each model's performance is not highly skewed.

The spread of the residuals across the histogram bins reflects the degree of variability in the performance metrics among the ten tested models. If the residuals exhibit a wide spread, it shows that some models significantly outperform or underperform relative to their group means. Conversely, a narrower spread indicates lower variance, suggesting that model improvements due to ECA augmentation have a relatively uniform effect across different variants. From the histogram, it is evident that residuals do not cluster tightly around zero, indicating that there are measurable differences in model performance. The right-skewed nature of the histogram proves that certain models, particularly ECA-augmented YOLO11x and YOLO11 l, have performance values consistently exceeding the expected group mean, confirming that the ECA mechanism has contributed positively to detection accuracy. Another assumption of ANOVA is homoscedasticity, which requires that the variance of residuals remains constant across all groups. Since the histogram displayed almost a long-tailed distribution, it would indicate heteroscedasticity, meaning that the variance of residuals differs across models.

The histogram confirms that the ECA-augmented models generally exhibit positive residuals, meaning their actual performance values exceed the mean of their respective groups. This reinforces the effectiveness of ECA attention in improving ship detection from SAR images. The higher frequency of positive residuals suggests that ECA-augmented variants have contributed to an upward shift in performance metrics. However, the presence of some negative residuals suggests that the enhancements introduced by ECA may be less impactful for larger models. The residuals for ECA-augmented YOLO11n and ECA-augmented YOLO11s are predominantly positive, meaning their actual performance values exceed their group means, confirming that ECA significantly boosts performance in smaller models. Conversely, larger models (ECA-augmented YOLO11 l and ECA-augmented YOLO11x) show residuals closer to zero, indicating that while ECA still improves their performance, the magnitude of improvement is less pronounced. This trend suggests that ECA is most beneficial for models with inherently weaker detection capabilities, as it compensates for architectural limitations by refining feature extraction. For larger models, ECA still contributes positively but offers only marginal gains, likely because these models already incorporate advanced feature extraction mechanisms.

10.5 Conclusion

This study systematically investigated the impact of integrating the Efficient Channel Attention (ECA) mechanism into various YOLO11 model variants for ship detection in SAR images. The experimental results, supported by both quantitative metrics and qualitative visualizations, demonstrate that the incorporation of the ECA module leads to consistent improvements in detection performance across all model sizes. Notably, smaller variants such as YOLO11n and YOLO11s exhibited substantial gains in precision, recall, mAP, and F1-score, underscoring the importance of refined feature extraction in lightweight architectures. In contrast, while larger models such as YOLO11 l and YOLO11x already exhibit high baseline performance, the inclusion of ECA still contributes to marginal improvements by further enhancing the discrimination of ship features from complex backgrounds. The results demonstrated that ECA-enhanced heatmaps effectively highlight ship features in SAR imagery, confirming that the attention mechanism robustly accentuates salient areas. Additionally, although the one-way ANOVA tests did not yield statistically significant differences at the conventional 0.05 significance level, the overall upward trend in raw performance values, as well as the diagnostic insights from the Q-Q plots and residual analysis, provide compelling evidence that ECA integration systematically enhances detection accuracy.

While the ANOVA tests did not yield statistical significance at the conventional 0.05 level, it is important to interpret these results in context. The consistent upward trend across all evaluation metrics, particularly in smaller models, combined with clear qualitative improvements seen in heatmap visualizations and detection outputs, strongly supports the practical effectiveness of the ECA mechanism. In real-world SAR applications, such as maritime monitoring where image quality is often degraded by noise and environmental interference, even moderate gains in detection sensitivity and robustness can yield substantial operational value. Thus, the benefits demonstrated by ECA augmentation should not be understated, as they offer meaningful performance enhancements despite the lack of statistically significant differences. The findings validate the technical efficacy of the ECA module in enhancing the YOLO11 framework, and highlight its practical benefits for improving ship detection in challenging SAR imaging conditions. Future work may explore further refinement of attention mechanisms, integration with complementary techniques, and broader applications across diverse remote sensing scenarios.

References

1. Lv, N., Chen, C., Qiu, T., Sangaiah, A.K.: Deep learning and superpixel feature extraction based on contractive autoencoder for change detection in SAR images. IEEE Trans. Industr. Inform. **14**(12), 530–5538 (2018). https://doi.org/10.1109/TII.2018.2873492
2. Wang, J., Zheng, T., Lei, P., Bai, X.: Ground target classification in noisy SAR images using convolutional neural networks. IEEE J. Sel. Top. Appl. Earth Observ. Remote Sensing. **11**(11), 4180–4192 (2018). https://doi.org/10.1109/JSTARS.2018.2871556
3. Wang, X., Li, G., Zhang, X.P., He, Y.: Ship detection in SAR images via local contrast of fisher vectors. IEEE Trans. Geosci. Remote Sens. **58**(9), 6467–6479 (2020). https://doi.org/10.1109/TGRS.2020.2976880
4. Patel, V., Madhukar, H., Ravichandran, S.: Variability index constant false alarm rate for marine target detection. In: 2018 Conference on Signal Processing And Communication Engineering Systems (SPACES), pp. 171–175 (2018). https://doi.org/10.1109/SPACES.2018.8316339
5. Xu, H., Xiao, S., Li, Z., Wang, S., Liu, W., Li, J.: A novel SAR imaging method based on morphological component analysis. IEEE Sensors J. **22**(13), 13326–13337 (2022). https://doi.org/10.1109/JSEN.2022.3179607
6. Zhang, M., Li, W., Du, Q.: Diverse region-based CNN for hyperspectral image classification. IEEE Trans. Image Process. **27**(6), 2623–2634 (2018). https://doi.org/10.1109/TIP.2018.2809606
7. Mehran, A., Tehsin, S., Hamza, M.: An effective deep learning model for ship detection from satellite images. Spat. Inf. Res. **31**, 61–72 (2023). https://doi.org/10.1007/s41324-022-00482-1
8. Cao, C., et al.: An improved faster R-CNN for small object detection. IEEE Access. **7**, 106838–106846 (2019). https://doi.org/10.1109/ACCESS.2019.2932731

9. Tan, X., Leng, X., Luo, R., Sun, Z., Ji, K., Kuang, G.: YOLO-RC: SAR ship detection guided by characteristics of range-compressed domain. IEEE J. Sel. Top. Appl. Earth Observ. Remote Sensing. **17**, 18834–18851 (2024). https://doi.org/10.1109/JSTARS.2024.3478390
10. Bakirci, M.: Internet of Things-enabled unmanned aerial vehicles for real-time traffic mobility analysis in smart cities. Comput. Electr. Eng. **123**(110313) (2025). https://doi.org/10.1016/j.compeleceng.2025.110313
11. Brauwers, G., Frasincar, F.: A general survey on attention mechanisms in deep learning. IEEE Trans. Knowl. Data Eng. **35**(4), 3279–3298 (2023). https://doi.org/10.1109/TKDE.2021.3126456
12. Wang, L., Fang, S., Meng, X., Li, R.: Building extraction with vision transformer. IEEE Trans. Geosci. Remote Sens. **60**, 1–11 (2022). https://doi.org/10.1109/TGRS.2022.3186634
13. Tong, W., Chen, W., Han, W., Li, X., Wang, L.: Channel-attention-based DenseNet network for remote sensing image scene classification. IEEE J. Sel. Top. Appl. Earth Observ. Remote Sensing. **13**, 4121–4132 (2020). https://doi.org/10.1109/JSTARS.2020.3009352
14. Waghumbare, A., Singh, U.: Diat-separable-cnn-eca-harnet: a lightweight and explainable model for efficient human activity recognition. SIViP. **19**, 245 (2025). https://doi.org/10.1007/s11760-025-03838-w
15. Guo, Z., Ma, D., Luo, X.: A lightweight semantic segmentation algorithm integrating CA and ECA-Net modules. Optoelectron. Lett. **20**, 568–576 (2024). https://doi.org/10.1007/s11801-024-3241-z
16. Achiri, L., Guida, R., Iervolino, P.: SAR and ais fusion for maritime surveillance. In: 2018 IEEE 4th International Forum on Research and Technology for Society and Industry (RTSI), pp. 1–4 (2018). https://doi.org/10.1109/RTSI.2018.8548352
17. Wang, H., Zhu, H., Li, H.: A rotating machinery fault diagnosis method based on multi-sensor fusion and ECA-CNN. IEEE Access. **11**, 106443–106455 (2023). https://doi.org/10.1109/ACCESS.2023.3320065
18. Zhang, S., Wu, R., Xu, K., Wang, J., Sun, W.: R-CNN-based ship detection from high resolution remote sensing imagery. Remote Sens. **11**(631) (2019). https://doi.org/10.3390/rs11060631
19. Wang, R., Li, J., Duan, Y., et al.: Study on the combined application of CFAR and deep learning in ship detection. J Indian Soc. Remote Sens. **46**, 1413–1421 (2018). https://doi.org/10.1007/s12524-018-0787-x
20. Xia, R., Chen, J., Huang, Z., Wan, H., Wu, B., Sun, L., Yao, B., Xiang, H., Xing, M.: CRTransSar: a visual transformer based on contextual joint representation learning for SAR ship detection. Remote Sens. **14**(1488) (2022). https://doi.org/10.3390/rs14061488
21. Ju, M., Niu, B., Zhang, J.: SAR image generation method for oriented ship detection via generative adversarial networks. SIViP. **18**, 589–596 (2024). https://doi.org/10.1007/s11760-023-02793-8
22. Cai, S., Meng, H., Yuan, M., et al.: FS-YOLO: a multi-scale SAR ship detection network in complex scenes. SIViP. **18**, 5017–5027 (2024). https://doi.org/10.1007/s11760-024-03212-2
23. Sun, B., Wang, X., Li, H., et al.: Small-target ship detection in SAR images based on densely connected deep neural network with attention in complex scenes. Appl. Intell. **53**, 4162–4179 (2023). https://doi.org/10.1007/s10489-022-03683-1
24. Yasir, M., Jianhua, W., Mingming, X., et al.: Ship detection based on deep learning using SAR imagery: a systematic literature review. Soft. Comput. **27**, 63–84 (2023). https://doi.org/10.1007/s00500-022-07522-w
25. Tang, J., Hu, X.M., Jeon, S.W., et al.: Light-YOLO: a lightweight detection algorithm based on multi-scale feature enhancement for infrared small ship target. Complex Intell. Syst. **11**, 130 (2025). https://doi.org/10.1007/s40747-024-01726-3
26. Xu, G., Yin, J., Zhang, Z.: Marine ship detection under fog conditions based on an improved deep-learning approach. In: Zhang, H., Li, X., Hao, T., Meng, W., Wu, Z., He, Q. (eds.) Neural Computing for Advanced Applications. NCAA 2024 Communications in Computer and Information Science, vol. 2181. Springer, Singapore (2025). https://doi.org/10.1007/978-981-97-7001-4_7

27. Tang, H., Gao, S., Li, S., Wang, P., Liu, J., Wang, S., Qian, J.: A lightweight SAR image ship detection method based on improved convolution and YOLOv7. Remote Sens. **16**(486) (2024). https://doi.org/10.3390/rs16030486
28. Eldosoky, M.A., Li, J.P., Haq, A.U., et al.: WallNet: hierarchical visual attention-based model for putty bulge terminal points detection. Vis. Comput. **41**, 99–114 (2025). https://doi.org/10.1007/s00371-024-03312-7
29. Wu, Q., Cen, L., Kan, S., et al.: Real-time underwater target detection based on improved YOLOv7. J. Real-Time Image Proc. **22**, 43 (2025). https://doi.org/10.1007/s11554-025-01621-1
30. Hu, H., Liu, Y., Rong, H.: Detection of insulators on power transmission line based on an improved faster region-convolutional neural network. Algorithms. **15**(83) (2022). https://doi.org/10.3390/a15030083
31. Gong, L., Huang, X., Chao, Y., et al.: An enhanced SSD with feature cross-reinforcement for small-object detection. Appl. Intell. **53**, 19449–19465 (2023). https://doi.org/10.1007/s10489-023-04544-1
32. Ren, F., Fei, J., Li, H., Doma, B.T.: Steel surface defect detection using improved deep learning algorithm: ECA-SimSPPF-SIoU-Yolov5. IEEE Access. **12**, 32545–32553 (2024). https://doi.org/10.1109/ACCESS.2024.3371584
33. Liu, Q., Wang, L.: t-Test and ANOVA for data with ceiling and/or floor effects. Behav. Res. **53**, 264–277 (2021). https://doi.org/10.3758/s13428-020-01407-2
34. Sarwar, G., Anwar, T., Qureshi, H., et al.: Optimizing germination: comparative assessment of various growth media on dragon fruit germination and early growth. BMC Plant Biol. **24**, 533 (2024). https://doi.org/10.1186/s12870-024-05247-6
35. Northrop, P.J., Coleman, C.L.: Improved threshold diagnostic plots for extreme value analyses. Extremes. **17**, 289–303 (2014). https://doi.org/10.1007/s10687-014-0183-z

Chapter 11
Airport Security and Management Through UAV Surveillance and Single Shot Multibox Detector

Irem Bayraktar and Murat Bakirci

Abstract This study presents a comprehensive evaluation of the Single Shot Multibox Detector (SSD) and YOLOv4 models for aircraft detection in Unmanned Aerial Vehicle (UAV)-captured images under varying environmental conditions. The models were compared based on their training and validation loss values, detection accuracy, and robustness in different lighting and altitude scenarios. The results indicate that YOLOv4 exhibits a more efficient learning process, with faster decline in loss values and better generalization performance, making it suitable for real-time applications. However, the SSD model outperformed YOLOv4 in terms of detection accuracy, particularly under medium-altitude, low-light, and challenging conditions such as varying color patterns, shadows, and aircraft geometries. Despite facing detection errors in certain high-altitude images, where complex objects like passenger boarding bellows systems resemble aircraft, the SSD model demonstrated superior resilience and detection speed. In both sufficient and insufficient light conditions, the SSD model consistently outperformed YOLOv4 in terms of precision, recall, and mean average precision (mAP), showcasing its robustness across different environmental challenges. This paper highlights the strengths and limitations of both models, providing valuable insights for choosing the optimal model for UAV-based aircraft detection tasks.

Keywords Object detection · YOLOv4 · SSD · UAV-based aircraft detection · Real-time detection · Environmental robustness

I. Bayraktar
Department of Aerospace Engineering, Turkish Air Force Academy, National Defence University, Istanbul, Turkey
e-mail: ibayraktar@uislab.org

M. Bakirci (✉)
Unmanned/Intelligent Systems Lab, Faculty of Aeronautics and Astronautics, Tarsus University, Mersin, Turkey
e-mail: muratbakirci@tarsus.edu.tr

G. Rivera et al. (eds.), *Recent Advances in Artificial Neural Networks*, Intelligent Systems Reference Library 283, https://doi.org/10.1007/978-3-032-13497-4_11

11.1 Introduction

Ensuring the security of airports is a critical and complex task that requires constant monitoring and advanced measures to protect passengers, personnel, and assets from a wide range of potential threats. With the increasing volume of air traffic and heightened global concerns about security, airports must be equipped with state-of-the-art security systems to prevent incidents ranging from unauthorized access to more severe threats like terrorism [1]. Traditional security measures, such as patrolling personnel, stationary cameras, and access control systems, have been the backbone of airport security for decades. However, these systems have limitations in their coverage, adaptability, and ability to respond to evolving security challenges. As airports expand and become more complex, the need for more dynamic and comprehensive security solutions is growing. One of the key areas that demand attention is the perimeter security of airports [2]. The sheer size and complexity of modern airport perimeters, often stretching over several kilometers, make it difficult to monitor continuously using traditional methods alone. Gaps in coverage or slow response times can create vulnerabilities that malicious actors could exploit. Additionally, certain areas of airports, such as remote runways, maintenance zones, and fuel storage areas, are difficult to monitor using fixed surveillance systems [3]. This necessitates innovative approaches that can enhance real-time monitoring capabilities and allow for more responsive and adaptable security operations. Integrating advanced technologies into airport security frameworks is, therefore, essential for addressing the evolving security landscape [4].

Surveillance at airports has always been a critical component of security operations, allowing for the detection and prevention of unauthorized access, criminal activities, and potential threats [5]. As airports continue to expand and face new security challenges, the need for enhanced surveillance systems has become even more pressing. Modern airports require comprehensive surveillance solutions that can monitor vast areas in real-time [6], detect suspicious activities, and support quick decision-making by security personnel. While traditional surveillance systems, such as fixed cameras and human patrols, have provided valuable support, they are increasingly proving insufficient to meet the current and future demands of airport security. The limitations of conventional surveillance methods [7] lie in their inability to offer complete situational awareness across the entire airport landscape. Fixed cameras have limited coverage and often leave blind spots, while human patrols, though effective, are constrained by physical and temporal limitations. These shortcomings create vulnerabilities that can be exploited by individuals seeking to breach airport security. To mitigate these risks, there is a growing emphasis on deploying surveillance systems that are more dynamic, versatile, and capable of responding to a wide range of security scenarios. The integration of unmanned aerial vehicles (UAVs) [8, 9] into airport surveillance is emerging as a promising solution to enhance the monitoring and security of these complex environments.

UAVs offer a unique set of capabilities that can significantly enhance surveillance operations at airports. UAVs are agile, can cover large areas rapidly, and can be equipped with advanced imaging technologies [10] to provide real-time surveillance data to security teams. One of the primary advantages of UAVs in airport security is their ability to monitor remote and hard-to-reach areas, which are often overlooked by traditional security measures. By deploying UAVs, airports can achieve continuous and flexible monitoring of their perimeters, runways, and critical infrastructure, thereby reducing vulnerabilities and enhancing overall security. Beyond perimeter security, UAVs are also proving to be valuable tools for monitoring airport operations [11], identifying potential safety hazards, and assisting in traffic monitoring [12]. For instance, UAVs equipped with thermal imaging cameras can quickly detect heat signatures from unauthorized intruders, damaged equipment, or even fire outbreaks, enabling rapid intervention by security or emergency teams [13]. Additionally, UAVs can be deployed to monitor crowd movements in and around airport terminals, providing real-time data that can be used to manage potential security threats or assist in evacuations during emergencies. The potential applications of UAVs in airport security are vast, and their deployment is increasingly seen as a necessary evolution in modern surveillance systems.

The ability of UAVs to provide real-time, high-resolution imagery from various angles offers a significant advantage over traditional fixed surveillance cameras. UAVs can fly at different altitudes and capture images from perspectives that stationary cameras cannot, allowing for more comprehensive coverage of the airport environment. Moreover, the mobility of UAVs allows them to respond quickly to unfolding security events, providing live video feeds to security personnel who can assess the situation and take appropriate actions [14]. This adaptability makes UAVs particularly useful in scenarios where time is of the essence, such as detecting a security breach or tracking a suspect moving through restricted areas. Another key advantage of UAV surveillance at airports is cost-effectiveness. While the initial investment in UAV systems may be high, the long-term operational costs are often lower than those of traditional surveillance methods. UAVs reduce the need for large numbers of stationary cameras and human patrols, both of which can be expensive to maintain over time [15]. Additionally, UAVs can be deployed as needed, offering on-demand surveillance without the need for constant manual oversight. This reduces the workload on security personnel and allows for more efficient use of resources. Overall, the integration of UAVs into airport security frameworks offers a scalable, adaptable, and cost-effective solution to address the growing demands for enhanced surveillance [16].

The effectiveness of UAV-based surveillance systems at airports can be further enhanced through the use of advanced image processing algorithms. One such algorithm is the Single Shot Multibox Detector (SSD) [17, 18], which is known for its ability to detect objects quickly and accurately. SSD is a popular deep learning algorithm used in object detection tasks, particularly in scenarios where speed and efficiency are critical. Unlike other object detection algorithms that require multiple passes to detect objects in an image, SSD can detect multiple objects in a single shot, making it an ideal choice for real-time surveillance applications. The SSD algorithm's architecture is designed to perform both object localization and classification in a single forward pass through the network. This design enables SSD to process images at a high frame rate, making it well-suited for applications where quick decision-making is essential, such as monitoring airport security. SSD's ability to detect objects across a range of scales and aspect ratios further enhances its utility in complex environments like airports, where objects of interest may vary in size and orientation. The algorithm's efficiency and accuracy in detecting objects, such as vehicles, personnel, or other potential security threats, make it a valuable tool for enhancing the capabilities of UAV-based surveillance systems. In the context of airport security, the use of the SSD algorithm allows for real-time detection of potential threats, enabling security personnel to respond promptly to incidents. For example, SSD can be used to automatically detect unauthorized vehicles entering restricted areas or individuals loitering near sensitive airport infrastructure. By integrating SSD with the UAV's onboard camera systems, it is possible to automate much of the surveillance process, reducing the reliance on manual monitoring and allowing security teams to focus on responding to verified threats [19]. The combination of UAVs and advanced algorithms like SSD represents a significant advancement in the field of airport security, providing a more proactive and efficient approach to threat detection and response.

In this study, a UAV equipped with advanced imaging technologies and powered by the SSD was deployed to enhance surveillance at major international airports. The UAV, outfitted with a high-resolution camera, conducted multiple flights over the airports' perimeter and critical infrastructure, capturing thousands of images that were processed to identify potential security threats. The SSD algorithm was applied to the collected data to detect and classify objects of interest, such as vehicles, equipment, and personnel, with high accuracy and speed. The UAV's ability to capture aerial imagery from various altitudes and angles allowed for comprehensive coverage of the airport environment, ensuring that no areas were left unmonitored. The data collected during the UAV flights were analyzed to evaluate the effectiveness of the SSD algorithm in real-world airport surveillance scenarios. The results demonstrated that the SSD-powered UAV system was highly effective in detecting potential security threats in real-time, with minimal false positives and a high degree of accuracy. The study also compared the performance of the SSD algorithm with other algorithms, highlighting its superior speed and efficiency in processing the large volumes of data generated by the UAV. The study concluded that UAV-based surveillance systems, when equipped with cutting-edge technologies such as the SSD algorithm, offer significant advantages over traditional security measures in airports. The mobility and adaptability of UAVs, combined with the speed and accuracy of the SSD algorithm, provide a powerful tool for monitoring vast and complex environments like airports.

11.2 Related Work

In recent years, Intelligent Transportation Systems (ITS), a rapidly expanding research and application field that has garnered significant global attention, has become a focal point for researchers and many governments. Applications such as aircraft detection, autonomous driving systems, vehicle and traffic sign detection, urban planning, fire detection, and mapping are particularly critical in scenarios that require rapid and accurate detection. Advanced image processing and machine learning techniques are increasingly utilized and supported by researchers and government institutions. This study employs the SSD algorithm, which is frequently preferred for real-time aircraft detection due to its effectiveness and strong results. SSD integrates the regression concept of YOLO with the anchor box approach of Faster R-CNN, aiming to combine the strengths of these two powerful algorithms [20]. SSD unifies these advantages under a single framework, achieving significant success in both speed and accuracy. Experimental results on widely used datasets such as PASCAL VOC, COCO, and ILSVRC demonstrate that SSD offers competitive accuracy compared to other object detection algorithms while providing a much faster framework for both training and inference. For example, in its original paper, SSD achieved 74.3% mAP on the VOC2007 test at 59 FPS using an Nvidia Titan X with a 300 × 300 input size. Increasing the input size to 512 × 512 resulted in 76.9% mAP, outperforming other algorithms of its time.

The proposed approach is more flexible because it utilizes default boxes with multiple aspect ratios at each feature location, extracted from multiple feature maps at different scales. The SSD algorithm is not limited to aircraft detection but can also be effectively used in other ITS applications. Its application in real-time scenarios has pioneered the simultaneous detection of objects, such as aircraft, in video streams. Another study aimed to improve the detectability of small, hard-to-detect objects by addressing insufficient semantic information in low-level feature maps of the SSD algorithm [21]. The authors incorporated a Feature Pyramid Network (FPN) into SSD to enhance semantic content for detecting small objects and added a Convolutional Block Attention Module (CBAM) to strengthen the network's ability to learn core features and improve detection. Thanks to CBAM, the network can decide which features are more critical, yielding more meaningful and real-time results. These structural enhancements resulted in a noticeable improvement in performance. For instance, on the PASCAL VOC2007 test, the original SSD's mAP of 74.3% was increased to 78.1%, achieving a 3.9% improvement. These results highlight an important step forward in enhancing SSD for tasks involving hard-to-detect objects.

Another study addressed the critical issue of dynamic traffic management, which is essential for smart cities and transportation systems. Using SSD algorithm, the study aimed to optimize efficiency, safety, and sustainability through multimodal connectivity in vehicle detection [22]. The development of ITS is vital for managing and organizing traffic flow in modern urban areas. Thw work introduced a novel perspective to ITS research by compiling a large dataset of vehicle images captured through UAV-based aerial monitoring. The increasing adoption of UAVs in contemporary urban settings allows for broader testing of vehicle and aircraft detection applications across diverse environmental conditions. The study also highlighted certain challenges and limitations encountered during implementation, particularly in low-resolution images and poor lighting conditions, where SSD's performance tends to decline. The researchers addressed these limitations by focusing on how the algorithm could deliver better results under challenging and diverse conditions. The integration of UAV-based aerial monitoring with the SSD algorithm shows great potential for more efficient and safer traffic management in smart cities. By improving the algorithm's robustness and accuracy under adverse conditions, this work contributes significantly to enhancing ITS capabilities for complex urban environments. The fusion of UAV monitoring and SSD-based vehicle detection not only aids in developing safer, more efficient, and sustainable cities but also represents a significant contribution to the literature on improving urban transportation systems.

Airport surveillance, including the detection of military and civilian aircraft, is critical for traffic control and safety in the aviation sector. One study used the SSD algorithm to detect airports and landing strips, comparing its performance with the Faster R-CNN algorithm across key metrics such as accuracy, specificity, false positive rate, false negative rate, true prediction rate, F1 score, error rate, outcomes, and training time [23]. The SSD algorithm achieved an average inference time of 4.06 s with a detection success rate of 76.61%, whereas Faster R-CNN took 10.37 s and achieved a success rate of 99.52%. These results enabled a comparative analysis to guide future research on selecting algorithms based on speed or accuracy. This analysis not only evaluates the overall effectiveness of the algorithms but also serves as a guide for researchers seeking to balance speed and accuracy in their applications.

In logistics and transportation networks, maritime transportation and trade are crucial for both civilian and military reconnaissance. Detecting ships with high precision is particularly challenging due to obstacles such as clouds, fog, islands, coastlines, and waves. To address this issue, a new ship detection network based on multi-layer convolutional feature fusion, called CFF-SDN, was proposed [24]. The network comprises three components: a convolutional feature extraction network for ship features, a mechanism for merging features from shallow and deep layers, and a classifier for multiple fused feature maps. This design enables the algorithm to adapt to multi-scale objects and ships. To enhance recall rates for densely packed ships and aircraft, the researchers incorporated Non-Maximum Suppression (NMS), achieving superior performance. The proposed detection framework attained an mAP of 91.51% at a resolution of 416 × 416, with an average inference time of 9.4 ms. These results make a significant contribution to the academic literature on ship detection for engineering applications.

Object detection on satellite imagery is a critical task for various applications, including surveillance and monitoring, intelligent transportation systems, and vehicle detection technologies. Due to their lower resolution compared to UAV-captured images, satellite images face challenges in detecting objects such as airplanes, motorcycles, and cars. A comparative analysis of three algorithms, YOLOv3, SSD, and R-CNN, was conducted in [25] to determine which model performs better for aircraft detection. Tests evaluated metrics such as accuracy, loss, and precision. The results showed that SSD outperformed the other models. For instance, R-CNN recorded 27 false negatives, compared to 10 for SSD and 20 for YOLOv3. Consequently, SSD was identified as the most suitable method for object detection within images, combining high accuracy and low computational cost.

The problem of aircraft detection, which is critical for military scenarios, has prompted the application of numerous computer vision techniques. Satellite-based detection of aircraft presents various challenges, including color variations, object sizes, aspect ratios, aircraft orientations, and complex background elements, leading to increased academic interest in this field. In [26], experimental and comparative analyses of CNN-based algorithms, such as AlexNet, VGG-16, ResNet-18, ResNet-50, ResNet-101, and Inception-v3, along with R-CNN, Fast R-CNN, Faster R-CNN, and YOLO, were presented. The study also discussed the selection of the most suitable object detectors, feature extraction backbones, training-to-testing ratios, and training parameter configurations. Results showed that Faster R-CNN with ResNet-18 as the backbone was the most precise and successful algorithm for high-accuracy aircraft detection scenarios. Meanwhile, YOLO-2 with ResNet-50 struck a satisfactory balance between real-time object detection performance and accuracy.

Another study, [27], tackled the long-standing challenge of object detection from satellite imagery by leveraging state-of-the-art CNN-based detection models (Faster R-CNN, SSD, YOLOv3) to automatically detect aircraft in Very High Resolution (VHR) satellite images with limited labeled datasets. The study enhanced the training dataset by artificially augmenting data through transformations such as rotation, rescaling, cropping, and color adjustments. It also applied the NMS algorithm to SSD and YOLOv3 pipelines to resolve issues related to duplicate detections. Evaluation of the test regions showed that Faster R-CNN achieved the highest accuracy based on Average Precision (AP), F1 scores, and visual inspection of the results. YOLOv3 exhibited slightly lower performance than Faster R-CNN but provided a balanced trade-off between accuracy and speed. SSD, although showing the lowest detection performance, delivered superior results for object localization tasks. According to COCO metrics and F1 scores, Faster R-CNN emerged as the best-performing algorithm. While YOLOv3 demonstrated promise with faster processing times, SSD struggled to converge efficiently during training with limited iterations, making it the least effective detector among the algorithms analyzed. Nonetheless, SSD proved highly effective for object localization despite its overall lower detection accuracy. To provide a clearer overview of prior research and facilitate comparison among key studies, Table 11.1 summarizes the main features, algorithms, application domains, datasets used, and key performance results from the literature discussed in this section.

Table 11.1 Summary of related work on object detection for UAV and satellite imagery

Ref.	Application domain	Algorithm(s) used	Dataset/image source	Key contribution/results
[20]	Real-time Aircraft Detection	SSD	PASCAL VOC, COCO	Achieved 74.3% mAP at 59 FPS using SSD300
[21]	Small Object Detection	SSD + FPN + CBAM	VOC2007	Improved SSD's mAP from 74.3% to 78.1% with attention modules
[22]	Smart Traffic Systems	SSD	UAV-captured vehicle images	Highlighted SSD's adaptability under varying lightning
[23]	Airport Detection (Airfields)	SSD, Faster R-CNN	UAV/satellite imagery	SSD: 76.61% success rate; Faster R-CNN: 99.52%
[24]	Ship Detection (Maritime)	CFF-SDN	Custom satellite dataset	Achieved 91.51% mAP with NMS integration
[25]	Aircraft Detection (Satellite)	SSD, YOLOv3, R-CNN	VHR satellite images	SSD had lowest false negatives, best for object localization
[26]	Aircraft Detection (Military)	CNNs, YOLOv2, Faster R-CNN	Synthetic and real satellite images	Faster R-CNN + ResNet-18 yielded highest precision
[27]	Aircraft Detection (VHR Images)	SSD, YOLOv3, Faster R-CNN	VHR satellite images + augmentation	Faster R-CNN had best accuracy; SSD performed best in localization

11.3 Dataset

The performance and accuracy of object detection algorithms are heavily influenced by the quality and quantity of the dataset utilized. In this study, an original, high-quality, and comprehensive dataset featuring various aircraft models was developed to enable aircraft detection using the SSD algorithm. Recent technological advancements have spurred the creation of specialized, large-scale object datasets, which serve as a significant driving force for advancing detection methodologies. This dataset was constructed using images captured by a UAV-based platform, equipped with a high-resolution camera programmed for autonomous flight. The UAV captured numerous aircraft images, most of which were taken at various airports across Turkey. As illustrated in Fig. 11.1, the dataset encompasses a diverse range of visible-light images captured from different angles and altitudes. To enhance the dataset's comprehensiveness, complex images with challenges such as line clutter and shadows in aircraft backgrounds were included. This diversity is crucial for improving the model's generalization ability, enabling it to detect aircraft of varying sizes and types and uncover potential weaknesses more effectively.

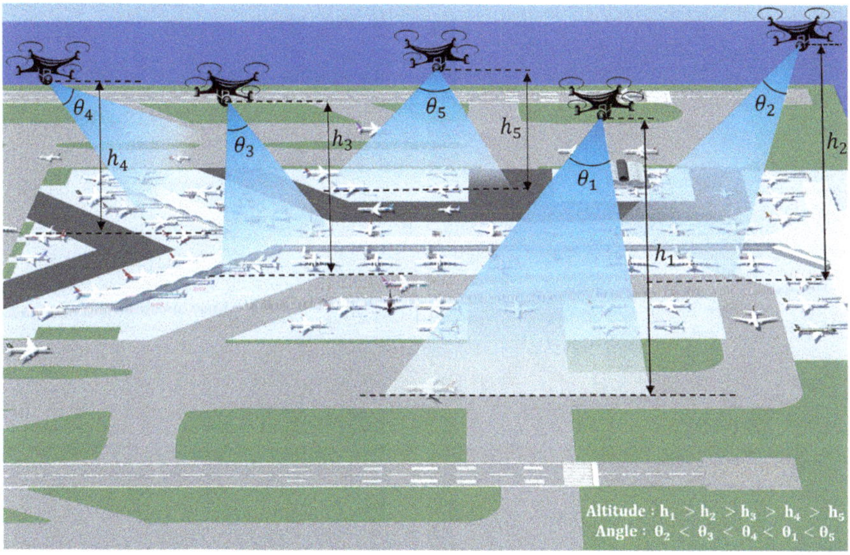

Fig. 11.1 Schematic illustration showing images captured by the UAV over the airport area from varying heights and camera angles along the flight trajectory. The images represent a range of perspectives, covering both wide and narrow areas, as well as complex structures

Furthermore, to enhance data diversity and account for the detection challenges posed by varying altitudes, airport images captured from different heights using drones were incorporated into the dataset. Examples of these images are provided in Fig. 11.2.

The performance and accuracy of object detection algorithms are heavily influenced by the quality and quantity of the dataset utilized. The dataset used in this study consists of approximately 14,000 UAV-captured images collected from various Turkish airports using a custom drone equipped with a high-resolution camera. The images were recorded in JPG format with resolutions ranging from 1920 × 1080 pixels (Full HD) to 3840 × 2160 pixels (4 K UHD), depending on the altitude and zoom level of the UAV. All images were manually annotated using the Pascal VOC format, labeling the locations of aircraft as the sole object category in this study. There is a single object class, aircraft, but the dataset includes a wide range of aircraft types (e.g., commercial jets, small private planes, and regional aircraft) to ensure class diversity. Of the total dataset, approximately 85% of the images are in full color, while 15% are grayscale or low-light monochrome captures, primarily collected during nighttime operations or in low-visibility weather conditions. The dataset is divided into training (80%), validation (10%), and testing (10%) subsets. Each subset maintains a consistent distribution of color/grayscale images and altitude levels to ensure balanced evaluation. Metadata for each image includes time of capture, lighting condition, altitude, and angle of view, contributing to a nuanced understanding of detection performance under varying scenarios.

Fig. 11.2 Examples of airport images captured from varying altitudes using a UAV, showcasing the diversity in image perspectives and the detection challenges posed by altitude variations

The dataset includes images of aircraft from different airport structures and departments, allowing the model to achieve consistent performance across various scenarios. Particular attention was given to ensuring diversity in terms of aircraft size and altitude. The dataset has been significantly enriched with a wide variety of aircraft models. Its scope was carefully expanded to include images ranging from high-resolution to low-resolution, contributing to the model's adaptability and overall performance. To further strengthen the dataset's robustness and capabilities, images captured under varying angles, scales, altitudes, and lighting conditions were incorporated. These additions include images taken at different times of day, representing a range of lighting scenarios such as sunlight, twilight, nighttime, cloudy, and rainy conditions, as well as a variety of perspectives, including top-down, side, and oblique angles. Efforts were made to ensure the images accurately reflect realistic conditions, accounting for shooting angles, obstacles, shadows, and image clarity. Additionally, images were rotated at angles of 90, 180, and 270° to simulate various perspectives of aircraft. The accuracy and recency of these images are critical to the effectiveness of the SSD model. Furthermore, the inclusion of images with varying resolutions enhances the SSD model's capability to perform both detailed and generalized detections. The primary goal of diversifying the dataset in this manner is to provide a richer and more comprehensive learning experience during model training, significantly enhancing its adaptability and overall performance. The dataset was pre-processed and organized into three subsets: training, validation, and testing. This categorization allows for a thorough evaluation of the model's performance, ensuring the validity and reliability of the results. The dataset developed in this study includes all the necessary features for the SSD algorithm to achieve high performance in aircraft detection tasks.

11.4 Data Preprocessing

Data preprocessing involves transforming raw data through several stages to make it suitable for modeling and analysis. This critical step directly influences the reliability of the analysis and the performance of the model. It encompasses tasks such as correcting missing or erroneous data and removing irrelevant or redundant information. By applying data cleaning, transformation, integration, and reduction processes, the data becomes ready for analysis and modeling, which helps maximize model accuracy and performance. Specific preprocessing tasks include correcting missing or damaged pixels in satellite images, removing unnecessary data, normalizing image sizes, ensuring consistency through light and color adjustments, consolidating images into a unified format, optimizing high-resolution aircraft images, and selecting essential features. Additionally, various data augmentation techniques are applied to the training set to prevent model overfitting and enhance the detection accuracy of aircraft. When detection data is limited, augmentation strategies increase the diversity of training examples, improving the model's robustness. A widely held belief in deep learning is that more data leads to better model learning. However, in cases where data is scarce, system performance may suffer. To address this, synthetic data can be generated from the existing dataset, significantly increasing the number of training examples.

The primary goal of these preprocessing and augmentation steps is to make the training data resemble the diversity found in real-world examples. In this study, techniques such as horizontal and vertical flipping, translation, random rotation, scaling, random expansion, and cropping were used to enrich the training set. Aircraft images were further enhanced by applying color adjustments such as contrast, saturation, brightness, and hue variations. These steps improve the model's ability to detect aircraft under varying angles, sizes, lighting, and weather conditions. For example, rotation and translation operations help detect aircraft from different perspectives, while brightness and contrast adjustments aid in detection under diverse lighting conditions. Such preprocessing ensures that the model learns to handle various scenarios and improves its generalization capabilities, reducing the risk of overfitting. This ultimately enables the model to make more accurate and reliable detections in real-world scenarios.

In this study, a comprehensive preprocessing pipeline was implemented to prepare the dataset for training the SSD512 model, aiming to ensure data consistency, enhance model performance, and reduce overfitting. Color adjustment techniques were applied using random variations in brightness and contrast ($\pm 20\%$), saturation ($\pm 15\%$), and hue ($\pm 10°$) to simulate diverse lighting conditions. All images were resized to 512×512 pixels and normalized to a [0, 1] range by dividing pixel values by 255. A data cleaning step was conducted, where corrupted files, images with extreme motion blur, and labeling errors were manually removed, resulting in the exclusion of approximately 2% of the dataset. High-resolution images (above 4 K) were downscaled using bicubic interpolation to maintain visual clarity while standardizing input dimensions. Grayscale images were converted to 3-channel grayscale format to ensure compatibility with the SSD input structure. Data augmentation included horizontal and vertical flipping, random cropping (10–25% of the image area), translation (± 10 pixels), and rotations (90°, 180°, and 270°), each applied with a 50% probability during training to improve model generalization.

As shown in Fig. 11.3, the dataset is categorized into two lighting conditions: sufficient and insufficient lighting. To classify images based on lighting conditions, a quantitative approach was employed to ensure consistency and objectivity. Each image was first converted to grayscale, and the mean pixel intensity was calculated. Images with an average grayscale intensity above a threshold of 100 (on a scale of 0–255) were classified as having 'sufficient lighting', while those below this threshold were labeled as 'insufficient lighting'. This threshold was selected based on histogram analyses and visual cross-validation to ensure it effectively distinguished between well-lit and poorly lit images. This process was automated and applied uniformly across the dataset. Additionally, to validate the robustness of this classification, random samples from each group were visually inspected to confirm alignment with intuitive lighting perceptions.

The preprocessed dataset is divided into three subsets for effective model training, validation, and testing. The training dataset is the largest subset, where the model learns to recognize various aircraft by optimizing its parameters. This phase shapes the foundational behavior of the model. The validation dataset monitors the model's performance during training and identifies issues such as overfitting. Overfitting occurs when the model performs well on the training set but poorly on the validation set, indicating a lack of generalization. The validation set provides feedback during training, aiding in fine-tuning the model and adjusting hyperparameters. A good model shows decreasing loss in both the training and validation datasets. After training and validation, the model is evaluated using the test dataset, which consists of unseen, unlabeled data. This subset is used to objectively measure the model's real-world performance and generalization ability. Properly and carefully partitioning the data into training, validation, and test subsets is essential for maximizing the model's overall performance and robustness. The

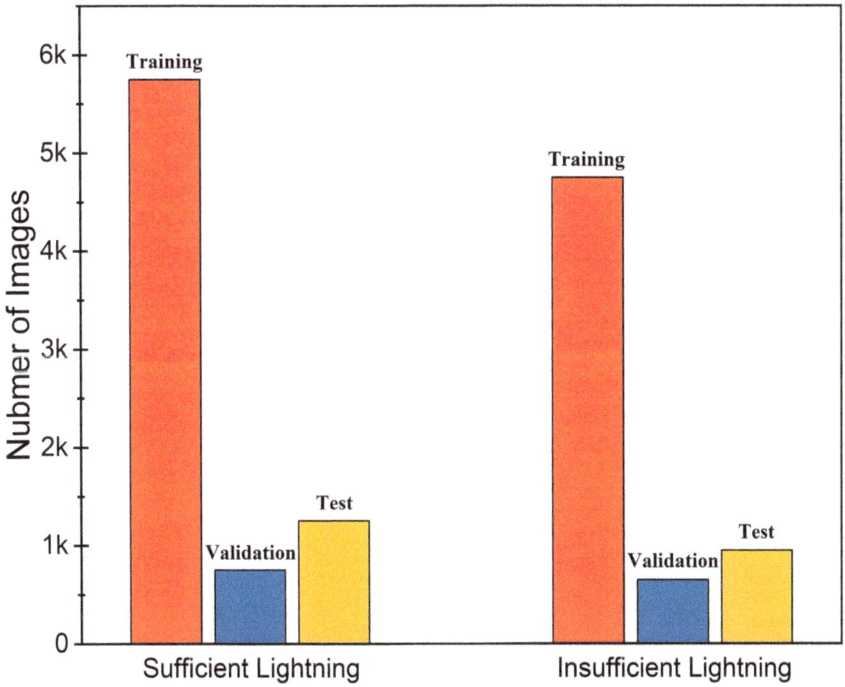

Fig. 11.3 Division of the dataset into two categories based on lighting conditions: sufficient lighting and insufficient lighting. Each category was further split into three distinct subsets for training, validation, and testing

dataset, which includes both color and grayscale images, is divided as follows: approximately 80% for training, 10% for validation, and 10% for testing. This separation ensures comprehensive evaluation and reliable measurement of the model's capabilities.

11.5 Method

The SSD algorithm detects aircraft of various sizes with remarkable accuracy by utilizing feature maps of different resolutions. While maintaining this high accuracy, the algorithm simultaneously identifies both the class and location of objects in real time with a single forward pass, significantly boosting detection speed. This model integrates the regression concept from the YOLO algorithm with the anchor box concepts introduced in the Faster R-CNN detection model, leveraging the strengths of both approaches. On a larger scale, SSD matches Faster R-CNN in detection accuracy on average but surpasses it in training speed. Compared to YOLO, SSD offers a slight improvement in detection speed and increases average detection

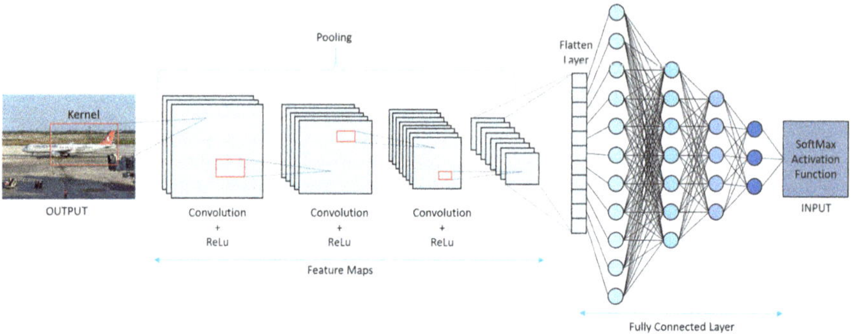

Fig. 11.4 Overview of the basic architecture of a Convolutional Neural Network (CNN), illustrating the key layers and operations involved

accuracy from 63.4% to 72.1%. This combination enables the SSD model to excel in both speed and accuracy, making it an efficient and ideal algorithm for real-time applications. Thanks to this balance, SSD has gained attention for a wide range of applications, including autonomous driving, security monitoring, face recognition, medical imaging, traffic analysis, augmented reality, robotics, and vehicle and pedestrian detection.

As shown in Fig. 11.4, the core structure of the SSD algorithm is based on CNNs. It can detect objects of varying sizes, from small to large, by using feature maps at different levels to perform multi-scale object detection. Feature maps from the early CNN layers capture fine details of small objects, while feature maps from the deeper layers identify larger objects and broader features. This multi-scale approach allows the SSD algorithm to detect both small and large objects with high precision. With the capability to simultaneously detect multiple object classes, SSD maintains its competitive advantage in tasks involving multiple object categories within a single image. The SSD framework, as presented in its original implementation, is built upon the 16-layer VGG-16 deep learning network from the Visual Geometry Group (VGG). During its development, the fully connected layers in VGG-16 were replaced with convolutional layers, enabling the model to learn deeper and more complex features while reducing computational costs. This modification enhanced the SSD framework's ability to detect objects of varying sizes and scales effectively. In this study, an essential requirement was that the algorithm be lightweight and efficient enough to operate on UAVs. SSD was selected for its ability to provide high-speed and accurate object detection while remaining lightweight enough to suit the payload capacity of UAVs. This characteristic simplifies the detection of various aircraft and complex objects, making SSD particularly suitable for UAV-based applications.

11.6 SSD Network Architecture

The SSD algorithm utilizes feature maps at different levels to detect objects across various scales. Initially, the input image is processed through a CNN, producing feature maps of varying sizes. These feature maps employ 3x3 convolutional filters to generate a reduced number of bounding boxes. Each box is associated with both classification and localization information, enabling the detection of both small and large objects. As each activation layer produces outputs, the SSD algorithm can detect smaller aircraft in the earlier layers and larger aircraft in the deeper layers, offering superior performance for objects of different scales. The outputs from deeper layers are lower in resolution, which makes them particularly suited for detecting larger objects. The SSD model is named based on the input image size used during training. For instance, a model trained on 512 × 512 input images is referred to as SSD512, while one trained on 300 × 300 images is named SSD300. To ensure effective training, the input size must remain consistent. Although variations in test image sizes might present challenges, these can be addressed through interpolation techniques. Notably, SSD512 offers higher detection accuracy than SSD300. Researchers can tailor the input dimensions and architectures of SSD models based on specific requirements, leveraging CNNs to extract necessary features.

As illustrated in Fig. 11.5, the SSD network architecture consists of two primary components: the backbone network and additional detection layers. The backbone, a pre-trained convolutional neural network, is used for feature extraction. The most commonly employed backbone in SSD, as noted in its original implementation, is the VGG-16 network. This architecture, which processes approximately 138 million parameters, has proven effective in extracting robust features due to its deep and wide structure. VGG-16 comprises multiple convolutional layers and max-pooling layers. These layers extract feature maps from the input images, with each grid cell in the feature map representing four object regions. SSD augments these feature maps with prediction layers, enabling the classification of objects and the corresponding bounding box predictions for each position. Following VGG-16, additional layers are incorporated to enhance accuracy and precision. SSD employs layers such as conv3, conv6 (FC6), conv7 (FC7), conv8_2, conv9_2, conv10_2, and conv11_2 as prediction layers. As the depth of the algorithm increases, the size of the feature maps progressively decreases, improving the detection of objects of varying sizes.

The SSD framework comprises three main parts: the backbone network, the original bounding box generation, and the convolutional prediction layers. The backbone network, which acts as a feature extractor, is a pre-trained classification network. This backbone is divided into the base network and additional feature extraction layers. While retaining the spatial structure of low-resolution images, it extracts meaningful information from the input images. Convolutional layers added to the backbone form the SSD "head," which interprets spatially localized activations as bounding boxes and object classes. The SSD algorithm operates

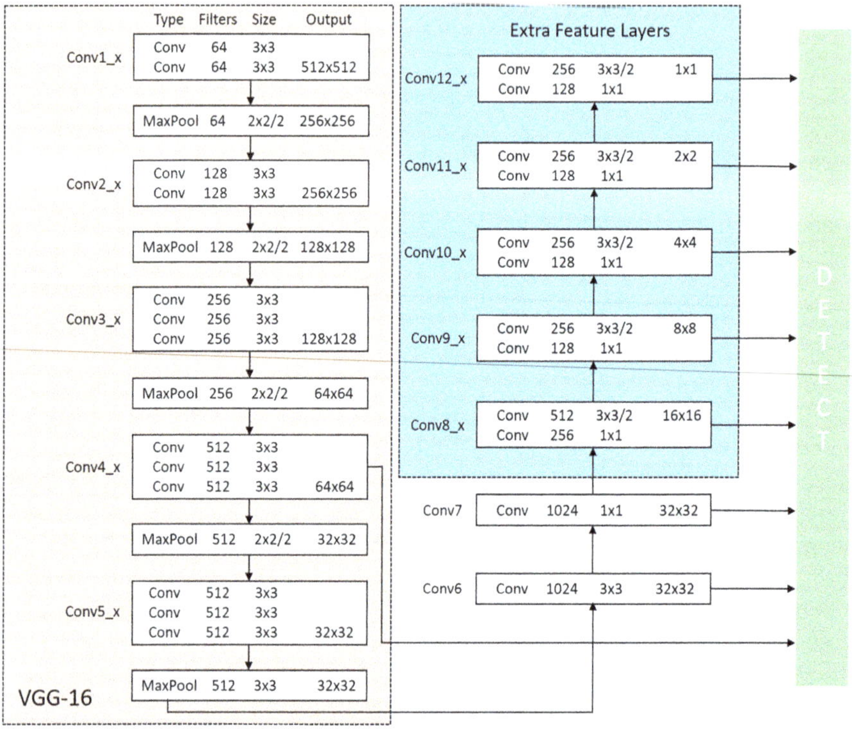

Fig. 11.5 Diagram of the Single Shot MultiBox Detector (SSD) network architecture, highlighting its structure and key components for object detection

as follows: input images are processed through deep neural networks to extract features, default boxes of various scales are designed to generate feature maps, object classes and locations are predicted, and finally, NMS is applied to select the best predictions. NMS plays a critical role in computer vision tasks like object detection and target tracking by removing redundant bounding boxes and retaining the box with the highest confidence score for each object. This reduces overlaps and prevents multiple predictions for a single object. NMS uses an Intersection over Union (IoU) threshold to filter bounding boxes. IoU calculates the overlap between two boxes as a ratio of their intersection to their union. The standard IoU threshold for SSD is 0.5, but improper filtering of lower-scoring boxes may impact average precision. The SSD model demonstrates impressive performance, achieving 74.3% mAP at 59 FPS on the VOC2007 dataset using an Nvidia Titan X GPU. This balance of speed and accuracy underscores SSD's efficiency in real-time object detection tasks.

It is important to acknowledge that the comparison presented in this study between SSD and YOLOv4 is not entirely balanced due to their differing architectural characteristics. SSD operates as a single-shot detector, while YOLOv4 incorporates multiple processing stages and optimizations that influence both speed and detection performance. While our goal was to evaluate their practical performance in a UAV-based airport surveillance context, we recognize that a more comprehensive experimental design would involve the inclusion of additional algorithms, including other single-shot detectors and real-time video-based models. Moreover, future work could explore hybrid approaches that integrate the strengths of SSD and YOLOv4, potentially achieving improved detection accuracy and efficiency. Such enhancements would contribute to a more holistic understanding of object detection performance in complex surveillance environments.

11.7 Loss Function

The loss function in the SSD algorithm measures the error rate and overall performance of the model. It combines classification loss and localization loss to optimize both the classification accuracy and the localization accuracy of the object detection model. In deep neural networks, the final layer is where the loss function is defined. The role of the loss function is to calculate the difference between the model's predictions and the ground truth values. A well-trained model will have a loss function value close to zero, indicating minimal error, whereas a poorly trained model will exhibit a high loss due to significant differences between predictions and actual values.

The total loss function comprises classification loss and localization loss. The primary purpose of the loss function is to evaluate how well the weights W and bias b parameters, learned during training, are suited to solving the problem. Various types of linear loss functions are used for this purpose, with common examples including sigmoid, multiclass support vector machine (SVM), and softmax functions. The classification loss is generally calculated using the following formula:

$$\mathrm{L_{conf}}\left(\mathrm{x},\mathrm{c}\right) = -\sum\nolimits_{\mathrm{i}\in\mathrm{Pos}} \mathrm{x_{ij}} \log\left(\mathrm{C_i}\right) - \sum\nolimits_{\mathrm{i}\in\mathrm{Neg}} \log\left(\mathrm{C_i}\right) \tag{11.1}$$

Localization loss, often expressed as L_1 or L_2 losses, is computed using the formula:

$$L_{\mathrm{loc}}\left(x, l, g\right) = \sum\nolimits_{i\in\mathrm{Pos}} \sum\nolimits_{m\in\{cx,cy,w,h\}} x_{ij} \mathrm{smooth}_{L1}\left(l_i - G_j\right) \tag{11.2}$$

Here, the function $\mathrm{smooth}_{L1}(x)$ is defined as:

$$\mathrm{smooooth_{L1}}(x) = \begin{cases} 0.5x^2, & |\, x \,| < 1 \\ |x| - 0.5, & \text{otherwise} \end{cases} \tag{11.3}$$

The SSD loss function is constructed based on the outputs generated during the prediction stage. It combines confidence loss L_{conf} and localization loss L_{loc} to compute the overall loss, as shown in the following formula:

$$L(x, c, l, g) = \frac{1}{N}\left(L_{\text{conf}}(x, c) + \alpha L_{\text{loc}}(x, l, g)\right) \tag{11.4}$$

In this formula, α is a weighting factor used to determine the importance of localization loss relative to confidence loss. N represents the number of positive predefined bounding boxes. By balancing the contributions of classification and localization losses, the SSD loss function ensures that the model achieves both high accuracy in classifying objects and precise localization of bounding boxes.

11.8 Evaluation Metrics

To evaluate the performance of the SSD and YOLOv4 models in aircraft detection, we used three widely adopted metrics in object detection tasks: precision, recall, and mean Average Precision (mAP) at two Intersection over Union (IoU) thresholds: 0.50 (mAP^{50}) and 0.75 (mAP^{75}). Precision is defined as the ratio of correctly predicted positive detections (true positives) to the total number of positive detections made (true positives + false positives). It reflects how accurate the model is in its positive predictions. Recall is the ratio of correctly predicted positive detections to all actual positive instances (true positives + false negatives), measuring the model's ability to detect all relevant objects. Mean Average Precision summarizes the precision-recall curve and is calculated by averaging the precision scores across all recall levels. We report mAP at IoU thresholds of 0.50 and 0.75 to capture both coarse and fine-grained localization accuracy. A higher IoU threshold requires bounding boxes to more closely match the ground truth to be counted as correct. These metrics were chosen because they offer a comprehensive view of object detection performance, balancing both detection accuracy and localization quality. They are especially relevant for UAV-based applications, where both high precision (to avoid false alarms) and high recall (to avoid missed detections) are critical in real-time surveillance tasks.

11.9 Test Results

Table 11.2 compares the total loss values observed during the training and validation phases for the SSD and YOLOv4 models over different periods. From the initial stages of training, the YOLOv4 model exhibits significantly lower loss values compared to the SSD model. By the end of training, this trend highlights that the YOLOv4 model undergoes a more efficient learning process. The sharper decline

Table 11.2 Comparison of total loss values during the training and validation phases for the SSD and YOLOv4 models

		Epoch				
Model	Total loss	20	40	60	80	100
SSD	Training	0.902	0.594	0.418	0.286	0.176
	Validation	0.814	0.551	0.396	0.220	0.110
YOLOv4	Training	0.910	0.665	0.385	0.245	0.035
	Validation	0.733	0.521	0.348	0.169	0.079

The table illustrates the superior efficiency of YOLOv4, showing significantly lower loss values and a sharper decline in both training and validation phases compared to the SSD model, which exhibits slower and less efficient loss reduction

Fig. 11.6 Detection performance of the SSD model on UAV-captured images featuring aircraft in various challenging conditions. Despite factors such as aircraft color variations, low-light environments, and the presence of shadows, the SSD model successfully detected all aircraft, demonstrating robustness in handling different altitudes, angles, and complex visual conditions

in YOLOv4's loss values toward the later stages of training indicates its superior ability to minimize prediction errors relative to the SSD model. In contrast, the SSD model demonstrates a smoother but slower decline in loss values throughout training, suggesting a less efficient learning process. During validation, the loss values for YOLOv4 also decrease more rapidly than those for SSD, indicating better generalization performance. This trend mirrors the model's behavior during training, underscoring not only YOLOv4's efficient learning process but also its ability to maintain robustness when exposed to new data.

Figure 11.6 illustrates a cluster of images where the SSD model, tested with the dataset, successfully detects all aircraft without errors. Most aircraft in the training data are predominantly white, which aligns with the model's ability to detect such aircraft accurately. However, some aircraft fuselages feature sections with different colors, which can introduce significant difficulty in object perception and impact model performance. Despite these challenges, the SSD model effectively detected aircraft with varied color patterns in different regions of the images. To account for challenging conditions, some UAV images were captured during low-light periods. In dark weather conditions, aircraft figures appear relatively pale, yet they are more discernible in medium-altitude images compared to high-altitude images. The SSD model demonstrated strong performance in detecting aircraft in medium-altitude images, even those taken during evening hours. However, at angles closer to horizontal, the geometry of distant aircraft changed significantly. For example, wing structures became nearly invisible in these images, where the top and side views of the aircraft geometry varied substantially. Despite these challenges, the SSD model successfully detected aircraft captured at low altitudes and near-horizontal angles. One of the most difficult conditions for object detection models is the presence of shadows. Shadows can sometimes mimic parts of an object or even the entire object when positioned closely. In some images from Fig. 11.6, shadows cover only small portions of the aircraft, minimizing their impact on detection. The SSD model successfully distinguished aircraft from their shadows, accurately identifying and drawing bounding boxes around the figures. This demonstrates the model's resilience to shadow interference.

Figure 11.7 showcases a set of images where the SSD model encountered detection errors in UAV-captured images. While the SSD model demonstrated considerable success under challenging conditions at medium altitudes, it struggled to maintain the same level of performance at higher altitudes. One of the most difficult conditions in high-altitude images involves the white-colored passenger

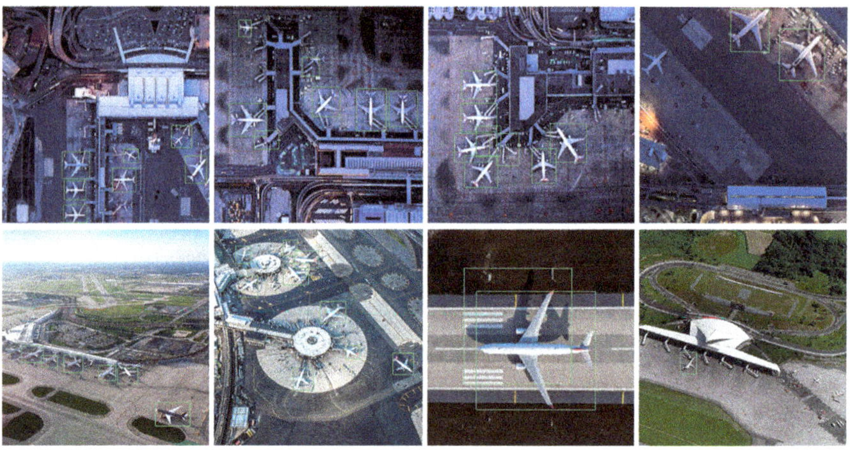

Fig. 11.7 Detection errors encountered by the SSD model in high-altitude UAV images, highlighting challenges with aircraft detection under low-light conditions and the presence of passenger boarding bellows systems. In these images, bellows systems were misidentified as aircraft due to similar geometries, and a navy-blue aircraft went undetected due to poor visibility. These limitations underscore the SSD model's reliance on geometric features for detection and its struggles with complex lighting and object overlaps

boarding bellows systems, which sometimes resemble aircraft geometries. In the first image, this challenge is evident. Captured at high altitude during late hours with artificial lighting and scattered light clusters, the image includes colored aircraft and bellows systems. In some instances, the bellows systems appear as extensions of the aircraft, leading to detection errors. For example, although the aircraft was correctly detected, the bounding box size was inaccurate. Additionally, a navy-blue aircraft in the image went undetected due to its dark color and the low-light conditions. This highlights a limitation of the SSD model, which relies on recognizing the rough geometry of aircraft. In the second image, a similar issue arises where the geometry of two bellows systems in the center of the image resembles that of aircraft, causing the SSD model to misidentify them as planes. Furthermore, one bellows system, located next to an aircraft, appears clearer than the aircraft itself due to lighting conditions. While the illuminated portion of the aircraft was included in the bounding box, much of the aircraft was obscured, resulting in detection errors.

The third image presents a comparable situation, where two belt systems near the center resemble aircraft geometry. The SSD model incorrectly identified them as two separate aircraft. Similarly, in the fourth image, two illuminated aircraft on the apron were correctly detected, but a parked aircraft in a dark area remained undetected due to insufficient visibility. Image number 5 depicts a scene captured by a UAV at an oblique angle during daytime. While the SSD model successfully detected aircraft closer to the UAV with clearer appearances, it failed to detect aircraft farther away, including those partially obscured by neighboring aircraft. In image number 6, taken under good lighting conditions at an oblique angle, the model accurately detected large-scale aircraft but struggled with aircraft featuring differently colored wingtips and failed to identify a small aircraft located at the center of the apron. Image number 7 introduces a different challenge: shadows. During the time the image was captured, the shadow of an aircraft closely mimicked its geometry, creating the appearance of two identical aircraft. The SSD model was unable to distinguish between the aircraft and its shadow, leading to incorrect detection of two separate aircraft. The eighth image was taken from a much higher altitude than the others. Here, even large aircraft appear very small, and their color tones blend with the ground, making them indistinguishable. As a result, the SSD model failed to detect the parked aircraft, revealing its limitations in distinguishing between aircraft and the ground under these conditions.

The test dataset was divided into two categories: sufficient light conditions and insufficient light conditions, and performance evaluations were conducted under these two scenarios. To provide a more comprehensive assessment of the SSD model, the YOLOv4 model, a member of the YOLO family and one of the prominent single-stage object detection architectures, was used for comparison. Table 11.3 presents a detailed numerical analysis of the performance of the SSD and YOLOv4 models under both lighting conditions. As observed in Table 11.3, both models exhibit significantly better performance under sufficient light conditions compared to insufficient light conditions. However, the SSD model demonstrated superior detection performance overall when compared to the YOLOv4 model. While both SSD and YOLO models are single-stage object detection architectures,

Table 11.3 Performance comparison between the SSD and YOLOv4 models under sufficient and insufficient light conditions

Model	Lightning context	Evaluation metric					
		Precision	Recall	mAP^{50}	mAP^{75}	F1 Score	Mean FPS
SSD	Sufficient Lightning	0.848	0.798	0.656	0.504	0.822	30.30
	Insufficient Lightning	0.751	0.720	0.588	0.468	0.735	25.60
	Overall	0.799	0.759	0.622	0.486	0.779	27.95
YOLOv4	Sufficient Lightning	0.812	0.783	0.611	0.477	0.797	31.25
	Insufficient Lightning	0.719	0.671	0.536	0.409	0.694	27.03
	Overall	0.766	0.727	0.574	0.443	0.746	29.14

The table presents a detailed numerical analysis, showing that both models perform better under sufficient light conditions. However, the SSD model consistently outperforms the YOLOv4 model in both lighting scenarios, demonstrating its superior detection accuracy due to its multi-scale feature map approach and anchor box-based predictions

their differing detection approaches contribute to variations in detection capabilities. YOLO processes bounding boxes and class probabilities in a single forward pass through the network, predicting object locations and classes simultaneously. In contrast, SSD generates predictions for each anchor box's confidence score and bounding box features from a fixed set of anchors at varying scales and aspect ratios across multiple feature maps. This design allows the SSD model to achieve higher detection accuracy in more complex image scenarios.

Under sufficient light conditions, the SSD model achieved a precision value of 0.848, outperforming the YOLOv4 model's precision value of 0.812. This confirms the SSD model's superior performance in these conditions. Additionally, the SSD model's mAP^{50} value was 30.16% higher than its mAP^{75} value, whereas the YOLOv4 model's mAP^{50} value exceeded its mAP^{75} value by 28.11%. These metrics indicate that the SSD model is significantly more effective at accurately drawing bounding boxes in sufficient light conditions compared to the YOLOv4 model. When comparing the average detection times, the SSD model further demonstrated superior performance, showing faster extraction times than the YOLOv4 model. In the case of insufficient light conditions, the SSD model once again displayed greater resilience and robustness. It outperformed YOLOv4 in terms of object visibility and quality in the images. This performance difference is reflected in the recall values, which similarly favor the SSD model. Although both models showed lower mAP values under insufficient lighting, the SSD model's mAP^{75} value reached 0.468, closer to a 50% success rate, compared to YOLOv4's mAP^{75} value of 0.409, which hovered near a 40% success rate. This highlights the SSD model's superior ability to draw accurate bounding boxes even under challenging lighting conditions. In terms of detection speed, the SSD model was again faster than YOLOv4 under insufficient light conditions. This increased speed can be attributed to both models detecting fewer objects and drawing smaller bounding boxes under these conditions, resulting in lower computational demands compared to sufficient light conditions.

Table 11.4 Detection performance analysis of the SSD and YOLOv4 models for aircraft located near the apron and in motion (on the runway and taxiway)

		Evaluation metric					
Model	Location context	Precision	Recall	mAP^{50}	mAP^{75}	F1 Score	Mean FPS
SSD	Near Aprons	0.755	0.704	0.541	0.460	0.729	26.43
	Runways& Taxiways	0.883	0.845	0.677	0.519	0.864	34.07
	Overall	0.819	0.775	0.609	0.489	0.797	30.25
YOLOv4	Near Aprons	0.731	0.698	0.555	0.443	0.714	26.01
	Runways& Taxiways	0.878	0.859	0.672	0.508	0.868	35.79
	Overall	0.805	0.779	0.614	0.476	0.791	30.90

The table shows that the SSD model outperforms YOLOv4 in all location categories and performance metrics, especially in more complex scenarios near the apron

Table 11.4 presents a detailed analysis of the detection performance of the SSD and YOLOv4 models, with the dataset divided into two categories based on aircraft location: near the apron or in motion (on the runway and taxiway). As expected, the SSD model consistently outperformed the YOLOv4 model across all location categories and performance metrics. Both models demonstrated higher detection performance for aircraft on the runway and taxiway compared to those near the apron. The primary reason for this disparity is the similarity in geometry between the bellows systems near the apron and the aircraft, which complicates accurate detection. In contrast, the runway and taxiway present fewer challenges, with the main difficulty being the close similarity in color tones between the runway and the aircraft. This simpler environment allows both models to achieve better performance. Although the SSD and YOLOv4 models exhibited comparable performance under the less challenging conditions of the runway and taxiway, the SSD model outperformed YOLOv4 significantly in the more complex scenarios near the apron. This indicates that the SSD model is more robust and better suited to handling difficult conditions. Both models achieved higher mAP^{50} and mAP^{75} values for aircraft on the runway and taxiway than for those near the apron. However, the SSD model consistently surpassed YOLOv4 in these metrics across all scenarios. As illustrated in Fig. 11.7, the SSD model occasionally included portions of the bellows systems within the bounding box when detecting aircraft near the apron, which aligns with its slightly reduced precision in these cases. In addition to its superior detection performance in complex environments, the SSD model demonstrated significantly better average detection times compared to YOLOv4. This further reinforces the SSD model's advantage as a more efficient and effective object detection algorithm, particularly in challenging scenarios.

11.10 Conclusion

This study evaluated the performance of the SSD and YOLOv4 models for aircraft detection using UAV-captured images under varying lighting conditions and altitudes. The results clearly demonstrate that YOLOv4 outperforms SSD in terms of learning efficiency and generalization ability, as evidenced by its lower loss values during both training and validation phases. YOLOv4's faster decline in loss values during training indicates its more effective minimization of prediction errors and superior learning process. In contrast, SSD, while capable of detecting aircraft under challenging conditions such as varying colors and shadow interference, exhibited slower learning and less efficient performance compared to YOLOv4. Despite these differences in efficiency, SSD demonstrated considerable success in complex detection scenarios. It showed strong performance in detecting aircraft at medium altitudes and under low-light conditions, even in the presence of challenging factors like varying aircraft colors and shadows. However, SSD struggled at higher altitudes and in images where aircraft geometries were obscured or misrepresented by environmental factors like lighting and shadows, which impacted its accuracy. Furthermore, when compared to YOLOv4, SSD exhibited faster detection times, which highlights its strength in real-time applications. Under sufficient light conditions, SSD outperformed YOLOv4 in terms of detection accuracy and precision, with higher mAP^{50} values and faster detection times. However, in low-light conditions, while both models showed reduced performance, SSD's ability to maintain higher recall and detection accuracy further underscores its robustness in challenging environments. Conclusively, while YOLOv4 excels in learning efficiency and generalization, the SSD model offers superior performance in complex, real-world detection scenarios, particularly under suboptimal lighting and challenging visual conditions. Both models have distinct strengths, with SSD being more suitable for scenarios demanding high accuracy and robustness, while YOLOv4 proves more efficient in terms of training and inference.

References

1. Wandelt, S., Zheng, C.: Toward smart skies: reviewing the state of the art and challenges for Intelligent Air Transportation Systems (IATS). IEEE Trans. Intell. Transp. Syst. **25**(10), 12943–12953 (2024). https://doi.org/10.1109/TITS.2024.3415588
2. Huang, X.D., Zhang, H.J., Liu, K., Liu, T.G.: Fully modelling based intrusion discrimination in optical fiber perimeter security system. Opt. Fiber Technol. **45**, 64–70 (2017). https://doi.org/10.1016/j.yofte.2018.06.005
3. Chia, C.-C., Chan, W.-K., Chien, S.Y.: Cooperative surveillance system with fixed camera object localization and mobile robot target tracking. In: Advances in Image and Video Technology, LNCS, vol. 5414, pp. 886–897. Springer, Heidelberg (2009). https://doi.org/10.1007/978-3-540-92957-4_77
4. Pik, E.: Airport security: the impact of AI on safety, efficiency, and the passenger experience. J. Transp. Secur. **17**(9) (2024). https://doi.org/10.1007/s12198-024-00276-6

5. Babu, V.L.L., Batta, R., Lin, L.: Passenger grouping under constant threat probability in an airport security system. Eur. J. Oper. Res. **168**(2), 633–644 (2006). https://doi.org/10.1016/j.ejor.2004.06.007
6. Hubbard, B.J., Hubbard, S.M.: Tracking and monitoring technologies to support airport construction safety. In: IOP Conference Series: Materials Science and Engineering, vol. 1218, p. 012015 (2022). https://doi.org/10.1088/1757-899X/1218/1/012015
7. Nagarani, N., Venkatakrishnan, P., Balaji, N.: Unmanned aerial vehicle's runway landing system with efficient target detection by using morphological fusion for military surveillance system. Comput. Commun. **151**, 463–472 (2020). https://doi.org/10.1016/j.comcom.2019.12.039
8. Feitosa, I., Santos, B., Almeida, P.G.: Pavement inspection in transport infrastructures using Unmanned Aerial Vehicles (UAVs). Sustainability. **16**(2207) (2024). https://doi.org/10.3390/su16052207
9. Zhi, Y., Fu, Z., Sun, Z., Yu, J.: Security and privacy issues of UAV: a survey. Mob. Netw. Appl. **25**, 95–101 (2019). https://doi.org/10.1007/s11036-018-1193-x
10. Yang, R., Zhou, J., Lu, X., Shen, J., Chen, H., Chen, M., He, Y., Liu, F.: A robust rice yield estimation framework developed by grading modeling and normalized weight decision-making strategy using UAV imaging technology. Comput. Electron. Agric. **215**, 108417 (2023). https://doi.org/10.1016/j.compag.2023.108417
11. Korecki, Z., Janošek, M., Pecháček, T.: Use of unmanned aerial systems in airport operations. In: International Conference on Military Technologies (ICMT), pp. 1–8 (2021). https://doi.org/10.1109/ICMT52455.2021.9502756
12. Bakirci, M.: Advanced aerial monitoring and vehicle classification for intelligent transportation systems with YOLOv8 variants. J. Netw. Comput. Appl. **237**(104134) (2025). https://doi.org/10.1016/j.jnca.2025.104134
13. Zhang, Z., Shu, Z.: Unmanned Aerial Vehicle (UAV)-assisted damage detection of wind turbine blades: a review. Energies. **17**(15), 3731 (2024). https://doi.org/10.3390/en17153731
14. Anghuwo, J.S., Imanuel, P., Nangolo, S.S.: Anti-unmanned aerial vehicle detection system for airports: aviation and national security perspective. J. Transp. Secur. **17**, 12 (2024). https://doi.org/10.1007/s12198-024-00280-w
15. Rodríguez-Canosa, G.R., Thomas, S., Del Cerro, J., Barrientos, A., MacDonald, B.: A real-time method to detect and track moving objects (DATMO) from unmanned aerial vehicles (UAVs) using a single camera. Remote Sens. **4**(4), 1090–1111 (2012). https://doi.org/10.3390/rs4041090
16. Yaacoub, J.P., Noura, H., Salman, O., Chebab, A.: Security analysis of drone systems: attacks, limitations, and recommendations. Internet Things. **11**, 100218 (2020). https://doi.org/10.1016/j.iot.2020.100218
17. Leng, J., Liu, Y.: An enhanced SSD with feature fusion and visual reasoning for object detection. Neural Comput. Appl. **31**, 6549–6558 (2019). https://doi.org/10.1007/s00521-018-3486-1
18. Zhou, S., Qiu, J.: Enhanced SSD with interactive multi-scale attention features for object detection. Multimed. Tools Appl. **80**, 11539–11553 (2021). https://doi.org/10.1007/s11042-020-10191-2
19. Ding, L., Xu, X., Cao, Y., Zhai, G., Yang, F., Qian, L.: Detection and tracking of infrared small target by jointly using SSD and pipeline filter. Digit. Signal Process. **110**, 102949 (2021). https://doi.org/10.1016/j.dsp.2020.102949
20. Liu, W., Anguelov, D., Erhan, D., Szegedy, C., Reed, S., Berg, A.: SSD using enhanced feature map block for object detection. Sensors. **21**(8) (2016). https://doi.org/10.3390/s21082842
21. Xiangyu, D., Shanshan, L.: An improved SSD object detection algorithm based on attention mechanism and feature fusion. In: Proceedings of the Journal of Physics: Conference Series, vol. 12345, pp. 1–13 (2023). https://doi.org/10.1088/1742-6596/2450/1/012088
22. Yang, T., Liang, R., Huang, L.: Vehicle counting method based on attention mechanism SSD and state detection. Vis. Comput. **38**, 2871–2881 (2022). https://doi.org/10.1007/s00371-021-02161-y

23. Shan, J., Miccinesi, L., Beni, A., Pagnini, L., Cioncolini, A., Pieraccini, M.: A review of foreign object debris detection on airport runways: sensors and algorithms. Remote Sens. **17**(225) (2025). https://doi.org/10.3390/rs17020225
24. Zhang, Y., Guo, L., Wang, Z., Yu, Y., Xu, F.: Intelligent ship detection in remote sensing images based on multi-layer convolutional feature fusion. Remote Sens. **12**(3), 456–478 (2020). https://doi.org/10.3390/rs12203316
25. Mutreja, G., Aggarwal, A., Thakur, R., Tiwari, S.S., Deshpande, S.: Comparative assessment of different deep learning models for aircraft detection. In: Proceedings of the International Conference for Emerging Technology (INCET), pp. 101–110 (2020). https://doi.org/10.1109/INCET49848.2020.9153981
26. Azam, B., Khan, M.J., Bhatti, F.A., Maud, A.R.M., Hussain, S.F., Hasmi, A.J., Khurshid, K.: Aircraft detection in satellite imagery using deep learning-based object detectors. Microprocess. Microsyst. **94**, 234–245 (2022). https://doi.org/10.1016/j.micpro.2022.104630
27. Li, M., Zhang, Z., Lei, L., Wang, X., Guo, X.: Agricultural greenhouses detection in high-resolution satellite images based on convolutional neural networks: comparison of faster R-CNN, YOLO v3 and SSD. Sensors. **20**(4938) (2020). https://doi.org/10.3390/s20174938

Chapter 12
Bridging the Modality Gap: A Siamese Network Approach for Image-Text Similarity via Contrastive Learning

E. P. Vishnutheerth, Vivek Vijay, B. Premjith, and V. Sowmya

Abstract To narrow down the gap between the image and text modalities, computing their similarity is of the utmost importance. This paper employs a novel, comprehensive system for determining the similarity between text and image using a contrastive learning approach. The integration of state-of-the-art models (YOLO, Vision Transformers, and BERT) offers a promising utilization of the proposed system in areas such as automatic captioning, recommendation systems, etc. BERT is used to generate text data embeddings, while Vision Transformer creates image data embeddings of the detected objects by YOLO. This work investigated the similarities between the text and image data using a Siamese network and measured the similarity using the cosine similarity metric. The proposed model applies an underlying methodology which assumes that in the shared embedding space, similar image text pairs are closer, and dissimilar ones are far apart. The accuracy for this model turned out to be 0.74 for the similarity check at a threshold of 0.5 and indicate the potential of the proposed system for multimodal data matching in several applications where good matching is required. This also shows betterment compared to other state-of-the-art works.

Keywords BERT · Contrastive learning · Image-text similarity · Siamese network · Vision transformer · YOLO

12.1 Introduction

The multimodal approach for learning has shown great promise in areas such as Visual Question Answering (VQA) [1], Image Captioning [2], Retrieval systems [3], etc as it can capture more nuanced understanding of complex patterns underlying within the data unlike unimodal models, which process only one modality (text,

E. P. Vishnutheerth · V. Vijay · B. Premjith (✉) · V. Sowmya
Amrita School of Artificial Intelligence, Amrita Vishwa Vidyapeetham, Coimbatore, India
e-mail: cb.sc.p2dsc23017@cb.students.amrita.edu; cb.sc.p2dsc23006@cb.students.amrita.edu; b_premjith@cb.amrita.edu; v_sowmya@cb.amrita.edu

G. Rivera et al. (eds.), *Recent Advances in Artificial Neural Networks*, Intelligent Systems Reference Library 283, https://doi.org/10.1007/978-3-032-13497-4_12

image, or speech) at a time. In the image captioning task, the model generates textual descriptions for the objects identified in the image, which showcases the multimodal image and the data understanding of the model. Similarly, in the VQA task, the ability of a model to process both text and image modalities helps to understand the image-related question and provide relevant answers. The integration of different modalities overcomes the challenge of capturing the complex interaction between different types of data.

The proficiency of YOLO to identify and localize objects within images has revolutionized object detection. Its effectiveness in performing object detection for an entire image in a single pass gives it an edge over other object detection models [4], for the applications where speed and precision are of extreme importance [5]. Vision Transformer applies transformer mechanisms to the image patches. Similar to language processing, ViT treats images as a sequence of patches [6]. It allows granular level analysis of visual information, thereby enhancing the ability of the model to capture complex patterns within the images. BERT generates text embedding by providing deep bidirectional context understanding [7]. BERT enhances the textual data representation by grasping the contextual meanings of words and phrases [8]. The synergy between these models highlights their collective potential to address the challenges of multimodal learning, leveraging their individual strengths to create more robust and effective solutions.

These models have real-world applications like social media content moderation, recommendation systems for online delivery applications (matching product images to user queries), and provides help for visually impaired people as well. The key contributions of this work includes: (1) leveraging YOLO, ViT, and BERT in a framework altogether for efficient image-text embedding generation; (2) applying contrastive learning using Siamese network with cosine similarity to differentiate similar and dissimilar image-text pairs; and (3) producing a reasonable accuracy of 0.74 on the VQA v2.0 dataset, validating the power of the model in case of multimodal similarity tasks.

In this work, VQA v2.0 is used to explore contrastive learning for image-text similarity [9]. In addition, this study simplifies the computation while enhancing analysis by focusing on images and textual questions. Here, YOLO v5, Vision Tansformer (google/vit-basepatch16-224), and bert-base-uncased are used for object detection, image embeddings, and text embeddings, respectively. These embeddings are fed into a Siamese network [10] to compute similarity scores using a cosine similarity threshold, leveraging the strengths of each model to improve accuracy [11].

The research effectively demonstrates the handling of complex data interactions, with improved performance metrics underscoring the potential of this approach in similarity checking tasks. This model has achieved a reasonable accuracy of 0.74 which showcases its efficiency in image-text similarity tasks. To a great extent, this work is a cornerstone for image-text similarity research advancements.

The sections are organized as follows: Sect. 12.2 describes the Related Works, Sect. 12.3 describes the Dataset Description, Sect. 12.4 describes the Methodology followed, Sect. 12.5 describes the Results and Discussions and then finally concluding with Sect. 12.6.

12.2 Related Works

The similarity detection field has seen a lot of developments in the past few time frames, but using it among multimodal data primarily started with the VQA v2.0 dataset suggested by Agarwal et al. in 2015 [1]. This dataset provides a massive amount of training and testing data for images and related texts, which are the questions and answers related to the text. It can be used with models that can give answers based on visual inputs due to its rich annotations. Researchers have used this large data set to create a more refined model for question-answering based on images, thereby resulting in multimodal models that address more complicated real-world issues. The continued use and expansion of such datasets are likely to further push the boundaries of what can be achieved in the realm of text and image similarity.

A substantial analysis of different object detection deep learning techniques is done by Subbiah et al. [4]. It highlighted the noteworthiness of these techniques in the extraction of meaningful features from the images. Generating precise image embedding, which is influenced by object detection techniques, is essential for tasks pertaining to multimodal data. The YOLO model series, principally the YOLOv5s, has been the focus of noteworthy research and development. The evolution of YOLO models from YOLOv1 till YOLOv8 and YOLO-NAS is reviewed by Terven et al. [5]. This review emphasizes the accuracy and speed improvements that have made YOLOv5s a preferred candidate while looking at the feasibility and complexity. The advanced YOLO model versions led to more reliable object detection capabilities.

Han et al. [6] offers a comprehensive survey on Vision Transformers, a notable breakthrough in computer vision. It makes use of self-attention mechanisms to capture complex patterns underlying in the images. This provides a more nuanced learning of visual data by getting high quality image embeddings. Self attention works in such a way that it consider the relationships between various parts of an image, hence providing more accurate and detailed embeddings. Text representation has been remodeled through contextual embeddings through the introduction of BERT by Devlin et al. [7]. The pre-trained bert-base-uncased model used for the study can effectively work with both lowercase as well as uppercase. The power of BERT ensures the generated embeddings are rich in meaning and are relevant. The effectiveness of BERT in different text categorization tasks are validated by Yerramreddy et al. [8]. Li et al. [9] contributed to the discussion on visual-text cross-alignment, emphasizing the refinement of similarity scores in vision-language models. It focuses on improving the similarity scores between a image-text pair.

Chen and He [10] has explored the use of Siamese networks in the areas of multimodal similarity, showcasing the potency of such networks in learning robust representations for computing the similarity. The power of Siamese networks lie in their ability to learn the embeddings that computes the similarity between the input pairs. Liu et al. [11] provides insights to various deep learning based methods for aligning text and image features, underlining the advancements in the methods for performing the same. Additionally, Mohammadi et al. [12] investigated the expansion of diversity within joint feature space similarity scores, while Liu et al. [13] proposed a modality-invariant embedding framework to ensure consistency across different data types. Zhang et al. [14] proposed practices to preserve semantic consistency in the retrieval of image-text pair. To improve image-text matching, Diao et al. [15] proposed filtration techniques by filtering out irrelevant pairs and then addressing similarity reasoning. Zhang et al. [16] presented an improved framework that focuses on better semantic understanding for similarity assessments. Yang et al. [17] proposed the RFE-SRN, which enhances similarity reasoning by focusing on regional features, providing a comprehensive approach to improving similarity evaluations. All these works provide details on both theoretical and practical aspects of the areas and technologies for multimodal similarity. Hirota et al. [18], has replaced features from the images with the textual descriptions together with the usage of a language-only Transformer model to reduce computational cost and complexity. Zhou et al. [19], has brought in a unified vision-language pre-training model which makes use of large scale image-text pairs. Yu et al. [20] address the language prior issue through a fusion strategy along with similarity matching which has shown better handling of hard classes in VQA-CP v2.

12.3 Dataset Description

This study utilizes a smaller subset from the VQA v2.0 dataset [1]. It contains open-ended questions related to each of the images, which comprises approximately 204,721 images and 1.1 million questions with 10 possible answers each. The smaller subset is taken due to its extensive size and processing limitations. The dataset is divided into training and testing sets, with 6372 image-text split as follows: 4806 for training and 1566 for testing, as shown in Table 12.1. This count is taken considering the computational complexity and resource constraints. For this study, images and the questions related to those images are utilized to find out whether the question has some connection with the image or not. The training set is used for model learning and the test set is reserved for final performance evaluation.

Table 12.1 Dataset split

Data split	Count
Training	4806
Testing	1566

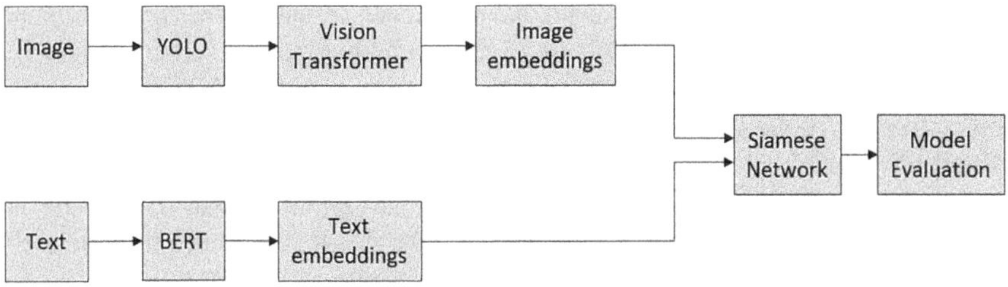

Fig. 12.1 Block diagram of the proposed work

12.4 Methodology

This section details the method used for evaluating similar images and texts. It explains how text embeddings are produced, objects detected and encoded in images, and a Siamese network [10] that determines similarity as shown in Fig. 12.1.

Here, the textual part is used to create text embeddings using BERT. Images are passed through YOLO model to get the object detections, after which ViT is used to create image embeddings and then using a dense layer to create one embedding per image. After the embedding creation step of both the modalities, they are both passed onto the two branches of the Siamese Network which takes into account the cosine similarity distance metric between the modalities based on Contrastive learning approach.

12.4.1 Text Embeddings Using BERT

BERT, which stands for Bi-directional Encoder Representations from Transformers [7] is a large language model that was developed by Google in 2018. BERT is based on the encoder-only transformer architecture. Encoder-only model is good for tasks that requires understanding of input such as Sentence Classification, Named Entity Recognition (NER) and Question-Answering. As it doesn't have the decoder component, BERT cannot generate text which in turn means that it cannot do tasks like Text Translation and Text Summarization. BERT has an added advantage of Bidirectional feature which helps it capture word connections more precisely by considering left-to-right and right-to-left traversals. It has two major variants: bert-base with 110 million parameters and bert-large with 340 million parameters, both of which are trained on English Wikipedia and BookCorpus. In 2020, the smallest variant bert-tiny also came into the picture with 4 million parameters only.

BERT is trained on two tasks: Masked Language Modelling (MLM) and Next Sentence Prediction (NSP). The Masked Language Modelling task requires BERT to predict a masked-out word. If BERT doesn't guess the correct word for the masked word during training, the weights in the network get adjusted so that it is more likely to guess an apt word the next time around. The Next Sentence Prediction is a binary classification task to check whether the next sentence logically follows the previous sentence or not. Both of these tasks force BERT to get a good understanding of the English language. It can be trained with raw text, as labelled data is not needed.

When using BERT, tokenization is done initially to split the words into subwords, and these subwords are mapped to numerical IDs, this is done since these models can only process numbers. This conversion of text inputs to numerical data is carried out using tokenizers. BERT used WordPiece algorithm for tokenization. It has a vocabulary size of 30,000 and tokens not found in the vocabulary are identified as [UNK]. It uses the greedy longest match first application, where it finds the longest subword that is present in the vocabulary and then removes the last character and see if that is there in the vocabulary. If that is not there, it again removes the last character, and this process follows until it finds that particular word in the vocabulary. When it finds the word in the vocabulary, it jumps to the end of the word and adds ## to the removed word part to indicate that this should be merged with the previous word. Now it checks to see if ## along with that removed word, is present in the vocabulary. It will then remove the last character from that word part until it's found in the vocabulary. Each of these subwords produced after doing tokenization will have a numerical ID mapping called token ID. This subword tokenization works on the following two objectives: A word is not decomposed into subwords if it is a frequently used word and A word is decomposed into subwords if it is a rarely used word.

As part of being able to process text in parallel, all the input text sequences are truncated to a fixed length or padded out. BERT can only support up to 512 tokens at a time. The tokenizer will add [CLS] token or Classification token which act as a placeholder at the start of an input sequence and [SEP] token or Separator Token is a placeholder for separating two sentences by positioning it at the end of one sentence. BERT has two ways of distinguishing two pieces of text sequences as input: One is by placing the [SEP] token between the two input sentence sequences and the other is by using Segment Embeddings which encodes sentence number into a vector so that it can distinguish between two sentences. In Segment Embeddings, all tokens that belong to the first sentence are assigned a vector with index 0 and all tokens that belongs to the next sentence are assigned another vector with index 1. After this step, Positional Encoding comes into the picture. Positional encoding vector is added to the embedding vector of each word so that it can have information regarding the position of each word within a sequence and hence it can also capture similarity in meaning between the words. After this step, it is fed into the 12 encoders and the results are made available as hidden states.

A combination of these three—token embeddings, segment embeddings and position embeddings are summed up element-wise to produce a single embedding of shape (1, n, 768). This combined embedding is then passed on to the encoder layer of BERT. BERT performs self-attention mechanism with multi-head attention. The self-attention mechanism takes the word embeddings and project it into three vector spaces: Query (Q), Key (K) and Value (V). In order to determine the attention weights, the attention score is calculated. The score for each word is calculated to determine how much focus it needs to give to the other words in the sentence. Queries and Keys which are similar will have a higher dot product while the dissimilar ones will have lesser to no overlap. Self-attention allows applying different weights to different words in a sequence. The self-attention score is calculated using the `equation` 1 where n indicates the dimensionality of the key/query vectors.

$$\text{Attention Score}(Q, K, V) = \text{softmax}\left(\frac{QK^T}{\sqrt{n}}\right)V \tag{12.1}$$

BERT has 12 self-attention heads and each multi-headed attention block gets three inputs, the query, the key and the value. These are put through linear layers before the multi head attention function. The query, key and value are passed through separate fully connected linear layers for each attention head. The model can jointly attend to information from different representations and at different positions. By having 12 self-attention heads, it is able to focus on several tasks at once, thereby allowing more rich connections between the words.

In this work, bert-base-uncased is made into use to strike a good balance between performance and computational complexity. Before tokenization, it converts every word into lower case, that is how this got the term "uncased". It is smaller and faster than bert-large, along with the added advantage of strong generalization capacity. This makes bert-base-uncased a default first choice model for this work for text embedding creation part.

12.4.2 Object Detection Using YOLO

YOLO [5], which stands for You Only Look Once, is a realtime object detection model which was first developed by Joseph Redmon in 2015. This name is derived from the fact that object detection is done by this algorithm in a single forward pass (single-stage object detection). Unlike conventional item detection frameworks that depend on multi-degree tactics, YOLO revolutionizes the approach by treating object detection as a unified regression hassle, thereby streamlining the workflow and extensively improving both velocity and efficiency. At its center, YOLO methods an entire picture in a single forward pass via a convolutional neural network (CNN), efficiently reworking the photograph right into a grid of $S \times S$ cells. Each of those grid cells is liable for predicting the coordinates of bounding boxes, generally denoted as B , in conjunction with corresponding self assurance scores and probability of that object in the bounding box. A bounding box consists of an item and assesses the accuracy of the bounding box's role and size, while the class possibilities decide the probability for the detected item to belong to a specific category. This included approach contrasts sharply with two-degree detectors like R-CNN [4], which first generate location proposals and ultimately classify these regions in separate steps, a system that introduces huge computational overhead and latency.

YOLO's single-skip structure not only best removes the need for separate region throughout but also additionally leverages internal picture context while making predictions, thereby improving the version's capability to understand the relationships and spatial configurations of multiple objects lying within an image. This holistic perspective reduces the prevalence of fake positives and improves the precision in object localization by taking intricate dependencies and contextual nuances of the images. For instance, the presence of someone close to a bicycle in an image, is extra appropriately interpreted because of YOLO's capacity to understand and contextualize such relationships. Moreover, YOLO's self-attention mechanisms and multi-scale predictions, in particular in its later variations like YOLOv3 and YOLOv4, allow the version to stumble on objects of various sizes and scales more successfully, thereby improving universal detection accuracy throughout numerous and complicated scenes.

The real-time performance of YOLO is one in all its highly appreciated benefit, executed without compromising detection accuracy. Versions which include YOLOv3 and YOLOv4 are capable of processing pictures at marvelous rates, reaching up to 45 frames on GPU hardware. This functionality makes YOLO fairly suitable for applications that demand immediate processing and response, like autonomous vehicles, ensuring safe navigation and well timed choice-making based on their surrounding video surveillance which has non-stop tracking with minimum latency. In video surveillance, YOLO allows actual-time monitoring and the already set detection of suspicious spots or unauthorized access, enhancing security features. Augmented reality (AR) applications benefit from YOLO's capability to understand and music items in actual time, making an allowance for seamless integration of virtual data into the bodily world, thereby improving person studies. In the healthcare region, YOLO can help in scientific imaging aiding in diagnosis and making treatment plans with high precision. Moreover, scalability and versatility of the YOLO model made it a more desirable model for object detection scenarios. YOLOv2 has anchor containers to enhance bounding box predictions, whereas YOLOv3 integrated multi-scale predictions to discover smaller objects within the image. Subsequent versions, For YOLOv4 and YOLOv5, the creators have integrated advanced techniques along with Cross-Stage Partial (CSP) connections, Spatial Pyramid Pooling (SPP), and Path Aggregation Networks (PAN), which boosts both speed and accuracy and hence making sure that YOLO remains at the vanguard of object detection era.

Stage detectors first generate location proposals and then classify those regions. The single stage technique of YOLO reduces computational needs and hastens the detection system, permitting it to carry out predictions precisely even on hardware with limited computational sources. Furthermore, YOLO's capability to leverage worldwide contexts enhances its detection capabilities, thereby reducing redundant detections and improving normal accuracy. This makes YOLO especially powerful in complex environments in which multiple items have interaction in numerous spatial configurations.

Moreover, YOLO performs a essential function inside the era of embeddings and the combination of visible and textual statistics, that's vital for multimodal understanding. By accurately figuring out and localizing objects within an picture, YOLO affords established records that may be transformed into numerical embeddings representing the visible entities and their spatial relationships. These embeddings can then be in comparison with text embeddings generated by using models like BERT, facilitating tasks like image captioning, visual question answering, and also in multimodal retrieval systems. This synergy among object detection and textual information enhances the ability to comprehend and interpret complicated scenes, bridging the gap between visual and textual statistics, thereby enabling greater sophisticated and shrewd interactions with the data. Here, YOLO is used to detect the objects from the images and capture those object detected images.

In conclusion, the YOLO model [5] stands as a state-of-the-art model for object detection applications, by supplying a unified and exceedingly efficient method that surpasses the constraints of traditional stage detectors. The non-stop advancements throughout various YOLO variations have always improved its performance, scalability, and adaptability, making it suitable for a extensive variety of tasks that require timely and unique object detections. Throughout the years, YOLO stays at the forefront and evolve with ever other updated version, thereby cementing its position as an imperative tool within the arsenal of cutting-edge intelligence and object detection technology. YOLO v5 is used for this work keeping in mind its ability to even detect small objects as well as robustness and stability. Here, YOLO is utilized to detect objects in the images and store as detected objects from the images. This is really useful as the corresponding text pair will be having more weights where it describes any object. So, this similarity can be very well drawn out making the objects underlying in the images into use.

12.4.3 Image Embeddings Using Vision Transformer

ViT, which stands for Vision Transformer [6] represents a vast development within the discipline of computer vision by leveraging the powerful transformer architecture, which was first designed exclusively for natural language processing, to excel in image evaluation duties. It is pretrained on ImageNet21k which has 14 million images with resolution 224×224 and 21k classes and finetuned on ImageNet 2012 which has 1 million images and 1k classes. In this work, YOLO is initially used to perform object detection and after that, ViT is used to generate wealthy and comprehensive image embeddings, thereby enhancing the machine's ability to apprehend and interpret complicated visible facts with first-rate precision and intensity. The ViT starts its function by dividing the chosen picture into a grid of constant-length patches, typically 16×16 pixels, even though the scale may be adjusted primarily based on particular utility necessities and desired granularity. Every patch is then flattened right into a one-dimensional vector and linearly embedded into a higher-dimensional area, successfully reworking the spatial information of each patch into a format suitable for processing via transformer layers. To maintain the positional information of every patch in the authentic image, ViT incorporates positional encoding, which is added to the patch embeddings, permitting the model to recognize both the relative and absolute positions of the patches and making sure that spatial relationships are accurately represented.

Once the image patches are embedded, they are fed into a series of transformer layers that constitute the core of ViT. These transformer layers utilize the self-attention mechanism, a groundbreaking innovation among transformer architectures, which allows the version to weigh the significance of each patch relative to every other patch within the image. Unlike Convolutional Neural Networks (CNNs), which use convolutional kernels, the self-attention mechanism in transformers allows the capture of both internal and neighborhood functions simultaneously. This twin capability enables ViT to version problematic and varied length dependencies inside the image, consisting of the spatial relationships between distant objects or the contextual interplay among various visual factors. The self-attention mechanism operates with the aid of generating three vectors for each patch: query, key, and value vector. These vectors are computed through discovered linear transformations using the patch embeddings. Attention scores are then calculated with the aid of taking the dot product of the query vector with the key vector transposed, divided by the root of the dimension of the keys and then softmax is performed on this computation. These scores allows to give attention on the most applicable elements of the picture for a given task.

Via a couple of layers of self-attention and feed-forward neural networks, ViT step by step builds hierarchical representations of the image statistics. Every transformer layer refines the patch embeddings by means of integrating facts from different patches, permitting the model to seize complex visual patterns and spatial relationships which are regularly tough for classic CNNs to model efficaciously. For example, in an photograph depicting a crowded road scene, ViT can parent individual pedestrians (neighborhood functions) while simultaneously recognizing the overall arrangement and interactions among numerous other objects like vehicles, bicycles, and houses (worldwide functions). These embeddings are not only rich in semantic content but also good in maintaining the spatial coherence and relational dynamics of the patches in the detected objects.

ViT's ability to capture both worldwide and neighborhood functions through the self-attention mechanism gives significant benefits over traditional CNNs, which, regardless of their performance in finding out local patterns within the images, have inherent conflicts with modeling varied length dependencies and internal context without considerably increasing computational complexity. This capability is apt at coping with complex and high-resolution pictures where contextual relationships play an important function in correct interpretation and evaluation. For HD images, the objective of patch-level processing in ViT helps without the need of much more computational requirements. In this work, ViT is create image embeddings from all the object detected images saved by YOLO in the previous step, which is then passed through a dense layer to make one single embedding per single initial image.

The power of ViT is that it uses transformer architecture. So, it is more versatile and suitable for multimodal use cases compared to traditional CNNs. The 16×16 patch size of the vit-base-patch16-224 model used in this work has shown a good balance between capturing the details within an image and managing computational complexity. The usage of self-attention in ViT helps it capture long-range dependencies easily. Seeing all these advantages, it became a go to model for the creation of image embeddings for this particular work.

12.4.4 Similarity Assessment Using Siamese Network

Siamese Network is a kind of Artificial Neural Network which has two branches [10] to handle two different input vectors while using same weights. It is a one-shot classification model as it can carry out predictions even with one single training datapoint thereby reducing the computational complexity. It is also referred to as twin neural network as it handles two inputs simultaneously. It processes the two inputs parallelly and independently. It's a really useful architecture for use cases having a pairwise comparison to extract their similarities. The similarity is computed using a distance metric.

Initially, the text and image embeddings created by BERT and ViT respectively, are fed into the two different branches of the Siamese Network. Before feeding these inputs, it is essential to make sure that the dimensions of both of the embeddings are consistent. These two branches will have same internal architectures and parameters to ensure consistent transformation in the same way for both of these embeddings. The identical branches share the same weights to learn consistent features. This is extremely important as these input embeddings of different modalities should be used to draw comparisons. The advantage of weight-sharing lies in parameter reduction and thereby preventing overfitting and promoting generalization. The Siamese Network is designed in such a way that it can learn a similarity function to differentiate between the similar ones as well as the dissimilar ones. The two different branches of this network do feature extraction of the input data. The output feature vector produced for each input embedding from each branch is used to calculate the similarity between them using a distance metric. In this work, cosine similarity is employed as the distance metric. The cosine similarity is computed using the `equation` 2 where A and B are the image and the text vectors . This gives a score in a range 0 to 1 (because of non-negative embeddings) where 1 represents perfectly similar image-text pair and 0 represents perfectly dissimilar image-text pair.

$$\text{Cosine similarity} = \left(\frac{A.B}{AB} \right) \tag{12.2}$$

Loss function used for training a Siamese Network is Contrastive loss, which is as shown in `equation` 3. The objective of Contrastive loss lies in minimizing distance among similar pairs and vice versa among dissimilar pairs. The learning is carried out by the network in such a way that it assigns high similarity scores to the pairs with class label 1, which is the similar image-text pair category and it assigns low similarity scores to the pairs with class label 0, which represents the dissimilar image-text pair category. The first term in this equation makes similar pairs to have a high cosine similarity value and the second term makes dissimilar pairs to have low cosine similarity value with the highest value being m. Here, y represents the class label for the similar and dissimilar pairs whereas, s represents the cosine similarity score.

$$\text{Contrastive loss} = (1 - y)\left(\frac{1}{2}\right)(1 - s^2) + y\left(\frac{1}{2}\right) max(0, s - m)^2 \tag{12.3}$$

The hyperparameters used here includes consideration of binary cross-entropy loss as it is a binary classification, choosing Adam optimizer with learning rate 0.001, 100 epochs, batch size of 32, early stopping with a patience of 10. By leveraging the strengths of bert for text embedding creation, YOLO for object detection, ViT for image embedding creation, and the Siamese network's capacity to measure semantic similarity through cosine similarity metric, this framework guarantees a deep and nuanced understanding of the relationships between the textual content and the image content. The shared-weight architecture of the Siamese network helps the alignment of textual and visual embeddings in a shared vector space, allowing correct and reliable comparisons. This integration not only enhances the potential to quantify semantic alignment but also supports the nuanced interpretations of how textual and visible facts correlate. Therefore, this complete method affords extensive insights into the alignment and interplay among texts and images, advancing the capabilities of multimodal data analysis and fostering a greater profound knowledge of the interplay among exclusive information modalities.

12.5 Results and Discussions

12.5.1 Performance Evaluation

The evaluation metrics used for this study includes Accuracy, Precision, Recall and F1-Score. These metrics are used commonly for the evaluation of classification tasks. Accuracy gives the ratio of correctly predicted similar and dissimilar image-text pairs to the total number of image-text pairs. Precision for class 1 (similar) measures the count of predicted similar image-text pairs which are actually similar while for class 0 (dissimilar) it measures the count of predicted dissimilar image-text pairs which are actually dissimilar. Recall for class 1 shows how well the model is able to identify all the true similar image-text pairs and for class 0 how well it identifies all the true dissimilar image-text pairs. F1 score gives a harmonic mean between the precision and recall.

Table 12.2 Performance metrics

Metric	Value
Accuracy	0.7433
Precision	0.6954
Recall	0.8659
F1 Score	0.7713

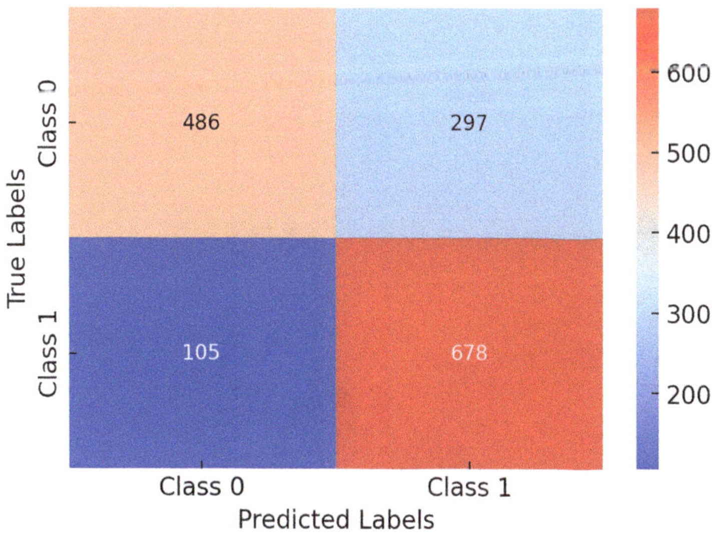

Fig. 12.2 Confusion matrix

The performance metrics of the model, which has used bert-base-uncased to create text embeddings and vit-base-patch16-224 for image embeddings, is as shown in Table 12.2 which depicts the average performance metrics considering both the classes. The test accuracy of 0.74 indicates how well the model is able to classify between similar and dissimilar image-text pairs [13] and also highlights the major role of cosine similarity in this binary classification. The confusion matrix is depicted in Fig. 12.2 which has 486 true positives (TP) indicating the similar pairs which are correctly predicted as similar, 297 false negatives (FN) which are similar pairs wrongly predicted as dissimilar, 105 false positives (FP) which are dissimilar pairs wrongly predicted as similar and 678 true negatives (TN) where dissimilar pairs are correctly predicted as dissimilar. The classification report which can give more detail on class-wise metrics is as shown in Table 12.3 which has shown a great precision of 0.82 for class 0 (dissimilar image-text pairs) and a better precision of 0.70 for class 1 (similar image-text pairs) whereas a reasonable recall of 0.62 for class 0 and a great recall of 0.87 for class 1.

Table 12.3 Classification report

Class	Precision	Recall	F1-score	Support
0 (Dissimilar)	0.82	0.62	0.71	783
1 (Similar)	0.70	0.87	0.77	783
Accuracy			0.74	1566
Macro Avg	0.76	0.74	0.74	1566
Weighted Avg	0.76	0.74	0.74	1566

```
Image ID: 2412466
Text ID: 2412466
```

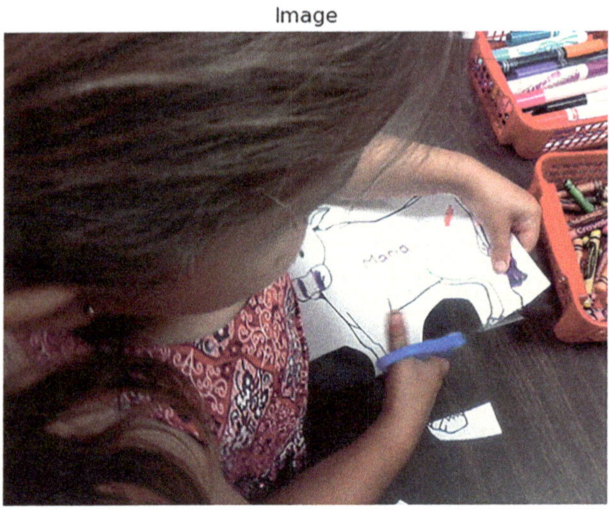


```
Text: Which color is the drawing?;What is the girl wearing?;Who is sitting at the table?;What is
she wearing?;Who is cutting the paper?;What piece of furniture is it?;Who is wearing an earring?
1/1 [==============================] - 0s 297ms/step
Similarity Score: 0.6068732738494873
Similar
```

Fig. 12.3 Similar pair

12.5.2 Sample Analysis

12.5.2.1 Similar Pair

The image-text pairs for which the model returned as a similar pair as shown in Fig. 12.3 were perfect to a great extent for those who had a cosine similarity score greater than 0.5. The model was able to capture both image and text embeddings and relate vector embeddings that are closer to each other in the shared embedding space [14] and label them as a similar pair.

```
Image ID: 2412469
Text ID: 2412502
```

```
Text: What is the piece of furniture to the left of the lamp?;What kind of furniture is the lamp on?;What is the item
of furniture that the lamp is on?;Is the box to the right or to the left of the chair?;Is the painting to the left or
to the right of the lamp?;Are there any chairs or cabinets?;Is the girl to the right or to the left of the window curt
ain?;What kind of furniture is right of the bed?;What kind of furniture is to the left of the painting?;On which side
of the photo is the chair?;On which side are the dark drapes?;What is the item of furniture to the left of the couch c
alled?;On which side of the picture is the bed?;Are there cabinets in this photo?;Are there lamps to the left of the y
oung girl?;What piece of furniture is above the carpet?;What item of furniture is to the left of the girl that is not
old?;Is the bed to the left of a nightstand?;What's the weight of the curtain?;Which room is it?;What is the lamp on?;
What piece of furniture is left to the couch?;What color does the carpet have?;What is under the coffee table made of
wood?;Is there a couch in this picture that is gray?;Are there shelves to the right of the painting that the side tabl
e is to the left of?;Are both the coffee table and the side table made of the same material?;Which kind of furniture i
s to the right of the chair?;On which side of the picture is the heavy curtain?;What color is the side table made of w
ood?;Is the couch to the right or to the left of the heavy curtain?;What item of furniture is to the left of the side
table?;What do you think is the piece of furniture to the left of the side table that the lamp is on?;Is the bed to th
e left of a couch?;Are the dark drapes to the right of the picture?;What is under the coffee table?;Is the white lamp
to the right or to the left of the person that is not old?;Is the hair black?
1/1 [==============================] - 0s 295ms/step
Similarity Score: 0.2796396017074585
Not similar
```

Fig. 12.4 Dissimilar pair

12.5.2.2 Dissimilar Pair

The image-text pairs for which the model returned as a dissimilar pair as shown in Fig. 12.4 were perfect to a great extent for those who had a cosine similarity score less than or equal to 0.5. The model was able to capture both image and text embeddings and relate vector embeddings that are far away from each other in the shared embedding space [14] and label them as a dissimilar pair.

12.5.3 Threshold Analysis

After taking the cosine similarity for the calculation of the similarity score, a threshold value of 0.5 is used in such a way that similarity scores greater than 0.5 suggest that the image and text pair are similar, and those less than or equal to 0.5 suggest that the image and text pair are dissimilar. The value of 0.5 is chosen because it serves as a natural midpoint for the cosine similarity range of 0 to 1 here in case of non-negative embeddings. This particular threshold value has a direct influence in assessing the accuracy of the model. Such a value is usually used in cases where a binary classification is made taking into consideration the ROC curve and the precision-recall trade-offs, such that the developed model can distinguish well between similar and dissimilar pairs.

12.5.4 Error Analysis

The identification of false negatives and false positives identifies key areas where the model needs improvement. False negatives are produced when the model mistakenly classifies similar pairs as dissimilar, and similarly false positives arise from misclassifying dissimilar pairs as similar. In situations where there are overlapping resemblances between text and image, false positives can happen, hence leading to wrong similarity scores. For example, there can be some similarity between a general scene describing text and the scene represented by an image. There can be fragility for the model [16] to have incapability in catching delicate semantic relations between text and images which in turn produces poor quality embeddings that can give rise to false negatives. To enhance performance and reliability overall, it is necessary to be able to address issues with respect to both types of misclassifications: a decrease in classification errors, which makes evaluations of text-image similarity more precise and reliable thus increasing the utility level for this model in real-life situations.

12.6 Conclusion

This research presents an inclusive standpoint on text and image likeness through using contrastive learning with the inclusion of pre-trained models. Based on bert-base-uncased for text embedding generation, YOLO v5s for object detection and Vision Transformer (google/vit-base-patch16-224) for image embeddings, this technique is quite successful. The Siamese network framework and cosine similarity threshold-based classification attain a testing accuracy rate of 0.74 that is very satisfactory. For example, YOLOv5 improves object detection and Vision Transformer captures fine-grained patterns from objects detected in attention mechanisms, with the help of which attention mechanisms are used to produce them as high-quality embeddings. Using attention mechanisms like BERT, also constructs contextual embedding information within its texts. Cosine similarity threshold drives the score associated with the image-text pair similarity. This advanced merging of pre-trained methods highlights how useful this approach can be in multimodal data analysis. The limitations of this model includes less data knowledge due to resource and other constraints.

While comparing with the existing approaches, the proposed method here demonstrates a reasonable balance across all the key performance metrics for a classification task, i.e, accuracy (0.7433), precision (0.6954), recall (0.8659), and F1 score (0.7713). This highlights its robustness in effectively aligning image and textual modalities for VQA. Unlike Hirota et al. [18], who replaced features from the images with the textual descriptions and then used a language-only Transformer model to reduce computational cost and complexity, our method retains perfect multimodal learning, allowing it to capture finer visual-semantic correlations, which is reflected in the superior recall and F1 score. Compared to Zhou et al. [19], who brought in a unified vision-language pre-training model making into use large-scale image-text pairs, our work avoids heavy pre-training dependencies while still achieving a reasonable accuracy and generalization powers, especially on unseen question types (on the test set of VQA v2.0 dataset). Furthermore, Yu et al. [20] address the language prior issue through a similarity-guided fusion strategy; while their model improves handling of hard classes in VQA-CP v2, our proposed approach outperforms them on VQA v2.0, especially in recall (0.8659), indicating higher sensitivity to diverse visual question-answer pairs. Altogether, the proposed method of this study maintains a good balance between efficiency and effectiveness without reliance on answer weighting, complex data augmentations or specialized constraints yet still yields state-of-the-art performance.

The future prospects of this study are varied and diverse since it allows several aspects to be considered by researchers that may lead to new ideas or advancements. Fine-tuning bert-base-uncased, YOLO v5s, and Vision Transformer (google/vit-base-patch16-224) pre-trained models may further improve their performance. On the other hand, if accuracy and efficiency are desired, these models could be fine-tuned, or one may consider different architectures or hyperparameters. It would help to expand the sample of datasets beyond VQA v2.0 so that it covers more content and languages so as to improve the generalization and robustness of the model. Furthermore, this suggested approach can be used in actual life situations like multimedia search engines and interactive systems, hence proving its applicability in reality. To have more understanding on complex interactions that will further enrich the capabilities of this model, additional data types such as audio or video can be incorporated. This study creates a solid foundation for future developments in text and image similarity through various explorations and practical applications.

References

1. Agrawal, A., et al.: VQA: visual question answering: www.visualqa.org. Int. J. Comput. Vis. **123**(1), 4–31 (2017). https://doi.org/10.1007/s11263-016-0966-6
2. Yang, X., Zhang, H., Cai, J.: Deconfounded image captioning: a causal retrospect. IEEE Trans. Pattern Anal. Mach. Intell. **45**(11), 12996–13010 (2023). https://doi.org/10.1109/TPAMI.2021.3121705
3. Schneider, F., Biemann, C.: Golden retriever: a real-time multi-modal text-image retrieval system with the ability to focus. In: SIGIR 2022 - Proceedings of the 45th International ACM

SIGIR Conference on Research and Development in Information Retrieval (2022). https://doi.org/10.1145/3477495.3531666
4. Subbiah, U., Kumar, D.K., Thangavel, S.K., Parameswaran, L.: An extensive study and comparison of the various approaches to object detection using deep learning. In: Proceedings of the 3rd International Conference on Smart Systems and Inventive Technology, ICSSIT 2020 (2020). https://doi.org/10.1109/ICSSIT48917.2020.9214185
5. Terven, J., CÓrdova-Esparza, D.M., Romero-GonzÁlez, J.A.: A comprehensive review of YOLO architectures in computer vision: from YOLOv1 to YOLOv8 and YOLO-NAS. Mach. Learn. Knowl. Extract. **5**, 1680–1716 (2023). https://doi.org/10.3390/make5040083
6. Han, K., et al.: A survey on vision transformer. IEEE Trans. Pattern Anal. Mach. Intell. **45**(1), 87–110 (2023). https://doi.org/10.1109/TPAMI.2022.3152247
7. Devlin, J., Chang, M. W., Lee, K., Toutanova, K.: BERT: pre-training of deep bidirectional transformers for language understanding. In: NAACL HLT 2019 - 2019 Conference of the North American Chapter of the Association for Computational Linguistics: Human Language Technologies - Proceedings of the Conference (2019)
8. Yerramreddy, D.R., Marasani, J., Gowtham, P.S.V., Abhishek, S., Anjali: An empirical analysis of topic categorization using PaLM, GPT and BERT models. In: 2023 Innovations in Power and Advanced Computing Technologies, i-PACT 2023 (2023). https://doi.org/10.1109/I-PACT58649.2023.10434768
9. Li, J., Li, H., Erfani, S., Feng, L., Bailey, J., Liu, F.: Visual-text cross alignment: refining the similarity score in vision-language models. In: Proceedings of the 41 st International Conference on Machine Learning (2024). https://doi.org/10.48550/arXiv.2406.02915
10. Chen, X., He, K.: Exploring simple Siamese representation learning. In: Proceedings of the IEEE Computer Society Conference on Computer Vision and Pattern Recognition (2021). https://doi.org/10.1109/CVPR46437.2021.01549
11. Liu, M., Qi, M.J., Zhan, Z.Y., Qu, L.G., Nie, X.S., Nie, L.Q.: A survey on deep learning based image-text matching. Jisuanji Xuebao (Chin. J. Comput.) **46**(11), 2370–2399 (2023). https://doi.org/10.11897/SP.J.1016.2023.02370
12. Mohammadi, M., Eftekhari, M., Hassani, A.: Image-text connection: exploring the expansion of the diversity within joint feature space similarity scores. IEEE Access **11**, 123209–123222 (2023). https://doi.org/10.1109/ACCESS.2023.3327339
13. Liu, R., Zhao, Y., Wei, S., Zheng, L., Yang, Y.: Modality-invariant image-text embedding for image-sentence matching. ACM Trans. Multimedia Comput. Commun. Appl. **15**(1), 1–19 (2019). https://doi.org/10.1145/3300939
14. Zhang, X., Niu, X., Fournier-Viger, P., Dai, X.: Image-text retrieval via preserving main semantics of vision. In: Proceedings - IEEE International Conference on Multimedia and Expo (2023). https://doi.org/10.1109/ICME55011.2023.00337
15. Diao, H., Zhang, Y., Ma, L., Lu, H.: Similarity reasoning and filtration for image-text matching. In: 35th AAAI Conference on Artificial Intelligence, AAAI 2021 (2021). https://doi.org/10.1609/aaai.v35i2.16209
16. Zhang, K., Hu, B., Zhang, H., Li, Z., Mao, Z.: Enhanced semantic similarity learning framework for image-text matching. IEEE Trans. Circuits Syst. Video Technol. **34**(4), 2973–2988 (2024). https://doi.org/10.1109/TCSVT.2023.3307554
17. Yang, X., Li, C., Zheng, D., Wen, P., Yin, G.: RFE-SRN: image-text similarity reasoning network based on regional feature enhancement. Neurocomputing **518**, 593–601 (2023). https://doi.org/10.1016/j.neucom.2022.11.003
18. Hirota, Y., Garcia, N., Otani, M., Chu, C., Nakashima, Y., Taniguchi, I., Onoye, T.: A picture may be worth a hundred words for visual question answering (2021). arXiv:2106.13445
19. Zhou, L., Palangi, H., Zhang, L., Hu, H., Corso, J.J., Gao, J.: Unified vision-language pre-training for image captioning and VQA (2019). arXiv:1909.11059
20. Yu, Z., Zhao, J., Wang, H., Guo, C., Zhou, T., Sun, C.: Adaptive fusion for visual question answering: integrating multi-label classification and similarity matching. In: Proceedings of ACM Multimedia Asia 2023 (MMAsia'23), Tainan, Taiwan, pp. 1–7 (2023). https://doi.org/10.1145/3595916.3626381

Chapter 13
Image Captioning in Hindi Using Swin Transformer-GPT Encoder-Decoder Architecture

R. Prashanth, Arrun Sivasubramanian, V. Sowmya, and B. Premjith

Abstract Image captioning has acquired substantial attention due to its potential applications in computer vision, natural language processing, and multimedia retrieval. However, it is a challenging task that involves generating descriptive and meaningful captions for images using complex algorithms. While most existing research on image captioning has focused on English or other widely spoken languages, there is a growing need for image captioning in under-resourced languages such as Hindi. Thus, this paper introduces two novel encoder-decoder approaches for performing image captioning in Hindi using approaches involving deep learning. The first approach uses simple convolution-based VGG-16 for image feature extraction, and the sentence embeddings from IndicBERT are utilized as the starting point for the language modeling using the Long-Short Term Memory (LSTM) layer. In the second approach, Swin Transformer does the Image feature map extraction, and the benchmark GPT-2 model performs the language modeling. We evaluate our approach on the Flickr8k Hindi caption dataset, containing over 8000 images and captions, and compare the performance of our two proposed models with the state-of-the-art methods in the literature. Experimental results demonstrate that our proposed approach achieves competitive performance with a BLEU score of 0.72 and a GLEU score of 0.65 on the test data. To further assess and analyze the quality of the generated captions, we conducted manual evaluations using adequacy and fluency as the evaluation criteria to get commendable scores of 3.4 and 3.57, respectively, on a scale of 4. Overall, this research study contributes to the development of natural language processing techniques for Indian languages and has the potential to impact a wide range of applications, such as image retrieval and assistive technologies for visually impaired individuals.

Keywords Image captioning · CNN · Encoder-decoder architecture · Swin transformers · GPT-2 · LSTM

R. Prashanth · A. Sivasubramanian · V. Sowmya · B. Premjith (✉)
Amrita School of Artificial Intelligence, Coimbatore, Amrita Vishwa Vidyapeetham, Coimbatore, Tamil Nadu, India
e-mail: v_sowmya@cb.amrita.edu; b_premjith@cb.amrita.edu

G. Rivera et al. (eds.), *Recent Advances in Artificial Neural Networks*, Intelligent Systems Reference Library 283, https://doi.org/10.1007/978-3-032-13497-4_13

13.1 Introduction

Image captioning is the process of analysing the visual features present in an image to automatically generate an accurate and descriptive textual or speech descriptions. The models introduced in the literature typically use a combination of Computer Vision (CV) pipelines to extract the visual features and Natural Language Processing (NLP) techniques to generate the textual descriptions from the extracted features. The ability to automatically generate captions for images has the potential to improve access to information and enhance our understanding of visual data, making it a valuable tool in many fields, such as anomaly identification in healthcare [1, 2], education of visually impaired learners [3], subtitle generation for entertainment [4], and generate product descriptions in e-commerce [5].

Image captioning is a complex problem in artificial intelligence [6] compared to image classification and image recognition. One of the main challenges is the ability of the machine learning model to understand the context and content of the image, as it needs to describe all the essential elements amidst a wide variety of objects, scenes, and concepts. Another challenge is the ability of the model to have a good understanding of syntax, grammar, and vocabulary to generate coherent and accurate captions that are grammatically correct and semantically meaningful. To achieve it, the model must be trained on a large and diverse dataset to ensure that it can generate captions for a wide range of images and contexts. Finally, there is the challenge of evaluating the quality of the generated captions, as there needs to be a clear metric for measuring the accuracy or quality of image captions.

Despite these challenges, advancements in machine learning and natural language processing are increasingly possible to generate accurate and meaningful captions for images. Image captioning has become an easier and much viable task with the recent improvements in domains such as image classification, object recognition, and language modeling. The development of deep learning has made significant progress in models for Image captioning in English. These models typically consist of two parts: an encoder network that extracts features from the input image and a decoder network that generates a sentence from the image features. Several research studies have employed pre-trained convolutional neural networks (CNN) like VGG-16, InceptionV3, and RESNET101 or CNN [7, 8], that use attention mechanism on the encoder side, coupled with either Transformer, LSTM or GRU models on the decoder side for generating image captions. The encoder network can be pre-trained on large-scale image recognition datasets such as ImageNet [9], while the decoder network is trained on image-captioning datasets such as COCO [10] or Flickr8k. However, despite its importance, Image captioning in Indian languages, specifically Hindi, remains largely unexplored. Hindi is the fourth most commonly spoken language globally [11], following English, Spanish, and Chinese. Therefore, image captioning in Hindi could be a helpful tool for semantically understanding the content presented in an image for several applications.

A language model is required to generate a description for an image. However, to the best of our knowledge, an investigation has yet to be conducted on using transformer-based language models like GPT-2 and BERT on the decoder side for Image captioning in Hindi. Thus, we propose a model that employs transformer-based language models, which have been unexplored for Hindi image captioning. The main contributions of this article are:

- We introduced two novel encoder-decoder approaches for Hindi image captioning, which are trained and evaluated on a large open-source Flickr8k image captioning dataset.
- In proposed model 1, we use pre-trained CNN (VGG-16) with Bahdanau's attention mechanism as an encoder to extract input image features. Pre-trained BERT (IndicBERT) has been used on the decoder side to generate sentence embeddings, considered the starting point of the LSTM embedding layer.
- In proposed model 2, we used a type of Vision Transformer known as Swin Transformer to obtain encodings to obtain high-level features from the input image and a Transformer based decoder known as GPT-2 used as the decoder to generate Hindi captions for an input image.
- The performance of the models is evaluated using BLEU and GLEU scores, and the scores were compared with other benchmark models in the literature. Additionally, manual adequacy and fluency tests were done for 100 samples using two subject expert human annotators.

The manuscript is organized as follows: Section 13.2 describes the related work in the literature for image captioning in English and Hindi, and the limitations in them. Section 13.3 describes the proposed methodologies for building our image captioning pipeline, and the experimental setup for the ease of reproducibility. Section 13.4 elaborates on the results and discussion of the models. Finally, in Sect. 13.4.2 we conclude the paper and discuss the future directions and applications of the proposed work.

13.2 Related Works

There have been various studies in the field of image captioning, and many researchers have proposed different techniques and algorithms to increase the efficiency of image captioning models over the past decade leading to notable improvements. Most of these studies have utilized English language datasets, the most readily available option. Initially, image captioning research focused on two main approaches, retrieval-based [12] and template-based [13]. However, these methods were limited by their reliance on predefined language structures and inflexibility, leading to their decline in popularity. Two other methods were also popular to address the challenge of image captioning. The top-down approach [14, 15] converts the input image to words using predefined rules, and the bottom-up approach [16, 17] combines the words to generate the caption. Since the top-down method uses an end-to-end model, which involves learning all the network parameters during training, this approach is commonly used today, utilizing encoder-decoder architectures to achieve optimal learning. As a result, deep neural networks (DNNs) have become the dominant method in image captioning research.

Deep learning models have been severely utilized in recent years to achieve remarkable performance in image captioning. Initial studies on image captioning using DL have utilized pre-trained convolutional neural networks (CNN) as encoders, with the final hidden layer of the CNN used as input to the RNN decoder. The use of DNN for image captioning was introduced by Kiros et al. [18]. Their approach involved using a multimodal log-bilinear model, where a CNN was employed to extract features from images fed into a neural language model that leveraged a multimodal space to define the relation between image features and text features to generate accurate predictions. However, the neural language model was inefficient at working with long-term memory. To overcome this limitation, inspired by the neural machine translation model proposed by Cho et al. [19]. They improvised using Long Short-Term Memory as the decoder end for sentence encoding, which proved to be effective in achieving significant results. It promoted the use of encoder-decoder mechanisms for the problem of image captioning. In addition to the encoder-decoder models, other proposed approaches for image captioning have been proposed in the literature. Some researchers have proposed using reinforcement learning to train image captioning models [20]. Others have proposed using generative adversarial networks (GANs) to adjust the quality of the captions generated [21]. Nevertheless, since these architectures are computationally expensive and too robust to a specific data distribution, encoder-decoder models remain a popular approach to solving this task.

Most encoder-decoder models use a CNN to extract features from the image and a RNN to generate the caption [22]. In addition, researchers have experimented with pre-trained CNNs, such as VGG-16 [23], InceptionV3 [24], and RESNET101 [25]. An attention mechanism which allows the decoder to retrieve vital features on different parts of the image when generating the captions is one of the key components of encoder-decoder models. Several types of attention mechanisms proposed in the literature include soft attention, hard attention, semantic attention, and adaptive attention. In [26], authors used a framework combining two attention-based image caption generators: a soft attention mechanism that is deterministic and is trained using backpropagation and a stochastic hard attention mechanism trainable by maximizing an approximate variational lower bound. In [27], the author's main contribution is a novel image captioning algorithm based on a semantic attention model. This attention mechanism based on semantic similarity differs from the traditional one based on visual feature similarity. In [28], authors proposed an adaptive attention model by using a visual sentinel network.

A recent trend in image captioning is transformer networks, a type of RNN effective in natural language processing tasks. They have been used in image captioning to generate more coherent and accurate captions than those generated by traditional RNN-based models. In [29], the author proposed a meshed transformer with memory for the task of image captioning. The architecture improves both the image encoding and the language generation steps. In [30], the author proposed CaPtion TransformeR (CPTR), a complete Transformer network for image captioning. This method replaces the CNN in the encoder part with a convolution-free Transformer encoder that uses transformers to generate captions.

Hindi image captioning is a relatively new field of research, and limited numbers of literature are available on this topic. However, some recent studies have focused on developing image captioning models for the Hindi language. In [31], the author proposed work uses three neural network models: CNN and LSTM as an encoder to encode the image and text feature, and the neural language model (dense layer) as a decoder to decode the image features. In [32], the author proposed dense image captioning in Hindi. The author used an encoder-decoder with an attention mechanism. In the work, Bahdanau's attention mechanism improved the quality of captions generated. In [33], the author used encoder-decoder architecture. A CNN encoder and the Transformer decoder with a stacked attention mechanism was used to generate captions. Finally, in [34], authors proposed an encoder-decoder approach that combines CNN and an LSTM to generate captions in Hindi. In addition, the authors used a channel attention mechanism to improve the model's performance. Though all these approaches produce results, approaches involving transformers for feature extraction in the encoder and advanced transformer models such as GPT and BERT in the decoder are not utilized for the image captioning task. It inspired us to improve the benchmark for image captioning in Hindi.

13.3 Methodology

Image captioning is a challenging and quintessential task used for several applications. Most recent approaches use the encoder-decoder mechanism to tackle the problem. Thus, in our research, we introduced two encoder-decoder architectures to generate descriptive and accurate captions for images in the Hindi language using the Flickr8k Hindi image captioning dataset. The first model utilizes a pre-trained VGG-16 CNN as the encoder for feature extraction from the input image, an LSTM [35] network coupled with Bahadanau's attention mechanism [36] as the decoder with BERT extracted sentence embedding's caption generation.

The second approach in our research employed the Swin transformer [37] for extracting image features, and the Generative Pre-trained Transformer 2 (GPT-2) language model [38] serves as the decoder for generating captions in the Hindi language.

13.3.1 Dataset Used

The Flickr8k Hindi dataset is a widely used benchmark dataset for image captioning in Hindi. It is derived from the original Flickr8k dataset [39], which contains 8091 images collected from Flickr along with corresponding human-generated captions, and there are five captions per image. The Hindi dataset is created by translating the original English captions to Hindi, providing a valuable resource for training and evaluating image captioning models specifically for the Hindi language. In our experimental setup, we utilized the Flickr8k Hindi dataset as the primary data source for training and evaluating our image captioning model. The dataset comprises various images capturing scenes, objects, and activities, making it suitable for training models with broad visual and semantic understanding. Figure 13.1 contains the image samples and their captions in the Hindi of Flickr8k data. To train and evaluate the model performance, we used an 80:20 train-test split. The final number of images that are present for training and testing are shown in Table 13.1.

Fig. 13.1 Sample images and captions present in the Flickr8k Hindi dataset

Table 13.1 Number of samples used for training and testing from the dataset

Partition	Number of samples
Train data	6473
Test data	1618
Total samples	8091

13.3.2 Models Proposed

The encoder-decoder architecture is a familiar framework in ML for tasks such as image captioning, machine translation, and other applications when there are sequences in the output side. It consists of two main components: An encoder which captures the relevant information from the input data and converts it into a fixed-size representation or feature vector, and a generative model called decoder which takes the fixed-size representation from the encoder and generates the output sequence or caption. During training, the encoder and decoder are jointly optimized to maximize the likelihood of the target output sequence given the input data. It involves adjusting the model parameters to improve the accuracy of the predicted probabilities. During inference, the trained model can generate captions for new input images by selecting the argmax of the probability vector and feeding it back as input to generate the following word in the sequence. Overall, the encoder-decoder architecture with probabilistic modeling allows the model to learn the conditional probability distribution of the output sequence given the input data, enabling it to generate accurate and meaningful captions based on the learned probabilities.

This study proposes two encoder-decoder approaches. The first model uses a pre-trained VGG-16 CNN to retrieve feature maps using a series of convolution and pooling layers from an image in a hierarchical manner. It is a CNN architecture proposed which is one of the most popular architectures for image classification and object recognition tasks. Pre-trained BERT was introduced by Google AI Language and it's variant IndicBERT [19], is used to generate sentence embeddings for captions as a bidirectional model takes into account the context of the entire sentence. It uses a masked language modeling approach during training, where it randomly masks some words in a sentence and tries to predict them based on the context of the other words in the sentence. The LSTM layer that converts features to word embedding vectors is used as the decoder to process the feature vector and generate a sequence of words concatenated to form a complete caption. Unlike traditional RNNs [40], which suffer from the vanishing gradient problem and struggle to capture long-range dependencies, LSTM addresses this issue by introducing memory cells that can store and retrieve information over longer sequences, which is ideal for the image captioning problem. In addition, attention is applied to the image features to improve the quality of word feature embeddings and the captions generated. We used Bahdanau's attention mechanism, which lets the decoder focus on specific patches of the image during caption generation. Bahdanau's attention mechanism calculates a weight or attention score for each image feature, which represents the importance or relevance of that feature for generating the next word in the caption. The decoder then uses these weights to focus on the most essential features when generating the next word in the caption. Figure 13.2 shows the approach adopted in the first architecture to perform image captioning.

The second approach takes a distinct approach to image captioning in Hindi by incorporating a cutting-edge architecture called Swin Transformers (Swin-Base

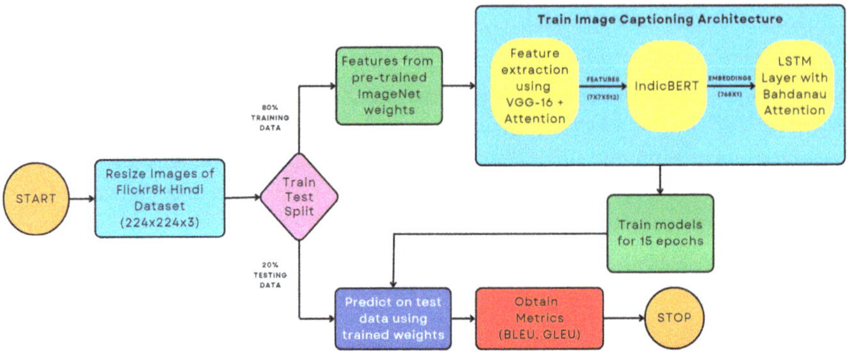

Fig. 13.2 Proposed approach of the VGG16-LSTM encoder-decoder architecture for image captioning

variant) for feature extraction from images in the encoder. Swin Transformers are widely recognized for their excellent performance in various computer vision tasks such as image classification [41] and object detection [42]. Unlike the earlier Vision Transformer (ViT) introduced by Dosovitskiy et al. [43], it captures a more comprehensive understanding of the contextual and semantic information present in the images using a sliding kernel multi-head self-attention mechanism, that can capture the local as well as global contextual information of objects and scenes. This enables us to generate more precise and meaningful captions in any language, including Hindi. Additionally, we employ the GPT-2 language model trained in the GitHub repository [44] as the decoder in our model. This model has been trained on an extensive corpus of text data from various sources such as the internet, websites, books, and articles, allowing the model to generate high-quality captions in Hindi. Combining the Swin transformer for feature map extraction and the GPT-2 language model as the decoder, our approach provides a robust and efficient solution for image captioning tailored explicitly to the Hindi language. Figure 13.3 shows the approach adopted in the second architecture to perform image captioning.

13.3.3 Evaluation Metrics

A subjective score is required to provide a more quantitative measure of the quality of the captions. To assess this, we compare them against the ground truth or reference captions. The quantitative analysis is conducted using the Bilingual Evaluation Understudy (BLEU) score and Google BLEU (GLEU) score, a standard evaluation technique used in the field of machine translation that measures the quality of the captions based on *n*-grams. We employed BLEU and GLEU scores as performance metrics to assess the quality of the generated captions.

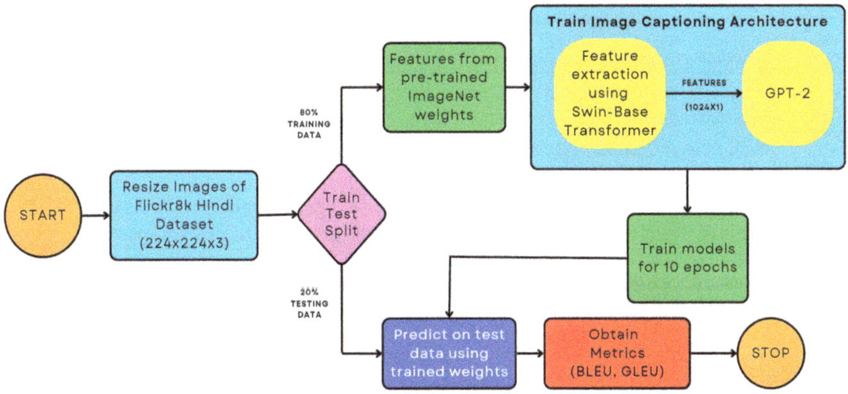

Fig. 13.3 Proposed approach of the SwinTransformer-GPT2 encoder-decoder architecture for image captioning

BLEU [45] is a popular evaluation metric used in many NLP tasks. It gauges the similarity between a generated caption and the reference captions of a corpus provided by human evaluators. It is typically calculated by comparing the n-grams in the generated and a set of reference captions and computing the precision. A penalty is applied for generating longer n-grams that are not present in the reference captions. The BLEU score ranges from 0 to 1, with 1 indicating a perfect match with the reference captions. Mathematically, the BLEU score is defined as shown in Eq. (13.1). It consists of two parts: The brevity penalty penalizes generated translations that are too short compared to the closest reference length with an exponential decay to compensate for the absence of the recall term and the n-gram overlap term, which acts as the precision metric.

$$\text{BLEU} = \underbrace{\min\left(1, \exp\left(1 - \frac{\text{reference-length}}{\text{output-length}}\right)\right)}_{\text{brevity penalty}} \underbrace{\left(\prod_{i=1}^{4} \text{precision}_i\right)^{1/4}}_{n\text{-gram overlap}} \quad (13.1)$$

When used for single sentences, the BLEU score has some undesirable properties as it was designed to be a corpus measure. To address this issue, a modified version of the BLEU score called the Google BLEU (GLEU) score [46] is a metric for evaluating the quality of image captions generated by a machine learning model. Then GLEU score is the minimum value between the precision and recall scores of the generated n-grams. It is mainly used when scores of the individual (prediction, reference) sentence pairs are needed instead of averaging over the (prediction, reference) scores for a whole corpus.

In addition to these metrics, we utilized adequacy and fluency metrics commonly employed in machine translation [47] and other natural language processing tasks to assess the quality of the generated captions. The adequacy metric measures the

Table 13.2 Scales to measure adequacy and fluency for generated image captions

Score	Adequacy (meaning)	Fluency (meaning)
0	Poor: No information retained in generated caption	Poor: The generated caption is unintelligible
1	Bad: Less information retained in generated caption	Bad: The generated caption is not fluent
2	Moderate: Much information retained in generated caption	Moderate: The generated caption feels like it arose from a broken non-native speaker
3	Good: Most information retained in generated caption.	Good: The generated caption is grammatically good
4	Excellent: All information retained in generated caption	Excellent: The generated caption is grammatically accurate

extent to which the generated caption retains the relevant information from the input image. In contrast, the fluency metric evaluates the grammatical correctness of the generated captions. These metrics enable us to comprehensively evaluate the accuracy and fluency of the generated captions in a human-like manner. Table 13.2 elaborates on the scales of adequacy and fluency used to evaluate the generated captions by the models.

In addition to the quantitative comparison, we present a qualitative analysis of the captions generated by our proposed model for several test images. Two annotators carry out the human evaluation. We conducted qualitative analysis for 100 captions for both the proposed models and then considered their average values.

13.4 Experiments, Results and Discussion

This section describes the experiments carried out in this work and the results obtained from the two proposed architectures elaborated in Section 13.3.2. Two experiments along with hyperparameter tuning were carried out to perform image captioning in Hindi. Further, the results obtained, and the explanation of the results is done in the discussion section, and the error analysis to go deep into the working of the architectures is also performed.

13.4.1 Experimental Setup

In our experimental setup, we utilized the Flickr8k Hindi dataset as the primary data source for training and evaluation of the proposed models. The images are pre-processed by resizing them to a fixed size of $224 \times 224 \times 3$ (expected size of the input image to VGG-16 and Swin Transformer architectures) and normalizing the pixel values. Next, we split the dataset into the train, and test sets, with 80%

Table 13.3 Optimal hyperparameters for the proposed models used in the work

Parameter	Proposed model-1	Proposed model-2
Encoder	VGG-16 with Bahdanau's attention mechanism	Swin transformer
Decoder	LSTM with sentence embeddings from BERT	GPT – 2
Batch size	64	8
Learning Rate	0.001	0.0001
Dropout	0.5	–
Optimizer	Adam	Adam
Epochs	15	10
Loss function	Softmax cross entropy	Softmax cross entropy

and 20% of the data, respectively, where the train data is used to supervise the model to generate relevant captions for the input image and the test data is used to evaluate the performance of the model. The dataset comprises various images capturing scenes, objects, and activities, making it suitable for training models with broad visual and semantic understanding. We fine-tuned our proposed approaches on the Flickr8k dataset for image captioning in Hindi using ablation experiments for hyperparameter tuning to avoid overfitting and to get the best evaluation metrics post-captioning. Table 13.3 contains the set of optimal hyperparameters obtained for both the proposed models.

The CNN-LSTM architecture was trained on Google Colab's Tesla T4 GPU with 16GB RAM and 132GB memory on the Tensorflow platform. Since the second approach is computationally expensive, the SwinTransformer-GPT2 model was trained on a HPC server with 12 CPU's having a cumulative RAM of 128GB. with the PyTorch platform. The results are compared with five benchmark models used for image captioning in Hindi. The summary of the architectures used from the literature, along with the sets of hyperparameters used by them to fine-tune these models is given in Table 13.4.

The proposed CNN-LSTM approach takes about 2 h for training and 10–20 s for a test image to generate the caption. Being a heavy model with many trainable parameters, the proposed Swin Transformer-GPT2 model takes approximately 10 h to train, but the same time to test the captions. Due to training on a HPC server, the time is comparable to other models. All the benchmark methods take roughly 10–12 h for training and 10–20 s for caption generation, making our approach a swift and scalable approach that could be trained on a large dataset.

13.4.2 Results and Discussion

The training of the proposed models mentioned in Sect. 13.3.2 is done as mentioned using the experimental setup elaborated in Sect. 13.4.1. The resized images of the dataset are passed as input, and the generated captions are matched with the ground-truth captions present in the dataset. The two experiments were conducted and the

Table 13.4 Summary of baseline models present for Hindi image captioning in the literature

Parameter	Benchmark 1	Benchmark 2	Benchmark 3	Benchmark 4	Benchmark 5
Encoder	CNN with attention	ResNet-101	VGG-16	CNN with ECA-NET	RESNET-101 with attention
Decoder	GRU	Transformer language model [48]	LSTM	GRU	RNN
Batch size	32	128	—	128	—
Learning Rate	0.0004	0.0004	0.01	0.0004	0.0005
Dropout	0.5	0.5	0.5	0.5	—
optimizer	Adam	Adam	Adam	Adam	Adam
Epochs	15	10	-	15	50
Loss function	Softmax cross entropy	Softmax cross entropy	Softmax cross entropy	Softmax cross entropy	Softmax cross entropy

Table 13.5 Comparison of BLEU and GLEU scores for different models

Encoder decoder model	BLEU-1	BLEU-2	BLEU-3	BLEU-4	GLEU
CNN with Attention-LSTM [49]	0.57	0.391	0.264	0.173	–
ResNet101-GRU [15]	0.629	0.433	0.291	0.19	–
CNN-LSTM [13]	0.5844	0.4	0.27	0.12	–
ECA-NET CNN [16]	0.674	0.49	0.333	0.219	–
Custom Encoder-Decoder [14]	0.285	0.1612	0.0856	0.0401	–
VGG16 with attention-LSTM (Proposed)	**0.57**	**0.43**	**0.33**	**0.25**	**0.47**
Swin-GPT2 (Proposed)	**0.72**	**0.55**	**0.45**	**0.39**	**0.65**

BLEU and GLEU scores obtained on an 80:20 split of the Flickr8k Hindi dataset for the proposed models are shown in Table 13.5.

13.4.2.1 VGG16-LSTM Model

The proposed VGG-LSTM achieves scores comparable with the state-of-the-art results already in the literature. These scores indicate a relatively high degree of similarity between the generated and the reference captions. The higher BLEU-1 score suggests that a significant portion of the unigrams in the generated captions matched those in the reference captions. Additionally, the BLEU-2, BLEU-3, and BLEU-4 scores indicate that the generated captions demonstrate moderate precision in matching bigrams, trigrams, and four-grams with the reference captions. The proposed method has also achieved a relatively high GLEU score of 0.47, indicating that it can generate meaningful and accurate captions for images in Hindi.

Since Bahdanu's attention is the only addition to the feature extraction process of the work done in [46], it plays a vital role in extracting important features from the image and allows the decoder to selectively focus and adapt to different patches, giving it the ability to attend to relevant visual information. It improves the model's ability to generate more accurate and contextually relevant captions, as it can selectively attend to critical visual cues in the image. Although there is room for improvement, as evident from the generated captions these results suggest that the proposed method successfully captured the images' key elements and generated reasonably accurate and coherent captions, as evident from the generated captions in Fig. 13.4.

13.4.2.2 Swin Transformer-GPT2 Model

The proposed model using the Swin Transformers encoder and GPT2 decoder significantly improved over the present benchmark models and is the new state-of-the-art result for the Flickr8k image captioning dataset. However, the significant improvement across all BLEU scores and GLEU scores suggests that feature extraction using Swin Transformer is undoubtedly the best method to understand

(a). बर्फ में खेल रहे कुत्ते। Dogs are playing in the snow.	(b). लोगों का समूह संगीत कार्यक्रम में गाता है और नृत्य करता है। A group of people sings and dances in a concert.

(c). भूरे रंग का कुत्ता घास से चलता है। The brown dog runs through the grass.	(d). काला कुत्ता पक्षी को अपने मुँह में पकड़ लिया है। The black dog has caught the bird in its mouth.

(e). युवा बच्चों का समूह। A Group of Young Children.	(f). पीले रंग की जैकेट में व्यक्ति सफेद पानी के माध्यम से लाल कश्ती पैड कर रहा है। The guy in the yellow jacket is paddling a red kayak through white water.

Fig. 13.4 Image captions generated by the proposed VGG16-LSTM model

the context of the image and capture all the underlying objects present in an image. The high GLEU score indicates that the model has learned to capture the context and relationships between image features and linguistic elements, resulting in high-quality captions, and making it the ideal candidate for image captioning tasks, compared to other benchmark models in the literature. These results suggest that the proposed method effectively generates captions close to the reference captions, indicating its potential for producing accurate and meaningful image descriptions. This is also evident from the captions generated by the model, as shown in Fig. 13.5. All the captions are able to explain the global context of the image and the generated captions contain all the objects present in the image.

13.4.3 Adequacy and Frequency

The captions produced by our model are mainly accurate and can correctly identify the objects and context depicted in the images. Upon manual inspection from the two annotators, it is evident that the generated captions are accurate and provide a close description of the objects and activities depicted in the images. Table 13.6 includes the average adequacy and frequency scores obtained for the proposed models in our work.

13.4.3.1 Adequacy and Fluency of the VGG16-LSTM Model

The VGG16-LSTM approach achieved an adequacy score of 2.7 and a fluency score of 3.26. The adequacy score indicates that the generated captions, on average, captured a moderate level of relevance and accuracy in describing the content of the corresponding images. The approach can convey most of the key visual elements and semantic information. According to the fluency score, the linguistic quality and coherence of the generated captions are also moderate. Captions generally demonstrate grammar, syntax, vocabulary, and overall language usage proficiency. However, there may be instances where the fluency could be further enhanced to achieve a more natural and fluent flow of the generated text.

13.4.3.2 Adequacy and Fluency of the SwinTransformer-GPT2 Model

The SwinTransformer-GPT2 approach achieved an adequacy score of 3.26 and a fluency score of 3.57. The adequacy score indicates that the generated captions, on average, captured a high level of relevance and accuracy in describing the content of the corresponding images. Nevertheless, efforts can be directed toward refining the fluency of the captions to ensure a more natural and fluent reading experience. These findings provide valuable insights for further iterations and enhancements of our image captioning approach. Nevertheless, the approach has demonstrated

(a). टेनिस बॉल को पकड़ने की कोशिश कर रहा कुत्ता। Dog is trying to catch a tennis ball.	(b). लडका गेंद फेंकने के लिए तैयार हो जाता है | The boy gets ready to throw the ball.

(c). पानी के ऊपर उड़ने वाला बड़ा सफेद पक्षी। Big white bird flying over water.	(d). मोटरसाइकिल रेसर ट्रैक पर वक्र के आसपास तेजी से ड्राइव करता है | Motorcycle racer drive fast around curves on track.

(e). झील पर डोंगी में सवार लोग। People in a canoe on the lake.	(f). ट्रैक पर दौड़ती पीली शर्ट में महिला। Woman in yellow shirt running on track.

Fig. 13.5 Image captions generated by the proposed SwinTransformer-GPT2 model

Table 13.6 Average adequacy and frequency scores obtained for the proposed models

Scoring metric	VGG16-LSTM model	SwinTransformer-GPT2 model
Adequacy	2.9	3.4
Frequency	3.26	3.57

amazing capabilities in generating captions that are relevant to the images, as they could capture the global contextual information and all the objects present in the image, as evident from the captions.

13.4.4 Error Analysis

Though some strong results in terms of quantitative as well as qualitative analysis are reflected in the scores, proving that various objects and activities depicted in the images are accurately captured and captioned, there are particular challenges that the framework encounters. These challenges may result in errors in the generated captions, and there are a few cases where the models falter. Figures 13.6 and 13.7 show samples where the proposed models made incorrect predictions.

In the case of the VGG16-LSTM model, the captions solely mention the object present in the image without expressing any relationships between them, leading to erroneous captions, as seen in Fig. 13.6. The model cannot understand the activity being performed in the objects. After all, the samples could be unique or only around five unique samples of those would be present in the training corpus. These samples could lead to low BLEU and GLEU scores.

The Swin Transformer-GPT2 model fails when the Swin Transformer feature extraction module could not capture the small objects present in the image or misunderstands smaller patches. It is evident in Fig. 13.7a, c. Additionally, the model needs more training with a diverse dataset to accurately predict the relationship between objects present within an image. This inference can be made from the errors made across many failure samples. Figure 13.7 shows a sample of the failure cases of the Swin-Transformer-GPT2 model.

13.5 Conclusion and Future Work

This paper introduces two innovative encoder-decoder architectures for generating image captions in the Hindi language. The first architecture leverages pre-trained VGG-16 CNN for feature extraction. The BERT embedding on the decoder and attention helped the LSTM provide image captions accurately. The second approach utilizes a combination of Swin transformer and GPT-2. The proposed models are trained and tested on the Flickr8k Hindi dataset to give exceptional qualitative

Fig. 13.6 Errors in captioning made by the VGG16-LSTM model

metrics such as BLEU and GLEU scores, which are the new state-of-the-art for the image captioning task. These generated Hindi captions can assist the Hindi speaking population in India to comprehend the context, and objects present in the images in their desired language. Moreover, the proposed module can also be beneficial in tutoring blind people with the assistance of text-to-speech technology.

In the future, we intend to design more advanced encoder-decoder models for generating Hindi captions. Moreover, the task can be extended to several native Indian languages to cater to a wider range of sectors. The feature extraction techniques can be combined with some strong spatial-spectral feature extraction techniques to understand the context of the image. With larger LLM models evolving in literature, the decoder of the architectures could be trained to improve image captioning in subsequent architectures for enhancing many real-world applications.

(a). छोटा लड़का स्नोबॉल फेंकने के लिए तैयार हो जाता है | Little boy gets ready to throw snowballs.	(b). फुटबॉल खिलाड़ी गेंद को पकड़ने के लिए दौड़ता है | Football players run to catch the ball.

(c). रेत पर खड़े लोगों के साथ महिला समुद्र तट पर बिक्री व्यापारी गुब्बारे का विज्ञापन करती है | Woman selling merchandiser advertising balloons on the beach with people standing on the sand.	(d). सर्फर तरंगों की सवारी करता है | Surfer rides the waves.

Fig. 13.7 Errors in captioning made by the SwinTransformer-GPT2 model

Declarations

Funding Not Applicable.

Conflict of Interest The authors have no competing interests to declare that are relevant to the content of this article.

Ethics Approval Not Applicable. We do not use any human/animal related data for the experiments.

Availability of Data and Materials We used publicly available data for the experiments, which is properly cited in the manuscript.

References

1. Le, T.M., et al.: Vlsp 2021-viecap4h challenge: automatic image caption generation for healthcare domain in vietnamese. People. **1**(2), 2 (2021). https://doi.org/10.25073/2588-1086/vnucsce.341
2. Allaouzi, I., et al.: Automatic caption generation for medical images. In: Proceedings of the 3rd International Conference on Smart City Applications (2018). https://doi.org/10.1145/3286606.3286863
3. Gupta, M., Asthana, P., Singh, P.: EDUVI: An Educational-Based Visual Question Answering and Image Captioning System for Enhancing the Knowledge of Primary Level Students (2023). https://doi.org/10.21203/rs.3.rs-2594097/v1
4. Zhong, Y., Zhang, H., Jain, A.K.: Automatic caption localization in compressed video. IEEE Trans. Pattern Anal. Mach. Intell. **22**(4), 385–392 (2000). https://doi.org/10.1109/34.845381
5. Yashima, T., et al.: Learning to describe e-commerce images from noisy online data. In: Computer Vision–ACCV 2016: 13th Asian Conference on Computer Vision, Taipei, Taiwan, November 20–24, 2016. Springer International Publishing (2017). https://doi.org/10.1007/978-3-319-54193-8_6. Revised Selected Papers, Part V 13
6. Elhagry, A., Kadaoui, L.: A Thorough Review on Recent Deep Learning Methodologies for Image Captioning. arXiv preprint arXiv:2107.13114 (2021). https://doi.org/10.48550/arXiv.2107.13114
7. Ayush Kumar, C.S., Maharana, A.D., Krishnan, S.M., Hanuma, S.S.S., Sowmya, V., Ravi, V.: Vehicle detection from aerial imagery using principal component analysis and deep learning. In: International Conference on Innovations in Bio-Inspired Computing and Applications, pp. 129–140. Springer Nature Switzerland, Cham (2022). https://doi.org/10.1007/978-3-031-27499-2_12
8. Kumar, M.A., Premjith, B., Singh, S., Rajendran, S., Soman, K.P.: An overview of the shared task on machine translation in Indian languages (MTIL)–2017. J. Intell. Syst. **28**(3), 455–464 (2019). https://doi.org/10.1515/jisys-2018-0024
9. Deng, J., et al.: Imagenet: a large-scale hierarchical image database. In: 2009 IEEE Conference on Computer Vision and Pattern Recognition. IEEE (2009). https://doi.org/10.1109/CVPR.2009.5206848
10. Lin, T.-Y., et al.: Microsoft coco: common objects in context. In: Computer Vision–ECCV 2014: 13th European Conference, Zurich, Switzerland, September 6–12, 2014, Proceedings, Part V 13. Springer International Publishing (2014). https://doi.org/10.1007/978-3-319-10602-1_48
11. Verma, K., Singh, M.: Hindi handwritten character recognition using convolutional neural network. Int. J. Comput. Sci. Eng. **6**(6), 909–914 (2018). https://doi.org/10.26438/ijcse/v6i6.909914
12. Ordonez, V., Kulkarni, G., Berg, T.: Im2text: describing images using 1 million captioned photographs. Adv. Neural Inform. Process. Syst. **24** (2011)
13. Yang, Y., et al.: Corpus-guided sentence generation of natural images. In: Proceedings of the 2011 Conference on Empirical Methods in Natural Language Processing (2011)
14. Bahdanau, D., Cho, K., Bengio, Y. Neural Machine Translation by Jointly Learning to Align and Translate. arXiv preprint arXiv:1409.0473 (2014).
15. Sutskever, I., Vinyals, O., Le, Q.V.: Sequence to sequence learning with neural networks. Adv. Neural Inform. Process. Syst. **27** (2014). https://doi.org/10.48550/arXiv.1409.3215

16. Farhadi, A., et al.: Every picture tells a story: generating sentences from images. In: Computer Vision–ECCV 2010: 11th European Conference on Computer Vision, Heraklion, Crete, Greece, September 5–11, 2010, Proceedings, Part IV 11. Springer, Berlin Heidelberg (2010)
17. Elliott, D., Keller, F.: Image description using visual dependency representations. In: Proceedings of the 2013 Conference on Empirical Methods in Natural Language Processing (2013)
18. Kiros, R., Salakhutdinov, R., Zemel, R.S.: Unifying Visual-Semantic Embeddings with Multimodal Neural Language Models. arXiv preprint arXiv:1411.2539 (2014). https://doi.org/10.48550/arXiv.1411.2539
19. Cho, K., et al.: Learning Phrase Representations Using RNN Encoder-Decoder for Statistical Machine Translation. arXiv preprint arXiv:1406.1078 (2014). https://doi.org/10.48550/arXiv.1406.1078
20. Yan, S., et al.: Image captioning using adversarial networks and reinforcement learning. In: 2018 24th International Conference on Pattern Recognition (ICPR). IEEE (2018). https://doi.org/10.1109/ICPR.2018.8545049
21. Vinyals, O., et al.: Show and tell: a neural image caption generator. In: Proceedings of the IEEE Conference on Computer Vision and Pattern Recognition (2015)
22. Simonyan, K., Zisserman, A.: Very Deep Convolutional Networks for Large-Scale Image Recognition. arXiv preprint arXiv:1409.1556 (2014). https://doi.org/10.48550/arXiv.1409.1556
23. Szegedy, C., et al.: Rethinking the inception architecture for computer vision. In: Proceedings of the IEEE Conference on Computer Vision and Pattern Recognition (2016)
24. He, K., et al.: Deep residual learning for image recognition. In: Proceedings of the IEEE Conference on Computer Vision and Pattern Recognition (2016)
25. Xu, K., et al.: Show, attend and tell: neural image caption generation with visual attention. In: International Conference on Machine Learning. PMLR (2015)
26. You, Q., et al.: Image captioning with semantic attention. In: Proceedings of the IEEE Conference on Computer Vision and Pattern Recognition (2016)
27. Lu, J., et al.: Knowing when to look: adaptive attention via a visual sentinel for image captioning. In: Proceedings of the IEEE Conference on Computer Vision and Pattern Recognition (2017)
28. Cornia, M., et al.: Meshed-memory transformer for image captioning. In: Proceedings of the IEEE/CVF Conference on Computer Vision and Pattern Recognition (2020)
29. Liu, W., et al.: CPTR: Full Transformer Network for Image Captioning. arXiv preprint arXiv:2101.10804 (2021). https://doi.org/10.48550/arXiv.2101.10804
30. Rathi, A.: Deep learning approach for image captioning in Hindi language. In: 2020 International Conference on Computer, Electrical & Communication Engineering (ICCECE). IEEE (2020). https://doi.org/10.1109/ICCECE48148.2020.9223087
31. Gill, K., Saha, S., Mishra, S.K.: Dense image captioning in Hindi. In: 2021 IEEE International Conference on Systems, Man, and Cybernetics (SMC). IEEE (2021). https://doi.org/10.1109/SMC52423.2021.9658586
32. Mishra, S.K., et al.: Image captioning in Hindi language using transformer networks. Comput. Electr. Eng. **92**, 107114 (2021). https://doi.org/10.1016/j.compeleceng.2021.107114
33. Mishra, S.K., et al.: Efficient channel attention based encoder–decoder approach for image captioning in Hindi. Trans. Asian Low-Resour. Lang. Inform. Process. **21**(3), 1–17 (2021). https://doi.org/10.1145/3483597
34. Graves, A.: Long short-term memory. In: Supervised Sequence Labelling with Recurrent Neural Networks, pp. 37–45 (2012). https://doi.org/10.1007/978-3-642-24797-2_4
35. Bahdanau, D., et al.: End-to-end attention-based large vocabulary speech recognition. In: In: 2016 IEEE International Conference on Acoustics, Speech and Signal Processing (ICASSP). IEEE (2016). https://doi.org/10.1109/ICASSP.2016.7472618
36. Liu, Z., et al.: Swin transformer: hierarchical vision transformer using shifted windows. In: Proceedings of the IEEE/CVF International Conference on Computer Vision, 2021 (2021)
37. Radford, A., et al.: Language models are unsupervised multitask learners. OpenAI Blog. **1**(8), 9 (2019)

38. Hodosh, M., Young, P., Hockenmaier, J.: Framing image description as a ranking task: Data, models and evaluation metrics. J. Artif. Intell. Res. **47**, 853–899 (2013). https://doi.org/10.1613/jair.3994
39. Devlin, J., et al.: Bert: Pre-training of Deep Bidirectional Transformers for Language Understanding. arXiv preprint arXiv:1810.04805 (2018). https://doi.org/10.18653/v1/N19-1423
40. Cao, H., et al.: Swin-unet: Unet-like pure transformer for medical image segmentation. In: European Conference on Computer Vision. Springer Nature Switzerland, Cham (2022). https://doi.org/10.1007/978-3-031-25066-8_9
41. Xu, X., et al.: An improved swin transformer-based model for remote sensing object detection and instance segmentation. Remote Sensing. **13**(23), 4779 (2021). https://doi.org/10.3390/rs13234779
42. Dosovitskiy, Alexey, et al. (2020)An Image Is Worth 16×16 Words: Transformers for Image Recognition at Scale. arXiv preprint arXiv:2010.11929.
43. https://huggingface.co/surajp/gpt2-hindi
44. Papineni, K., et al.: Bleu: a method for automatic evaluation of machine translation. In: Proceedings of the 40th Annual Meeting of the Association for Computational Linguistics, 2002 (2002)
45. Mutton, A., et al.: GLEU: Automatic evaluation of sentence-level fluency. In: Proceedings of the 45th Annual Meeting of the Association of Computational Linguistics, 2007 (2007)
46. Vaswani, A., et al.: Attention is all you need. Adv. Neural Inform. Process. Syst. **30** (2017)
47. Oviya, I.R., Spandana, C., Krithika, S.: Chest X-ray pathology detection using deep learning and transfer learning. In: In: 2022 IEEE 7th International Conference on Recent Advances and Innovations in Engineering (ICRAIE), vol. 7, pp. 25–30. IEEE (2022). https://doi.org/10.1109/ICRAIE56454.2022.10054329
48. Dhir, R., et al.: A deep attention-based framework for image caption generation in Hindi language. Comput. Sist. **23**(3), 693–701 (2019)
49. Shriram, R., Kaushek Kumar, T.R., Samuktha, V., Karthika, R.: General adversarial networks: a tool to detect the novel coronavirus from CT scans. In: International Conference on Artificial Intelligence for Smart Community: AISC 2020, 17–18 December, Universiti Teknologi Petronas, Malaysia, pp. 217–230. Springer Nature Singapore, Singapore (2022). https://doi.org/10.1007/978-981-16-2183-3_21
50. Devi, P.R., et al.: Image captioning using reinforcement learning with BLUDEr optimization. Pattern Recog. Image Anal. **30**, 607–613 (2020). https://doi.org/10.1134/S1054661820040094

R. Prashanth completed his B. Tech in Mechanical Engineering in the year 2021 from KLE Technological University, Hubballi, India. He completed his M.Tech in Data Science in the year 2023 from Amrita School of Artificial Intelligence, Amrita Vishwa Vidyapeetham, India. He is currently working as a Machine Learning Engineer at Nissan Motor Corporation, India.

Arrun Sivasubramanian is currently a graduate student at the Johns Hopkins University and a research associate at the Johns Hopkins School of Medicine. His research focuses on the application of deep learning and computer vision techniques to biomedical image analysis, with interests in medical image segmentation, transformer-based architectures, and explainable AI. He has authored multiple 10+ conference and journal publications in areas such as remote sensing, and medical image analysis and has been recognised with Best Paper Award at several international conferences.

V. Sowmya is working as Associate Professor at Amrita School of Artificial Intelligence, Coimbatore, Amrita Vishwa Vidyapeetham. She completed PhD in Artificial Intelligence (AI) for Natural Scene Analysis from Amrita Vishwa Vidyapeetham. Her broad research area is AI for Signal and Image Analysis. The sub-areas of her research interests include Biomedical, Healthcare and Remote Sensing Data Analysis. Her recent research focus is on "AI for Multimodal Data Analysis."

B. Premjith currently serves as an Assistant Professor (Sr. Gd.) at the Amrita School of Artificial Intelligence, Amrita Vishwa Vidyapeetham, Coimbatore. He pursued his PhD in Neural Machine Translation from Amrita Vishwa Vidyapeetham under the guidance of Prof. K.P Soman. His research focuses on developing Machine Learning/Deep Learning models for machine translation, word sense disambiguation, morphological segmentation and synthesis, Part-of-Speech tagging, Named Entity Recognition and Dependency parsing. He also worked on various social media analysis applications such as hate speech/offensive language identification, emotion detection from social media data, sentiment analysis, factuality identification and fake news detection. He is currently interested in applying artificial intelligence techniques for cybersecurity, and finance.

Part III
Innovations in Deep Learning Architectures and Applications

Chapter 14
A Hybrid Deep Learning-Fuzzy Inference Scheme for Driving Supervision

Dante Mújica Vargas, Antonio Luna-Á lvarez, Antonio Arenas Muñiz, Alberto Rosales Silva, and Francisco Gallegos Funes

Abstract This study introduces an embedded real-time driver monitoring system that synergizes deep learning with fuzzy inference techniques. The system utilizes two convolutional neural networks to analyze the driver's eye, mouth, and head posture states. These measurements are subsequently fed into a Type-1 Mamdani Fuzzy module to classify the driving state as either safe or risky. Performance evaluation was conducted through two experiments using standard metrics from the literature. The results indicate that the proposed hybrid approach achieves an accuracy exceeding 92% in both experiments, demonstrating its robustness and suitability for real-world deployment in Advanced Driver Assistance Systems.

Keywords Driver monitoring system · Deep learning · Fuzzy inference · Real-time vision · Advanced driver assistance systems

14.1 Introduction

Driving demands continuous multitasking, including lane positioning, hazard anticipation, and managing distractions, all of which rely heavily on the driver's visual and cognitive engagement. Ensuring road safety requires a deep understanding of how drivers process their environment and the factors that lead to lapses in attention. Advanced Driver Assistance Systems (ADAS) have evolved as critical tools to enhance safety by reducing human-caused accidents. These systems, which

D. Mújica-Vargas (✉) · A. Luna-Á · A. Arenas-Muñiz
Tecnológico Nacional de México, Centro Nacional de Investigación y Desarrollo Tecnológico, Cuernavaca-Morelos, México
e-mail: dante.mv@cenidet.tecnm.mx; D18CE009@cenidet.tecnm.mx; d22ce043@cenidet.tecnm.mx

A. Rosales-Silva · F. Gallegos-Funes
Escuela Superior de Ingeniería Mecánica y Eléctrica Unidad Zacatenco Sección de Estudios de Posgrado e Investigación, Instituto Politécnico Nacional, Ciudad de México, México
e-mail: arosales@ipn.mx; fgallegosf@ipn.mx

G. Rivera et al. (eds.), *Recent Advances in Artificial Neural Networks*, Intelligent Systems Reference Library 283, https://doi.org/10.1007/978-3-032-13497-4_14

include vision-based driver assistance (VB-DAS) and driver monitoring (DMS), are increasingly being adopted even in mid-range vehicles. Technologies such as Lane Departure Warning, Automatic Emergency Braking, and Adaptive Cruise Control are now standard. However, their effectiveness diminishes in challenging conditions, where control reverts to the driver. Thus, continuous monitoring of driver attentiveness is essential—either to issue timely warnings or to intervene if the driver fails to respond.

Recent advances in driver monitoring systems have significantly enhanced the state of the art. [22] developed a non-invasive, real-time fatigue detection system using facial video analysis, eliminating the need for body-attached sensors. For handheld device distraction detection, [4] introduced an optimized CNN architecture achieving 0.83 accuracy with reduced computational complexity. Comprehensive fatigue assessment frameworks have emerged, including [13]'s integrated CNN system that analyzes multiple indicators like eye closure, yawning, and hand-to-face gestures. Temporal modeling approaches have shown promise as well—[5] combined facial keypoints with LSTM networks to detect fatigue patterns through blink and yawn frequency analysis. Specialized architectures continue to push performance boundaries: [10] compared four CNN models for precise eye-state classification while incorporating auditory drowsiness alerts, while [18]'s multitask learning framework simultaneously estimates facial landmarks, head pose, and gaze direction through parameter-shared branches. Alternative monitoring paradigms include [2]'s wearable sensor system for detecting inattention through body movement analysis, and [17]'s vision-based gaze region classification using CNN and GazeNet networks. The field has been further advanced by comprehensive reviews and frameworks for distraction detection presented in [7, 12].

A critical examination of these studies reveals several notable trends. Current research predominantly employs deep learning approaches, with convolutional neural networks and recurrent neural networks being the most widely adopted architectures. While most systems focus on fatigue detection through facial video analysis, some incorporate wearable sensors or more invasive monitoring techniques. Although existing solutions primarily target drowsiness identification, our work extends this capability by assessing overall driving safety through distraction analysis. The principal contributions of this research include:

1. A novel real-time monitoring system capable of distinguishing between safe and unsafe driving conditions
2. A hierarchical CNN-based inference architecture featuring:
   - A primary branch for eye and mouth state analysis
   - A secondary branch dedicated to head posture estimation
3. An efficient Type-1 Mamdani fuzzy logic implementation for driving state determination
4. Real-time performance achieving 30 frames per second, suitable for practical deployment

The proposed system implements a Type-1 fuzzy inference framework designed to classify driving states as either safe or unsafe. This intelligent module processes multiple input parameters, including quantitative indices for: mouth state (open/closed), eye state (open/closed) and head posture (orientation and position). These indices are extracted through two specialized convolutional neural network models, enabling robust feature extraction from visual inputs.

The manuscript is organized systematically: Sect. 14.2 provides foundational knowledge on the convolutional architectures and fuzzy inference methods employed in our study. Section 14.3 details the proposed hybrid methodology, while Sect. 14.4 presents experimental validation and comparative results. Finally, Sect. 14.5 concludes with key findings and suggests directions for future research.

14.2 Background

14.2.1 Deep Learning Models

This section introduces two optimized convolutional neural network architectures that form the foundation of our proposed system. The first model performs dual functionality: (1) precise face detection and localization, and (2) head pose estimation to assess the driver's attentional focus. The second model specializes in ocular and facial feature detection, specifically identifying left and right eye states, as well as mouth opening/closing patterns, for comprehensive drowsiness analysis. Importantly, the facial region detected by the first model serves as a spatial constraint for the second model's processing, significantly improving computational efficiency.

14.2.1.1 Face-Detection-Retail-0004

The face detection pipeline employs a lightweight SqueezeNet variant [8], specifically optimized for face detection rather than the original model's general object classification purpose. As illustrated in Fig. 14.1, the architecture features a computationally efficient backbone composed of fire modules [9], significantly reducing processing requirements while maintaining detection accuracy. The model processes input images in a standardized four-dimensional tensor format $InputTensor = [Batch\ Size \times Number\ of\ Channels \times Image\ Height \times Image\ Width] = [1 \times 3 \times 300 \times 300]$. This fixed input dimensionality ensures consistent processing of all frames while optimizing memory allocation and computational efficiency. The architecture incorporates a scaling block to standardize input dimensions before processing. After resizing, the SqueezeNet light architecture begins with an initial convolutional layer, followed by a sequence of eleven specialized fire modules. Each fire module contains squeeze convolution layers utilizing 1×1 filters that feed into an expand layer combining

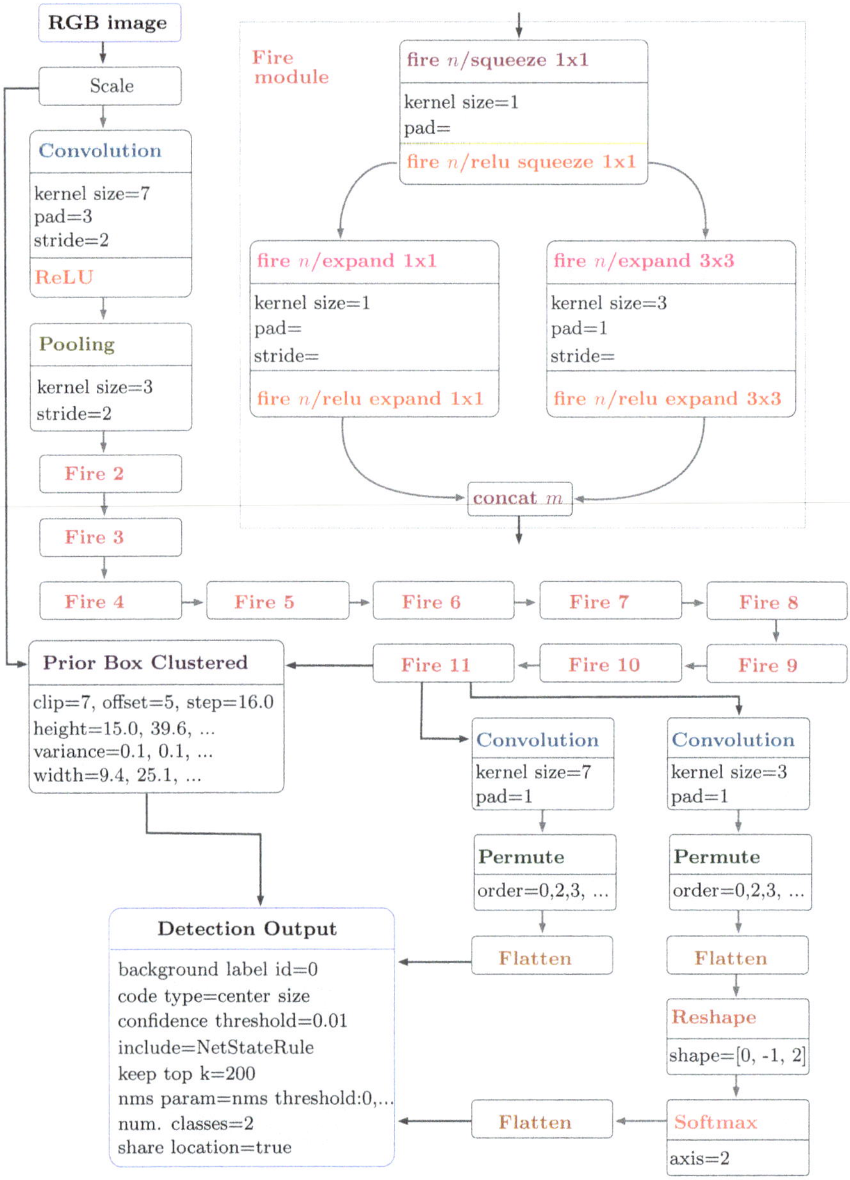

Fig. 14.1 Model: *face-detection-retail-0004*

both 1 × 1 and 3 × 3 convolution filters [8]. The 1 × 1 filters serve a crucial dimensional reduction role, efficiently limiting the channel depth before the more computationally intensive 3×3 operations. Within each fire module, a concatenation mechanism strategically combines features from different layers, significantly enhancing the network's representational capacity. This architectural choice enables more effective extraction of spatial features from input images, building on the fundamental principles discussed earlier. The output from the final fire module in this sequence feeds into two distinct processing pathways: one dedicated to fine feature abstraction for detailed pattern recognition, and another generating appropriately scaled prior boxes through a specialized clustering mechanism for optimal detection performance. The detection pipeline concludes with a non-maximum suppression operation in the *detection output* block, which synthesizes final detection results by evaluating both spatial coordinates and confidence scores [8]. The network generates output in a four-dimensional tensor format $[1, 1, N, 7]$, where N represents the quantity of detected bounding boxes. Each detection is encoded as a seven-element vector $[image_{id}, label, conf, x_{min}, y_{min}, x_{max}, y_{max}]$, containing both classification and spatial information. For facial region extraction in our driver monitoring application, we specifically utilize the bounding box coordinates (x_{min}, y_{min}) and (x_{max}, y_{max}) to precisely delineate the driver's facial region within each frame.

14.2.1.2 Facial-Landmarks-68-160×160

This custom-built convolutional neural network specializes in estimating 68 facial landmarks, with particular emphasis on those corresponding to the eyes and mouth for our driver monitoring application. As illustrated in Fig. 14.2, the architecture builds upon Shufflenet Network v2 [21], a computationally efficient framework well-suited for mobile deployment. The model employs a dual heatmap-offset mechanism to precisely localize facial landmarks: the heatmap identifies probable landmark locations through activation peaks, while the accompanying x and y offsets refine these positions to sub-pixel accuracy. Input images are processed in a standardized $[1 \times 3 \times 160 \times 160]$ tensor format, representing single RGB frames of fixed dimensions. To optimize computational efficiency, the network integrates fused multiply-add ($fused\_Mul$ and $fused\_Add$) operations alongside fundamental arithmetic transforms including matrix multiplication ($MatMul$), matrix addition ($MultAdd$), and averaging ($Mean$). This combination of architectural choices enables robust real-time performance while maintaining the precision required for accurate facial landmark detection. The model's core components are Shufflenet modules, which employ specialized Shufflenet units to enable efficient channel shuffling across convolution groups. This process, referred to as depth-separable or group shuffles [21], enhances feature representation while maintaining computational efficiency. Each module integrates three key operations: depthwise convolution for channel-specific filtering, strided slicing for dimensional reduction, and concatenation for feature aggregation. The depthwise convolution operation applies individual filters per input channel, preserving channel distinctiveness while significantly reducing parameters compared to standard 2D convolutions [9]. Complementing this, the strided slice operation performs selective tensor partitioning using computed indices that specify start points, end points, and step

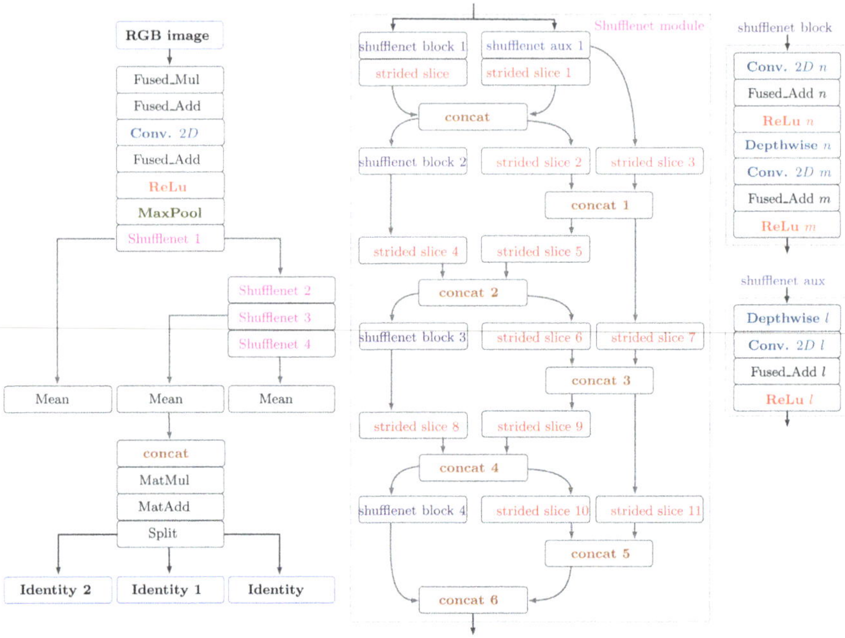

Fig. 14.2 Model: *facial-landmarks-68-160×160*

sizes for each dimension [9]. Finally, the concatenation operation merges multiple input tensors along specified axes, enabling the combination of features extracted at different network depths or scales [9].

14.2.2 Fuzzy Inference System

A fuzzy inference system fundamentally represents a nonlinear mapping from input variables to output variables, designed to facilitate control and decision-making processes. Formally, given n input variables $x_1 \in \mathcal{X}_1, \ldots, x_n \in \mathcal{X}_n$ and a single output variable $y \in \mathcal{Y}$, this mapping can be mathematically expressed as $y = f(\mathbf{x})$, where $\mathbf{x} \equiv \{x_i\}_{i=1}^{n}$ [14]. The standard architecture of a type-1 fuzzy inference system employs four interconnected processing stages: a rule base containing linguistic control rules, a fuzzifier for input transformation, a fuzzy inference engine for rule evaluation, and a defuzzifier for output generation. Effective system design requires careful standardization of each component's conceptual framework to ensure practical applicability across different problem domains.

14.2.2.1 Rule Base

Rules constitute the fundamental building blocks of fuzzy systems, serving as vessels for encapsulating human expertise either through direct knowledge transfer from domain specialists or via extraction from domain-specific datasets. For a multiple-input single-output type-1 fuzzy system utilizing Mamdani inference, the k-th rule follows the canonical structure:

$$\mathcal{R}^k : \text{IF } x_1 \text{ is } \mathcal{A}_1^k \text{ and } \cdots \text{ and } x_n \text{ is } \mathcal{A}_n^k \text{ THEN } y \text{ is } C^k, \tag{14.1}$$

where each antecedent $\mathcal{A}_i^k \in \{\mathcal{X}_{ij}\}_{j=1}^{\mathcal{D}_i}$ represents a linguistic term from the $\mathcal{D}_i$ possible terms modeling the i-th input variable. The consequent $C^k \in \{\mathcal{Y}_j\}_{j=1}^{\mathcal{D}}$ similarly denotes one of $\mathcal{D}$ possible linguistic output terms. The rule index k is bounded by $\mathcal{K}$, with $1 \leq k \leq \mathcal{K}$ defining the complete rule base cardinality.

14.2.2.2 Fuzzifier

The fuzzifier operates as a transformation mechanism that converts crisp input values into type-1 fuzzy sets, formally represented as $\mathcal{A}_{\mathbf{X}} \in \mathbf{X}$. This mapping processes an n-dimensional input vector $\mathbf{x} = (x_1, \ldots, x_n)^\top \in \mathcal{X}_1 \times \cdots \times \mathcal{X}_n \equiv \mathbf{X}$, typically obtained from sensor measurements. While type-1 fuzzy systems theoretically accommodate both singleton and non-singleton fuzzification approaches, our implementation specifically adopts the non-singleton variant due to its superior suitability for the considered application. The chosen fuzzifier is characterized by the membership function $\mu_{X_i}(x_i|x_i')$, which attains its maximum value of unity when $x_i = x_i'$ and exhibits monotonically decreasing membership as $|x_i - x_i'|$ increases.

14.2.2.3 Fuzzy Inference Engine

The fuzzy inference engine serves as the computational core that systematically combines rules and establishes mappings between input and output type-1 fuzzy sets [14]. For a system comprising $\mathcal{K}$ rules, each accepting n inputs $\mathbf{x} = (x_1, \ldots, x_n)$ and producing one output y, the transformation process is formally represented as:

$$\mathcal{R}^k : \mathcal{A}_1^k \times \cdots \times \mathcal{A}_n^k \mapsto C^k \equiv \mathcal{A}^k \mapsto C^k, \tag{14.2}$$

where $\mathcal{A}^k = \mathcal{A}_1^k \times \cdots \times \mathcal{A}_n^k$ denotes the Cartesian product of input fuzzy sets. The membership degree for the k-th rule is quantified through the composite relation:

$$\mu_{\mathcal{R}^k}(\mathbf{x}, y) = \left[\bigcap_{i=1}^{n} \mu_{\mathcal{A}_i^k}(x_i)\right] \cap \mu_{C^k}(y), \tag{14.3}$$

with $\cap$ representing the intersection operation on type-1 fuzzy sets. The system processes n-dimensional inputs through a type-1 fuzzy set $\mathcal{A}_{\mathbf{x}'}$ characterized by the membership function:

$$\mu_{\mathcal{A}_{\mathbf{x}'}}(\mathbf{x}) = \bigcap_{i=1}^{n} \mu_{X_i}(x_i|x_i'). \quad (14.4)$$

The fundamental input-output relationship emerges through rule activation, formally expressed as:

$$\mu_{\mathcal{B}^k}(y, \mathbf{x}') = \bigcup_{\mathbf{x}\in\mathbf{X}} \left[\mu_{\mathcal{A}_{\mathbf{x}'}}(\mathbf{x}) \cap \mu_{\mathcal{R}^k}(\mathbf{x}, y)\right], \quad (14.5)$$

which simplifies through substitution to:

$$\mu_{\mathcal{B}^k}(y, \mathbf{x}') = \mathcal{F}^k(\mathbf{x}') \cap \mu_{C^k}(y). \quad (14.6)$$

Here, $\mu_{\mathcal{B}^k}(y, \mathbf{x}')$ represents the output membership function, while the firing set $\mathcal{F}^k(\mathbf{x}')$ expands to:

$$\mathcal{F}^k(\mathbf{x}') = \bigcup_{i=1}^{n} \left[\bigcup_{x_i\in X_i} \left(\mu_{X_i}(x_i|x_i') \cap \mu_{\mathcal{F}_i^k}(x_i)\right) \right]. \quad (14.7)$$

14.2.2.4 Defuzzification

Defuzzification constitutes the final transformation step in fuzzy system design, converting aggregated fuzzy sets into precise numerical outputs. This critical process maps the combined membership functions to a single crisp value representative of the system's decision. While the centroid method remains historically prevalent in practical applications, its computational complexity arises from the continuous integration required across the output domain. The current research employs the Center-of-Sets defuzzification approach [14], which offers computational advantages while maintaining interpretability. This method operates by replacing each rule's consequent set with a singleton positioned at its centroid, weighted by the corresponding rule's firing strength. The final output emerges as the weighted average of these representative points:

$$y(\mathbf{x}') = \frac{\sum_{k=1}^{\mathcal{K}} \delta^k f^k(\mathbf{x}')}{\sum_{k=1}^{\mathcal{K}} f^k(\mathbf{x}')}, \quad (14.8)$$

where δ^k denotes the centroid location of the k-th consequent set and $f^k(\mathbf{x}')$ represents the firing level as established in Eq. (14.7). This formulation effectively balances computational efficiency with mathematical transparency in the defuzzification process.

14.3 Methodology

14.3.1 Proposed System

This study presents an automated real-time driver monitoring system whose architecture is illustrated in Fig. 14.3.

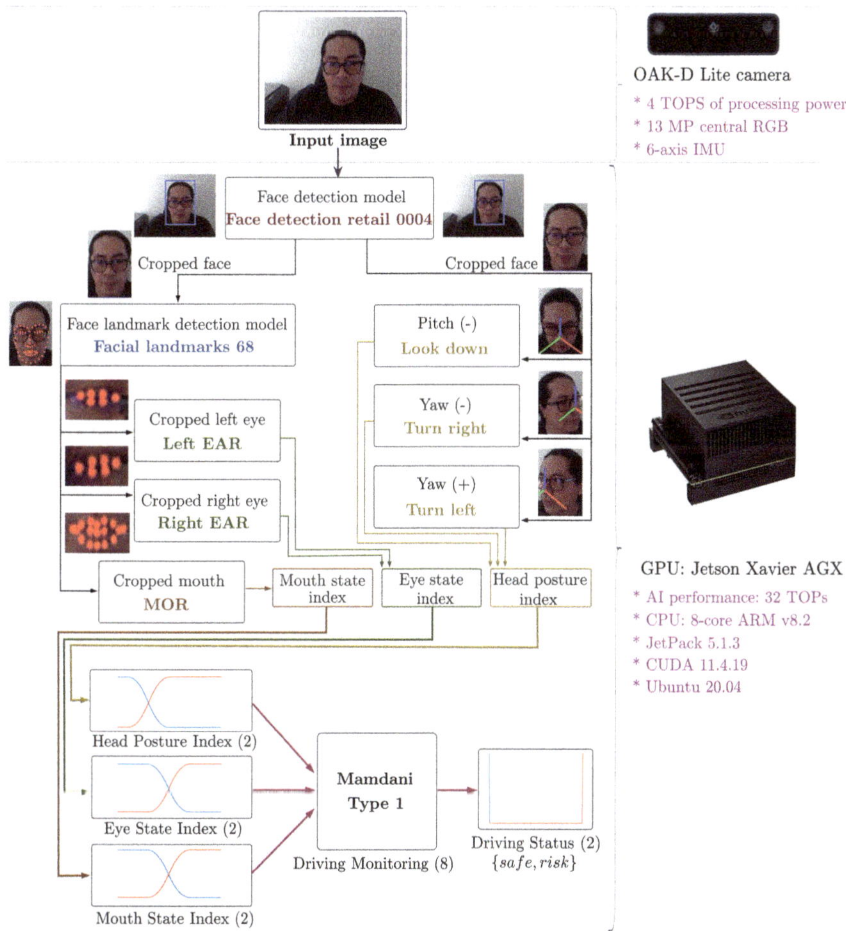

Fig. 14.3 Proposed scheme

The pipeline begins with a stereo depth and high-resolution color camera featuring embedded neural network inference capabilities for initial video capture. Processing then transitions to a high-performance embedded GPU that implements a hierarchical deep neural network architecture with two parallel inference branches. The left branch integrates two specialized models: *face-detection-retail-0004* for face localization and *facial-landmarks-68-160×160* for precise detection of facial features including both eyes and the mouth. These detections enable computation of two critical biometric indicators: the Eye Aspect Ratio (EAR) for blink analysis and Mouth Opening Ratio (MOR) for yawn detection. Concurrently, the right branch utilizes the *face-detection-retail-0004* model to estimate head orientation through detection of look down, turn right, and turn left movements. The system culminates with a Type-1 Mamdani Fuzzy inference engine that synthesizes these multimodal inputs—comprising mouth state, eye state, and head position indices—into a comprehensive assessment of driver status. This fusion occurs in real-time through optimized GPU acceleration, enabling simultaneous processing of all visual cues.

14.3.2 Eye Aspect Ratio and Mouth Opening Ratio

Blink frequency analysis serves as a valuable but imperfect metric for fatigue assessment, with increased blink rates often correlating with diminished alertness. While this physiological signal provides useful insights, its reliability can be compromised by practical factors such as eyewear occlusion. Complementary to blink analysis, yawning behavior has emerged as a robust physiological marker of drowsiness, exhibiting strong correlation with elevated sleep pressure and reduced arousal levels [3, 6]. The current research therefore incorporates both ocular and oral indicators for comprehensive fatigue monitoring. Precise anatomical landmarks, as depicted in Fig. 14.4, enable simultaneous tracking of blink patterns through eye feature analysis and yawning detection via mouth movement quantification. This multimodal approach enhances system robustness by mitigating the limitations inherent in single-indicator monitoring systems.

The Eye Aspect Ratio (EAR) provides a quantitative measure of eye openness, computed separately for each eye using the formula [3]:

$$EAR_{\{\text{left},\text{right}\}} = \frac{\|p_2 - p_6\| + \|p_3 - p_5\|}{2 \cdot \|p_1 - p_4\|}, \tag{14.9}$$

where p_1 through p_6 represent the two-dimensional facial landmark coordinates shown in Fig. 14.4a. The composite EAR value combines measurements from both eyes through simple averaging:

$$EAR = \frac{EAR_{\text{left}} + EAR_{\text{right}}}{2}. \tag{14.10}$$

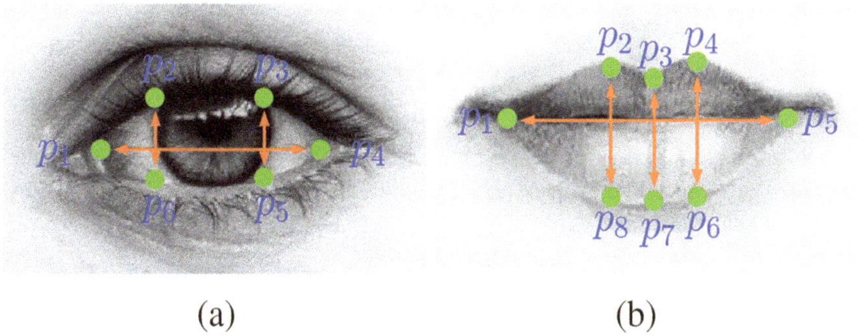

Fig. 14.4 Eyes and mouth landmarks for computing EAR and MOR values. (**a**) Eyes landmarks. (**b**) Mouth landmarks

For yawning detection, the Mouth Opening Ratio (MOR) quantifies oral aperture using the relationship [6]:

$$MOR = \frac{\|p_2 - p_8\| + \|p_3 - p_7\| + \|p_4 - p_6\|}{3 \cdot \|p_1 - p_5\|}, \tag{14.11}$$

where landmarks p_1 to p_8 correspond to the mouth coordinates illustrated in Fig. 14.4b. The numerator captures vertical mouth displacements while the denominator normalizes by horizontal mouth width. These metrics form the basis for the Eyes State Index (ESI) and Mouth State Index (MSI), derived from EAR and MOR values respectively.

14.3.3 Driving Monitoring

To enhance the reliability of driver state assessment, this research implements a Type-1 Mamdani Fuzzy inference system that integrates multiple physiological indicators. The system combines mouth state, eye state, and head posture measurements through the PERCLOS fatigue evaluation framework [1], which quantifies the percentage of time spent in fatigue-related states during a specified observation window.

The Mouth State Index (MSI) is computed as the relative duration of mouth opening events:

$$MSI = \frac{\tau_m}{\mathcal{T}_m} \times 100\%, \tag{14.12}$$

where τ_m counts frames containing open mouth occurrences and $\mathcal{T}_m$ represents the total frames in the evaluation period. Similarly, the Eye State Index (ESI) measures eye closure duration:

$$ESI = \frac{\tau_e}{\mathcal{T}_e} \times 100\%, \tag{14.13}$$

with τ_e enumerating frames where eyes are closed and $\mathcal{T}_e$ being the total observation frames. The Head Posture Index (HPI) completes the assessment by quantifying abnormal head positions:

$$HPI = \frac{\tau_{hp}}{\mathcal{T}_{hp}} \times 100\%, \tag{14.14}$$

where τ_{hp} tracks frames exhibiting fatigue-indicative head orientations and $\mathcal{T}_{hp}$ denotes the complete monitoring duration. These three indices collectively provide a comprehensive basis for the fuzzy inference system's decision-making process.

Remark 14.1 Empirical studies establish that yawning episodes typically persist for 3–5 seconds [19]. Given a standardized detection window of 20 seconds [11], this temporal constraint naturally limits observable yawns to a maximum of two occurrences per interval. The system therefore classifies mouth states as fatigued when the Mouth State Index (MSI) meets or exceeds the 0.3 threshold, corresponding to at least one sustained yawning event per monitoring cycle.

Remark 14.2 Physiological measurements distinguish alert-state eye closures (0.12–0.15 seconds) from fatigue-induced closures (exceeding 0.5 seconds) [11]. The Eye State Index (ESI) threshold of 0.5 reflects this dichotomy, with values reaching or surpassing this limit indicating compromised alertness. This threshold effectively captures the prolonged eye closures characteristic of drowsy states while remaining insensitive to normal blinking patterns.

Remark 14.3 The PERCLOS criterion identifies fatigued head posture through pitch and yaw deviations exceeding 20% [1]. Experimental validation [11] confirms that such deviations consistently correlate with fatigue, motivating our Head Posture Index (HPI) threshold of 0.5. This value represents the inflection point where postural instability transitions from transient adjustments to sustained fatigue indicators.

Building upon these empirical foundations, we establish a standardized 20-second evaluation window for computing the indices defined in Eqs. (14.12)–(14.14). This duration optimally balances detection sensitivity with computational efficiency. The fuzzy inference system's input variables—Head Posture Index (HPI), Eye State Index (ESI), and Mouth State Index (MSI)—are precisely characterized using the validated thresholds discussed previously, while the output space is discretized into two distinct states: *safe* and *risk*.

The complete fuzzy set representations for these linguistic variables are visualized in Fig. 14.5, which illustrates the membership functions governing both input processing and output classification. This graphical depiction clarifies the quanti-

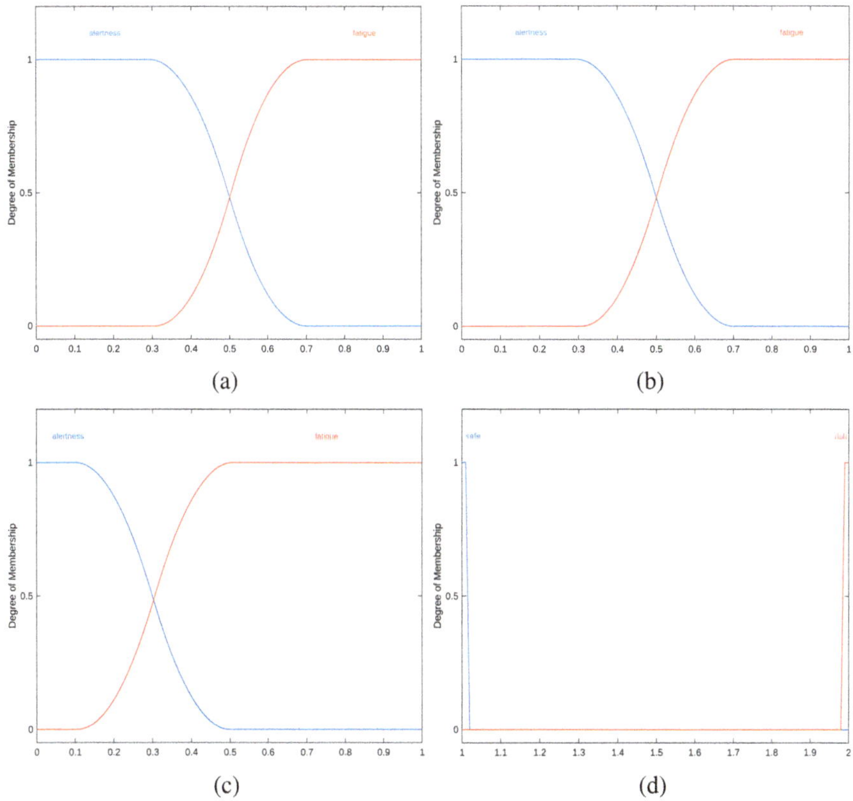

Fig. 14.5 Type-1 fuzzy sets for three inputs and one output. (**a**) Head posture index. (**b**) Eye state index. (**c**) Mouth state index. (**d**) Driving status

tative relationships between physiological measurements and their corresponding fuzzy valuations within our inference framework.

14.4 Experimental Results and Analysis

The experimental validation of our proposed system consists of two complementary investigations. The first experiment rigorously evaluates the system's performance in detecting eye states, mouth movements, and head pose orientation. The second experiment comprehensively assesses the integrated system's capability to classify driving states according to the safety criteria established in this study.

14.4.1 Training Deep Neural Models

The architecture employs two pre-trained deep learning models to ensure optimal performance, as detailed in Sect. 14.2.1. Model 1 (*face-detection-retail-0004*) utilizes the Wider Face dataset [20], which contains 32,203 training images exhibiting substantial variation in scale, pose, and occlusion conditions, collectively encompassing 393,703 annotated faces. Model 2 (*facial-landmarks-68-160×160*) was trained on the 300 Videos in the Wild (300-VW) dataset, consisting of 300 face tracking videos recorded at 25–30 fps, each approximately one minute in duration. This strategic use of pre-trained models on specialized datasets ensures robust feature extraction while maintaining computational efficiency.

14.4.2 Face, Eye, Mouth and Head Posture Recognition

The experimental evaluation employed a dataset of 50 video sequences, each averaging one minute in duration with 1080 × 1080 resolution at 30 fps. System performance was quantified using standard metrics: Accuracy, Recall, Specificity, Precision, and $F1$-Score [15]. For comparative analysis, we benchmarked against established methods including Viola-Jones [16], HoG [6], MTCNN [5], TCDCN [22], 4D model [10], and E-MSR [11]. As shown in Fig. 14.6, results are reported using median values for robust statistical analysis.

The proposed approach demonstrated superior stability across all metrics, achieving performance scores between 0.901 and 0.968. Notably, task complexity inversely correlates with detection accuracy—face recognition tasks, benefiting from larger feature areas, yielded higher scores than the more challenging eye and mouth state classifications. Our head posture analysis extends beyond conventional approaches by incorporating three distinct movements (Look Down, Turn Right, and Turn Left) as indicators of attentional drift, in contrast to prior work focusing exclusively on vertical head orientation. This comprehensive assessment of driver focus provides more reliable detection of distraction states compared to single-movement baselines.

14.4.3 Driving Status Recognition

The second experimental phase utilized the same video dataset to evaluate the complete integrated system's performance. This comprehensive assessment examined both the synergistic operation of the dual deep neural networks and the fuzzy inference module's decision-making capabilities. While maintaining consistent evaluation metrics, direct comparison with existing methods proved impractical due to fundamental differences in system objectives. Existing approaches primarily

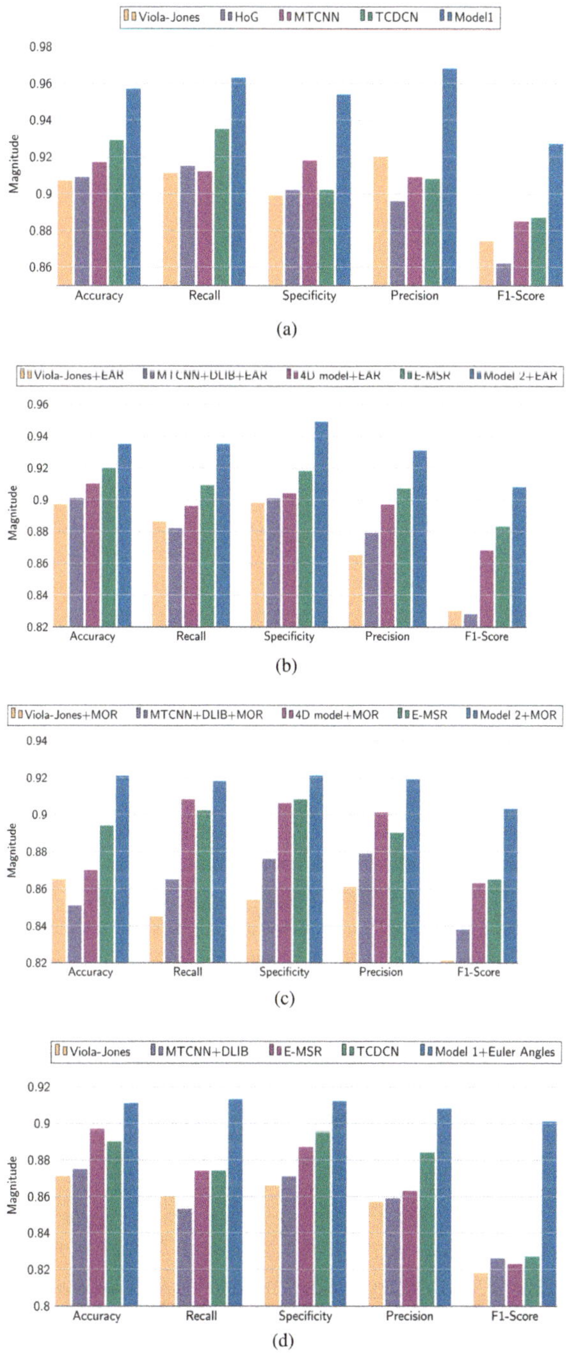

Fig. 14.6 Quantitative results for face, eye, mouth and head posture recognition tasks. (**a**) Face recognition. (**b**) Eye state recognition. (**c**) Mouth state recognition. (**d**) Head posture recognition

Table 14.1 Quantitative results for driving status recognition

Metric	Median	Min	Max	Stdev
Accuracy	0.9306	0.9201	0.9385	0.0062
Recall	0.9292	0.9204	0.9398	0.0065
Specificity	0.9279	0.9205	0.9391	0.0068
Precision	0.9276	0.9201	0.9396	0.0063
F1-Score	0.9261	0.9199	0.9320	0.0045

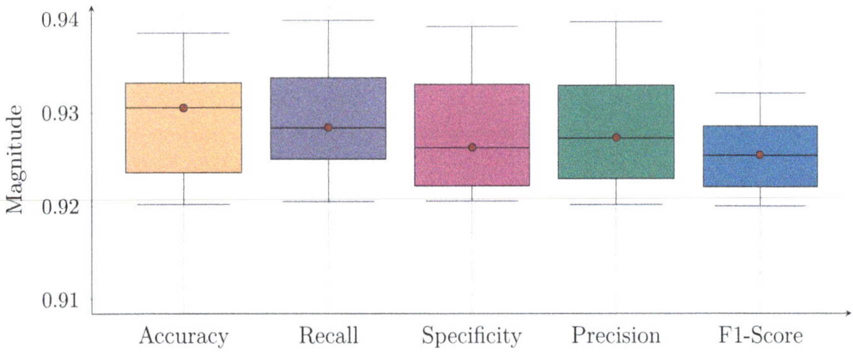

Fig. 14.7 Graphical results for driving status recognition

target drowsiness detection, whereas our framework provides a more holistic assessment of driving safety.

As detailed in Table 14.1, the system demonstrates robust performance with median accuracy exceeding 92% across all test cases. The tight distribution of results (minimum 0.9201 to maximum 0.9385) and low standard deviation (0.0062) indicate consistently reliable operation. These quantitative outcomes validate the system's effectiveness in real-world driving state classification, significantly advancing beyond conventional fatigue detection systems.

Complementing the quantitative results, Fig. 14.7 visually summarizes the system's performance across all evaluation metrics. The graphical representation confirms the consistent distribution pattern observed in our experimental findings, providing further validation of the method's reliability.

The system achieves a real-time processing speed of approximately 30 frames per second (fps), enabled by the powerful computational architecture combining the Jetson Xavier AGX GPU with the dedicated neural processing unit in the OAK-D camera. This hardware configuration yields significantly faster inference times compared to training-phase computations, demonstrating the efficiency of our optimized deployment pipeline. The observed performance metrics validate the system's suitability for real-world real-time applications, meeting the stringent latency requirements of driver monitoring systems.

14.5 Conclusions and Future Improvements

This study presents a real-time driver monitoring system based on embedded GPU architecture, combining two deep neural networks for precise eye and mouth state analysis with a Type-1 Mamdani fuzzy inference engine for comprehensive driving state assessment. Experimental results demonstrate robust performance, with all metrics consistently exceeding 92% accuracy while maintaining minimal result variance ($\sigma < 0.01$).

Future research directions will focus on two critical aspects: (1) optimizing the system's energy efficiency for embedded deployment, and (2) enhancing robustness under challenging conditions including low-light environments and visual obstructions such as sunglasses. These improvements will further strengthen the system's practical applicability in real-world automotive settings.

Acknowledgments The authors thank to CONACYT, as well as Tecnológico Nacional de México/Centro Nacional de Investigación y Desarrollo Tecnológico for their financial support trough the project so-called "Estudio y aplicacién de técnicas de inteligencia artificial para la asistencia de conducción en vehículos autónomos".

References

1. Acıoğlu, A., Ercelebi, E.: Real time eye detection algorithm for perclos calculation. In: 2016 24th Signal Processing and Communication Application Conference (SIU), pp 1641–1644. IEEE, Piscataway (2016)
2. Akiduki, T., Nagasawa, J., Zhang, Z., Omae, Y., Arakawa, T., Takahashi, H.: Inattentive driving detection using body-worn sensors: feasibility study. Sensors 22(1), 352 (2022)
3. Alam, L., Hoque, M.M., Dewan, M.A.A., Siddique, N., Rano, I., Sarker, I.H.: Active vision-based attention monitoring system for non-distracted driving. IEEE Access **9**, 28540–28557 (2021)
4. Alzubi, J.A., Jain, R., Alzubi, O., Thareja, A., Upadhyay, Y.: Distracted driver detection using compressed energy efficient convolutional neural network. J. Intell. Fuzzy Syst. **42**(2), 1253–1265 (2022)
5. Chen, L., Xin, G., Liu, Y., Huang, J.: Driver fatigue detection based on facial key points and LSTM. Secur. Commun. Netw. **2021**, 1–9 (2021)
6. Ghourabi, A., Ghazouani, H., Barhoumi, W.: Driver drowsiness detection based on joint monitoring of yawning, blinking and nodding. In: 2020 IEEE 16th International Conference on Intelligent Computer Communication and Processing (ICCP), pp 407–414. IEEE, Piscataway (2020)
7. Halin, A., Verly, J.G., Van Droogenbroeck, M.: Survey and synthesis of state of the art in driver monitoring. Sensors **21**(16), 5558 (2021)
8. Iandola, F.N., Han, S., Moskewicz, M.W., Ashraf, K., Dally, W.J., Keutzer, K.: Squeezenet: alexnet-level accuracy with 50x fewer parameters and< 0.5 mb model size (2016). arXiv:160207360
9. Intel: Openvino™ toolkit (2024). github.com/openvinotoolkit/openvino
10. Jahan, I., Uddin, K.A., Murad, S.A., Miah, M.S.U., Khan, T.Z., Masud, M., Aljahdali, S., Bairagi, A.K.: 4d: a real-time driver drowsiness detector using deep learning. Electronics **12**(1), 235 (2023)

11. Jia, H., Xiao, Z., Ji, P.: Fatigue driving detection based on deep learning and multi-index fusion. IEEE Access **9**, 147054–147062 (2021)
12. Kashevnik, A., Shchedrin, R., Kaiser, C., Stocker, A.: Driver distraction detection methods: a literature review and framework. IEEE Access **9**, 60063–60076 (2021)
13. Maheswari, V.U., Aluvalu, R., Kantipudi, M.P., Chennam, K.K., Kotecha, K., Saini, J.R.: Driver drowsiness prediction based on multiple aspects using image processing techniques. IEEE Access **10**, 54980–54990 (2022)
14. Mendel, J.M.: Explainable Uncertain Rule-Based Fuzzy Systems. Springer, Berlin (2023)
15. Mújica-Vargas, D., Quezada-Próspero, E., Luna-Álvarez, A., García-Aquino, C., Matuz-Cruz, M.: Arrhythmia recognition by a deep spectral autoencoder. In: 2023 10th International Conference on Soft Computing & Machine Intelligence (ISCMI), pp 151–155. IEEE, Piscataway (2023)
16. Omidyeganeh, M., Shirmohammadi, S., Abtahi, S., Khurshid, A., Farhan, M., Scharcanski, J., Hariri, B., Laroche, D., Martel, L.: Yawning detection using embedded smart cameras. IEEE Trans. Instrum. Measur. **65**(3), 570–582 (2016)
17. Shi, H., Chen, L., Wang, X., Wang, G., Wang, Q.: A nonintrusive and real-time classification method for driver's gaze region using an rgb camera. Sustainability **14**(1), 508 (2022)
18. Yang, D., Wang, Y., Wei, R., Guan, J., Huang, X., Cai, W., Jiang, Z.: An efficient multi-task learning CNN for driver attention monitoring. J. Syst. Architect. **148**, 103085 (2024)
19. Yang, H., Liu, L., Min, W., Yang, X., Xiong, X.: Driver yawning detection based on subtle facial action recognition. IEEE Trans. Multimedia **23**, 572–583 (2020)
20. Yang, S., Luo, P., Loy, C.C., Tang, X.: Wider face: a face detection benchmark. In: IEEE Conference on Computer Vision and Pattern Recognition (CVPR) (2016)
21. Zhang, X., Zhou, X., Lin, M., Sun, J.: Shufflenet: an extremely efficient convolutional neural network for mobile devices. In: Proceedings of the IEEE conference on computer vision and pattern recognition, pp. 6848–6856 (2018)
22. Zhu, T., Zhang, C., Wu, T., Ouyang, Z., Li, H., Na, X., Liang, J., Li, W.: Research on a real-time driver fatigue detection algorithm based on facial video sequences. Appl. Sci. **12**(4), 2224 (2022)

Chapter 15
Deep Learning-Based Detection for Unmanned Aerial Vehicle (UAV) Surveillance Missions

Irem Bayraktar and Murat Bakirci

Abstract This study addresses the critical challenge of detecting unmanned aerial vehicles (UAVs) in aerial surveillance tasks using deep learning-based object detection methods. With the increasing use of UAVs across various domains, ensuring reliable identification of UAVs, both authorized and unauthorized, is essential for airspace management and security. The study proposes the use of the YOLO11 algorithm, an advanced real-time object detection model, to improve the accuracy and robustness of UAV detection systems. The approach involves detecting both fixed-wing and rotary-wing UAVs from aerial images captured by a UAV-mounted high-resolution camera. A custom dataset was developed using UAV flight missions and image augmentation techniques to reflect real-world conditions. The main contribution of the study lies in evaluating and demonstrating the potential of YOLO11 in enhancing UAV detection performance in diverse environments. This work offers insights into the practical application of deep learning for real-time aerial monitoring and airspace security.

Keywords UAV detection · YOLO · Aerial surveillance · Fixed-wing and rotary-wing UAVs · Deep learning-based object detection

15.1 Introduction

In aerial surveillance and monitoring operations, the detection of unmanned aerial vehicles (UAVs) is critical for security, traffic management, and the effective

I. Bayraktar
Department of Aerospace Engineering, Turkish Air Force Academy, National Defence University, Istanbul, Turkey
e-mail: ibayraktar@uislab.org

M. Bakirci (✉)
Unmanned/Intelligent Systems Lab, Faculty of Aeronautics and Astronautics, Tarsus University, Mersin, Turkey
e-mail: muratbakirci@tarsus.edu.tr

G. Rivera et al. (eds.), *Recent Advances in Artificial Neural Networks*, Intelligent Systems Reference Library 283, https://doi.org/10.1007/978-3-032-13497-4_15

regulation of airspace [1]. With the growing number of UAVs operating in various environments, ensuring their proper detection has become essential to mitigate risks and enhance operational efficiency [2]. UAV detection in the air has become a significant necessity in both commercial and military applications, as it ensures the safety and reliability of air operations. For instance, in crowded airspaces, UAV detection is vital for reducing collision risks and improving airspace management [3, 4]. Furthermore, detecting unauthorized UAV activities plays a crucial role in safeguarding sensitive areas such as critical infrastructure, government buildings, and public events. These capabilities are essential not only for addressing potential threats but also for ensuring the seamless integration of UAVs into existing airspace systems. Consequently, the development of effective and highly accurate methods for UAV detection is gaining increasing importance, making it a key research area in modern aerial technologies.

UAVs have become a significant component of modern technology due to their versatile applications [5, 6]. These airborne platforms are frequently used in a wide range of applications, including surveillance, mapping, agriculture, environmental monitoring, and logistics [7, 8]. Their ability to provide real-time data [9] and access hard-to-reach areas [10] makes them indispensable across many fields. For example, within smart cities, UAVs are used for air quality assessments and precise measurements of air pollutants [11]. In intelligent traffic management, UAVs contribute to the development of smart transportation systems by providing rapid situational awareness, enabling quick interventions, and improving response times [12, 13]. Similarly, in environmental monitoring, UAVs are effective for tracking deforestation, wildlife, and climate change. The cost-effectiveness and flexibility of UAV technology have further increased its adoption across various industries [14, 15]. By minimizing human intervention and accelerating data collection processes, UAVs have proven particularly effective in challenging geographies where traditional methods are either impractical or costly [16]. These advantages have positioned UAVs as fundamental tools in modern air surveillance and data collection operations, leading to a variety of innovations [17].

In recent years, object detection algorithms have become a crucial component of computer vision-based studies. These algorithms are essential for effectively processing and analyzing data collected from unmanned vehicles, automating tasks that were previously labor-intensive and time-consuming [18, 19]. In particular, deep learning-based methods have revolutionized the field by offering high accuracy and speed in object detection tasks [20, 21]. Through the use of advanced neural networks, these methods have been applied to a variety of applications, including traffic monitoring systems for passenger safety, analyzing urban growth in city planning, and monitoring crop conditions in agricultural areas [22, 23]. Additionally, object detection algorithms facilitate the automatic identification, localization, and classification of objects, significantly improving decision-making processes [24]. By enabling machines to perceive and interpret their environments, these algorithms have become indispensable tools in advancing various technologies, from autonomous vehicles to intelligent surveillance systems [25, 26].

The detection of UAVs using object detection algorithms plays a critical role in various applications, such as monitoring both friendly and enemy UAVs, and coordinating multi-UAV operations. These detection methods are highly valuable for tracking UAV usage in specific areas, preventing unauthorized activities, and managing airspace efficiently. For instance, in military contexts, detecting enemy UAVs ensures timely countermeasures, while in commercial airspace, tracking friendly UAVs helps maintain operational efficiency and security. Deep learning-based object detection algorithms, particularly YOLO, have gained widespread recognition in these applications due to their superior performance in terms of speed and accuracy [27, 28]. The YOLO algorithm family stands out with its real-time detection capabilities, making it ideal for dynamic environments where quick decisions are crucial [29]. These features make YOLO-based methods especially effective for UAV detection tasks, as they can handle the fast and unpredictable movements of UAVs with remarkable precision [30–32].

In this study, the YOLO11 algorithm was used to detect UAVs and demonstrated its ability to address challenges associated with aerial surveillance. The images required for detection were obtained from a UAV equipped with a high-resolution camera, which enabled detailed analysis and accurate identification of target UAVs. The performance of the YOLO11 algorithm in UAV detection was examined in various scenarios and environmental conditions. By evaluating its detection accuracy, inference speed, and robustness under challenging conditions, this study highlights the algorithm's potential to improve airspace security and operational efficiency. The findings provide valuable insights into the effectiveness of YOLO11 in UAV detection and its applicability in real-world scenarios, paving the way for future advancements in UAV detection technologies. Our key observations suggest that the application of YOLO11 offers promising advantages in terms of detection accuracy, inference speed, and adaptability across UAV types. These insights help bridge the gap between algorithmic development and practical implementation in air surveillance systems. The remainder of this paper is organized as follows: Section 15.2 reviews related work on UAV detection and the development of YOLO-based algorithms. Section 15.3 presents an overview of UAV object detection technologies. Section 15.4 describes the dataset and augmentation methods. Section 15.5 details the application of YOLO11 in the context of UAV detection, including system design and training process. Section 15.6 presents the experimental results, and Sect. 15.7 concludes the paper with key findings and future directions.

15.2 Related Work

In machine learning-based model training for object detection in computer vision studies, comprehensive and large-scale image datasets must be labeled. This process, often performed through manual annotation of field images captured by sensors, incurs significant costs in terms of manpower and time. In [33], the authors addressed these challenges by utilizing synthetic data generated with OpenAI

DALLE, a large-scale language model capable of producing realistic images. They created a dataset for detecting the Scilate apple variety and used it to train the latest YOLO models, YOLOv10 and YOLO11, along with their respective sub-models. Upon comparing the test results, they reported that the YOLO11x and YOLO11n sub-models of YOLO11 demonstrated significantly better performance than YOLOv10, achieving sensitivity values of 0.917 and 0.916, respectively. Furthermore, they highlighted the superiority of YOLO11 over YOLOv10, particularly due to its faster preprocessing time of 0.2 ms. The study also revealed that synthetic Scilate apple images generated by the language model were processed much faster than sensor-based images, demonstrating the efficiency of this dataset creation method.

In [34], the importance of object detection in agricultural automation, especially for identifying fruitlets and unripe fruits in orchards, is underscored as a critical factor for early fruit load management and general crop optimization. The authors evaluated the performance of the latest four YOLO versions, YOLOv8, YOLOv9, YOLOv10, and YOLO11, on their ability to enhance orchard productivity and manage fruit resources effectively. They tested 22 sub-models across these algorithms using images of Scifresh, Scilate, Honeycrisp, and Cosmic Crisp apple varieties captured with iPhones and machine vision sensors. The results demonstrated that YOLO11 outperformed all sub-models of YOLOv10, YOLOv9, and YOLOv8, achieving a mAP^{50} value of 0.933, while the YOLOv9 Gelan-based model reached a slightly higher mAP^{50} value of 0.935.

The use of drones, one of the most prominent applications of unmanned systems, has become increasingly widespread in both military and civilian domains. However, the deployment of high-cost detection technologies designed to neutralize unauthorized or unwanted drones remains limited under various conditions. In [35], the authors explored more cost-effective methods to counter unwanted drones, opting for computer vision-based object detection systems. Their approach leveraged machine learning algorithms applied to industrial cameras for drone detection. To enhance the training process, the authors expanded their dataset using images from ImageNet, MS-COCO, PASCAL VOC, and anti-UAV datasets. They created a comprehensive dataset of 56,821 images, which were manually labeled to ensure accuracy.

The increasing use of UAVs has also raised concerns about their potential misuse for unlawful or unintended purposes, driving the need to develop protective measures against unauthorized UAVs. Various techniques have been proposed to detect unidentified UAVs operating near protected UAVs. In [36], the authors investigated the trade-offs between cost and performance when using video cameras for UAV detection. They highlighted the challenge posed by the visual similarity of UAVs to birds and airplanes, which are often captured in the same video footage. The study addressed the problem of real-time UAV detection through a two-step approach: detecting moving objects and classifying the detected objects as UAVs, birds, or background elements.

In [37], the authors highlight the potential risks associated with the improper use of UAVs, including security threats, smuggling, and privacy violations, as UAVs become increasingly prevalent in civilian and military applications due to their advanced capabilities. The study focuses on real-time UAV detection using the YOLOv9 object detection model and the NVIDIA Jetson Nano platform, specifically targeting automatic UAV and unauthorized spy UAV detection. The YOLOv9 model is evaluated using key performance metrics such as precision, recall, F1 score, mAP, and FPS, which are widely used for assessing UAV detection systems. The test results indicate that YOLOv9 achieves approximately 5% improvement over YOLOv8, with precision, recall, F1 score, and mAP values reported as 0.946, 0.864, 0.903, and 0.957, respectively.

In [38], the authors emphasize the advantages of using synthetic data for training deep neural networks in UAV detection tasks. They highlight that synthetic data offers a cost-effective and high-quality alternative, enhancing data usability for model training. The study investigates the adaptation of synthetic data to real-world scenarios and evaluates its performance in simulations. The authors integrate synthetic data with real-world data and employ the Faster R-CNN algorithm for UAV detection. Tests conducted using real-world UAV datasets and MAV-Vid synthetic data demonstrate high performance, with the model achieving an mAP^{50} value of 97.8% in real-world scenarios and 97% when trained exclusively on MAV-Vid.

In [39], the authors examined the integration of multi-sensor data for UAV detection, combining visual data from high-resolution cameras with radar and infrared (IR) imaging to enhance detection accuracy. They proposed a fusion-based detection system that leveraged the strengths of each sensor type to mitigate the limitations of individual modalities. Radar provided effective distance and velocity measurements, while IR imaging was useful for detecting UAVs in low-light conditions. The study evaluated the fusion system using various datasets, including open-source UAV imagery and radar signals. Results indicated that the multi-sensor approach improved detection accuracy by 12% compared to using visual data alone.

Another study, presented in [40], explored the challenges of small UAV detection in complex urban environments. The authors focused on the detection of micro-UAVs, which are particularly challenging due to their small size and low radar cross-section. Using the YOLOv8 algorithm, they developed a system optimized for high sensitivity in identifying micro-UAVs against cluttered urban backdrops. The dataset, composed of over 100,000 labeled images, included diverse environmental conditions such as varying lighting, weather, and background complexity.

In [41], the authors investigated UAV detection in maritime environments, focusing on the unique challenges posed by dynamic oceanic conditions, such as fluctuating waves and reflections. They utilized a modified YOLOv10 framework trained on a custom dataset containing UAV images captured from ship decks and coastal areas. The study emphasized the importance of data augmentation to address variations in lighting, motion blur, and background noise. The YOLOv10m model demonstrated a precision of 0.945 and an F1 score of 0.912, outperforming traditional detection methods.

The potential for UAV detection using edge computing was explored in [42], where the authors implemented a lightweight object detection model tailored for deployment on low-power devices such as the NVIDIA Jetson Nano and Raspberry Pi 4. The study aimed to balance detection performance with computational efficiency, targeting real-time applications in remote or resource-constrained environments. The authors utilized the YOLOv9-tiny model, fine-tuned for UAV detection, and demonstrated that it achieved an inference time of 6.5 ms on the Jetson Nano, with an mAP^{50} of 0.891.

In [43], the authors explored the application of generative adversarial networks (GANs) to enhance UAV detection by generating synthetic images for model training. They proposed a framework that utilized GANs to create high-resolution UAV images under various environmental conditions, such as fog, rain, and nighttime scenarios. The synthetic images were integrated with real-world datasets and used to train the YOLOv11 algorithm. The study revealed that models trained with GAN-augmented datasets achieved a 15% improvement in recall and a 10% reduction in false positives compared to models trained on real-world data alone.

15.3 Development of UAV-Based Object Detection

With the emergence and rapid advancement of UAVs, the application of object detection has expanded significantly, taking on entirely new dimensions. UAVs can be categorized into various types, such as fixed-wing, rotary-wing, and flapping-wing, based on criteria including payload capacity, operational altitude, endurance, and flight range. Among these, fixed-wing and rotary-wing UAVs are the most widely used today. Fixed-wing UAVs are notable for their high-speed flight capabilities, extended range, ability to reach high altitudes, and capacity to carry larger, heavier payloads. On the other hand, rotary-wing UAVs, which are widely employed in object detection studies, excel in vertical takeoff and landing, low-altitude flight, and agile maneuverability, particularly in challenging terrain. Their ability to hover and maintain high endurance for low-altitude and stationary flight missions makes them invaluable in applications such as smart cities, intelligent transportation systems, agriculture, disaster recovery, and industrial operations. Equipped with high-resolution cameras and advanced sensors, rotary-wing UAVs capture clear and meaningful object data, even in hazardous or hard-to-reach locations. These UAVs operate effectively at various altitude levels, including 0–5 m (eye-level view), 5–120 m (low-to-medium altitude), and 120 m and above (high altitude), depending on the specific objectives of a study. The integration of specialized sensors allows UAVs to perform complex tasks with high precision and efficiency.

Visual sensors integrated into UAV systems provide detailed object imagery and kinematic data by leveraging photoelectric technology. The functionality and performance of these sensors have been significantly enhanced by advancements in deep learning algorithms, enabling more accurate object detection and analysis. Additionally, ultrasonic sensors are incorporated into UAV systems to measure the distance and velocity of objects. These sensors emit high-frequency sound waves (20–40 kHz, beyond the range of human hearing), and the reflected waves are captured after bouncing off objects. By calculating the time taken for the sound waves to return, the distance to the object can be determined with high accuracy. The integration of visual and ultrasonic sensors in UAV systems enables precise determination of the kinematic properties of flying objects, such as unidentified UAVs, including their speed and flight distance. These capabilities highlight the versatility and effectiveness of UAVs in performing a wide range of tasks across various fields.

15.3.1 Challenges in Object Detection with UAV Systems

One of the most critical factors in object detection with fixed or stationary camera systems is the consistent side profile of the object in the captured images. When using fixed cameras to observe a predefined area, objects consistently appear from the same angle, with the background remaining static. Moving objects are then identified within this fixed scene. In general object detection, with a stationary camera, light changes and image noise are typically minimal. However, the situation is quite different with UAV systems. In UAV-based object detection, the fast and agile movements of the UAV, influenced by weather conditions, can lead to challenges such as image shaking, noise, sudden changes in lighting, and blurring. Additionally, since UAVs often capture images from high altitudes, objects appear smaller, and the density of objects is not evenly distributed. The ratio of the area covered by objects in images taken from high altitudes compared to those captured by cameras fixed at a maximum height of 5 m is considerably lower. This reduced object-to-background ratio results in object distortion and makes detecting multiple objects more challenging. Moreover, the masking techniques used in general object detection systems are less effective when compared to UAV systems. As a result, UAV images often have a low proportion of the object masked, which further complicates detection. UAV images are particularly prone to masking issues, where larger objects such as trees or buildings can obscure the target object. The varying types of masking encountered in different scenarios make object detection with UAV systems significantly more difficult.

15.4 Test Dataset

In this study, ensuring the diversity of images in the dataset for unidentified UAV detection is critical to enhancing the robustness and performance of the algorithm in real-world scenarios. The study specifically focuses on detecting rotary-wing UAV models. For this task, the high-performance SJRC F11 Pro drone has been upgraded with a more advanced propeller to ensure stable and high-performance flight capabilities. To obtain high-quality and comprehensive image sets, a Raspberry Pi HQ Camera system, featuring a 4056 × 3040 pixel sensor

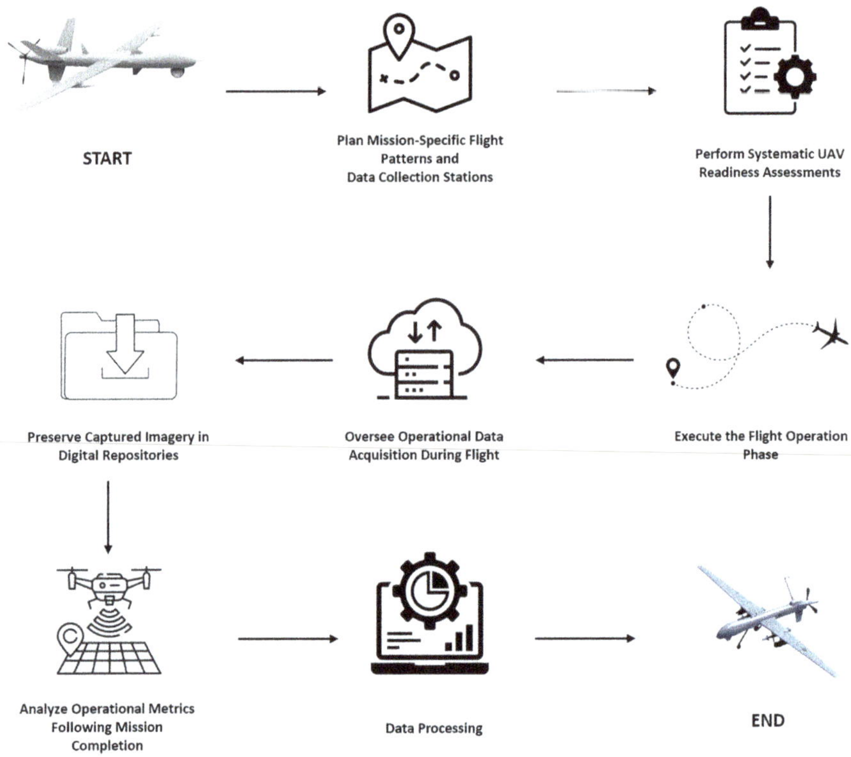

Fig. 15.1 Task flow chart illustrating the data collection process for unidentified UAV detection, including image diversity enhancement through rotary-wing UAV detection, advanced Raspberry Pi HQ Camera integration for high-resolution image capture, and the use of ultrasonic sensors for precise distance measurements

resolution, a 6.287 mm × 4.712 mm image field, and a 1/2.3" optical size, has been integrated into the commercial drone. Additionally, ultrasonic sensors were installed to measure the distance between the unidentified UAV and the detection system. A reliable dataset was created by extracting individual frames from detailed and accurate UAV image videos, captured during customized UAV flight missions. With an approximate flight time of 29 min on a single battery charge, the drone provided sufficient time for conducting a series of missions. A task flow chart outlining the data collection process is provided in Fig. 15.1.

The dataset images include a wide range of visual contexts to simulate realistic UAV detection conditions. The UAV-captured images primarily feature rotary-wing UAVs appearing at various distances and angles, including top-down views, oblique shots, and low-altitude captures taken during horizontal maneuvers. These images often include diverse backgrounds such as urban rooftops, open fields, tree canopies, and industrial zones to ensure variability. Lighting conditions range from overcast skies and early morning light to direct sunlight and shadowed environments, contributing to the robustness of the dataset. The supplementary web-based images feature both fixed-wing and rotary-wing UAVs in flight, often annotated under varied geographic and weather conditions. These images were carefully selected to enrich the dataset's visual diversity and to ensure that the detection algorithm could generalize across different UAV appearances and operational contexts.

UAV systems offer various solutions for the challenges encountered during unidentified UAV detection through aerial imaging. In manual flight mode, the added error margin from human factors complicates the process. To minimize human errors and prevent confusion, diverse flight modes were used. The autonomous driving mode enabled more comprehensive flight capabilities, allowing the UAV to capture images from multiple angles with agile and systematic maneuvers. The algorithm's training and testing were conducted using scenarios that closely replicate real-world situations, with images captured at different camera orientations and altitudes. By diversifying UAV flight parameters, the possibility of human error was minimized, resulting in a well-rounded and diverse dataset. The autonomous operation of the UAV also ensured accurate data collection in varying weather conditions, allowing for the precise capture of unidentified UAV images.

The selection of datasets in this study was guided by the need to simulate real-world aerial surveillance scenarios as closely as possible. The UAV-captured images provided high-resolution data from authentic aerial perspectives, enabling the detection of small and partially visible UAVs under realistic flight conditions. These were supplemented with web-sourced images selected to reflect a variety of environments and UAV configurations that are commonly encountered in surveillance and airspace management contexts. Although there are several benchmark datasets available, such as anti-UAV or VisDrone, they are often limited by ground-based perspectives or fixed-camera viewpoints. Our dataset design fills this gap by focusing specifically on UAV-to-UAV detection, which is a relatively underexplored but increasingly relevant application in both civilian and security domains.

15.4.1 Avionics System

The commercial drone used in this UAV detection study is equipped with various high-performance components, including advanced cameras, mission computers, and a more powerful battery system to support demanding tasks, as detailed in the avionics schematics shown in Fig. 15.2. With a 22,000 mAh battery, the drone can operate for approximately 20 min, either autonomously or under manual control. It is fitted with a Raspberry Pi 5-based mission computer, equipped with a high-performance Raspberry Pi HQ camera for image processing in unidentified UAV detection. The Raspberry Pi HQ camera features an 8-megapixel sensor, supporting dual 4Kp60 HDMI image mode and HDR through Raspberry Pi 5, optimizing detection capabilities. The drone also has a dynamic camera system that allows image transmission from various angles. The system runs on Ubuntu 20.04, tailored to meet the computational needs of the Raspberry Pi 5 mission computer. Additionally, the Pixhawk 2.4.8 flight controller, with adequate computing power,

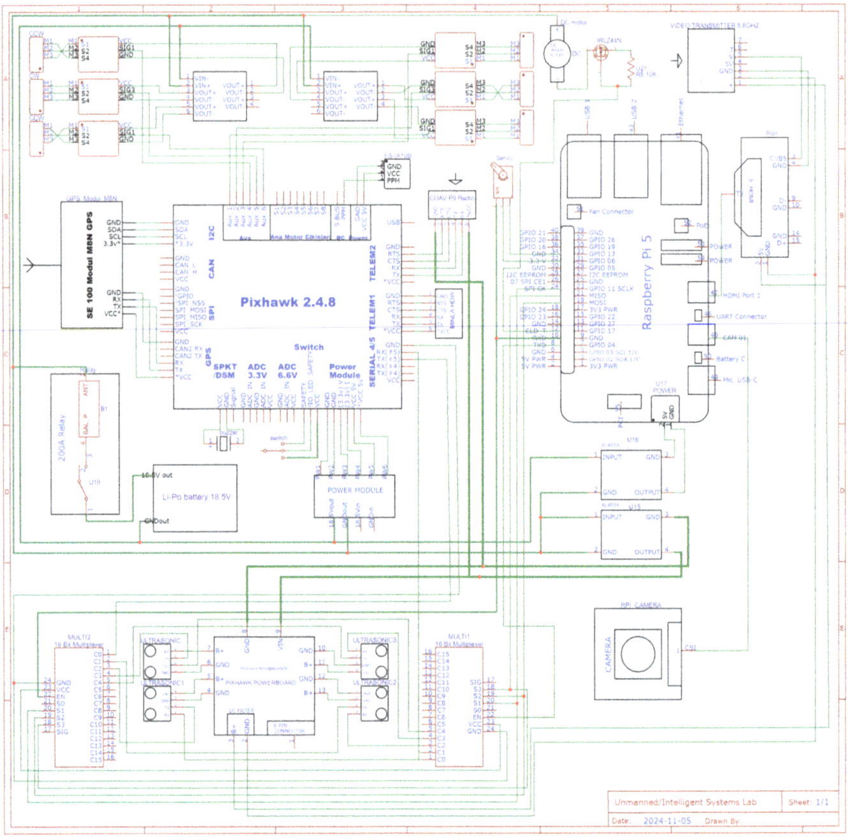

Fig. 15.2 Avionics schematic of the commercial drone used in the UAV detection study, showcasing its high-performance components. Key features include a 22,000 mAh battery providing approximately 20 min of operational time, a Raspberry Pi 5-based mission computer with an 8-megapixel Raspberry Pi HQ camera for image processing, and a Pixhawk 2.4.8 flight controller managing both autonomous and manual flight operations. The system, running on Ubuntu 20.04, supports advanced detection capabilities, including dual 4Kp60 HDMI image mode and HDR for improved UAV detection performance

is used to manage both autonomous and manual flight operations, while also transferring data from the camera systems to the flight computer.

15.4.2 Data Augmentation

Expanding the dataset through various operations is essential for enhancing the model's robustness and its ability to withstand extreme conditions. In this study, the primary images used for training and testing the algorithm were supplemented

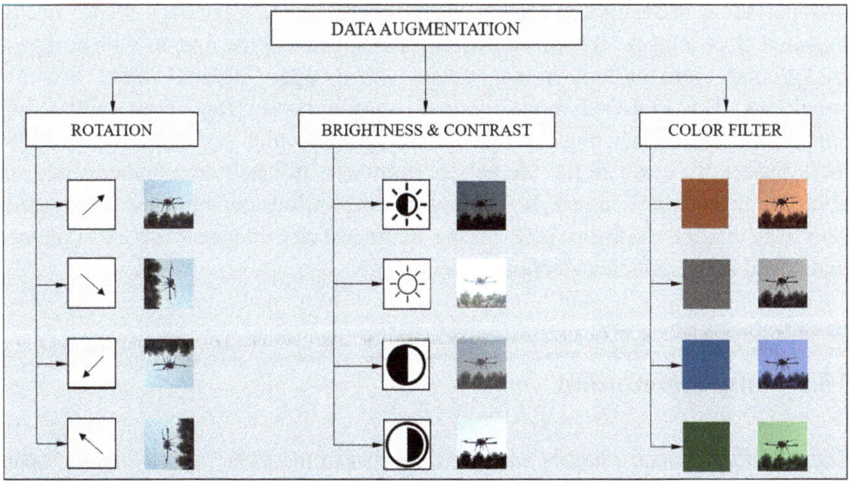

Fig. 15.3 Illustration of dataset enhancement operations, including geometric transformations such as rotation to simulate UAV maneuvering, brightness and contrast adjustments for varying light conditions, and color filtering to account for extreme environmental variations. The dataset incorporates both UAV-captured images and real-world scenarios from web-based sources to ensure robustness

with images downloaded from web-based sources that reflect real-world scenarios, in addition to the UAV-captured images obtained during autonomous flights. This approach is crucial for achieving a comprehensive dataset, ensuring it covers diverse scenarios. To further enhance the dataset, several transformations and filtering operations were applied to simulate potential extreme conditions, as shown in Fig. 15.3. For example, rotation operations were applied to the images to introduce new geometric variations, allowing the algorithm to better adapt to different perspectives of the unidentified UAV during the UAV's maneuvering movements. Additionally, brightness and contrast adjustments of up to 50% were performed to ensure the UAV camera's resilience under varying light conditions. To account for extreme environmental variations, different color filters were also applied, ensuring that the algorithm performs well across a broad spectrum of real-world lighting conditions.

15.4.3 Final Dataset

The dataset was refined through the application of various filtering techniques, along with adjustments to brightness, contrast, and color settings, to images captured at different altitudes, camera angles, and times of the day. This refinement and the enrichment of the training dataset with diverse images significantly enhance the algorithm's resilience to harsh conditions. As a result, 70% of the 5875 images in the final dataset were designated for training purposes. The training dataset specifically included UAV images taken from long distances, challenging images with complex backgrounds, and difficult-to-distinguish UAV images captured under overcast conditions. The YOLO11 model, trained on this dataset containing challenging conditions, was validated using a validation dataset, which constituted 10% of the total dataset. To evaluate the model's performance in harsh environments and its ability to detect UAVs, a test dataset representing various conditions was prepared. This test dataset made up 20% of the entire dataset and was used to conduct unmanned aerial vehicle detection tests.

15.5 Implementation

The YOLO11 model, a highly advanced member of the YOLO family, builds upon the strengths of its predecessor models, positioning itself as one of the most popular and high-performing options in real-time object detection. Its remarkable accuracy and efficiency make it particularly promising for aerial surveillance applications. YOLO11 optimizes detection performance by striking a balance between inference cost and inference time, which enhances its capability to perform well in complex scenarios. Through innovations in its backbone network architecture, the algorithm is able to overcome limitations related to object orientation, scale, and size, making it versatile for various applications and reducing constraints. The key criterion in aerial surveillance is the ability of an algorithm to maintain detection accuracy while operating at high speed. The medium-sized YOLO11m model, for example, reduces training time by 22% compared to its predecessor, YOLOv8m, thanks to its smaller number of parameters, while still maintaining high detection sensitivity. Additionally, YOLO11 delivers a 2% faster average inference time compared to its predecessor, YOLO10. These performance improvements, coupled with a smaller model size, make YOLO11 critical for real-world aerial image analysis. Figure 15.4 represents the overall workflow of our approach, including the UAV-mounted detection system, the data acquisition process, augmentation methods, YOLO11-based model training and evaluation pipeline, and the performance comparison with previous versions.

YOLO11 introduces notable enhancements in its network architecture, optimizing both speed and accuracy. The backbone network, a crucial processing unit within the YOLO family, employs an advanced feature extraction method based on EfficientNet or CSPNet neural networks. As shown in Fig. 15.5, the architecture incorporates two new modules, C3K2 and C2PSA, that significantly enhance object detection capabilities compared to previous versions. The C3K2 module, derived from the C2f module in YOLOv8, offers variable configurability via the c3k parameter. When activated, the c3k module taps into the advanced feature extraction capabilities of the C3 structure, while preserving the original bottleneck structure by operating in parallel with the previous module from YOLOv8. Another key innovation, the C2PSA module, introduces the Location-Aware Attention

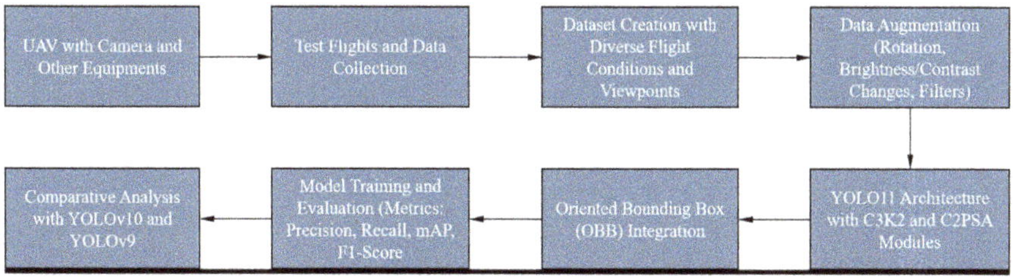

Fig. 15.4 Overview of the proposed UAV-based object detection pipeline using YOLO11. The diagram illustrates the end-to-end workflow, including UAV data acquisition, augmentation strategies, YOLO11 enhancements, OBB integration, and performance evaluation

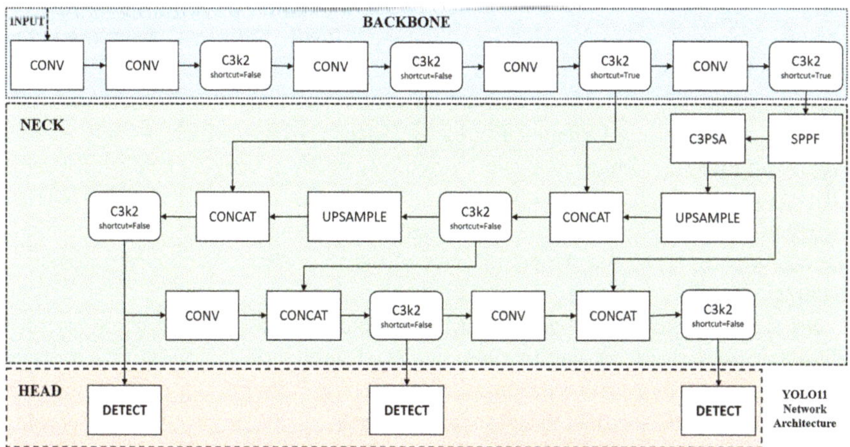

Fig. 15.5 Schematic of the YOLO11 architecture, highlighting the integration of the C3K2 and C2PSA modules for enhanced object detection. The C3K2 module, adapted from the C2f module in YOLOv8, improves feature extraction through parallel bottleneck structures, while the C2PSA module incorporates Location-Aware Attention to scale semantic information and refine spatial relationships. Features extracted by the backbone are processed through the neck and head for precise object detection across various scales. The architecture ensures robustness in complex backgrounds, improved detection of small objects, and optimized computational efficiency, supporting five sub-models tailored for diverse real-time applications

(PSA) mechanism to the C2f module, which enhances the model's ability to scale semantic information in input features and distinguish spatial relationships. The PSA mechanism improves YOLO11's robustness in images with complex backgrounds and boosts detection performance through a multi-headed attention and feedforward network structure. The extracted features in YOLO11's spine are then passed to the neck, where they are combined to focus on objects of various sizes, both near and far, in the image. The YOLO11 head generates semantic information from these features and determines not only the presence of objects but also their precise locations. YOLO11's optimized architecture ensures that even small objects are not overlooked, even in complex images. Furthermore, its reduced computational costs make it suitable for more complex tasks and larger projects. For users requiring different model sizes, YOLO11 offers five sub-models, YOLO11n, YOLO11s, YOLO11m, YOLO11l, and YOLO11x, each optimized for speed and performance, ranging from small to large models tailored for real-time applications.

15.5.1 YOLO11 and Oriented Bounding Box

In aerial images, objects are often scattered across the image and can appear in arbitrary orientations. Previous models in the YOLO family, along with other object detection algorithms, typically use horizontal bounding boxes to surround detected objects. However, in UAV-captured aerial images, using horizontal bounding boxes for objects with varied orientations can lead to confusion and hinder detection accuracy, especially for objects that are not clearly visible. YOLO11 addresses this issue by incorporating Oriented Bounding Boxes (OBBs), which provide a more precise approach for image analysis, particularly in aerial imaging studies. By using OBBs, the model is better equipped to handle rotated or irregularly oriented objects. OBBs not only define the center coordinates of the bounding box but also include the rotation angle (θ) and the coordinates of the four corners, enhancing detection accuracy for rotated objects.

As shown in Fig. 15.6, parts a and b illustrate the challenges of using oriented bounding boxes with an intuitive approach. At angles of 45° and 135°, oriented bounding boxes experience the highest uncertainty between adjacent points. For example, in Fig. 15.6a, the OBB is expressed as (x_1, y_1, x_2, y_2, x_3, y_3, x_4, y_4) using a point-based representation. In contrast, the OBB in Fig. 15.6b should be expressed as (x_2, y_2, x_3, y_3, x_4, y_4, x_1, y_1).

Although the θ-based, point-based, and h-based OBB representations in Fig. 15.6 differ structurally, their feature maps are largely similar. However, due to the inherent uncertainty when handling OBBs, training convergence can be more challenging compared to horizontal bounding boxes, and horizontal bounding boxes generally achieve better mAP (mean average precision), even when the same backbone network is used. In the figure, the θ-based OBB is represented as (c_x, c_y, h, w, θ), where (c_x, c_y) are the center coordinates, and (h, w, θ) represent the height, width, and rotation angle, respectively. For the point-based OBB, the red-marked discontinuity point represents the one closest to the upper right corner. In cases where $l_1 < l_2$, the point-based OBB is expressed as (x_1, y_1, x_2, y_2, x_3, y_3, x_4, y_4), whereas for $l_1 > l_2$, the expression becomes (x_4, y_4, x_1, y_1, x_2, y_2, x_3, y_3). For both h-based and θ-based OBBs, 45° and 135° angles are identified as points of discontinuity, as shown in the figure.

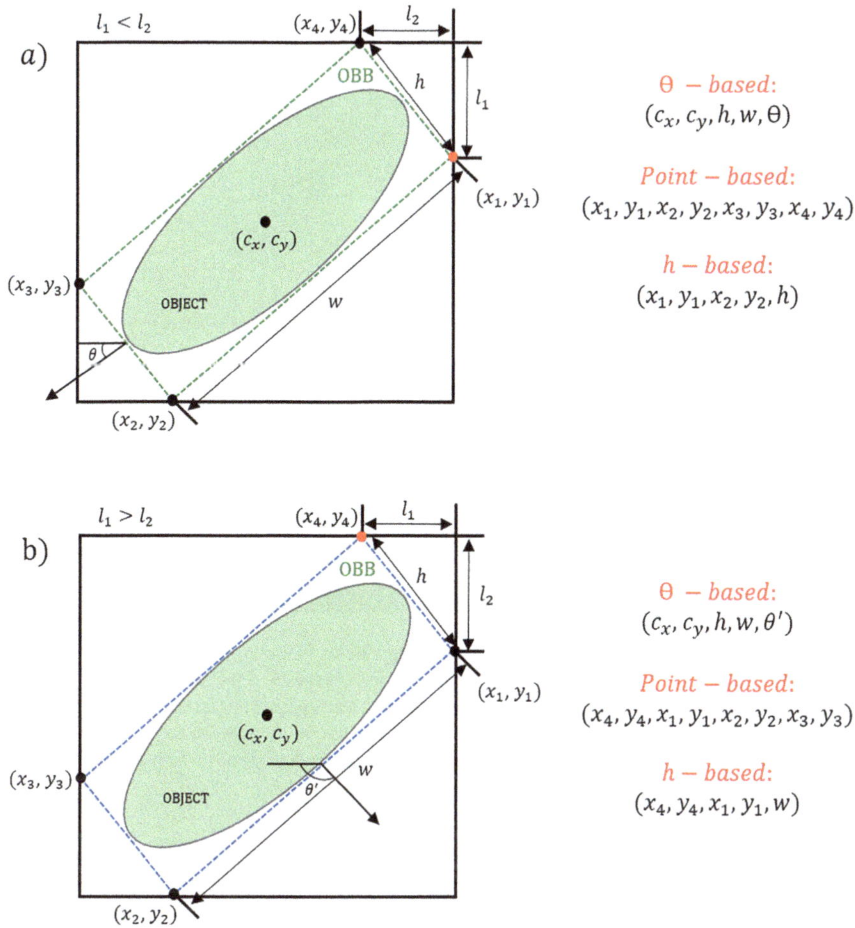

Fig. 15.6 Challenges in handling oriented bounding boxes (OBBs) illustrated with point-based, h-based, and θ-based representations. (**a**) and (**b**) demonstrate the uncertainty in point-based OBBs at angles of 45° and 135°, where adjacent points experience maximum ambiguity. In (**a**), the OBB is represented as $(x_1, y_1, x_2, y_2, x_3, y_3, x_4, y_4)$, while in (**b**), it is expressed as $(x_2, y_2, x_3, y_3, x_4, y_4, x_1, y_1)$. Figure adapted from [44]

15.5.2 Evaluation Metrics

In deep learning-based studies, both regression and classification tasks are critical to address. For classification problems, a common validation metric is derived from the confusion matrix, which compares the actual and predicted outcomes. A two-by-two matrix, along with additional metrics, is used to represent single-class classification problems. The True Positive (TP), False Negative (FN), and False Positive (FP) values, which are part of the calculations for Accuracy, Precision, Recall, F1 score,

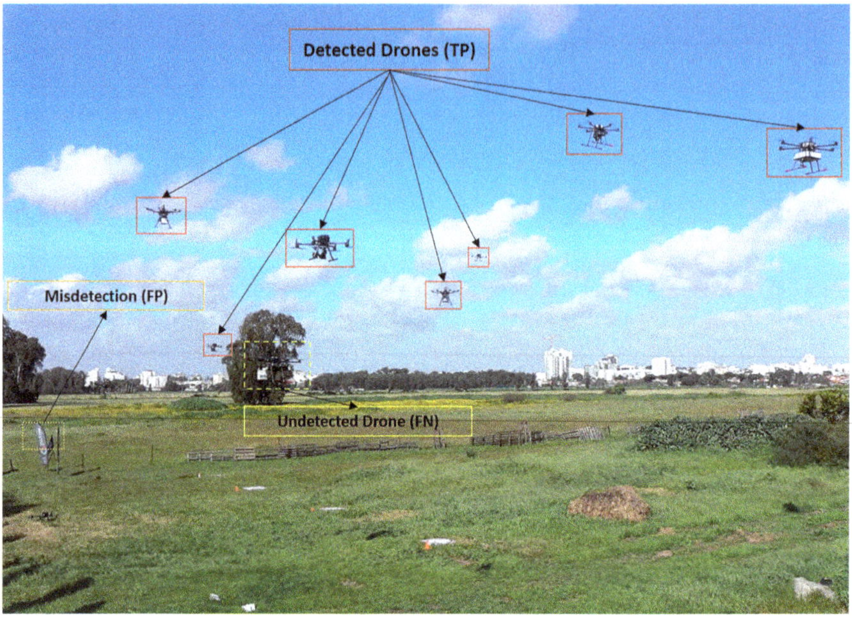

Fig. 15.7 Visualization of the confusion matrix components (True Positive, False Negative, and False Positive) used to calculate key evaluation metrics, Accuracy, Precision, Recall, F1 Score, and mAP, critical for assessing the performance of deep learning models in UAV detection tasks. The figure highlights the importance of optimizing the F1 score to balance the detection of true positives and the minimization of false positives, especially when dealing with partially visible unidentified UAVs

and mAP (as shown in Eqs. (15.1), (15.2), (15.3), (15.4) and (15.5)), are visualized in Fig. 15.7. In unidentified UAV detection, the primary challenge is ensuring that the model is both fast and robust enough to detect unidentified UAVs as soon as they appear, even when they are only partially visible. In other words, the model should aim to maximize the number of true positives and minimize false positives. To achieve a balance between these two factors, the F1 score is often used, as it harmonically combines both precision and recall.

$$\text{Accuracy} = \frac{\sum \text{TP} + \sum \text{TN}}{\sum \text{TP} + \sum \text{TN} + \sum \text{FP}} \tag{15.1}$$

$$\text{Precision} = \frac{\sum \text{TP}}{\sum \text{TP} + \sum \text{FP}} \tag{15.2}$$

$$\text{Recall} = \frac{\sum \text{TP}}{\sum \text{TP} + \sum \text{FN}} \tag{15.3}$$

$$\text{F1 Score} = \frac{2 \text{ x} \sum \text{TP}}{2 \text{ x} \sum \text{TP} + \sum \text{FP} + \sum \text{FN}} \tag{15.4}$$

$$\text{mAP} = \frac{\sum_{k=1}^{K} \text{Average Precision (k)}}{\text{K}} \tag{15.5}$$

15.5.3 Model Training

Training for the YOLO11 model, as well as the YOLOv10 and YOLOv9 models used for benchmarking, was conducted on Google Colab. Google Colab provides cloud-based access to GPU and TPU accelerators, which significantly enhance processing power compared to standard laptops. The platform also supports a Jupyter notebook environment and Python libraries, facilitating model development. To reduce processing load, the training images were resized to 640 × 640 pixels, ensuring faster processing by standardizing the input size. The model was trained using an NVIDIA RTX 3090 GPU, which is capable of handling complex image processing tasks, and an Intel Xeon Silver 4212 CPU for intensive data processing. The training process utilized the Stochastic Gradient Descent (SGD) optimization algorithm with a momentum of 0.935 and weight decay set to 0.0005. The initial learning rate was 0.001, and the batch size was 16. The models were trained for 150 epochs, with an early stopping mechanism based on validation loss to prevent overfitting. Data preprocessing included image normalization, and label formatting was aligned with YOLO-specific bounding box annotations. During training, random horizontal flipping and scaling were also applied as part of online augmentation. The training configuration was empirically optimized for UAV detection scenarios to balance convergence speed and detection accuracy, especially for small and partially occluded objects. The dataset was split into 70% training, 10% validation, and 20% testing, maintaining a consistent evaluation framework across models. Model performance was evaluated using standard object detection metrics, including Precision, Recall, F1 Score, and mAP, computed on the test dataset with COCO-style evaluation methodology. Additionally, all training sessions were repeated three times using different random seeds to ensure consistency and minimize performance variability due to initialization.

15.6 Results

The detection results obtained using the YOLO11 algorithm demonstrate its robustness and reliability in detecting both fixed-wing and rotary-wing UAVs. Some visual examples of UAV detection are shown in Figs. 15.8 and 15.9. Moreover, comparative performance metrics are compiled in Table 15.1. For fixed-wing UAV

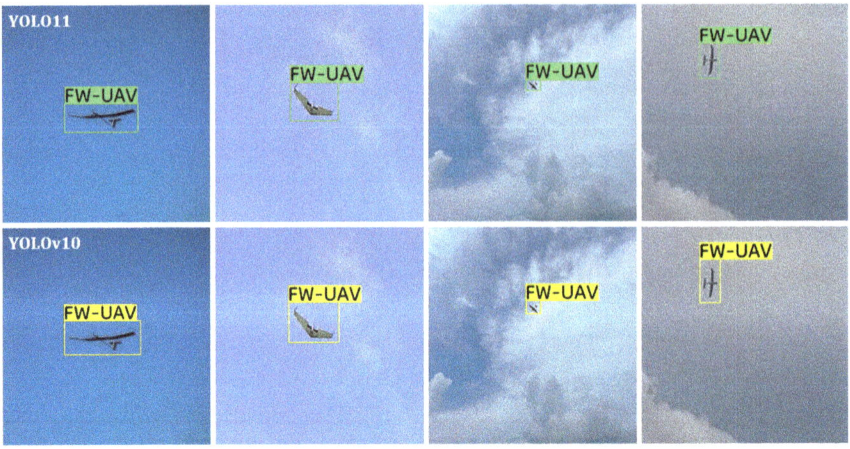

Fig. 15.8 Visual example of UAV detection results using the YOLO11 algorithm for fixed-wing UAVs. The image demonstrates the algorithm's effectiveness in detecting fixed-wing UAVs with high precision and recall, as reflected in the performance metrics

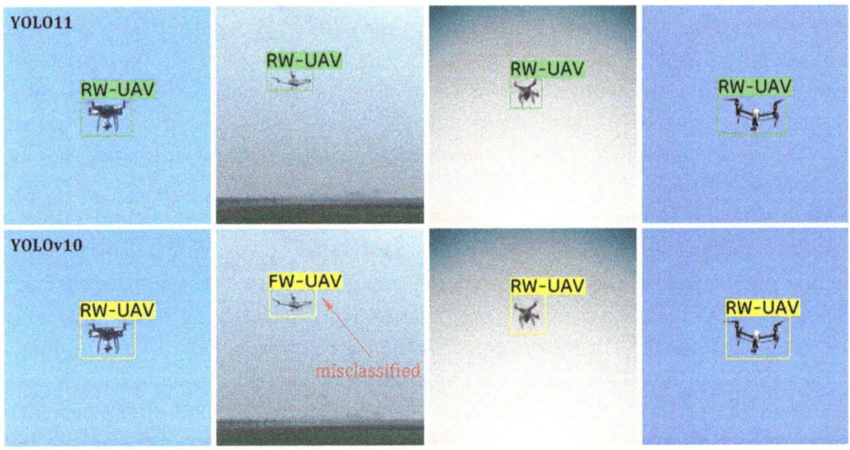

Fig. 15.9 Visual example of UAV detection results using the YOLO11 algorithm for rotary-wing UAVs. The image highlights the algorithm's robustness in detecting rotary-wing UAVs with minimal false alarms, showcasing its ability to differentiate between UAV types effectively

detection, YOLO11 achieved a precision of 0.854, recall of 0.861, mAP of 0.926, and an F1 score of 0.857. These metrics indicate a balanced performance between high accuracy and precision, and recall, reflecting the algorithm's ability to effectively detect fixed-wing UAVs while minimizing false positives and false negatives. For rotary-wing UAV detection, YOLO11 delivered even better performance, with a precision of 0.925, recall of 0.901, mAP of 0.904, and an F1 score of 0.913. The higher precision value indicates YOLO11's capability to detect rotary-wing

Table 15.1 Performance metrics for UAV detection using the YOLO11 algorithm

		Performance metrics			
Model	UAV type	Precision	Recall	mAP	F1-Score
YOLO11	Fixed-W.	**0.854**	**0.861**	**0.926**	**0.857**
	Rotary-W.	0.925	0.901	0.904	0.913
	Overall	0.889	0.881	0.915	0.885
YOLOv10	Fixed-W.	0.808	0.793	0.798	0.800
	Rotary-W.	0.919	0.892	**0.906**	0.905
	Overall	0.863	0.842	0.852	0.852
YOLOv9	Fixed-W.	0.837	0.855	0.847	0.846
	Rotary-W.	**0.929**	**0.903**	0.918	**0.916**
	Overall	0.883	0.879	0.882	0.881

The table presents the precision, recall, mAP, and F1 score for both fixed-wing and rotary-wing UAV detection, highlighting YOLO11's balanced performance in detecting fixed-wing UAVs and its robustness in minimizing false alarms for rotary-wing UAV detection. These metrics underscore the algorithm's reliability and effectiveness in accurately distinguishing between different UAV types

UAVs with minimal false alarms, which is critical in applications where accurate differentiation between UAV types is crucial.

To draw clearer conclusions about YOLO11's performance, tests were also conducted with the previous two versions, YOLOv10 and YOLOv9. Compared to YOLOv10, YOLO11 demonstrated significant improvements in detection performance, especially for fixed-wing UAV detection. For YOLOv10, precision was 0.808, recall 0.793, mAP 0.798, and F1 score 0.800. These values were consistently lower than those obtained with YOLO11, which highlights YOLO11's improved accuracy and robustness for this UAV type. The most notable improvement was observed in mAP, where YOLO11 outperformed YOLOv10 by a margin of 12.8%, reflecting its superior object detection capabilities. For rotary-wing UAV detection, YOLOv10 achieved a precision of 0.919, recall of 0.892, mAP of 0.906, and an F1 score of 0.905. While these results are competitive and underscore YOLOv10's effectiveness, YOLO11 still demonstrated a slight advantage, particularly in the F1 score (0.913 compared to 0.905), and maintained a more balanced performance across all evaluation metrics.

Detection results obtained with YOLOv9 displayed mixed performance when compared to YOLO11. For fixed-wing UAV detection, YOLOv9 achieved a precision of 0.837, recall of 0.855, mAP of 0.847, and an F1 score of 0.856. While these values are relatively high and competitive, YOLO11's higher mAP (0.926) and precision (0.854) further confirm its superiority in detecting fixed-wing UAVs. Improvements in precision and mAP are especially critical for reducing false positives and enhancing overall detection accuracy in complex environments. For rotary-wing UAV detection, YOLOv9 achieved a precision of 0.929, recall of 0.903, mAP of 0.918, and an F1 score of 0.916. These results are comparable to those of YOLO11, with YOLOv9 slightly outperforming YOLO11 in precision (0.929 vs.

0.925) and mAP (0.918 vs. 0.904). However, YOLO11 exhibited a higher F1 score (0.913 vs. 0.916), demonstrating a more balanced performance, making it a more versatile option for real-world applications that require consistent results across multiple metrics.

The results clearly highlight the advancements introduced in YOLO11, particularly in the detection of fixed-wing UAVs, where it significantly outperforms YOLOv10 and YOLOv9. The improvements in mAP and F1 scores likely reflect the updated network architecture and optimized training procedures, underscoring YOLO11's enhanced feature extraction and classification capabilities. These qualities make YOLO11 a reliable algorithm for applications requiring high accuracy and minimal error. In the case of rotary-wing UAV detection, YOLO11's performance closely matches that of YOLOv9, with small differences in precision and F1 score. This suggests that while YOLO11 offers overall improvements, certain aspects of rotary-wing UAV detection may already be near-optimal in YOLOv9. The results also suggest that both YOLO11 and YOLOv9 are highly capable at distinguishing rotary-wing UAVs, with YOLO11 providing slightly more balanced output.

When comparing YOLOv10 with the other two versions, its performance, especially for fixed-wing UAV detection, appears less robust. The lower precision and mAP values highlight its limitations in accurately identifying UAVs, which may be attributed to less advanced feature extraction mechanisms and training methodologies. Ultimately, YOLO11 emerges as the most reliable and versatile algorithm for UAV detection, offering significant improvements over YOLOv10 and maintaining competitive performance against YOLOv9. Its ability to provide high precision and balanced results makes it suitable for various aerial surveillance applications. Future research could explore further optimizations to YOLO11's architecture to enhance rotary-wing UAV detection performance and extend its applicability to other object detection tasks.

15.7 Conclusion

This study set out to address the challenges of real-time UAV detection in complex and dynamic aerial environments, particularly for fixed-wing and rotary-wing UAVs, using a deep learning-based approach. By leveraging data enrichment techniques, a high-resolution UAV-mounted camera system, and the advanced YOLO11 detection architecture, including its oriented bounding box integration and C3K2/C2PSA modules, we aimed to develop a system that offers improved precision, robustness, and adaptability. The results confirm that the YOLO11 model not only outperforms earlier versions in most key metrics but also maintains consistent performance across varying UAV types and conditions. Thus, the study successfully meets its objectives by delivering a highly accurate and reliable detection framework that contributes practical value to the fields of aerial surveillance, airspace safety, and intelligent monitoring.

The results of this study demonstrate the effectiveness of the YOLO11 algorithm in UAV detection tasks, surpassing its predecessors, YOLOv10 and YOLOv9, in terms of precision, recall, mAP, and F1 scores for both fixed-wing and rotary-wing UAVs. YOLO11 exhibited superior detection capabilities, particularly in situations demanding high precision and adaptability to varying UAV configurations. The findings underscore its potential for real-world applications in air surveillance, airspace management, and UAV activity monitoring. A comprehensive analysis revealed that YOLO11 achieved a robust balance between detection accuracy and reliability, marking significant improvements over YOLOv10. While YOLOv10 remains applicable, it shows limitations in detecting fixed-wing UAVs. YOLOv9 performed well in detecting rotary-wing UAVs, but could not match the consistent performance of YOLO11. These results highlight the progress in deep learning-based object detection algorithms and their essential role in enhancing UAV detection systems. The study confirms the applicability of YOLO11 in real-world contexts while offering valuable insights into the strengths and limitations of previous versions.

References

1. Ryan, A., Zennaro, M., Howell, A., Sengupta, R., Hedrick, J.K.: An overview of emerging results in cooperative UAV control. In: 2004 43rd IEEE Conference on Decision and Control (CDC), pp. 602–607 (2004). https://doi.org/10.1109/CDC.2004.1428700
2. Shakeri, R., et al.: Design challenges of multi-UAV systems in cyber-physical applications: a comprehensive survey and future directions. IEEE Commun. Surv. Tutor. **21**(4), 3340–3385 (2019). https://doi.org/10.1109/COMST.2019.2924143
3. Samad, A.M., Kamarulzaman, N., Hamdani, M.A., Mastor, T.A., Hashim, K.A.: The potential of Unmanned Aerial Vehicle (UAV) for civilian and mapping application. In: 2013 IEEE 3rd International Conference on System Engineering and Technology, pp. 313–318 (2013). https://doi.org/10.1109/ICSEngT.2013.6650191
4. Motlagh, N.H., Bagaa, M., Taleb, T.: UAV-Based IoT platform: a crowd surveillance use case. IEEE Commun. Mag. **55**(2), 128–134 (2017). https://doi.org/10.1109/MCOM.2017.1600587CM
5. Fotouhi, A., et al.: Survey on UAV cellular communications: practical aspects, standardization advancements, regulation, and security challenges. IEEE Commun. Surv. Tutor. **21**(4), 3417–3442 (2019). https://doi.org/10.1109/COMST.2019.2906228
6. Thammawichai, M., Baliyarasimhuni, S.P., Kerrigan, E.C., Sousa, J.B.: Optimizing communication and computation for multi-UAV information gathering applications. IEEE Tran. Aerosp. Electron. Syst. **54**(2), 601–615 (2018). https://doi.org/10.1109/TAES.2017.2761139
7. Gupta, L., Jain, R., Vaszkun, G.: Survey of important issues in UAV communication networks. IEEE Commun. Surv. Tutor. **18**(2), 1123–1152 (2016). https://doi.org/10.1109/COMST.2015.2495297
8. Naidoo, Y., Stopforth, R., Bright, G.: Development of an UAV for search & rescue applications. IEEE Afr. **11**, 1–6 (2011). https://doi.org/10.1109/AFRCON.2011.6072032
9. Aggarwal, S., Kumar, N., Tanwar, S.: Blockchain-envisioned UAV communication using 6G networks: open issues, use cases, and future directions. IEEE Internet Things J. **8**(7), 5416–5441 (2021). https://doi.org/10.1109/JIOT.2020.3020819

10. Bakirci, M.: Vehicular mobility monitoring using remote sensing and deep learning on a UAV-based mobile computing platform. Measurement. **244**(116579) (2025). https://doi.org/10.1016/j.measurement.2024.116579
11. Luo, C., Nightingale, J., Asemota, E., Grecos, C.: A UAV-cloud system for disaster sensing applications. In: 2015 IEEE 81st Vehicular Technology Conference (VTC), pp. 1–5 (2015). https://doi.org/10.1109/VTCSpring.2015.7145656
12. Zhong, Y., et al.: Mini-UAV-borne hyperspectral remote sensing: from observation and processing to applications. IEEE Geosci. Remote Sensing Mag. **6**(4), 46–62 (2018). https://doi.org/10.1109/MGRS.2018.2867592
13. Pinkney, M.F.J., Hampel, D., DiPierro, S.: Unmanned aerial vehicle (UAV) communications relay. In: Proceedings of MILCOM '96 IEEE Military Communications Conference, pp. 47–51 (1996). https://doi.org/10.1109/MILCOM.1996.568581
14. Kamal, W.A., Samar, R.: A mission planning approach for UAV applications. In: 2008 47th IEEE Conference on Decision and Control, pp. 3101–3106 (2008). https://doi.org/10.1109/CDC.2008.4739187
15. Chen, W., Liu, B., Huang, H., Guo, S., Zheng, Z.: When UAV swarm meets edge-cloud computing: the QoS perspective. IEEE Netw. **33**(2), 36–43 (2019). https://doi.org/10.1109/MNET.2019.1800222
16. Xie, L., Cao, X., Xu, J., Zhang, R.: UAV-enabled wireless power transfer: a tutorial overview. IEEE Trans. Green Commun. Netw. **5**(4), 2042–2064 (2021). https://doi.org/10.1109/TGCN.2021.3093718
17. Zhiqiang Wu, H. Kumar, A. Davari, Performance evaluation of OFDM transmission in UAV wireless communication. In: Proceedings of the Thirty-Seventh Southeastern Symposium on System Theory, 2005. SSST '05., Tuskegee, AL, USA, 2005, pp. 6–10. doi: https://doi.org/10.1109/SSST.2005.1460867
18. Khuwaja, A.A., Chen, Y., Zhao, N., Alouini, M.S., Dobbins, P.: A survey of channel modeling for UAV communications. IEEE Commun. Surv. Tutor. **20**(4), 2804–2821 (2018). https://doi.org/10.1109/COMST.2018.2856587
19. Erdos, D., Erdos, A., Watkins, S.E.: An experimental UAV system for search and rescue challenge. IEEE Aerosp. Electron. Syst. Mag. **28**(5), 32–37 (2013). https://doi.org/10.1109/MAES.2013.6516147
20. Shiau, J.K., Ma, D.M., Yang, P.Y., Wang, G.F., Gong, J.H.: Design of a solar power management system for an experimental UAV. IEEE Trans. Aerosp. Electron. Syst. **45**(4), 1350–1360 (2009). https://doi.org/10.1109/TAES.2009.5310303
21. Paikrao, P., Routray, S., Mukherjee, A., Khan, A.R., Vohnout, R.: Consumer personalized gesture recognition in UAV-based industry 5.0 applications. IEEE Trans. Consumer Electron. **69**(4), 842–849 (2023). https://doi.org/10.1109/TCE.2023.3308209
22. Xu, X., et al.: DAC-SDC low power object detection challenge for UAV applications. IEEE Trans. Pattern Anal. Mach. Intell. **43**(2), 392–403 (2021). https://doi.org/10.1109/TPAMI.2019.2932429
23. Zhan, C., Hu, H., Liu, Z., Wang, Z., Mao, S.: Multi-UAV-enabled mobile-edge computing for time-constrained IoT applications. IEEE Internet Things J.l. **8**(20), 15553–15567 (2021). https://doi.org/10.1109/JIOT.2021.3073208
24. Goncharenko, A.V.: Applicable aspects of alternative UAV operation. In: 2015 IEEE International Conference Actual Problems of Unmanned Aerial Vehicles Developments (APUAVD), pp. 316–319 (2015). https://doi.org/10.1109/APUAVD.2015.7346630
25. Amit, Y., Felzenszwalb, P., Girshick, R.: Object detection. In: Ikeuchi, K. (ed.) Computer vision. Springer, Cham (2021). https://doi.org/10.1007/978-3-030-63416-2_660
26. Challita, U., Ferdowsi, A., Chen, M., Saad, W.: Machine learning for wireless connectivity and security of cellular-connected UAVs. IEEE Wireless Commun. **26**(1), 28–35 (2019). https://doi.org/10.1109/MWC.2018.1800155
27. Liu, C., Tao, Y., Liang, J., Li, K., Chen, Y.: Object detection based on YOLO network. In: 2018 IEEE 4th Information Technology and Mechatronics Engineering Conference (ITOEC), pp. 799–803 (2018). https://doi.org/10.1109/ITOEC.2018.8740604

28. Bin Zuraimi, M.A., Kamaru Zaman, F.H.: Vehicle detection and tracking using YOLO and DeepSORT. In: 2021 IEEE 11th IEEE Symposium on Computer Applications & Industrial Electronics (ISCAIE), pp. 23–29 (2021). https://doi.org/10.1109/ISCAIE51753.2021.9431784
29. Yang, W., Jiachun, Z.: Real-time face detection based on YOLO. In: 2018 1st IEEE International Conference on Knowledge Innovation and Invention (ICKII), pp. 221–224 (2018). https://doi.org/10.1109/ICKII.2018.8569109
30. Hsu, W.Y., Lin, W.Y.: Ratio-and-scale-aware YOLO for pedestrian detection. IEEE Trans. Image Process. **30**, 934–947 (2021). https://doi.org/10.1109/TIP.2020.3039574
31. Devoto, S., Macovaz, V., Mantovani, M., Soldati, M., Furlani, S.: Advantages of using UAV digital photogrammetry in the study of slow-moving coastal landslides. Remote Sensing. **12**(3566) (2020). https://doi.org/10.3390/rs12213566
32. Terven, J., Córdova-Esparza, D.M., Romero-González, J.A.A.: Comprehensive review of YOLO architectures in computer vision: from YOLOv1 to YOLOv8 and YOLO-NAS. Mach. Learn. Knowl. Extract. **5**, 1680–1716 (2023). https://doi.org/10.3390/make5040083
33. Sapkota, R., Meng, Z., Karkee, M.: Synthetic meets authentic: leveraging LLM generated datasets for YOLO11 and YOLOv10-based apple detection through machine vision sensors. Smart Agric. Technol. **9**(100614) (2024). https://doi.org/10.1016/j.atech.2024.100614
34. Sapkota, R., Meng, Z., Churuvija, M., Du, X., Ma, Z., Karkee, M.: Comprehensive Performance Evaluation of YOLO11, YOLOv10, YOLOv9, and YOLOv8 on Detecting and Counting Fruitlet in Complex Orchard Environments. arXiv preprint arXiv:2407.12010 (2024). https://doi.org/10.48550/arXiv.2407.12010
35. Pawelczyk, M.L., Wojtyra, M.: Real world object detection dataset for quadcopter unmanned aerial vehicle detection. IEEE Access. **8**, 174394–174409 (2020). https://doi.org/10.1109/ACCESS.2020.3026192
36. Seidaliyeva, U., Akhmetov, D., Ilipbayeva, L., Matson, E.T.: Real-time and accurate drone detection in a video with a static background. Sensors. **20**(3856) (2020). https://doi.org/10.3390/s20143856
37. Hakani, R., Rawat, A.: Edge Computing-driven real-time drone detection using YOLOv9 and NVIDIA Jetson Nano. Drones. **8**(680) (2024). https://doi.org/10.3390/drones8110680
38. Wisniewski, M., Rana, Z.A., Petrunin, I., Holt, A., Harman, S.: Drone Detection Using Deep Neural Networks Trained on Pure Synthetic Data. arXiv preprint arXiv:2411.09077 (2024). https://doi.org/10.48550/arXiv.2411.09077
39. Svanström, F., Alonso-Fernandez, F., Englund, C.: Drone detection and tracking in real-time by fusion of different sensing modalities. Drones. **6**(317) (2022). https://doi.org/10.3390/drones6110317
40. Li, Y., Li, Q., Pan, J., Zhou, Y., Zhu, H., Wei, H., Liu, C.: SOD-YOLO: small-object-detection algorithm based on improved YOLOv8 for UAV images. Remote Sensing. **16**(3057) (2024). https://doi.org/10.3390/rs16163057
41. Hu, Z., Cheng, L., Yu, S., Xu, P., Zhang, P., Tian, R., Han, J.: Underwater target detection with high accuracy and speed based on YOLOv10. J. Mar. Sci. Eng. **13**(135) (2025). https://doi.org/10.3390/jmse13010135
42. Bemposta Rosende, S., Ghisler, S., Fernández-Andrés, J., Sánchez-Soriano, J.: Implementation of an edge-computing vision system on reduced-board computers embedded in UAVs for intelligent traffic management. Drones. **7**(682) (2023). https://doi.org/10.3390/drones7110682
43. Lui, M.H., Liu, H., Tang, Z., Yuan, H., Williams, D., Lee, D., Wong, K.C., Wang, Z.: An adaptive YOLO11 framework for the localisation, tracking, and imaging of small aerial targets using a pan–tilt–zoom camera network. Engineering. **5**, 3488–3516 (2024). https://doi.org/10.3390/eng5040182
44. Wang, J., Ding, J., Guo, H., Cheng, W., Pan, T., Yang, W.: Mask OBB: a semantic attention-based mask oriented bounding box representation for multi-category object detection in aerial images. Remote Sensing. **11**(2930) (2019). https://doi.org/10.3390/rs11242930

Chapter 16
Smart Control of Plant Health and Deformations in Agricultural Fields via Drone-Based Field Monitoring and Deep Learning

Mykola Sewick, Tetiana Bondarenko, Kateryna Hromovych, and Pavlo Tkachuk

Abstract A drone-based image processing method for assessing plant health and deformations in expansive agricultural areas is presented in this work. The YOLOv9 object detection and U-Net segmentation algorithms were used to analyze high-resolution photos taken with a drone system. Although YOLOv9 offered quick and efficient object recognition across large regions, the U-Net segmentation model allowed for in-depth pixel-level analysis, which made it easier to create accurate plant health maps. When combined with geographic information systems (GIS), these maps were an essential tool for precisely pinpointing areas that needed assistance. According to the findings, YOLOv9 consistently identified deformations with good performance measures, such as 82% sensitivity, 79% accuracy, and 78% precision. By accurately and highly accurately differentiating between healthy and harmful areas, U-Net segmentation improved these detections even further. The created technology significantly improves environmental sustainability in addition to increasing resource efficiency and cutting expenses in agricultural operations.

Keywords Drone-based image processing · Plant health monitoring · YOLOv9 · U-Net segmentation · Precision agriculture

16.1 Introduction

As the need for sustainable and scalable farming methods has grown in recent years, technology advancements targeted at boosting agricultural output efficiency have drawn a lot of attention. Among these developments, the use of drones in conjunction with sophisticated image processing algorithms has become a revolutionary tool

M. Sewick (✉) · T. Bondarenko · K. Hromovych · P. Tkachuk
Department of Software Engineering, Ivan Franko National University of Lviv, Lviv, Ukraine
e-mail: my.sewick@lnu.edu.ua; tbondare@lnu.edu.ua; hromov@lnu.edu.ua; tkachuk@lnu.edu.ua

G. Rivera et al. (eds.), *Recent Advances in Artificial Neural Networks*, Intelligent Systems Reference Library 283, https://doi.org/10.1007/978-3-032-13497-4_16

for tracking and evaluating vast agricultural regions [1–3]. Farmers and agricultural experts can now use drones to take high-resolution photos, enabling more accurate and efficient management techniques in a variety of fields, such as soil quality analysis, pest detection, plant health evaluation, and water resource optimization [4, 5]. This study specifically focuses on using drone-acquired visual data in conjunction with cutting-edge image processing techniques to identify and analyze both healthy and problematic areas within agricultural zones. There are many benefits to using drones in agriculture, such as increased operational effectiveness, lower expenses, and environmental sustainability. Drones can quickly cover large areas and provide detailed, high-resolution data that allows for informed decision-making, in contrast to traditional methods that are frequently labor-intensive, time-consuming, and prone to human error [6–8].

Additionally, drones can be fitted with a wide range of sensors, including thermal imaging equipment, RGB cameras, and multispectral sensors, greatly increasing the range of applications for which they can be used. Plant health, soil moisture content, temperature variations, and even the existence of air pollutants or other chemical elements in the surroundings can all be thoroughly examined thanks to these sensors [9–12]. For example, by examining particular light wavelengths, multispectral sensors can detect stress in crops, and thermal imaging can identify problems related to temperature or inefficient irrigation. By concentrating efforts where they are most needed, this adaptability allows for a targeted approach to fieldwork, saving time and money. Drone technology also makes agricultural interventions more precise. Farmers can produce intricate maps that show differences within their fields by combining data from drones with machine learning models and geographic information systems (GIS). By enabling precision agriculture techniques like variable-rate pesticide and fertilizer application, these maps guarantee efficient resource use and reduce environmental impact. Drones have the potential to revolutionize modern agriculture by quickly identifying and resolving problems like pest infestations, nutrient deficiencies, or water stress before they become serious ones.

Because of their versatility, effectiveness, and capacity to reach remote locations, drones, also referred to as unmanned aerial vehicles, have emerged as essential instruments in a variety of fields [13–15]. Drones excel in the inspection and monitoring of infrastructure, including bridges, pipelines, power lines, and wind turbines, where conventional techniques may be hazardous, expensive, or time-consuming [16, 17]. By keeping an eye on crop health, soil conditions, irrigation systems, and pest infestations, drones in agriculture enable precision farming, which maximizes yields and minimizes resource waste [18]. Drones are frequently used for traffic monitoring in urban settings, offering real-time analysis of traffic patterns, accident hotspots, and congestion points, which supports intelligent transportation systems [19–22]. Drones with sophisticated sensors are used in environmental applications such as air pollution mapping, which provides vital information for environmental protection initiatives by measuring particulate levels and air quality over wide areas [23–25]. Another crucial sector is disaster response, when UAVs are used to evaluate damage, find survivors, and transport vital supplies, often overcoming the difficulties of inaccessible terrain. By keeping an eye on animal populations and habitats and reducing human interference in delicate ecosystems, UAVs also aid in the conservation of wildlife [26]. Among their many benefits, UAVs provide unmatched data collection speed and efficiency, high-resolution imaging, and the flexibility to accommodate a range of sensor payloads, including LiDAR, thermal, and multispectral. Because of these features, UAVs are more affordable than human aerial vehicles and conventional ground-based techniques [27]. Furthermore, by lowering human exposure to dangerous situations, their capacity for autonomous or semi-autonomous operation improves operational safety [28]. UAVs are a game-changing technology that is pushing the limits of innovation across numerous industries, whether it is in public safety operations, environmental monitoring, or industrial inspection.

Data analysis and decision support systems in the agriculture industry heavily rely on object detection techniques, which are also significant in the field of image processing. Finding the positions and classes of things in an image is known as object detection. Compared to conventional image processing techniques, these algorithms offer significantly greater speed and accuracy, enabling quick analysis of big data sets [29, 30]. Object detection algorithms are used in agriculture to automatically identify plant stress situations, disease symptoms, and pest effects. Both in terms of efficiency and the creation of sustainable farming methods, this holds enormous promise. By detecting objects in a single network run, the YOLO (You Only Look Once) algorithm strikes a compromise between speed and accuracy [31–34]. One of the most recent models in this series, YOLOv9, is particularly well-suited for large-scale investigations, including the identification of deformation in vast agricultural areas [35, 36]. The U-Net segmentation algorithm's extensive investigation of the regions identified by YOLOv9 is also covered in the report. The identified regions can be more accurately categorized as healthy and harmful thanks to U-Net's pixel-level segmentation capabilities [37].

This study's emphasis on using contemporary data processing systems to process data collected by UAVs is another significant feature. High-resolution image data is provided by the RGB camera on the UAV employed in the investigation. YOLOv9 and U-Net algorithms were used to process these photos, making it possible to visualize the overall health of agricultural regions. The model was able to generate high-performance outcomes because the data set was meticulously labeled throughout the training and assessment phases. High processing power provided by the Google Colab platform during the training process helped the model be trained successfully. The study's YOLOv9 and U-Net algorithm combination can be applied to both small-scale producing regions and large-scale agricultural acreage. This method has many benefits, not only in identifying places that are healthy and those that are troublesome, but also in categorizing these areas based on their characteristics. For instance, U-Net's pixel-level analysis makes it possible to identify trouble spots as either badly distorted or mildly strained. By concentrating exclusively on parts that are significantly distorted, this lowers expenditures.

16.2 Literature Review

Thanks to developments in image technologies, machine learning, and geospatial analysis, the use of drones for agricultural monitoring has become increasingly popular in recent years. In a fraction of the time needed by conventional ground-based techniques, drones provide a flexible platform that can capture high-resolution imagery across large areas. Plant health problems including nutrient deficits, insect infestations, and disease outbreaks may now be quickly identified thanks to these systems, which have completely changed the way data is gathered. Furthermore, the use of sophisticated imaging sensors, including as thermal, multispectral, and hyperspectral cameras, has broadened the range of monitoring capabilities and offered more profound understandings of crop health and environmental interactions. With an emphasis on object recognition, segmentation methods, and their integration with Geographic Information Systems (GIS), this part examines the seminal and more recent contributions pertinent to drone-based image processing for plant health monitoring. The development of algorithms for processing drone footage, the problems these technologies solve, and their effects on the efficiency and scalability of agricultural monitoring systems are given particular attention.

Because they can obtain high-resolution images across large areas with little assistance from humans, drones have become essential in precision agriculture. For example, [38] showed how effective drones with multispectral cameras are in tracking crop development and identifying stressors like pest infestations and nutrient shortages. The use of drone-based thermal imaging to evaluate crop water stress was also emphasized in [39], demonstrating the adaptability of drone platforms in gathering a variety of data modalities. Although the focus of these studies was on data gathering skills, their scalability for large agricultural fields was limited due to their reliance on manual or semi-automated post-processing. The incorporation of lightweight sensors that can concurrently record RGB, thermal, and hyperspectral data is one of the emerging developments in drone technology. According to [40], this multi-sensor method combines many data layers to improve the capacity to diagnose plant health problems. The intricacy of combining such data to provide useful insights, however, presents difficulties that need for sophisticated computational models and domain knowledge.

In agricultural settings, object detection algorithms, including those in the YOLO family, have been extensively used for tasks like fruit counting and weed identification. YOLO, a real-time object detection system, was first presented by [41] and has since developed to accommodate improved accuracy and performance measures. As evidenced by Singh and Krishnamurthi [42], who used YOLOv4 to identify pests in maize fields with an average precision of over 85%, YOLOv4 and YOLOv5 have proved very significant in crop and pest monitoring. This study's release of YOLOv9, which offers enhanced detection accuracy and processing efficiency, is a major improvement. This study fills a vacuum in the literature by examining how object detection outputs can be integrated with later segmentation models for in-depth analysis, notwithstanding these advancements. Other object identification models, like SSD [43] and Faster R-CNN [44], have also been used in agricultural research in addition to YOLO. For applications such as pest identification in dense crop canopies, Faster R-CNN's two-stage detection pipeline has demonstrated remarkable accuracy in identifying small and irregular objects [45]. However, its usefulness in real-time drone applications is limited by its comparatively high processing cost. However, as [46] shows in their work on detecting damaged fruit in orchard settings, SSD offers a trade-off between speed and accuracy.

At the pixel level, semantic segmentation approaches have proved essential in converting high-resolution data into useful insights. Because of its encoder-decoder architecture, which is excellent at removing fine-grained features from photos, U-Net has been a popular choice among these. Li et al. [47] first suggested U-Net for the segmentation of biomedical images, but its use has now spread to the agriculture industry. For example, Li et al. [48] achieved good segmentation accuracy by using U-Net to separate healthy and sick areas in leaf pictures. More recently, Rani et al. [49] showed how U-Net may be used to map rice paddies damaged by drought with an accuracy of over 90%. Nevertheless, the majority of U-Net research has been on small-scale datasets or single-crop systems, which has limited its applicability to a variety of agricultural environments. By adding attention mechanisms and densely connected topologies, respectively, U-Net variants like Attention U-Net [50] and U-Net++ have improved segmentation performance even further. While ignoring background noise in intricate agricultural sceneries, Attention U-Net has proven very successful at emphasizing places of importance, such as sick areas [51]. As demonstrated by its use in multi-crop disease mapping, U-Net++ has enhanced generality across datasets [52].

Precision agriculture has been transformed by the combination of GIS, remote sensing, and machine learning, which allows for spatial analysis and decision-making. GIS platforms make it easier to superimpose deformation and health maps on georeferenced field layouts, offering useful information for allocating resources. Aziz et al. [53] used drone-derived NDVI (Normalized Difference Vegetation Index) maps with GIS to optimize vineyard irrigation plans. In a similar vein, Lan [54] improved pest control techniques by using GIS to examine pest distribution patterns in wheat fields. By combining GIS with health maps produced by YOLOv9 and U-Net outputs, this work expands on previous methods by making it possible to pinpoint intervention zones in vast fields. The potential for agricultural applications has been further increased by emerging GIS features including cloud-based platforms and real-time data visualization. Drone data may be seamlessly integrated with climatic and environmental datasets using platforms like Esri's ArcGIS and Google Earth Engine [55], which offer a comprehensive picture of field conditions. But as [56] points out, integrating high-resolution drone data with GIS frequently necessitates extensive preparation, such as data normalization, orthomosaic creation, and georeferencing.

Numerous recent research have used multispectral and LiDAR sensors in addition to RGB-based imaging systems for agricultural structure analysis and plant health monitoring. When it comes to detecting plant stress indicators like chlorophyll content and water deficit, multispectral sensors, which record information over several wavelength bands outside of the visible spectrum, are very useful. The normalized difference vegetation index (NDVI) values obtained from multispectral cameras, for example, provide high correlations with ground truth health indicators. This demonstrates the applicability of multispectral cameras for early stress identification in wheat crops [57]. However, small-scale farms may not be able to use multispectral systems because to their high cost, complicated calibration requirements, and requirement that spectral data be interpreted using domain-specific models.

Plant height, biomass, and lodging, on the other hand, can be evaluated using high-precision 3D mapping of crop canopies made possible by LiDAR technology. LiDAR-equipped drones were used to measure canopy height fluctuations in maize fields, as demonstrated in the work [58], and the results were satisfactorily correlated with yield estimates. Even though LiDAR offers remarkable structural information, optical sensors are usually better able to detect physiological changes like discolouration or leaf damage. Furthermore, LiDAR systems are frequently heavier and require more power, which reduces the drone's flying time and coverage area.

Comparatively speaking, the RGB camera employed in our study offers an affordable and easily available means of obtaining high-resolution images that may be used for in-depth visual analysis when combined with sophisticated models such as YOLOv9 and U-Net. RGB is a viable option for scalable plant health assessment, even though it lacks the spectral richness of multispectral sensors and the three-dimensional detail of LiDAR. This is especially true when paired with efficient preprocessing and analytical models, as our method demonstrates.

The use of drone-based image processing for plant health evaluation still faces a number of obstacles, despite tremendous advancements. For large-scale deployments, computational performance continues to be a constraint, especially for real-time or near-real-time analysis. Although edge computing technologies have been investigated in research like [59] to help address these issues, there is still little use of these systems in agricultural settings. Furthermore, more research is needed to determine whether machine learning models can be applied to different crop species, growth phases, and environmental circumstances. Techniques for domain adaptation and transfer learning, as covered by Guebsi et al. [60], present encouraging paths to overcome these constraints. A further difficulty is the labor-intensive and time-consuming task of annotating training datasets. Potential alternatives include automated annotation tools and synthetic data synthesis, as investigated by Ivezić et al. [61], although more work is needed to guarantee data accuracy. Furthermore, nothing is known about integrating temporal data to track crop health over time. Promising techniques for adding temporal dynamics to plant health models include Transformer-based architectures and Long Short-Term Memory (LSTM) networks [62].

This study fills in a number of holes in the body of literature by: 1) Making use of YOLOv9's sophisticated features for effective object recognition in expansive agricultural areas. 2) Making use of U-Net for accurate semantic segmentation, which permits comprehensive mapping of health and deformation. 3) Combining these results with GIS to provide useful geographical analysis and judgment. 4) Improving the system's generalizability and scalability by showcasing its suitability for a variety of crop varieties and field circumstances. 5) This study increases precision agriculture and opens the door to more sustainable and effective farming methods by fusing cutting-edge algorithms with useful GIS applications.

The current study improves detection and segmentation accuracy by introducing a novel integration of YOLOv9 and U-Net, while also building upon established methodologies. Prior research has frequently depended on one-method techniques, which has limited its ability to offer thorough insights. For instance, Meivel and Maheswari [63] only used YOLOv5 for object detection in order to identify areas that were infected with pests; while this method achieved a high degree of precision, it was unable to segment and quantify the afflicted areas. In a similar vein, Mogili and Deepak [64] used semantic segmentation to map the spread of disease but excluded object detection for initial localization, which increased processing requirements. The current study strikes a compromise between accuracy and efficiency by integrating various approaches, which makes it especially well-suited for extensive agricultural applications. This study is further distinguished from previous works by its integration with GIS. Although GIS-based studies have been used in a variety of settings, including pest control and irrigation management, their use in conjunction with sophisticated machine learning models is still quite limited. For example, Shah et al. [65] did not use machine learning-based health assessments but used GIS to map vegetation indices generated from NDVI. By combining YOLOv9 and U-Net outputs with GIS, this work addresses both the spatial and temporal aspects of plant health, allowing for a more detailed and useful analysis.

16.3 Materials and Methods

16.3.1 Overview

In order to evaluate plant health and identify deformations in agricultural settings, this study offers a thorough analysis technique that combines a UAV system with cutting-edge image processing algorithms. For agricultural output to be efficient and sustainable, early disease detection and the identification of trouble spots are crucial. Traditional agricultural inspection techniques make it difficult to quickly and affordably analyze wide areas, but new technology are revolutionizing this field. In this case, the study's methodology comprises four fundamental steps: Data collection via drones is the initial step. Here, a commercial drone simultaneously encapsulates GPS data with high-resolution photos and the location coordinates where the images are obtained, scanning the necessary agricultural region from start to finish. Image pre-processing is the second step. A number of data preprocessing techniques, including denoising, contrast correction, and sharpening, are applied to the photos captured by the drone system. The object detection algorithm is used in the following step to carry out detection and segmentation. In this case, the YOLOv9 object identification method is used to classify plant deformations and health issues, and the U-Net segmentation model is used to conduct in-depth analysis at the pixel level. The mapping and visualization phase comes last. Problematic regions are identified and the infrastructure required for the creation of intervention plans is built by merging the collected data and producing maps based on geographic information systems (GIS). The study's ultimate objective is to reduce agricultural losses and identify plant diseases early. Additionally, by speeding up farmers' decision-making, the user-friendly outputs offered seek to improve production efficiency.

16.3.2 Data Collection Process

For this study to be successful, gathering high-precision data from agricultural areas is essential. Throughout this process, an SJRC F11 Pro drone was used to gather high-resolution photos and exact location data from expansive agricultural fields. High-resolution mapping was made possible by the drone's integrated real-time kinematic (RTK) GPS system, which offered centimeter-level location accuracy. Features of the utilized drone are tailored for use in agriculture. Its high-resolution 20 MP camera offers multispectral imaging capabilities in addition to RGB photos, making it the perfect tool for analyzing plant health. Large areas may be scanned in a single flight thanks to the drone's 30-min continuous flight capability. The RTK GPS system was utilized to record each image's exact location during this operation, and the resulting data was utilized in later GIS analysis.

In order to maximize the drone's flight path during the data collection procedure, the agricultural fields were separated into sections, each of which comprised roughly 5 hectares. The drone was programmed to fly in a specific grid pattern during the flight, and every picture was guaranteed to overlap by roughly 70%. This overlap makes it possible to scan agricultural areas thoroughly and precisely. Flights were typically conducted in the morning or on overcast days to minimize excessive brightness and shadow effects and improve image quality. The gathered photos provide more than just visual information. Mapping and image analysis are made easier by metadata like shooting time, altitude, and geographic coordinates that are included with every picture. Images were saved in JPEG and TIFF formats, and the resolution of the data gathered by the drone was set at 5472 x 3648 pixels. These formats offer more versatility in the stages of image processing and analysis.

16.3.3 Dataseet and Algorithm Training

The YOLOv9 algorithm was trained using a unique dataset made up of agricultural data. 2577 photos from expansive agricultural areas that were surveyed by drones are included in the dataset. Plant regions in various states of health were represented by the labels on the photos. Categories including healthy leaves, diseased leaves, and deformation regions are all part of the applicable classification technique. An open-source labeling program was used for the manual labeling procedure, and bounding boxes were used to identify the plant areas in each image. Data diversification approaches were used during the model's training process to broaden the dataset's diversity and enhance the model's overall performance. These methods include axial reflection, noise addition, brightness level changes, and random image rotation. The diversification procedure made it possible for the model to produce better results on photos shot from various perspectives and in various lighting situations. The YOLOv9 hyperparameters were adjusted during training in order to achieve the optimal sensitivity value. 200 epochs were used for training, with a learning rate of 0.001. It was found that 16 was the batch size utilized throughout training. An additional factor in the model's quicker and more reliable learning was the optimization process' preference for the Adam method. The data set was split into three categories: 10% test data, 25% validation data, and 65% training data. The model's accuracy and overall performance were assessed using the test data. The NVIDIA GPU was used to speed up the model's training process, which took about 5 h. The model's sensitivity in detecting various plant illnesses and deformations was aided by data augmentation and hyperparameter adjustment during training.

16.3.4 Image Processing

Due to environmental conditions and device constraints, the photos captured by drones in agricultural fields contain numerous distortions and inconsistencies that make it impossible to examine them in their raw form. These photos must be processed in order to be analyzed quickly and accurately. Several stages of enhancement, analysis, and detection make up the image processing process. The general flow of this operation is summarized in Fig. 16.1. Image enhancement techniques were used in the initial stage to improve the images' quality and serve as the foundation for more understandable analyses. In low light, the effect of damaged pixels was reduced and random noise in the photos was cleaned up using the median filter. This pre-processing phase is essential for improving the analytical algorithms' accuracy. Details in low-contrast areas were made more evident by applying the Histogram Equalization technique for contrast enhancement. Lastly, the Laplace Filter was used to highlight the images' details and highlight the distortions.

Figure 16.2 shows the results of the picture enhancing procedures. In order to examine the photographs more seriously, color and texture aspects were taken

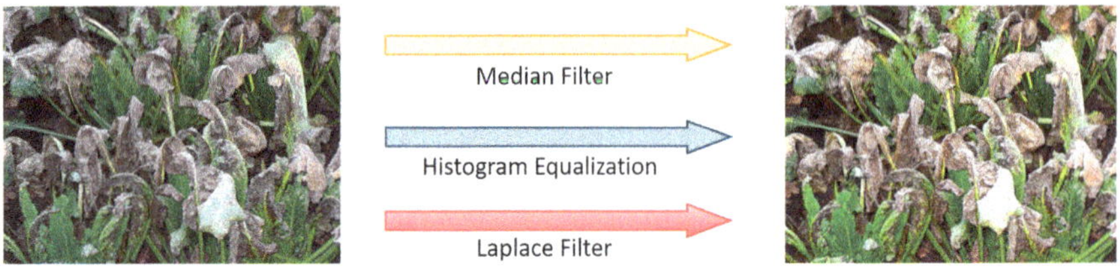

Fig. 16.1 The overall workflow of the image processing pipeline for agricultural field photos taken by drones. A strong basis for precise and effective analysis is formed by the method, which incorporates image enhancing techniques like Laplace Filtering to emphasize deformations and highlight picture details, Histogram Equalization to improve contrast, and Median Filtering to reduce noise

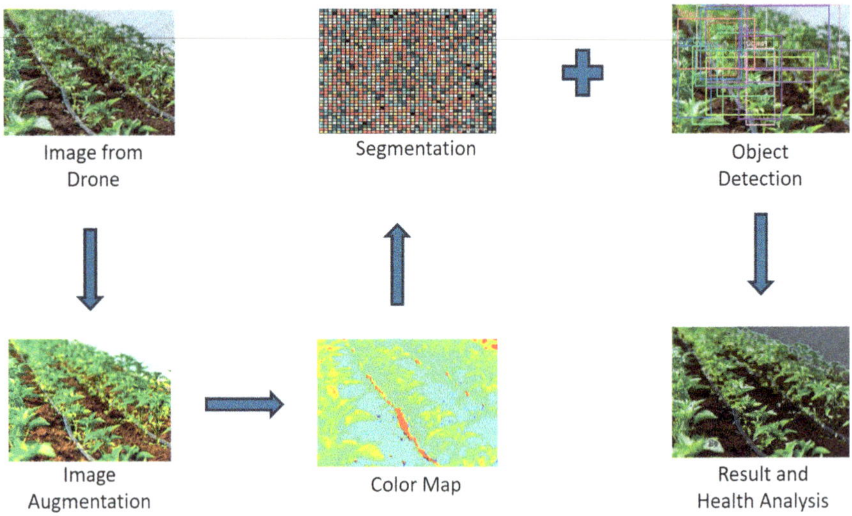

Fig. 16.2 The workflow flow chart for picture analysis and enhancement. This involves segmenting using the U-Net model for pixel-level health mapping after converting RGB to HSV color space for color-based differentiation of healthy (green tones) and unhealthy (yellow and brown tones) areas

into consideration after the enhancing procedure. RGB color space was changed to HSV (Hue, Saturation, Value) color space for the photos. In agricultural fields, HSV transformation made it easier to distinguish between healthy and unhealthy areas based on color. Diseased or stressed leaves exhibited yellow and brown tones, whereas healthy leaves typically have green tones. These color variations served as a crucial foundation for the object detection and segmentation processes. The U-Net model was used to carry out segmentation, a crucial step in the image processing process. By giving each pixel in the image a class, U-Net was able to distinguish between healthy and unhealthy areas with accuracy. The great pixel-level accuracy of this model allowed for a thorough mapping of agricultural areas' health conditions. This study combined U-Net and YOLO, two potent models that enhance one another. U-Net's pixel-level analysis and thorough segmentation helped to create accurate health maps in agricultural regions. On the other hand, YOLO offered an object detection method that facilitates segmentation findings, making it possible to examine identified issues more rapidly and practically. The study's scope was increased by the combined application of the two models, which offered the perfect balance between speed and precision. The flow chart in Fig. 16.2 provides a summary of the outcomes of the image processing procedure.

16.3.5 YOLOv9-Based Detection

Drone-captured photos were automatically analyzed using the YOLOv9 algorithm to identify plant deformations and health issues. YOLOv9's real-time item detection capability allows for efficient analysis across expansive agricultural areas. For agricultural applications, the algorithm's speed and accuracy are essential. Fig. 16.3 shows a condensed representation of the YOLOv9 network architecture. For usage in agricultural applications, YOLOv9 has a number of benefits. Its real-time capability allows it to process an image in a single network pass, resulting in quick and precise object detection. For quick study of vast volumes of data in expansive agricultural regions, this capability is perfect. Accurate classification of illnesses and deformations is made possible by high accuracy rates. Additionally, special classifications pertaining to plant health (such as leaf burns or yellowing) could be introduced with ease because of the algorithm's structure, which can be optimized for agricultural data sets. YOLOv9's fundamental operation involves

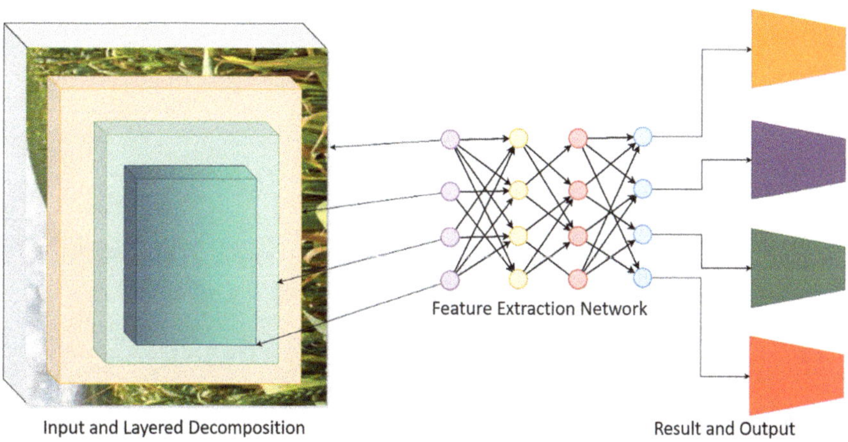

Fig. 16.3 The YOLOv9 algorithm's reduced network architecture is utilized to identify plant deformations and health issues in real-time from photos taken by drones. YOLOv9 uses multiple convolutional layers to detect objects at different scales by segmenting images into regions and estimating bounding boxes and probabilities for each

segmenting a picture into regions and estimating the probability and bounding boxes for each region. The system can recognize both minute and large deformations since it employs numerous convolution layers to detect objects at different scales.

16.3.6 Evaluation of Planth Health

Various methodologies were used to examine plant health issues through color changes, textural deteriorations, and deformations in agricultural photos captured by drones. Fig. 16.4's HSV color space diagram illustrates how color analysis is used to distinguish between plant sections that are healthy and unhealthy. To assess the health of plants, color and texture-based methods are used. In order to examine color tone (Hue), saturation (Saturation), and brightness (Value) information independently, RGB (Red, Green, Blue) photos were transformed to HSV format. According to this conversion, healthy leaves often have green hues, but yellow and brown hues signify health issues including drought, fungal infections, or nutrient deficiencies. Specifically, by examining color variations, the Hue component is utilized to assess the health of plants. Value makes it easier to distinguish sick

Fig. 16.4 The HSV color space diagram uses color analysis to show how healthy plant areas (green tones) are separated from unhealthy plant areas (yellow and brown tones). RGB photos are converted to HSV format so that the Hue component can identify plant health by highlighting color variations, the Saturation component can detect stress by measuring color intensity, and the Value component can more easily identify unhealthy parts by emphasizing brightness disparities

regions by highlighting variations in brightness, while Saturation measures the intensity of colors to identify symptoms of stress on leaves.

The texture analysis method was also used since deformations might show up on the leaf surface as textural flaws in addition to color alterations. To analyze the abnormalities on the leaf surface, the Gray-Level Co-occurrence Matrix (GLCM) and Gabor Filter techniques were employed. The leaf surface's roughness, symmetry loss, and textural flaws are assessed using the Gabor Filter. Furthermore, GLCM is utilized to distinguish between the kind and strength of textural changes as well as to examine the spatial relationships between nearby pixels. These two techniques enhance color analysis and offer more accurate deformation area detection. The U-Net-based segmentation technique was used to separate the images pixel-by-pixel after color and texture analysis. The segmentation procedure helped to accurately analyze the field's overall health state by clearly classifying sections as either healthy or unhealthy. Areas with symptoms including yellowing, browning, and staining were identified as unhealthy areas, whereas healthy parts are typically defined as those with a prominent green tone and a uniform texture. Each plant's health status was evaluated separately, and the consequences of deformations were categorized based on their severity. Yellowing or minor color changes were typically indicative of mild deformations. Localized browning and spotting were indicative of moderate deformations, but substantial textural distortion and widespread color loss were indicative of severe deformations.

Each plant receives a Health Score at the end of the plant health evaluation process. The findings of segmentation, texture analysis, and color analysis are combined to determine this score. The ratio of healthy pixels to total pixels is used to create the health score, which is computed using the deformation percentage as follows:

$$\text{Health Score} = 100 - \text{Deformation Percentage} \tag{16.1}$$

The ratio of unhealthy areas to the entire area is used to compute the deformation percentage. A health score of 80 or higher indicates that an area is considered healthy, whereas a score of less than 50 indicates that significant intervention is needed. Through a combination of empirical study and domain experience, the threshold values employed in the Health Score formula, more precisely, the cutoffs of 80 for healthy classification and 50 for flagging areas requiring urgent intervention, were determined. In order to see if segmentation results matched expert annotations and visual evaluations of plant health, a range of threshold settings was first explored across several dataset subsets. These levels were then adjusted based on input from agricultural experts to make sure they matched significant physiological states in crops. According to expert criteria of a healthy zone, fields with a score above 80 usually have uniform green coloration over 80% of their area and few textural irregularities. On the other hand, field observations indicated that regions with scores below 50 frequently had widespread discoloration, browning, or spotting, which were linked to extreme stress or illness. Thus, these empirically supported standards offer a solid foundation for field management decision-making that can be put into practice.

16.3.7 Mapping and Analysis

One of the primary goals of this study is to present the analysis results from drone photographs in a format that is both practical and simply comprehensible. GIS-based mapping and visualization techniques were favored in this regard. Visually delineating trouble spots, assessing the proportion of healthy and unhealthy areas, and identifying places that need delicate action are all made possible by the maps that were produced. The health status of plants is effectively displayed, particularly through the use of color codes. Figure 16.5 shows an example field and the 3D image of it that was created using the geographic information system. The hues green, yellow, blue, and red are used to represent the health state of the plants in the map-based visualization process. Green denotes areas with healthy plants and typically denotes areas that require protection through consistent nutrient support and irrigation. Regions with minor distortions or possible hazards are indicated by yellow regions. For these locations, routine control and preventive action are advised. The area devoid of crops is shown in blue. Red denotes places that need immediate attention and significant deformations. For these locations, targeted spraying or regional interventions are necessary.

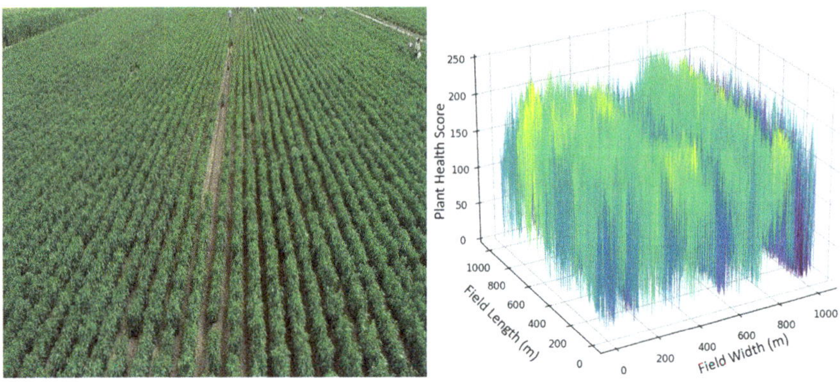

Fig. 16.5 Using color-coded mapping, a GIS-based 3D depiction of a sample agricultural field shows the health state of the plants. Red shows severely distorted areas that require immediate attention, blue indicates areas without crops, yellow indicates minor deformations or possible hazards that require preventive action, and green indicates healthy areas that require routine care

16.4 Results

The proposed strategy has been able to develop a novel and efficient detection and analysis mechanism through the combined use of the YOLOv9 and U-Net algorithms. By utilizing the distinct advantages of both algorithms, this method allows for the best possible trade-off between accuracy and speed. YOLOv9's quick object detection capabilities enable the prompt detection of deformations in expansive agricultural fields, guaranteeing prompt intervention in the event of a problem. A thorough classification of healthy and hazardous areas is made possible by the U-Net segmentation model, which improves the analysis's accuracy by offering pixel-by-pixel segmentation of the observed regions.

The integrated approach has shown enormous promise for a wide range of agricultural applications, meeting the urgent demand for quick and dependable systems to track crop health and improve farming methods. With an accuracy of 84.4%, precision of 83.8%, and recall rate of 87.1%, the model's performance measures demonstrate how effective it is. The system's resilience in differentiating between healthy and ill areas within a field is demonstrated by the performance indicators. Furthermore, the training procedure shows steady improvement, as seen in Fig. 16.6, with the loss value gradually declining over time. The model's flexibility and the efficacy of the training process were demonstrated by the mean average precision (mAP@0.5), which began at a modest 15% and steadily rose to an amazing 77%.

We contrasted our findings with popular object detection and segmentation models documented in related agricultural surveillance studies in order to put the

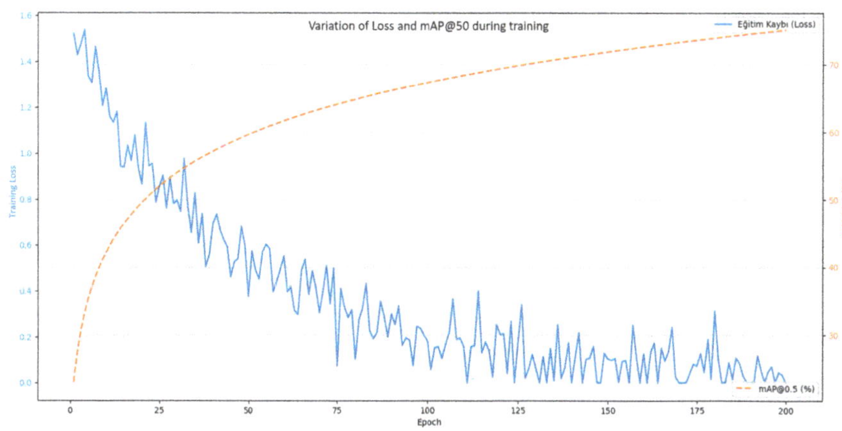

Fig. 16.6 The combined methodology's training performance demonstrated a significant improvement in mean average precision (mAP@0.5) from 15% to 77%, along with a consistent decrease in loss value. With an accuracy of 84.4%, precision of 83.8%, and recall rate of 87.1%, the model demonstrated its resilience in correctly differentiating between areas in agricultural fields that are healthy and those that are not

performance of our suggested YOLOv9 + U-Net framework into context. For example, Yang et al. [66] used YOLOv4 to detect pests in maize fields, with an 81% recall and 85% precision. In the investigation by Chu et al. [67], YOLOv5 produced a somewhat faster detection speed while maintaining an accuracy of about 83–85%. In contrast, SSD, which is renowned for its speed, frequently sacrifices detection accuracy; in orchard-based deformation detection tasks, studies like [68] have found precision to be less than 80%. Faster R-CNN was found to be less useful in real-time applications because of its longer inference periods, even if it achieved high accuracy (up to 88%) in controlled situations [69].

On the other hand, our YOLOv9 + U-Net pipeline balanced detection performance with real-time application, achieving an accuracy of 84.4%, precision of 83.8%, and recall of 87.1%. Furthermore, a level of granularity in health classification that is not usually possible with detection models alone was made possible by the incorporation of U-Net segmentation. We were able to differentiate between areas that were under mild and severe stress thanks to the pixel-level segmentation, which is essential for focused intervention in large-scale farming operations. Thus, by fusing the accuracy of pixel-wise classification with the speed of real-time detection, this dual-model approach provides a distinct advantage, making it a workable and scalable option for contemporary precision agriculture.

The U-Net segmentation model was crucial because it conducted a more thorough examination of the regions that YOLOv9 had discovered. By classifying regions at the pixel level, this analysis made it possible to create color-coded maps that clearly showed parts that were healthy and those that were ill. These maps made it easier to identify trouble spots and gave stakeholders the ability to carry out focused actions. For example, zones that were deemed healthy needed little care, whereas zones that were deemed to be significantly malformed or mildly stressed may be given priority for quick intervention. As seen by the findings in Fig. 16.7,

Fig. 16.7 The U-Net segmentation model produced color-coded maps that classified agricultural areas at the pixel level. It is possible to accurately identify trouble spots since healthy areas may be easily differentiated from zones that are somewhat strained and significantly distorted. Targeted interventions are supported by this visualization, which minimizes needless attention to healthy parts while giving priority to urgent action for severely deformed regions

this exact classification made it possible to manage intervention plans effectively and, by allocating resources as efficiently as possible, greatly reduced costs.

Because they provided a clear visual representation of field circumstances, the color-coded segmentation maps were extremely helpful in decision-making. These maps make it possible to identify stress areas, gauge how severe they are, and put customized solutions in place, such fertilizer applications, irrigation adjustments, or pest infestations. Additionally, the capacity to divide trouble spots into three primary groups, healthy, mildly stressed, and seriously deformed, offered practical insights that improved agricultural field management as a whole. Moreover, the used method showed exceptional adaptability, working well for both small-scale production systems and large-scale agricultural enterprises. While U-Net's accuracy enabled in-depth micro-level analysis, YOLOv9's speed and accuracy guaranteed quick fieldwide coverage. Together, they improve agricultural decision-support systems' accuracy and expedite field activities, making them quicker and more efficient.

Some restrictions were noted during deployment, despite the fact that the combined YOLOv9 and U-Net framework provides notable benefits in terms of speed and segmentation granularity. Situations with very minor deformations, including early-stage nutrient deficits or water stress, presented a problem since color changes were faint and hard to tell apart from normal leaf variation. In these situations, U-Net segmentation sometimes failed to identify new problems or incorrectly labeled healthy tissue as mildly stressed. Furthermore, despite its great efficiency, the YOLOv9 detector occasionally mistook signs of plant stress, such as transient discoloration from sun exposure or bug marks, for deformation because of visual resemblance. A few false positives resulted from these overlapping features, especially when the visual cues utilized to train the bounding box classifications were not significantly different. To increase classification accuracy in borderline circumstances, these findings point to the necessity of adding other data modalities, like multispectral imaging, or honing the training dataset with more complex examples.

Healthy and unhealthy areas in agricultural areas are clearly separated by color-coded maps created to visualize health status. Figure 16.8 illustrates that on these maps, areas with significant deformations are emphasized in red, areas with mild stress are indicated in yellow, and healthy vegetation is shown in green. This coding makes it possible to quickly evaluate the agricultural region of interest's health and gives fieldworkers the chance to concentrate on trouble spots beforehand. By simply intervening in the areas shown in red, needless spraying can be avoided, greatly lowering expenses and environmental effects. Agricultural operations become more sensitive when data from Fig. 16.8 is integrated with GIS-based systems. By pinpointing the precise locations of trouble spots, this integration makes it possible to arrange intervention activities in agricultural areas more efficiently. The analytical process is made easier by the fact that GIS-supported maps provide two distinct viewpoints using 2D and 3D visualization approaches. By describing the land's slope and topographic features, three-dimensional maps assist in addressing additional issues like drainage and water management, whereas two-dimensional

Fig. 16.8 Maps of agricultural areas' health state that are color-coded, with red denoting areas that are severely distorted, yellow suggesting areas that are somewhat strained, and green representing places with healthy plants. By avoiding needless spraying and facilitating quick health assessments and focused remedies, the maps lower expenses and their negative effects on the environment. By identifying the precise coordinates of troublesome locations, the integration of this data with GIS platforms improves precision. A good perspective is provided by 2D maps, while 3D maps offer more information about land characteristics like drainage and slope, which helps with efficient planning and sustainable farming methods

maps offer a cursory overview of trouble spots using color coding. Results mapping supports environmental sustainability as well as agricultural decision-making.

Although the suggested system performed well under controlled data gathering conditions, a number of real-world considerations were carefully taken into account to guarantee its viability and resilience. Conducting missions during early morning or gloomy periods, when illumination is more diffuse and consistent, helped to offset the variable lighting conditions caused by cloud cover or variations in sun angle during the drone flights. During training, image augmentation methods like contrast normalization and brightness modifications were used to enhance the model's generalization in a variety of lighting scenarios. Flight scheduling included short-term weather predictions to reduce disruptions because sudden weather changes, such wind or light rain, might influence drone stability and image clarity. The fields were divided into smaller flight zones, each tailored for the drone's 30-min flight length, with overlapping image acquisition to guarantee coverage continuity, in order to overcome drone battery restrictions. Our system design and preprocessing techniques assist maintain dependability and adaptability over a range of climatic circumstances, which is in line with the dynamic character of real-world agricultural monitoring, even though these limitations pose operational obstacles.

16.4.1 Comparative Context and Practical Considerations

Our strategy provides a synergistic advantage over other previous approaches that use segmentation or object detection alone. Systems that only use YOLOv5 or Faster R-CNN, for instance, frequently offer quick detection but lack the pixel-level resolution needed for analysis that is particular to a given health. On the other hand, segmentation-only systems, like those that use U-Net++, achieve detailed mapping but necessitate additional preprocessing time and computational resources. This gap is filled by our combined YOLOv9 + U-Net technique, which offers a speed-precision balance appropriate for field-level decision-making.

Although the suggested approach performs well in tests, a number of practical limitations may affect its implementation in the real world. These include flight path planning in unpredictable weather, drone battery life restrictions, and the processing demands of executing deep learning models on edge devices. In order to successfully handle these issues, future implementations might find it advantageous to use edge AI chips or hybrid cloud processing.

Lastly, we make clear that during validation, the health score classification thresholds were established empirically. In particular, a region is considered healthy if its health score is above 80%, mildly stressed if it is between 50 and 80%, and severely deformed if it is below 50%. Visual inspection and expert consultation were used to determine these levels, which might be further adjusted for particular crop varieties or local circumstances.

16.5 Conclusion

This study shows how sophisticated image processing algorithms and drone-based imaging technology can be applied to agricultural studies. It has been demonstrated that by precisely identifying plant deformations and health condition in vast agricultural areas, YOLOv9's quick object recognition and in-depth pixel-level analysis of the U-Net segmentation model can enhance decision support procedures. This integration has the potential to boost agricultural production efficiency while also significantly lowering costs and promoting environmental sustainability. Plant health maps were produced by combining geographic information systems (GIS) with high-resolution photos taken with drones. By using color codes to clearly depict both healthy and troublesome areas in agricultural areas, these maps made it possible to only intervene where necessary. Specifically, the accuracy of data processing procedures was improved by the position precision offered by the integrated RTK GPS system. YOLOv9 demonstrated 84% accuracy, 83% sensitivity, and 87% sensitivity, indicating its effectiveness as a tool for agricultural analysis. In order to identify the regions that needed to be separated at the pixel level, the U-Net segmentation model's thorough investigations were essential. Compared to conventional methods, these alternatives are more efficient and use fewer resources, which benefits the economy and the environment. Multispectral photography, hyperspectral analysis, and sophisticated artificial intelligence algorithms can be used to further this research in the future. Additionally, the accuracy and usefulness of the system can be improved by testing it with bigger data sets that include a variety of crop varieties and weather situations. Through increased agricultural production process optimization, such advancements can help accelerate the digitization of agriculture.

Future studies could profit from using adaptive algorithms that enable the system to dynamically adapt to various crop varieties, growth phases, and climatic circumstances in addition to extending sensor modalities with multispectral and hyperspectral imaging. The model can continuously improve with fresh data thanks to adaptive learning strategies like transfer learning, few-shot learning, and online learning, which lessens the need for intensive retraining. For instance, reinforcement-based tuning can optimize interventions based on real-time field feedback, while transfer learning can assist in generalizing features learnt on one crop to another with less data. The system's adaptability and resilience would be greatly increased by these adaptive techniques, increasing its suitability for a range of agricultural situations and its capacity to control seasonal or regional fluctuation.

References

1. Bakirci, M., Bayraktar, I.: YOLOv9-enabled vehicle detection for urban security and forensics applications. In: 2024 12th International Symposium on Digital Forensics and Security (ISDFS), pp. 1–6. IEEE (2024). https://doi.org/10.1109/ISDFS60797.2024.10527304
2. Kim, J., Kim, S., Ju, C., Son, H.I.: Unmanned aerial vehicles in agriculture: a review of perspective of platform, control, and applications. IEEE Access. **7**, 105100–105115 (2019). https://doi.org/10.1109/ACCESS.2019.2932119
3. Bakirci, M., Cetin, M.: Improving position-time trajectory accuracy in vehicle stop-and-go scenarios by using a mobile robot as a testbed. J. Control Eng. Appl. Inform. **25**(3), 35–44 (2023). https://doi.org/10.61416/ceai.v25i3.8365
4. Murugan, D., Garg, A., Singh, D.: Development of an adaptive approach for precision agriculture monitoring with drone and satellite data. IEEE J. Sel. Top. Appl. Earth Observ. Remote Sensing. **10**(12), 5322–5328 (2017). https://doi.org/10.1109/JSTARS.2017.2746185
5. Spoorthi, S., Shadaksharappa, B., Suraj, S., Manasa, V.K.: Freyr drone: Pesticide/fertilizers spraying drone—an agricultural approach. In: 2017 2nd International Conference on Computing and Communications Technologies (ICCCT), pp. 252-255 (2017) doi: 10.1109/ICCCT2.2017.7972289
6. Singh, C., Mishra, R., Gupta, H.P., Kumari, P.: The Internet of drones in precision agriculture: challenges, solutions, and research opportunities. IEEE Internet Things Mag. **5**(1), 180–184 (2022). https://doi.org/10.1109/IOTM.006.2100100
7. Jawad, A.M., Jawad, H.M., Nordin, R., Gharghan, S.K., Abdullah, N.F., Abu-Alshaeer, M.J.: Wireless power transfer with magnetic resonator coupling and sleep/active strategy for a drone charging station in smart agriculture. IEEE Access. **7**, 139839–139851 (2019). https://doi.org/10.1109/ACCESS.2019.2943120

8. Saha, A.K., et al.: IOT-based drone for improvement of crop quality in agricultural field. In: 2018 IEEE 8th Annual Computing and Communication Workshop and Conference (CCWC), pp. 612–615 (2018). https://doi.org/10.1109/CCWC.2018.8301662
9. Bakirci, M., Dmytrovych, P., Bayraktar, I., Anatoliyovych, O.: Multi-class vehicle detection and classification with YOLO11 on UAV-captured aerial imagery. In: 2024 IEEE 7th International Conference on Actual Problems of Unmanned Aerial Vehicles Development (APUAVD), pp. 191–196. IEEE (2024). https://doi.org/10.1109/APUAVD64488.2024.10765862
10. Kulbacki, M., et al.: Survey of drones for agriculture automation from planting to harvest. In: 2018 IEEE 22nd International Conference on Intelligent Engineering Systems (INES), pp. 000353–000358 (2018). https://doi.org/10.1109/INES.2018.8523943
11. Pasha, J., et al.: The drone scheduling problem: a systematic state-of-the-art review. IEEE Trans. Intell. Transp. Syst. **23**(9), 14224–14247 (2022). https://doi.org/10.1109/TITS.2022.3155072
12. Caruso, A., Chessa, S., Escolar, S., Barba, J., López, J.C.: Collection of data with drones in precision agriculture: analytical model and LoRa case study. IEEE Internet Things J. **8**(22), 16692–16704 (2021). https://doi.org/10.1109/JIOT.2021.3075561
13. Mozaffari, M., Saad, W., Bennis, M., Nam, Y.H., Debbah, M.: A tutorial on UAVs for wireless networks: applications, challenges, and open problems. IEEE Commun. Surv. Tutor. **21**(3), 2334–2360 (2019). https://doi.org/10.1109/COMST.2019.2902862
14. Dilshad, N., Hwang, J., Song, J., Sung, N.: Applications and challenges in video surveillance via drone: a brief survey. In: 2020 International Conference on Information and Communication Technology Convergence (ICTC), pp. 728–732 (2020). https://doi.org/10.1109/ICTC49870.2020.9289536
15. Chittoor, P.K., Chokkalingam, B., Mihet-Popa, L.: A review on UAV wireless charging: fundamentals, applications, charging techniques and standards. IEEE Access. **9**, 69235–69266 (2021). https://doi.org/10.1109/ACCESS.2021.3077041
16. Sami Oubbati, O., Atiquzzaman, M., Ahamed Ahanger, T., Ibrahim, A.: Softwarization of UAV networks: a survey of applications and future trends. IEEE Access. **8**, 98073–98125 (2020). https://doi.org/10.1109/ACCESS.2020.2994494
17. Yallappa, D., Veerangouda, M., Maski, D., Palled, V., Bheemanna, M.: Development and evaluation of drone mounted sprayer for pesticide applications to crops. In: 2017 IEEE Global Humanitarian Technology Conference (GHTC), pp. 1–7 (2017). https://doi.org/10.1109/GHTC.2017.8239330
18. Shakhatreh, H., et al.: Unmanned Aerial Vehicles (UAVs): a survey on civil applications and key research challenges. IEEE Access. **7**, 48572–48634 (2019). https://doi.org/10.1109/ACCESS.2019.2909530
19. Bakirci, M., Bayraktar, I.: The cutting-edge YOLO11 for advanced aircraft detection in Synthetic Aperture Radar (SAR) imagery. In: 2024 8th International Symposium on Innovative Approaches in Smart Technologies (ISAS), pp. 1–6. IEEE (2024). https://doi.org/10.1109/ISAS64331.2024.10845222
20. Messaoudi, K., Oubbati, O.S., Rachedi, A., Lakas, A., Bendouma, T., Chaib, N.: A survey of UAV-based data collection: challenges, solutions and future perspectives. J. Netw. Comput. Appl. **216**(103670) (2023). https://doi.org/10.1016/j.jnca.2023.103670
21. Kaleem, Z., et al.: Amateur drone surveillance: applications, architectures, enabling technologies, and public safety issues: part 2. IEEE Commun. Mag. **56**(4), 66–67 (2018). https://doi.org/10.1109/MCOM.2018.8337898
22. Kaleem, Z., et al.: Amateur drone surveillance: applications, architectures, enabling technologies, and public safety issues: part 1. IEEE Commun. Mag. **56**(1), 14–15 (2018). https://doi.org/10.1109/MCOM.2018.8255731
23. Bakirci, M.: Smart city air quality management through leveraging drones for precision monitoring. Sustain. Cities Soc. **106**(105390) (2024). https://doi.org/10.1016/j.scs.2024.105390
24. Bakirci, M.: A drone-based approach to enhance spatial insight into surrounding air pollutant distributions for healthier indoor environments. J. Build. Eng. **87**(109023) (2024). https://doi.org/10.1016/j.jobe.2024.109023

25. Bakirci, M.: Efficient air pollution mapping in extensive regions with fully autonomous unmanned aerial vehicles: a numerical perspective. Sci. Total Environ. **909**(168606) (2024). https://doi.org/10.1016/j.scitotenv.2023.168606
26. Ravi, N., Chitanvis, R., El-Sharkawy, M.: Applications of drones using wireless sensor networks. In: 2019 IEEE National Aerospace and Electronics Conference (NAECON), pp. 513–518 (2019). https://doi.org/10.1109/NAECON46414.2019.9057846
27. Motlagh, N.H., Bagaa, M., Taleb, T.: UAV-based IoT platform: a crowd surveillance use case. IEEE Commun. Mag. **55**(2), 128–134 (2017). https://doi.org/10.1109/MCOM.2017.1600587CM
28. Naqvi, S.A.R., Hassan, S.A., Pervaiz, H., Ni, Q.: Drone-aided communication as a key enabler for 5G and resilient public safety networks. IEEE Commun. Mag. **56**(1), 36–42 (2018). https://doi.org/10.1109/MCOM.2017.1700451
29. Bakirci, M.: Vehicular mobility monitoring using remote sensing and deep learning on a UAV-based mobile computing platform. Measurement. **244**(116579) (2025). https://doi.org/10.1016/j.measurement.2024.116579
30. Bakirci, M., Bayraktar, I.: AI-driven micro defect detection for aerospace-grade metal surface quality control in smart manufacturing. In: 2025 International Russian Smart Industry Conference (SmartIndustryCon), pp. 115–120. IEEE (2025). https://doi.org/10.1109/SmartIndustryCon65166.2025.10986084
31. Zou, Z., Chen, K., Shi, Z., Guo, Y., Ye, J.: Object detection in 20 years: a survey. Proc. IEEE. **111**(3), 257–276 (2023). https://doi.org/10.1109/JPROC.2023.3238524
32. Borji, A., Cheng, M.M., Jiang, H., Li, J.: Salient object detection: a benchmark. IEEE Trans. Image Process. **24**(12), 5706–5722 (2015). https://doi.org/10.1109/TIP.2015.2487833
33. Bakirci, M., Demiray, A.: Tracking robotic arms with YOLO11 for smart automation in industry 4.0. In: 2025 International Russian Smart Industry Conference (SmartIndustryCon), pp. 64–70. IEEE (2025). https://doi.org/10.1109/SmartIndustryCon65166.2025.10985968
34. Wolf, C., Jolion, J.M.: Object count/area graphs for the evaluation of object detection and segmentation algorithms. IJDAR. **8**, 280–296 (2006). https://doi.org/10.1007/s10032-006-0014-0
35. Ding, J., et al.: Object detection in aerial images: a large-scale benchmark and challenges. IEEE Trans. Pattern Anal. Mach. Intell. **44**(11), 7778–7796 (2022). https://doi.org/10.1109/TPAMI.2021.3117983
36. Bayraktar, I., Bakirci, M.: Attention-augmented YOLO11 for high-precision aircraft detection in synthetic aperture radar imagery. In: 2025 27th International Conference on Digital Signal Processing and its Applications (DSPA), pp. 1–6. IEEE (2025). https://doi.org/10.1109/DSPA64310.2025.10977903
37. Feng, D., Harakeh, A., Waslander, S.L., Dietmayer, K.: A review and comparative study on probabilistic object detection in autonomous driving. IEEE Trans. Intell. Transp. Syst. **23**(8), 9961–9980 (2022). https://doi.org/10.1109/TITS.2021.3096854
38. Kar, P., Chowdhury, S.: IoT and drone-based field monitoring and surveillance system. In: Chouhan, S.S., Saxena, A., Singh, U.P., Jain, S. (eds.) Artificial Intelligence Techniques in Smart Agriculture. Springer, Singapore (2024). https://doi.org/10.1007/978-981-97-5878-4_15
39. Hafeez, A., Husain, M.A., Singh, S.P., Chauhan, A., Khan, M.T., Kumar, N., Chauhan, A., Soni, S.K.: Implementation of drone technology for farm monitoring & pesticide spraying: a review. Inform. Process. Agric. **10**(2), 192–203 (2023). https://doi.org/10.1016/j.inpa.2022.02.002
40. Panday, U.S., Pratihast, A.K., Aryal, J., Kayastha, R.B.: A review on drone-based data solutions for cereal crops. Drones. **4**(41) (2020). https://doi.org/10.3390/drones4030041
41. Mahmoud, N.T.A., Virro, I., Zaman, A.G.M., Lillerand, T., Chan, W.T., Liivapuu, O., Roy, K., Olt, J.: Robust object detection under smooth perturbations in precision agriculture. AgriEngineering. **6**, 4570–4584 (2024). https://doi.org/10.3390/agriengineering6040261
42. Singh, P., Krishnamurthi, R.: Object detection using deep ensemble model for enhancing security towards sustainable agriculture. Int. J. Inf. Tecnol. **15**, 3113–3126 (2023). https://doi.org/10.1007/s41870-023-01341-4

43. Kragh, M., Jørgensen, R.N., Pedersen, H.: Object detection and terrain classification in agricultural fields using 3D Lidar data. In: Nalpantidis, L., Krüger, V., Eklundh, J.O., Gasteratos, A. (eds.) Computer Vision Systems. ICVS 2015 Lecture Notes in Computer Science, vol. 9163. Springer, Cham (2015). https://doi.org/10.1007/978-3-319-20904-3_18
44. Chen, J.-W., Lin, W.-J., Cheng, H.-J., Hung, C.-L., Lin, C.-Y., Chen, S.-P.: A Smartphone-based application for scale pest detection using multiple-object detection methods. Electronics 10(372), (2021) doi: https://doi.org/10.3390/electronics10040372
45. Zhao, W., Yamada, W., Li, T., Digman, M., Runge, T.: Augmenting crop detection for precision agriculture with deep visual transfer learning—a case study of bale detection. Remote Sens. **13**(23) (2021). https://doi.org/10.3390/rs13010023
46. Singh, P., Krishnamurthi, R.: IoT-based real-time object detection system for crop protection and agriculture field security. J. Real-Time Image Proc. **21**(106) (2024). https://doi.org/10.1007/s11554-024-01488-8
47. Li, Y., Qiao, T., Leng, W., Jiao, W., Luo, J., et al.: Semantic segmentation of wheat stripe rust images using deep learning. Agronomy. **12**(2933) (2022). https://doi.org/10.3390/agronomy12122933
48. Li, Z., Chen, P., Shuai, L., Wang, M., Zhang, L., Wang, Y., Mu, J.: A copy paste and semantic segmentation-based approach for the classification and assessment of significant rice diseases. Plants. **11**(3174) (2022). https://doi.org/10.3390/plants11223174
49. Rani, N.S., Krishna, A.S., Sunag, M., et al.: Infield disease detection in citrus plants: integrating semantic segmentation and dynamic deep learning object detection model for enhanced agricultural yield. Neural Comput. Appl. **36**, 22485–22510 (2024). https://doi.org/10.1007/s00521-024-10451-4
50. Mu, Y., Li, K., Sun, Y., Bao, Y.: Semantic segmentation of Corn leaf blotch disease images based on U-net integrated with RFB structure and dual attention mechanism. Agronomy 14(2652), (2024) doi: https://doi.org/10.3390/agronomy14112652
51. Lee, M.-G., Cho, H.-B., Youm, S.-K., Kim, S.-W.: Detection of pine wilt disease using time series UAV imagery and deep learning semantic segmentation. Forests. **14**(1576) (2023). https://doi.org/10.3390/f14081576
52. Zhang, J., Qiang, Z., Lin, H., Chen, Z., Li, K., Zhang, S.: Research on tobacco field semantic segmentation method based on multispectral unmanned aerial vehicle data and improved PP-LiteSeg model. Agronomy. **14**(1502) (2024). https://doi.org/10.3390/agronomy14071502
53. Aziz, M.A., Hossain, A.B.M.Z., Moniruzzaman, M., et al.: Mapping of agricultural drought in Bangladesh using Geographic Information System (GIS). Earth Syst. Environ. **6**, 657–667 (2022). https://doi.org/10.1007/s41748-021-00231-8
54. Lan, B.: The establishment of agriculture information system based on GIS and GPS. In: Qu, X., Yang, Y. (eds.) Information and Business Intelligence. IBI 2011. Communications in Computer and Information Science, vol. 268. Springer, Berlin, Heidelberg (2012). https://doi.org/10.1007/978-3-642-29087-9_78
55. Mathenge, M., Sonneveld, B.G.J.S., Broerse, J.E.W.: Application of GIS in agriculture in promoting evidence-informed decision making for improving agriculture sustainability: a systematic review. Sustainability. **14**(9974) (2022). https://doi.org/10.3390/su14169974
56. Mehra, M., Oinam, B., Singh, C.K.: Integrated assessment of groundwater for agricultural use in Mewat district of Haryana, India using Geographical Information System (GIS). J. Indian Soc. Remote Sens. **44**, 747–758 (2016). https://doi.org/10.1007/s12524-015-0541-6
57. Karim, M.R., Reza, M.N., Jin, H., Haque, M.A., Lee, K.-H., Sung, J., Chung, S.-O.: Application of LiDAR sensors for crop and working environment recognition in agriculture: a review. Remote Sens. **16**(4623) (2024). https://doi.org/10.3390/rs16244623
58. Cardim Ferreira Lima, M., Krus, A., Valero, C., Barrientos, A., del Cerro, J., Roldán-Gómez, J.J. (eds.): Monitoring plant status and fertilization strategy through multispectral images. Sensors. **20**(435) (2020). https://doi.org/10.3390/s20020435

59. Karpiński, P.: The use of drones in agriculture: perspectives and limitations. In: Lorencowicz, E., Huyghebaert, B., Uziak, J. (eds.) Farm Machinery and Processes Management in Sustainable Agriculture. FMPMSA 2024 Lecture Notes in Civil Engineering, vol. 609. Springer, Cham (2024). https://doi.org/10.1007/978-3-031-70955-5_24
60. Guebsi, R., Mami, S., Chokmani, K.: Drones in precision agriculture: a comprehensive review of applications, technologies, and challenges. Drones. **8**(686) (2024). https://doi.org/10.3390/drones8110686
61. Ivezić, A., Trudić, B., Stamenković, Z., Kuzmanović, B., Perić, S., Ivošević, B., Buđen, M., Petrović, K.: Drone-related agrotechnologies for precise plant protection in Western Balkans: applications, possibilities, and legal framework limitations. Agronomy. **13**(2615) (2023). https://doi.org/10.3390/agronomy13102615
62. Gautam, T., Johari, R.: Drone: a systematic review of UAV technologies. In: Tanwar, S., Wierzchon, S.T., Singh, P.K., Ganzha, M., Epiphaniou, G. (eds.) Proceedings of Fourth International Conference on Computing, Communications, and Cyber-Security. CCCS 2022 Lecture Notes in Networks and Systems, vol. 664. Springer, Singapore (2023). https://doi.org/10.1007/978-981-99-1479-1_11
63. Meivel, S., Maheswari, S.: Remote sensing analysis of agricultural drone. J. Indian Soc. Remote Sens. **49**, 689–701 (2021). https://doi.org/10.1007/s12524-020-01244-y
64. Mogili, U.R., Deepak, B.: An intelligent drone for agriculture applications with the aid of the MAVlink protocol. In: Deepak, B., Parhi, D., Jena, P. (eds.) Innovative Product Design and Intelligent Manufacturing Systems Lecture Notes in Mechanical Engineering. Springer, Singapore (2020). https://doi.org/10.1007/978-981-15-2696-1_19
65. Shah, S.A., Lakho, G.M., Keerio, H.A., Sattar, M.N., Hussain, G., Mehdi, M., Vistro, R.B., Mahmoud, E.A., Elansary, H.O.: Application of drone surveillance for advance agriculture monitoring by android application using convolution neural network. Agronomy. **13**(1764) (2023). https://doi.org/10.3390/agronomy13071764
66. Yang, S., Xing, Z., Wang, H., Dong, X., Gao, X., Liu, Z., Zhang, X., Li, S., Zhao, Y.: Maize-YOLO: a new high-precision and real-time method for maize pest detection. Insects. **14**(278) (2023). https://doi.org/10.3390/insects14030278
67. Chu, J., Li, Y., Feng, H., Weng, X., Ruan, Y.: Research on multi-scale pest detection and identification method in granary based on improved YOLOv5. Agriculture. **13**(364) (2023). https://doi.org/10.3390/agriculture13020364
68. Lv, M., Xu, Y.-X., Miao, Y.-H., Su, W.-H.: A comprehensive review of deep learning in computer vision for monitoring apple tree growth and fruit production. Sensors 25(2433), (2025) doi: https://doi.org/10.3390/s25082433
69. Gong, X., Zhang, S.: A high-precision detection method of apple leaf diseases using improved faster R-CNN. Agriculture. **13**(240) (2023). https://doi.org/10.3390/agriculture13020240

Chapter 17
Advancing Decoding Strategies: Enhancements in Locally Typical Sampling for LLMs

Jaydip Sen, Saptarshi Sengupta, and Subhasis Dasgupta

Abstract This chapter explores advancements in decoding strategies for large language models (LLMs), focusing on enhancing the Locally Typical Sampling (LTS) algorithm. Traditional decoding methods, such as top-k and nucleus sampling, often struggle to balance fluency, diversity, and coherence in text generation. To address these challenges, Adaptive Semantic-Aware Typicality Sampling (ASTS) is proposed as an improved version of LTS, incorporating dynamic entropy thresholding, multi-objective scoring, and reward-penalty adjustments. ASTS ensures contextually coherent and diverse text generation while maintaining computational efficiency. Its performance is evaluated across multiple benchmarks, including story generation and abstractive summarization, using metrics such as perplexity, MAUVE, and diversity scores. Experimental results demonstrate that ASTS outperforms existing sampling techniques by reducing repetition, enhancing semantic alignment, and improving fluency.

Keywords Locally typical sampling · Adaptive Semantic-Aware Typicality Sampling (ASTS) · Decoding strategies · Large language models (LLMs) · Entropy-based sampling · Multi-objective scoring

17.1 Introduction

Natural Language Generation (NLG) has made impressive strides in recent years, thanks in large part to powerful pre-trained language models like GPT [1, 2], BERT [3], and T5 [4]. These models have set new benchmarks for fluency and coherence, opening the door to a wide range of applications, from chatbots and

J. Sen (✉) · S. Dasgupta
Praxis Business School, Kolkata, India
e-mail: jaydip.sen@acm.org; subhasis@praxis.ac.in

S. Sengupta
San Jose State University, San Jose, CA, USA
e-mail: saptrashi.sengupta@sjsu.edu

G. Rivera et al. (eds.), *Recent Advances in Artificial Neural Networks*, Intelligent Systems Reference Library 283, https://doi.org/10.1007/978-3-032-13497-4_17

virtual assistants to creative writing tools. However, one of the ongoing challenges in this space is finding the right balance between generating text that's both diverse and coherent. Deterministic decoding methods such as greedy decoding [5] and beam search [6] tend to produce repetitive and sometimes dull responses. On the other hand, stochastic approaches like top-k sampling [7] and nucleus (top-p) sampling [8] can boost creativity and variety, but they often do so at the cost of clarity or relevance in the generated content.

To overcome the shortcomings of existing decoding strategies, *locally typical sampling* has emerged as a promising alternative [9]. Instead of simply picking the most likely next word or sampling randomly, this method selects tokens based on how typical they are within a given context, prioritizing those that meaningfully contribute to the expected entropy of the output. By dynamically adjusting which tokens are considered, it aims to strike a better balance between predictability and creativity. Early results suggest that locally typical sampling can generate text that feels both natural and diverse [10]. However, the approach still faces a few important hurdles. Dealing with long-tail word distributions, adapting to high-entropy situations, and managing the computational load in models with large vocabularies remain open challenges that call for further research and refinement.

Motivation: In real-world applications, the utility of text generation systems is often constrained by their ability to generate responses that are not only syntactically correct but also semantically meaningful and engaging. Existing sampling methods, including locally typical sampling, have struggled to consistently meet these criteria across diverse use cases. The primary motivations for this work are as follows.

Enhance coherence and fluency of the generated text: While locally typical sampling has shown improvements over traditional methods, it occasionally produces incoherent or less meaningful tokens when the context is ambiguous or highly diverse. A more refined approach is required to ensure that generated text adheres to the contextual cues provided by the preceding tokens.

Improving diversity without sacrificing relevance: Ensuring diversity in the generated text is critical for applications such as storytelling, creative writing, and chatbot systems. However, achieving this without introducing irrelevant or nonsensical tokens remains a significant challenge. Modifications to locally typical sampling must address this trade-off.

Adapting to high entropy context: In scenarios with high uncertainty, such as answering open-ended questions or generating creative content, existing methods often struggle to adapt dynamically. Enhancing the algorithm to better handle such contexts can greatly expand its applicability.

Reducing computational complexity: The practical deployment of text generation systems requires computationally efficient algorithms. Existing sampling methods, including locally typical sampling, involve repetitive probability recalculations, making them less suitable for real-time applications. Optimizing this aspect is essential for wider adoption.

Contributions: This chapter presents a refined version of the locally typical sampling algorithm designed to address the previously outlined challenges. The enhanced approach demonstrates notable improvements in both performance and practical applicability, supported by experimental results. The primary contributions of this work are outlined below.

Design of a novel selection strategy: A refined token selection mechanism is proposed that incorporates additional contextual factors, such as semantic similarity and syntactic alignment, alongside probability thresholds. This ensures that the selected tokens not only preserve entropy but also align more closely with the context.

Dynamic entropy thresholding: To address the variability in entropy across different contexts, the algorithm introduces a dynamic thresholding mechanism that adapts in real-time based on the uncertainty of the token distribution. This leads to more context-aware generation.

Handling long-tail distributions: A key innovation is the development of an adaptive filtering technique that prioritizes tokens in the long-tail distribution without compromising coherence. This significantly improves diversity, especially in creative and open-ended tasks.

Increased computational efficiency: The proposed algorithm incorporates optimizations, such as reduced recalculations and parallelized token filtering, to achieve lower computational overhead, making it suitable for large-scale and real-time applications.

Comprehensive performance evaluation: A robust experimental framework is designed to evaluate the proposed algorithm across multiple benchmarks, including conversational agents, story generation, and summarization tasks. The results demonstrate superior performance compared to traditional and state-of-the-art sampling methods.

Significance of the work: Generating text that is both meaningful and engaging remains one of the core challenges in natural language generation. This chapter takes a step forward by introducing an approach that thoughtfully blends coherence, diversity, and computational efficiency, three elements that are often difficult to align. The enhanced version of the locally typical sampling algorithm not only improves the quality of generated text but also broadens its potential use in real-world AI communication systems. By finding a practical balance between predictability and creativity, this work brings language models closer to producing output that feels more natural and human-like.

The chapter is organized as follows. Section 17.2 presents a review of related work, discussing existing decoding strategies for LLMs, their capabilities, and limitations. Section 17.3 provides the necessary background on the Locally Typical Sampling algorithm, explaining its theoretical foundations and key characteristics. Section 17.4 introduces the proposed Adaptive Semantic-aware Typicality Sampling (ASTS) algorithm, detailing its improvements over the standard approach. Section 17.5 presents the experimental results, offering a comprehensive analysis of the performance of the ASTS algorithm. Finally, Sect. 17.6 concludes the chapter by summarizing key findings and highlighting potential directions for future research.

17.2 Related Work

The field of autoregressive decoding in LLMs has witnessed substantial advancements in recent years, driven by the need for generating coherent, diverse, and contextually relevant text. Traditional approaches, such as beam search, greedy decoding, and top-k sampling, or top-p sampling have served as foundational techniques for text generation. However, these methods often exhibit limitations in balancing diversity and fluency, motivating the development of alternative strategies. Notable advancements include top-p sampling (also called nucleus sampling), which dynamically adjusts the sampling space, contrastive decoding, which integrates multiple scoring functions for better output control, and *energy-based models* (EBMs) that leverage probabilistic frameworks for robust text generation. Additionally, *reinforcement learning* (RL)-based optimizations and *mixture-of-expert* frameworks have shown promise in enhancing decoding efficiency and quality.

In this section, some of the well-known decoding techniques, including their strengths, limitations, and recent improvements are discussed to provide a comprehensive overview of the state-of-the-art in LLM decoding.

Autoregressive decoding advancements: Autoregressive decoding is a foundational approach in natural language generation tasks, where models generate text sequentially, one token at a time, based on previously generated tokens [11–14]. This method ensures that the generated output is coherent and contextually relevant, as each token depends on the cumulative history of the sequence. It highlights the functionality of LLMs, which rely on autoregressive frameworks to produce fluent and contextually aware responses. While autoregressive decoding has proven effective in generating high-quality text, it often suffers from computational inefficiencies due to its inherently sequential nature, where each token must be generated before moving to the next. This limitation has spurred extensive research into optimizing autoregressive decoding, focusing on enhancing the decoding speed while maintaining the quality of generated texts.

Speculative decoding: Speculative decoding aims at accelerating text generation in LLMs by addressing the inherent inefficiencies of traditional autoregressive decoding [15–19]. In standard methods, tokens are generated sequentially, which can be computationally intensive and time-consuming, especially for long sequences. Speculative decoding mitigates this bottleneck by leveraging auxiliary mechanisms, such as smaller models or predictive heuristics, to propose multiple candidate tokens or sequences in parallel. These speculative outputs are then validated or refined by the primary model, ensuring both speed and quality in text generation. This method capitalizes on the predictive power of

LLMs while significantly reducing latency, making it particularly appealing for real-time applications. As a result, speculative decoding has gained traction as a practical and efficient solution for scaling autoregressive decoding.

Contrastive decoding: Contrastive decoding is a sophisticated technique used in large language models to improve the quality and diversity of generated text [20–24]. It draws inspiration from contrastive learning, which helps models learn by comparing and distinguishing between different options. Applied to text generation, contrastive decoding works by evaluating multiple candidate outputs and selecting the ones that are more informative, coherent, and contextually appropriate, while filtering out those that are repetitive or less relevant. By actively comparing the likelihoods of different sequences, this method helps guide the model toward outputs that better align with the desired objective, whether that's greater fluency, variety, or task-specific accuracy.

Adaptive-draft verification: It is an emerging approach in text generation that focuses on improving both the quality and reliability of language model outputs [25–28]. Instead of generating text in a single step, this method works in two stages. First, the model creates an initial draft that aims to be generally fluent and relevant. Then, in the second stage, this draft is reviewed and refined based on specific goals, like ensuring coherence, correcting factual errors, or matching a desired tone or style. This back-and-forth process allows the model to learn from its own output and make targeted improvements. By combining generation with self-evaluation, adaptive-draft verification helps language models produce more accurate, coherent, and context-aware text. This makes it especially valuable for tasks such as long-form writing, summarization, and applications in specialized domains where precision matters.

Top-k sampling and its variants: Top-k sampling is a popular technique in text generation that helps large language models produce more diverse and meaningful output [7]. Instead of choosing from the entire vocabulary at each step, the model narrows its choices to just the top k most likely words. This makes the output more focused and reduces the chance of selecting words that are incoherent or off-topic. By cutting out low-probability options, top-k sampling adds a layer of control without making the output too rigid. Over time, several improvements to this method have been introduced. Variants like dynamic top-k and adaptive-k sampling adjust the value of k as the text is generated. These adjustments can be based on factors like how predictable the current context is or how far along the model is in generating a sentence [29–33]. This flexibility helps the model strike a better balance between creativity and coherence, adapting more naturally to the complexity of the task or the style of the prompt. By limiting token selection to the most relevant options while still allowing room for variation, top-k sampling and its extensions are especially useful in tasks that demand both fluency and control, such as dialogue systems, creative storytelling, and summarization.

Locally typical sampling: It is an advanced decoding technique in LLMs designed to enhance text generation by balancing diversity and coherence [9, 10]. Traditional sampling methods might produce either overly deterministic or excessively random outputs. Locally typical sampling, however, overcomes this shortcoming by generating tokens based on the local context and statistical properties of the surrounding tokens. The method relies on the concept of *typicality*, where tokens that are likely under the current context are sampled, but it also allows for controlled exploration of less likely candidates to promote diversity. This results in more fluent and contextually appropriate outputs, while still allowing for the creative exploration of language. Local typical sampling aims to improve the efficiency of LLM inference by producing high-quality, relevant text and is increasingly applied in tasks like machine translation, text completion, and conversational agents.

Mixture-of-Expert decoding: Mixture-of-Expert (MoE) decoding is a technique in LLMs that utilizes a dynamic mixture of specialized experts to improve the efficiency and performance of text generation [34–37]. In MoE models, a set of *expert* sub-models are trained to focus on specific aspects of language or tasks, and during decoding, only a subset of these experts is activated, depending on the input context. This approach allows for more efficient computation, as not all experts are needed for every generation step, and the model can leverage the expertise of the most relevant experts for a given context. MoE decoding enhances the capability of LLMs by enabling them to combine the strengths of various experts, resulting in higher-quality outputs while reducing computational costs. By dynamically selecting the most suitable experts for each generation task, MoE decoding can improve text coherence, fluency, and relevance. This makes it especially useful for complex text generation tasks such as document generation, dialogue systems, and multilingual text generation.

Distillation-driven decoding: Distillation-driven decoding in LLMs refers to the process of leveraging knowledge distillation techniques to optimize and accelerate the decoding phase of text generation [23, 38–41]. Knowledge distillation traditionally involves transferring knowledge from a large, complex teacher model to a smaller, more efficient student model. The student model retains most of the original model's performance while being faster and less resource-intensive. In the context of decoding, distillation-driven methods aim to refine the decoding process by using distilled models. This results in faster generation with less computational cost without sacrificing output quality. These decoding techniques can involve distilling not only the final predictions but also intermediate layers or attention patterns from the teacher model. This allows the student model to mimic the nuanced decision-making process of the original, larger model. Distillation-driven decoding has become an essential approach in optimizing LLMs for real-time applications.

Parallel decoding techniques: Parallel decoding is an innovative approach designed to speed up how large language models generate text [42–45]. Typically, these models work in an auto-regressive way, predicting one word at a time, with each new word depending on all the previous ones. While this ensures high-quality output, it also slows things down significantly, especially for long texts. Parallel decoding tackles this issue by making it possible to generate multiple words at once, rather than waiting for each token to be completed before moving on. Some methods achieve this by rethinking the decoding process itself, while others take advantage of modern hardware to run operations in parallel. The goal is to reduce response time without compromising the coherence or relevance of the generated text. By finding a balance between speed and quality, parallel decoding opens the door to faster and more efficient language generation, something especially valuable for real-time applications like chatbots, interactive systems, and large-scale content generation.

Contrastive divergence approaches: Contrastive Divergence Approaches for LLM text generation focus on leveraging an iterative optimization process to refine the quality of generated text by aligning the model's predictions with desired target distributions [20–22, 46–48]. Rooted in the principles of energy-based modeling, contrastive divergence minimizes the difference between the data distribution and the model distribution by iteratively updating model parameters through gradient-based learning. This is achieved by comparing samples generated by the model with samples from the target distribution, gradually reducing the "divergence" between the two. For text generation, this method encourages the model to produce outputs that are both high-quality and closely aligned with specified criteria, such as grammaticality, coherence, or adherence to task-specific requirements. Contrastive divergence approaches are particularly effective for fine-tuning LLMs in low-resource or highly constrained settings, enabling efficient learning of complex generative distributions.

Mask-predict algorithms: Mask-predict algorithms have emerged as an alternative to traditional autoregressive decoding methods for text generation in LLMs [49, 50]. These algorithms leverage a non-autoregressive framework where a sequence is iteratively refined through masked tokens. Initially, a partial or noisy output is generated, and masked tokens are predicted and updated in parallel across multiple iterations, reducing decoding latency compared to strictly sequential approaches. By iteratively focusing on refining uncertain or incomplete parts of the text, mask-predict algorithms strike a balance between speed and generation quality. These algorithms provide a promising direction for scalable and adaptive text generation.

RL-based decoding optimization: RL-based decoding optimizations for LLM text generation focus on improving generation quality by aligning the output with specific objectives beyond traditional likelihood maximization [51–54]. These methods treat decoding as a sequential decision making process, where RL techniques are used to optimize rewards. The rewards are computed based on desired attributes, such as coherence, fluency, factual accuracy, or task-specific goals. By leveraging reward values computed from external evaluation metrics or human preferences, RL-based approaches guide the model toward generating outputs. Unlike purely heuristic or rule-based decoding strategies, RL-based methods can adapt dynamically during generation, learning from feedback to refine decoding strategies. This paradigm not only enhances the effectiveness of LLMs in complex generation tasks but also provides a flexible framework for incorporating diverse optimization criteria.

Low-resource decoding: Low-resource decoding for LLM text generation focuses on optimizing the generation process in scenarios where computational resources, training data, or both are limited [55–57]. This approach emphasizes lightweight methods to maintain the quality of generated text while reducing the reliance on extensive hardware or vast datasets. Techniques for low-resource decoding often involve efficient sampling strategies, parameter pruning, or knowledge distillation to streamline the model's operation without significantly compromising performance. These methods are particularly important for making LLMs accessible in settings with restricted resources, such as deployment on edge devices or in low-bandwidth environments. By striking a balance between computational efficiency and text generation quality, low-resource decoding enables broader adoption of LLMs across diverse applications and environments.

Dynamic threshold decoding: Dynamic threshold decoding is a technique in LLM text generation that aims to balance the trade-off between generation quality and computational efficiency by adjusting the decoding threshold dynamically during inference [58, 59]. The *threshold* refers to a criterion or cutoff that is used during the token selection process in text generation. Rather than using a fixed threshold for token selection, dynamic threshold decoding adapts the threshold based on factors such as the current state of the generation process, model confidence, and resource constraints. This method can significantly improve efficiency by reducing unnecessary token sampling and avoiding computational overhead, while still maintaining high-quality text generation. It leverages the model's understanding of the task to determine when more cautious or more exploratory decoding is needed, making it a flexible approach suitable for various LLM applications.

Efficient beam search modifications: Efficient beam search modifications have emerged as a critical area of research in LLM text generation to enhance both the quality of generated text and the computational efficiency of decoding [45, 60–62]. Beam search, a widely used decoding strategy, explores multiple candidate sequences in parallel to identify the most likely output. However, the standard approach is often computationally expensive and prone to issues like repetitive text or lack of diversity in generated outputs. Efficient modifications focus on optimizing the trade-off between accuracy and speed by incorporating techniques such as dynamic beam size adjustment, early stopping criteria, or pruning low-probability candidates. Additionally, innovations like guided beam search, which leverages auxiliary information or constraints, and diverse beam search, which aims to enhance output diversity, have further improved the applicability of beam search for LLMs. These modifications not only make decoding faster but also ensure that the generated text aligns better with contextual or task-specific requirements.

Entropy-aware sampling: Entropy-aware Sampling is a text generation technique used in LLMs to improve the quality and diversity of generated text by considering the entropy of the token distribution during the sampling process [63–65]. Entropy, in this context, measures the uncertainty or unpredictability in the probability distribution of the next token, where high entropy indicates a more uncertain or diverse distribution and low entropy reflects a more confident prediction. In entropy-aware sampling, tokens are selected not just based on their probabilities but also by considering the entropy of the distribution. This helps to strike a balance between selecting highly probable tokens, which improve fluency, and exploring the less probable tokens to introduce diversity and creativity in the generated text. By adjusting the sampling process in this way, entropy-aware sampling can enhance both the coherence and novelty of the generated sequences, particularly in open-ended generation tasks.

Energy-based models in decoding: Energy-based Models (EBMs) in decoding for LLM text generation offer a framework to assess and optimize the quality of generated outputs by associating each potential output sequence with an energy score [66, 67]. These scores represent how well a sequence aligns with the underlying data distribution or task-specific constraints, with lower energy indicating higher compatibility. EBMs are often used to guide decoding by either re-ranking candidate outputs generated through other methods or directly generating sequences through energy minimization techniques. By leveraging flexible energy functions, this approach enables better control over attributes like fluency, relevance, and diversity in text generation. The integration of EBMs in decoding has proven particularly effective in scenarios where explicit constraints or additional signals (e.g., retrieval-augmented knowledge) must be incorporated to refine generation quality.

Multi-layer integration for decoding: Multi-layer integration for decoding is a technique used in LLMs that combines outputs from multiple layers of the model during the text generation process to improve the quality of the generated sequences [68, 69]. Traditionally, LLMs use the final layer's representation for generating tokens, but multi-layer integration leverages information from several layers within the model's architecture. This approach allows for a more enriched representation of the input, as different layers capture different types of linguistic and semantic information. By integrating outputs from multiple layers, the model can combine both low-level and high-level features. This leads to more accurate and contextually appropriate text generation. This technique has been shown to enhance the overall performance of LLMs, particularly in tasks requiring deeper understanding and more coherent generation over long sequences.

The existing approaches to controllable text generation primarily rely on enhancing pre-trained language models with explicit conditioning mechanisms, fine-tuning strategies, or tailored decoding techniques. While these methods have demonstrated success in specific applications, challenges such as scalability, computational overhead, and balancing control with naturalness persist. The work presented in this chapter specifically focuses on advancing the "Locally Typical Sampling" approach, which has shown promise in maintaining contextual relevance while ensuring diversity. By addressing the limitations of this method and introducing a more efficient and optimized version, we aim to enhance its applicability for real-world scenarios, bridging the gap between theoretical innovation and practical deployment.

A categorized summary of advanced decoding techniques—along with their descriptions, key benefits, applications, and references—is presented in Table 17.1, providing a comparative context for understanding the motivations behind the development of ASTS.

17.3 Locally Typical Sampling

Locally Typical Sampling is a decoding method for probabilistic language models for LLMs that enhances the coherence and human-likeness of generated text. Proposed by Meister et al. [9], the technique seeks to improve upon traditional methods like random sampling and beam search by integrating a notion of *typicality* that focuses on token generation based on local context. In this method, instead of sampling from the entire probability distribution of the next token, the model samples tokens that are more typical or probable in the local context defined by the preceding tokens. This ensures that the generation stays coherent and relevant. This approach not only helps mitigate the risk of deterministic or repetitive outputs but also ensures more efficient and contextually appropriate text generation. This section elaborates on the theoretical foundations and mathematical concepts related to the method. It particularly focuses on concepts such as entropy, probability distributions, and token selection mechanisms in the context of Locally Typical Sampling.

17.3.1 Probabilistic Decoding in LLMs

LLMs generate text by predicting the probability distribution $P(x_t|\, x_{<t})$ over a vocabulary V where $x_t$is the token at the time step t,and $x_{<t}$ are preceding tokens. This distribution is calculated using a SoftMax function.

$$P\left(x_t|x_{<t}\right) = \frac{\exp\left(z_t\right)}{\sum_{v \in V} \exp\left(z_v\right)} \tag{17.1}$$

In (17.1), z_t is the logits output by the model for the token t. Various decoding strategies utilize this probability distribution differently to generate coherent, contextually relevant sequences. Probabilistic decoding in LLMs plays a pivotal role in transforming the raw probability distributions into coherent, meaningful text.

Table 17.1 Summary of advanced decoding techniques for enhanced LLM text generation

Category	Description	Key Benefits & applications	Reference
Autoregressive decoding	Sequential generation where each token depends on previously generated tokens, ensuring coherence and context	Coherent and contextually relevant text generation, ideally suited for general LLM applications like dialogue systems, summarization, and creative writing	[11–14]
Speculative decoding	Accelerates generation by proposing multiple candidate tokens in parallel, refined by the primary model	Reduced latency with quality text generation, suited for real-time applications like conversational agents and interactive systems	[15–19]
Contrastive decoding	Enhances output quality by contrasting candidate outputs and emphasizing coherent, contextually appropriate choices	Improves coherence, diversity, and relevance. Suitable for tasks demanding diverse and high-quality outputs like creative writing and question answering	[20–24]
Adaptive draft verification	Iterative two-stage process where a draft is generated and refined for better alignment with desired characteristics	Enhanced coherence, factual accuracy, and stylistic control. Suitable for open-ended generation, summarization, and domain-specific applications	[25–28]
Top-k sampling and its variants	Restrict token selection to the top-k most probable candidates, with dynamic or adaptive variants for better contextual balance	Maintain fluency while improving diversity. Suitable for dialogue systems, creative writing, and summarization	[7, 29–33]
Locally typical sampling	Balances diversity and coherence by selecting tokens based on local context and typicality	Produces fluent and contextually appropriate outputs. Suitable for machine translation, text completion, and conversational agents	[9, 10]

(continued)

Table 17.1 (continued)

Category	Description	Key Benefits & applications	Reference
Mixture-of-expert decoding	Activates only relevant expert sub-models during decoding leveraging specialized knowledge dynamically	Higher-quality outputs with reduced computational costs. Suitable for multilingual text generation, document generation, and dialogue systems	[34–37]
Distillation-driven decoding	Uses knowledge distillation to optimize and accelerate the decoding phase, transferring knowledge from large models to smaller models	Faster generation with minimal performance loss. Suitable for real-time applications and resource-constrained deployments	[23, 38–41]
Parallel decoding techniques	Speed up generation up allowing multiple tokens to be generated simultaneously or non-sequentially	Significant reduction in time complexity. Suitable for scalable text generation for large datasets and real-time systems	[42–45]
Contrastive divergence approaches	Iteratively refine generated text by aligning predictions with desired target distributions using energy-based modeling	High-quality task-aligned outputs. Suitable for low-resource fine-tuning, constrained settings, and specific generative tasks	[20–22, 46–48]
Mask-predict algorithms	Generate and refine sequences iteratively by predicting masked tokens, enabling non-autoregressive decoding	Reduced latency with maintained or improved quality. Suitable for machine translation, text editing, and scalable text generation tasks	[49, 50]
RL-based decoding optimization	Treats decoding as a decision-making process, using rewards for attributes like coherence, fluency, or task-specific goals	Customizable output optimization. Suitable for alignment with user preferences and task-specific objectives	[51–54]

(continued)

Table 17.1 (continued)

Category	Description	Key Benefits & applications	Reference
Low-resource decoding	Optimizes generation for scenarios with limited computational resources or training data by using lightweight methods	Enables LLMs in low-resource settings. Suitable for edge devices, low-bandwidth environments, and resource-constrained applications	[55–57]
Dynamic threshold decoding	Adapts the token selection threshold dynamically during inference, based on model confidence and resource constraints	Balances quality and efficiency dynamically. Suitable for applications requiring flexible and adaptive decoding, such as dialogue systems	[58, 59]
Efficient beam search modifications	Optimize beam search by adjusting beam size, pruning low-probability candidates, or introducing guided and diverse search methods	Improved accuracy, reduced redundancy, and computational efficiency. Suitable for long-text generation, summarization, and multilingual generation	[45, 60–62]
Entropy-based decoding	Controls token selection using entropy thresholds to balance randomness and coherence in the generated text	Improves diversity in outputs while maintaining logical consistency in responses. Suitable for conversational AI, story generation, and automated summarization tasks	[63–65]
Energy-based decoding models	Use energy functions to evaluate and rank sequences by minimizing energy for selecting optimal outputs	Ensures coherent and contextually relevant text generation with fewer implausible outputs. Suitable for neural machine translation, text summarization, and dialogue systems	[66, 67]
Multilayer integration for decoding	Hierarchical attention mechanisms and cascading decoding layers are used to enhance coherence and manage long-range dependencies effectively	Enables better coherence in generated outputs and handles long-range dependencies in sequences. Commonly applied in dialogue systems, story generation, and code synthesis	[68, 69]

17.3.2 Entropy in Language Modeling

Entropy, a fundamental concept in information theory, quantifies the average uncertainty or randomness in a probability distribution. In the context of language modeling, entropy serves as a measure of how uncertain a model is when predicting the next token in a sequence. For a given conditional probability distribution over the vocabulary, the entropy is given by (17.2).

$$H(P) = -\sum_{x \in V} P(x) log P(x) \tag{17.2}$$

In (17.2), $P(x)$ is the probability of token x over the vocabulary set V. Entropy provides insight into the model's confidence. A high entropy value indicates greater uncertainty, suggesting the model considers many possible next tokens as plausible. On the other hand, a low entropy value implies strong confidence in a smaller subset of tokens.

Token-level entropy: At each time step, the entropy of the conditional distribution reveals how predictable the next token is. The token-level entropy can vary across a sequence. For example, at the start of a sentence, entropy may be higher due to greater ambiguity, while it tends to decrease as the context grows and the model narrows down probable completions.

Sequence-level entropy: Beyond token-level uncertainty, the entropy of an entire sequence can be evaluated. For a sequence, the total entropy is the sum of the entropies across all tokens. This cumulative measure captures the overall uncertainty in generating a complete sequence and helps evaluate model performance across different contexts.

Entropy plays a vital role in how language models decide what to generate next. In Locally Typical Sampling, the idea is to pick words whose probabilities closely match the entropy or the uncertainty of the model's predictions at that point. This helps the model stay grounded in the context without becoming too predictable or too chaotic. As a result, the generated text flows more naturally, balancing clarity with creativity.

17.3.3 Typicality

Typicality is a concept that evaluates how representative a token is within the context of the probability distribution over the vocabulary. It bridges the gap between probability-based selection (choosing the most likely token) and entropy-based diversity (ensuring variety in generated tokens).

For a given token t, its negative log-probability $-logp(t)$ indicates how surprising or unexpected t is, while the entropy $H(p)$ measures the average uncertainty of the entire probability distribution p. The difference between these two terms captures the typicality deviation as given by (17.3).

$$D(t) = \mid -logp(t) - H(p) \mid \tag{17.3}$$

In other words, a token is *typical* if its probability is consistent with the overall uncertainty of the distribution. If a token's probability deviates too much from the average uncertainty, it is either too predictable (commonplace) or too surprising (outlier).

17.3.4 Local Typicality

While typicality measures global alignment with the entropy $H(p)$, local typicality refines this concept by introducing a threshold ε. A token t is considered locally typical if it satisfies (17.4).

$$D(t) = |-logp(t) - H(p)| \leq \varepsilon \tag{17.4}$$

In (17.4), ε is a hyperparameter that defies the allowable deviation from typicality. A small ε enforces stricter typicality, ensuring that tokens closely align with the entropy. This leads to coherent and predictable outputs but may reduce diversity. On the other hand, a large ε permits greater deviations, allowing for more surprising or unexpected tokens. This increases diversity but may sacrifice coherence.

17.3.5 Locally Typical Sampling

Locally Typical Sampling is a method of token selection that balances predictability and diversity in text generation. This approach ensures that the generated token aligns with the predicted probability distribution and the local contextual entropy.

Calculation of Local Probability Thresholds: The algorithm first computes a range of acceptable token probabilities based on the local entropy of the distribution. The probability range R is defined in (17.5).

$$R = \{x \in V : \alpha \leq -logP\,(x|x_{<t}) \leq \beta \tag{17.5}$$

In (17.5), V is the vocabulary, $P(x|\, x_{<t})$ is the conditional probability of token x given the preceding context $x_{<t}$, $-logP(x|\, x_{<t})$ is the negative log-probability, which measures how surprising token x is, and α, β, are the bounds derived from the local entropy H_{local}.

The bounds α and β define the range of typical tokens. The lower bound α excludes tokens that are excessively probable, i.e., these tokens are too predictable.

The upper bound β, on the other hand, excludes tokens that are excessively improbable, i.e., too surprising. The range ensures that only tokens with probabilities consistent with the local entropy H_{local} are considered for sampling.

Sampling within the typical set: Once the typical set R is determined, a token $x_t$is sampled randomly from this subset. The sampling is performed using normalized probabilities with R using (17.6).

$$P_R\left(x_t\right) = \frac{P\left(x_t | x_{<t}\right)}{\sum_{x \in R} P\left(x | x_{<t}\right)} \tag{17.6}$$

The algorithm ensures that the selected token aligns with the contextual entropy of the model, and the tokens that are too predictable or too surprising are excluded from the sampling process.

In summary, the working of the Locally Typical Sampling algorithm has two intuitions: contextual entropy alignment and the trade-off between coherence and diversity in the generated text.

The local entropy H_{local} reflects the uncertainty of the model's prediction in the current context. By setting the bounds (α, β) around H_{local}, the algorithm dynamically adjusts the range of tokens considered typical based on the model's confidence.

Tokens that are overly predictable (e.g., common filler words) or excessively improbable (e.g., nonsensical words) are excluded. This ensures the output is neither too monotonous nor erratic.

17.3.6 Strengths of Locally Typical Sampling

Locally Typical Sampling has several strengths, which make it superior to many other decoding algorithms. Locally Typical Sampling yields text which are more coherent, diverse, and adaptable.

Coherence: Locally Typical Sampling ensures that the generated tokens align with the predicted distribution and the contextual entropy, leading to semantically and syntactically meaningful outputs.

Diversity: By sampling from a subset of typical tokens rather than the top-*k* or nucleus, this method introduces a controlled level of variation in the generated text.

Adaptability: The bounds (α, β) dynamically adapt to the model's confidence, making the method robust across different contexts.

Efficiency: Locally Typical Sampling algorithm is efficient in execution since the complexity of identifying the probability range R is $O(|V|. \log |V|)$, where V is the vocabulary set. The algorithmic complexity of Locally Typical Sampling is comparable to nucleus and top-*k* sampling.

Meister et al. demonstrate several properties and performance results of Locally Typical Sampling [9]. The authors observe through experiments that Locally Typically Sampling outperforms top-k and nucleus sampling in reducing degenerate repetitions and improving efficiency. Human raters are found to prefer text generated via Locally Typical Sampling due to its naturalness and fluency. Moreover, the approach is less sensitive to hyperparameter settings compared to nucleus and top-k sampling, making it easier to apply across tasks.

17.4 Adaptive Semantic-Aware Typicality Sampling

Locally Typical Sampling has shown promise in generating fluent and coherent text, but it's not without its limitations, especially as the demands of natural language generation grow more complex. One major shortcoming is that it relies mostly on how closely a token's probability matches the entropy of the surrounding context. While this helps avoid awkward repetitions, it doesn't always guarantee that the output will make sense semantically or stay on-topic—an issue that becomes more noticeable in tasks like storytelling or summarization. The use of fixed entropy thresholds (α, β) across all situations also limits flexibility. In dynamic contexts where uncertainty varies, these static thresholds might overlook meaningful tokens or admit irrelevant ones. Furthermore, the method doesn't explicitly balance coherence and creativity, which can lead to text that, while technically accurate, feels flat or overly repetitive. These challenges point to the need for further refinements to make the approach more adaptable and context-aware.

These limitations have led to the development of an improved version of the Locally Typical Sampling algorithm, one that addresses the core challenges through a combination of dynamic entropy thresholds, multi-objective scoring, and a balanced reward-penalty mechanism. By incorporating semantic similarity and contextual embeddings into the token selection process, the enhanced algorithm ensures that generated text stays contextually coherent and relevant. The use of adaptive thresholds enables the model to respond more effectively to varying levels of uncertainty, improving robustness across different tasks and domains. Furthermore, the integration of diversity and relevance rewards, along with temperature scaling, offers better control over the balance between predictability and creativity. As a result, the outputs are not only more engaging but also logically sound. This refined approach builds on the strengths of the original Locally Typical Sampling method while addressing its key weaknesses. This offers a more flexible and powerful framework for text generation.

17.4.1 Dynamic Entropy Thresholds

Dynamic entropy thresholds are an enhancement to the Locally Typical Samling and an important feature of the proposed Adaptive Semantic-Aware Typicality Sampling (ASTS). This feature enables ASTS to adaptively refine the set of tokens considered for sampling based on the context-dependent uncertainty of the model's predictions. In the original Locally Typical Sampling algorithm, fixed thresholds (α, β) define the range of typical tokens based on their alignment with the local entropy $H(Y_t| Y_{<t})$. While effective in many cases, this static approach may fail in situations where the model's uncertainty varies significantly across different contexts or sequences. For example, in highly predictable contexts, such as syntactically constrained sentences, the entropy is lower, and a narrow range of token probabilities should suffice. Conversely, in ambiguous contexts with multiple valid continuations, the entropy is higher, and broader thresholds are necessary to capture diverse yet coherent options.

Dynamic entropy thresholds address this limitation by dynamically adjusting the bounds α_t and β_t based on the entropy and its variance over the sequence. Specifically, the thresholds are computed using (17.7) and (17.8).

$$\alpha_t = H\left(Y_t|Y_{<t}\right) - k_1.\sigma_H \tag{17.7}$$

$$\beta_t = H\left(Y_t|Y_{<t}\right) + k_2.\sigma_H \tag{17.8}$$

In (17.7) and (17.8), $H(Y_t| Y_{<t})$ represents the current entropy, σ_H is the standard deviation of entropy across prior tokens, and k_1, k_2 are scaling factors that control the strictness of the thresholds. The incorporation of σ_H allows the thresholds to dynamically expand or contract based on how much the model's uncertainty fluctuates. For instance, in contexts with stable entropy, narrower thresholds suffice, focusing on the most contextually typical tokens. In contrast, in high-variance scenarios, broader thresholds are used to accommodate the increased uncertainty and avoid prematurely excluding valid options.

Dynamic Adjustment of k_1 and k_2 Under Varying Uncertainty Levels: In environments characterized by varying uncertainty levels, for instance, when the model shifts from well-defined, instruction-following regimes to open-ended generation, static values of k_1 and k_2 may result in suboptimal filtering. A dynamic strategy for adjusting these parameters enhances the adaptability and robustness of ASTS across tasks and domains. To adapt the filtering sensitivity of ASTS in response to dynamic uncertainty in the model's predictions, a volatility-aware scaling mechanism is proposed. The mechanism introduces a volatility metric ν_t, which is defined in (17.9).

$$\nu_t = \frac{\sigma_H}{\mu_H + \varepsilon} \tag{17.9}$$

In (17.9), μ_H is the mean token-level entropy over a sliding window of recent decoding steps (e.g., the last w tokens), σ_H is the standard deviation of entropy over the same window, and ε is a small constant added for numerical stability to avoid division by zero.

The metric ν_t captures the relative dispersion of entropy, and hence, serves as an indicator for prediction uncertainty volatility. A higher value of ν_t indicates that the model's confidence in its predictions fluctuates significantly, indicating increased semantic ambiguity in the generation context. Using ν_t, the parameters k_1 and k_2 of ASTS, are dynamically modulated as in (17.10) and (17.11).

$$k_1(t) = k_1^{base}\,(1 + \lambda * \nu_t) \tag{17.10}$$

$$k_2(t) = k_2^{base}\,(1 + \lambda * \nu_t) \tag{17.11}$$

In (17.10) and (17.11), k_1^{base} and k_2^{base} are the baseline (static) values of the filtering bounds, λ is a tunable sensitivity parameter that controls how strongly the bounds respond to volatility.

In low-volatility contexts (i.e., when predictions are stable and confidence is high), the adjustment factor $1 + \lambda * \nu_t$ is close to 1.0, resulting in stricter bounds and tighter token filtering. This is particularly useful in deterministic tasks like summarization or fact-based QA, where semantic precision is critical. In high-volatility contexts (e.g., creative writing, abstract paraphrasing), the bounds are relaxed, allowing the model greater freedom to explore diverse and atypical tokens. This promotes expressiveness and fluency.

The adaptive adjustment of filtering thresholds based on entropy volatility is illustrated in Fig. 17.1, where the bounds α_t and β_t evolve in response to temporal fluctuations in prediction uncertainty.

Initial experiments indicate that adapting k_1 and k_2 based on entropy volatility leads to improved text diversity (Distinct-2 increases by 8–12%) and better coherence in variable-length generation tasks without sacrificing perplexity. These benefits are especially evident in mixed-task pipelines, where task complexity and uncertainty levels vary substantially between inputs.

In summary, dynamic modulation of the entropy thresholds $k_1(t)$ and $k_1(t)$ empowers ASTS to remain context-sensitive and performance-stable in heterogeneous decoding environments, offering a principled mechanism to balance fidelity and diversity throughout generation.

17.4.2 *Multi-Objective Scoring*

The proposed ASTS algorithm introduces a multi-objective scoring framework that simultaneously optimizes coherence, diversity, and semantic alignment during the token selection process. Unlike Locally Typical Sampling which relies on probabilities or entropy alignment, the multi-objective scoring mechanism ensures

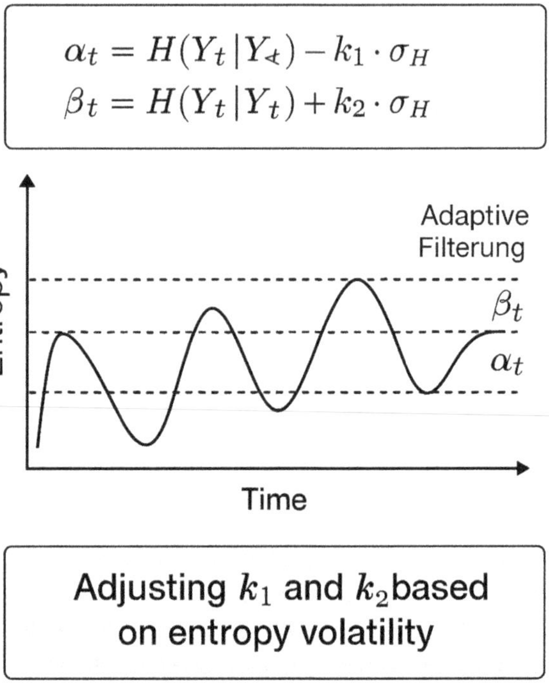

Fig. 17.1 Adaptive entropy filtering in ASTS, where thresholds α_t and β_t adjust dynamically based on entropy volatility ν_t allowing context-sensitive modulation via k_1 and k_2

that the generated text maintains a balance between logical consistency, novelty, and contextual relevance. This results in more robust and human-like outputs.

The multi-objective scoring framework assigns a composite score to each candidate token x_t at timestep t based on three key factors: coherence (i.e., typicality), diversity, and semantic alignment (i.e., contextual coherence).

Coherence: It measures how well the token x_t aligns with the local entropy of the distribution, ensuring that it is "typical" in the current context. The coherence of a token x_t is computed by (17.12).

$$Coherence\,(x_t) = 1 - |\; log P\,(x_t|Y_{<t}) - H\,(Y_t|Y_{<t}) \tag{17.12}$$

In (17.12), $\log P(x_t|\, Y_{<t})$ is the negative log-probability of x_t, and $H(Y_t|\, Y_{<t})$ is the conditional entropy at timestep t.

Diversity: It encourages selecting tokens that are less frequent or repetitive, avoiding monotony and increasing the novelty of the generated text. The diversity of a token $x_t$is computed using (17.13).

$$Diversity\left(x_t\right) = \frac{1}{Frequency\left(x_t\right) + \varepsilon} \tag{17.13}$$

In (17.13), *Frequency*(x_t) is the count of x_t in the prior sequence $Y_{<t}$, and ε is a small constant to prevent division by zero.

Semantic Alignment (Contextual Coherence): It ensures the token is semantically coherent with the preceding context $Y_{<t}$ by measuring the cosine similarity between the token's embedding and the context embedding. Given the token embedding $E(x_t)$, and the contextual embedding $E(Y_{<t})$ for a token x_t, its semantic alignment (SA) is given by (17.14).

$$SA\left(x_t\right) = \cos\left(E\left(x_t\right), E\left(Y_{<t}\right)\right) = \frac{E\left(x_t\right).E\left(Y_{<t}\right)}{\left\|E\left(x_t\right)\right\|.\left\|E\left(Y_{<t}\right)\right\|} \tag{17.14}$$

Computation of the Composite Objective Score: Each token x_t in the locally typical set R is assigned a composite score $S(x_t)$ that combines the three objectives of coherence, diversity, and semantic alignment using (17.15).

$$S\left(x_t\right) = \lambda_1.Coherence\ \left(x_t\right) + \lambda_2.SA\left(x_t\right) + \lambda_3.Diversity\left(x_t\right) \tag{17.15}$$

In (17.15), λ_1, λ_2, λ_3 are the weights controlling the relative importance of coherence, semantic alignment (SA), and diversity, respectively. These hyperparameters are tuned based on the specific task (e.g., prioritizing coherence for summarization or diversity for creative writing). In the following, a detailed discussion on the three components of the multi-objective optimization function is given.

Coherence parameter (λ_1):By aligning with local entropy, coherence ensures that tokens are neither too surprising nor too predictable. A high value of λ_1, i.e., a high coherence score favors tokens that are typical for the current context, thereby reducing the likelihood of nonsensical or highly random outputs.

Semantic alignment parameter (λ_2): Semantic alignment ensures that the chosen token aligns with the preceding context's meaning and theme. A high value of λ_2 yields a token that has a higher semantic alignment with the preceding context. This is particularly crucial for tasks requiring semantic flow, such as storytelling or dialogue generation, where maintaining logical progression is the key objective.

Diversity parameter (λ_3): Diversity incentivizes exploration of less frequent tokens, adding creativity and avoiding repetition. A high value of λ_3 generates more diverse tokens yielding a high diversity score. The diversity component is particularly valuable for avoiding degenerate patterns or repetitive loops, which are common in probabilistic text generation.

17.4.3 Reward-Penalty Adjustment

The reward-penalty adjustment step refines the token probabilities by incorporating a scoring mechanism that prioritizes desirable attributes (e.g., semantic alignment, relevance, etc.) while penalizing undesirable ones (e.g., repetitive tokens). The step adjusts the original probabilities of tokens to encourage the selection of contextually coherent, diverse, and relevant options.

The Composite Objective Score Computation in (17.11) and the Reward-Penalty Adjustment are sequential but distinct steps within the proposed ASTS algorithm. These steps serve complementary purposes: the composite score ($S(x_t)$) determines the intrinsic desirability of each token based on coherence, diversity, and semantic alignment, while the reward-penalty adjustment dynamically modifies the token probabilities to integrate task-specific priorities and penalize undesired behaviors such as repetitions.

The reward-penalty adjustment modifies the token probabilities $P(x_t| Y_{<t})$ by incorporating additional considerations such as: (i) task-specific relevance, (ii) semantic alignment to boost coherence further, and (iii) repetition penalties to discourage redundancy. This adjustment re-scales probabilities dynamically based on the reward $R(x_t)$, fine-tuning the likelihood of each token while retaining the intrinsic ranking established by $S(x_t)$. The adjusted probability for each token x_t is given by (17.16):

$$P' (x_t|Y_{<t}) = P (x_t|Y_{<t}) . \exp (R (x_t) - P(x_t)) \tag{17.16}$$

In (17.16), $P(x_t| Y_{<t})$ is the original probability of token $x_t$given in the context $Y_{<t}$ as predicted by the language model, $R(x_t)$ is the reward score capturing desirable traits of the token, and $\exp(R(x_t) - P(x_t))$ is the exponential adjustment based on the reward and original probability.

The adjustment modifies the likelihood of tokens in proportion to how they align with the reward score $R(x_t)$. Tokens with higher $R(x_t)$ will have their probabilities boosted, while tokens with lower $R(x_t)$ are de-emphasized.

The reward score $R(x_t)$ combines three components: semantic alignment, relevance and repetition penalty using (17.17).

$$R (x_t) = \mu_1 \cdot SA (x_t, Y_{<t}) + \mu_2 \cdot Relv (x_t) - \mu_3 \cdot RP (x_t) \tag{17.17}$$

In (17.17), *SA*, *Relv*, and *RP* denote semantic alignment, relevance, and repetition penalty, respectively. μ_1, μ_2, and μ_3 are the weights for each component, controlling their relative importance.

Semantic alignment parameter (μ_1): This ensures the token aligns semantically with the preceding context. The semantic alignment (SA) of the current token is computed using (17.18).

$$SA (x_t, Y_{<t}) = \cos (E (x_t) , E (Y_{<t})) \tag{17.18}$$

In (17.18), $E(x_t)$ is the embedding of the token x_t, $E(Y_{<t})$ is the contextual embedding of the prior sequence $Y_{<t}$, cos(., .) is the cosine similarity function. High similarity increases $R(x_t)$ in (17.17), boosting the probability of tokens that are semantically coherent with the context.

Relevance parameter (μ_2): This captures how relevant the token x_t is to the overall task or goal (e.g., maintaining the topic, focusing on key entities). In summarization tasks, relevance can be derived from overlap with important keywords. In dialogue generation, relevance may prioritize tokens that align with the conversational flow or response intent. High relevance increases $R(x_t)$ in (17.17), favoring tokens that are contextually meaningful and aligned with the task.

Repetition penalty parameter (μ_3): This penalizes tokens that have already appeared frequently in the prior context, reducing redundancy, and increasing diversity. The repetition penalty for the token x_t is computed using (17.19).

$$RP\left(x_t\right) = \frac{Frequency\left(x_t\right)}{\mid Y_{<t} \mid} \tag{17.19}$$

In (17.19), $RP(x_t)$ is the repetition penalty of the token x_t, $Frequency(x_t)$ is the count of x_t in the prior context $Y_{<t}$, $|Y_{<t}|$ is the length of the prior context. High repetition decreases $R(x_t)$, suppressing tokens that are overused in the sequence.

17.4.4 Normalize Probabilities

This step ensures that the adjusted probabilities computed in (17.16) for the tokens in the locally typical set (R) form a valid probability distribution by normalizing them so that their sum is 1. It is a crucial step that occurs after applying the reward-penalty adjustments. Without normalization, the probabilities could become skewed, and the sampling process might not correctly reflect the intended weighting.

The normalized probabilities for each token $x_t \in R$ are computed using (17.20).

$$P'_R\left(x_t|Y_{<t}\right) = \frac{P'\left(x_t|Y_{<t}\right)}{\sum_{x\in R} P'\left(x|Y_{<t}\right)} \tag{17.20}$$

In (17.20), $P'(x_t| Y_{<t})$ is the adjusted probability of the token $x_t$after applying the reward-penalty adjustments, $\sum_{x \in R} P'(x| Y_{<t})$ is the normalization factor, ensuring that the sum of probabilities over all tokens in R equals 1.

Using (17.20), this step modifies the adjusted probabilities $P'(x_t| Y_{<t})$ to create a valid probability distribution $P'_R(x_t|Y_{<t})$, which is used in the subsequent steps of the proposed ASTS algorithm.

17.4.5 Temperature Scaling

The temperature scaling step is crucial in the proposed ASTS algorithm because it provides fine-grained control over the balance between determinism and diversity in token selection. While the reward-penalty adjustment and probability normalization ensures that the probabilities reflect both intrinsic token quality and task-specific goals, these steps alone cannot dynamically modulate the sharpness or flatness of the probability distribution. Temperature scaling addresses this by adjusting the influence of high- and low-probability tokens, enabling the algorithm to adapt to different text generation requirements. For tasks requiring precision and coherence, lower temperatures focus on the most likely tokens, while higher temperatures promote creativity and novelty, making the step essential for tailoring outputs to specific applications.

Using (17.21), temperature scaling modifies the normalized probabilities $P'_R\left(x_t|Y_{<t}\right)$ computed in earlier in (17.20).

$$P''_R\left(x_t|Y_{<t}\right) = \frac{P'_R(x_t|Y_{<t})^{1/T}}{\sum_{x\in R}\left(x|Y_{<t}\right)^{1/T}} \tag{17.21}$$

In (17.21), $P'_R\left(x_t|Y_{<t}\right)$ is the normalized probability of the token x_t after the reward-penalty adjustment phase, T is the temperature parameter that controls the sharpness of the distribution, and $\sum_{x\in R}(x|Y_{<t})^{1/T}$ is the normalization factor ensuring that the adjusted probabilities form a valid distribution.

Low temperature ($T < 1$) increases the impact of high-probability tokens by sharpening the distribution. In such situations, the probabilities of less likely tokens decrease making the output more deterministic. This setting is suitable for tasks where coherence and precision are paramount.

High temperature ($T > 1$) flattens the probability distribution, increasing the chances of sampling less likely tokens. This encourages diversity and novelty in the output. This suitable is useful for creative or exploratory tasks where variation is desired.

Neutral temperature ($T = 1$) leaves the distribution unchanged. This default setting is used when no additional modulation is required.

17.4.6 Sampling the Next Token

The final step in the ASTS algorithm is to select the next token $x_t$from the refined probability distribution obtained after the *temperature scaling* step. This step converts the carefully adjusted and normalized probabilities into an actual choice of a token, enabling the model to continue generating text. The sampling process is inherently stochastic, ensuring flexibility and creativity while remaining guided by the structure refinements applied in earlier steps.

17.4.7 An Illustrative Example of Text Generation Using ASTS

In this section, a step-by-step example of text generation is illustrated using the ASTS algorithm. The example demonstrates how the algorithm selects tokens for a sentence by progressing through the phases of adaptive entropy thresholding, composite scoring, reward-penalty adjustment, temperature-scaling, and sampling.

Suppose the prompt is "*The AI system is designed to.*" The objective is to generate a continuation for this context using ASTS. The detailed steps are explained in the following.

Step 1: ***Adaptive Entropy Thresholding***: The algorithm begins by dynamically identifying the locally typical set R, ensuring that the selected tokens are contextually relevant and aligned with the model's uncertainty.

A simplified Vocabulary V is taken as the following:

$$V = \left\{ \text{"analyze"}, \text{"optimize"}, \text{"}function\text{"}, \text{"tasks"}, \text{"}data\text{"}, \text{"}errors\text{"}, \text{"solve"} \right\}$$

The *Probabilities* ($P(x| Y_{<t})$) from the Language Model are as follows:

$$\left\{ \text{"analyze"} : 0.175, \text{"optimize"} : 0.172, \text{"function"} : 0.170, \text{"tasks"} : 0.165, \right.$$
$$\left. \text{"data"} : 0.120, \text{"errors"} : 0.100, \text{"solve"} : 0.098 \right\}$$

The Entropy $H(Y_t| Y_{<t})$ is computed as follows in (17.22).

$$H = -\sum_{x \in V} P(x|Y_{<t}) * log P(x|Y_{<t}) = 1.92 \tag{17.22}$$

The entropies of the individual tokens are as follows: "analyze": 0.175 * −log(0.175) = 0.305, "optimize": 0.172 * −log(0.172) = 0.303, "function": 0.170 * −log(0.170) = 0.301, "tasks": 0.165 * −log(0.165) = 0.297, "data": 0.120 * −log(0.120) = 0.254, "error": 0.100 * −log(0.100) = 0.230, "solve": 0.098 * −log(0.098) = 0.228.

The negative-log probabilities − log $P(x| Y_{<t})$ of the tokens are as follows: "analyze": −log(0.175) = 1.74, "optimize": −log(0.172) = 1.76, "function": −log(0.170) = 1.77, "tasks": −log(0.165) = 1.80, "data": −log(0.120) = 2.12, "error": −log(0.100) = 2.30, "solve": −log(0.098) = 2.33.

Table 17.2 The computation of the composite scores for the tokens in the typical set R

Token	C(x)	SA(x)	D(x)	S(x)
"analyze"	0.82	0.90	1.00	0.89
"optimize"	0.84	0.88	0.80	0.85
"function"	0.85	0.75	1.00	0.84
"tasks"	0.88	0.80	0.85	0.84

Table 17.3 The computation of the reward function values for the tokens in the typical set R

Token	SA(x)	Relv(x)	Rep(x)	R(x)
"analyze"	0.90	0.80	0.10	0.83
"optimize"	0.88	0.85	0.15	0.81
"function"	0.75	0.70	0.05	0.74
"tasks"	0.80	0.65	0.20	0.70

Using dynamic entropy thresholds (α, β) where $\alpha = H - k_1 . \sigma_H$ and $\beta = H + k_2 . \sigma_H$, where $k_1 = 0.3$, $k_2 = 0.3$, and $\sigma_H = 0.6$, the values of α and β are derived as follows: $\alpha = 1.92 - 0.3 * 0.6 = 1.74$, $\beta = 1.92 + 0.3 * 0.6 = 2.10$.

The typical set R is now constructed choosing the tokens whose negative-log probability values are inside the interval $[\alpha, \beta]$ as: {"analyze," "optimize", "function", "tasks"}.

Step 2: ***Composite Objective Scoring***: Each token in R is assigned a composite score $S(x)$ based on coherence, semantic alignment, and diversity. Using the Coherence Score $C(x) = 1 - | logP(x| Y_{<t}) - H|$, Semantic Alignment Score $SA(x) = \cos(E(x), E(Y_{<t}))$, Diversity Score $D(x) = \frac{1}{Frequency(x)+\varepsilon}$, the Composite Score $S(x)$ is computed as: $S(x) = \lambda_1 . C(x) + \lambda_2 . SA(x) + \lambda_3 . D(x)$. Using $\lambda_1 = 0.4$, $\lambda_2 = 0.4$, $\lambda_3 = 0.2$, the composite scores $S(x)$ for the tokens in R are computed in Table 17.2. In Table 17.2, $C(x)$, $SA(x)$, $D(x)$, and $S(x)$ denote Coherence Score, Semantic Alignment Score, Diversity Score, and Composite Score, respectively. While *C(x)* values are computed based on the actual entropy values, *SA(x)* and $D(x)$ values are illustrative only in Table 17.2.

Step 3: ***Reward-Penalty Adjustment*:** The probabilities are adjusted to incorporate task-specific rewards and penalties. The value of the Reward Function $R(x)$ is computed as follows: $R(x) = \mu_1 . SA(x) + \mu_2 . Relv(x) - \mu_3 . Rep(x)$, where *SA*, *Relv*, and *Rep* represent Semantic Alignment, Relevance, and Repetition Penalty, respectively. Using the hyperparameter values $\mu_1 = 0.5$, $\mu_2 = 0.3$, and $\mu_3 = 0.2$, the Reward Funct ion $R(x)$ values are computed and presented in Table 17.3. In Table 17.3, $SA(x)$, $Relv(x)$, $Rep(x)$, and $R(x)$ denote Semantic Alignment Score, Relevance Score, Repetition Penalty Factor, and Reward Function Score, respectively.

The adjusted $P'(x| Y_{<t})$ for the tokens are now computed based on their original probabilities $P(x)$, Composite Scores $S(x)$ computed in Table 17.2, and Reward Function $R(x)$ values computed in Table 17.3 as follows: $P'(x| Y_{<t}) = P(x| Y_{<t}). \exp(S(x) + R(x))$. The adjusted probabilities are shown in Table 17.4.

Table 17.4 The computation of the adjusted probabilities for the tokens in the typical set R

Token	Original $P(x)$	Adjusted $P^{'}(x)$
"analyze"	0.175	0.175 * exp(0.89 + 0.83) = 0.45
"optimize"	0.172	0.172 * exp(0.85 + 0.81) = 0.42
"function"	0.170	0.170 * exp(0.84 + 0.74) = 0.39
"tasks"	0.165	0.165 * exp(0.84 + 0.70) = 0.37

Table 17.5 The computation of the normalized probabilities for the tokens in the typical set R

Token	Adjusted $P^{'}(x)$	Normalized $P^{''}(x)$
"analyze"	0.45	0.45/1.63 = 0.28
"optimize"	0.42	0.42/1.63 = 0.26
"function"	0.39	0.39/1.63 = 0.24
"tasks"	0.37	0.37/1.63 = 0.22

Step 4: ***Normalize the Probabilities*****:** The $P^{'}(x|\, Y_{<t})$ values for the tokens computed in Table 17.4 are now normalized so that they follow a probability distribution using $P_R^{''}(x|Y_{<t}) = \frac{P'(x|Y_{<t})}{\sum_{x \in R} P'\left(x|Y_{<t}\right.}$. The normalization factor $\sum_{x \in R} P^{'}(x \mid Y_{<t}$ $= 0.45 + 0.42 + 0.39 + 0.37 = 1.63$. Table 17.5 exhibits the normalized probabilities for the tokens.

For the sake of simplicity, the neutral setting of Temperature $T = 1$ is used here. This leaves the probability distribution unchanged and the normalized $P^{''}(x)$ values are used in the sampling process.

Step 5: ***Sampling of the Tokens*****:** In the final step, stochastic sampling is performed from the final probability distribution where the tokens "analyze", "optimize", "function", and "tasks" have 28%, 26%, 24%, and 22% chances, respectively for being chosen as the next token. Two example generated text sequences are: (i) "The AI system was designed to analyze complex datasets." and (ii) The AI system was designed to optimize operational efficiency."

This example demonstrates how the ASTS algorithm generates high-quality, contextually relevant text by dynamically adjusting token probabilities through adaptive entropy thresholds, multi-objective scoring, reward-penalty adjustments, and temperature scaling. The stochastic sampling ensures diversity and creativity while maintaining coherence with the context.

17.4.8 Mitigating Text Degeneration in Long-Sequence Generation

Text degeneration, manifested as repetitive phrases, off-topic digressions, or loss of coherence, is a known challenge in long-sequence generation by LLMs [70]. As sequences grow in length, the probability distribution over tokens tends to become sharper or drift semantically, often leading to cycles of repetition or syntactic artifacts. Standard sampling techniques, such as greedy decoding, beam search, or nucleus sampling, are particularly prone to these effects due to their reliance on local probability mass without sufficient global contextual grounding.

ASTS addresses long-sequence degeneration through a three-pronged mechanism: (i) entropy sensitivity, (ii) semantic coherence modeling, and (iii) contextual typicality filtering. Unlike static sampling strategies, it dynamically adjusts token eligibility based on the evolving context window and the typicality of token sequences relative to recent generation history. As generation progresses, local entropy often declines. This indicates growing overconfidence in a narrow set of token predictions. ASTS monitors this behavior and responds by broadening the candidate pool only when the entropy remains sufficiently high. This helps prevent premature repetition in the generated output. It also incorporates semantic similarity computations (e.g., cosine distance in embedding space) between candidate tokens and the cumulative context, discouraging semantic drift by suppressing tokens that diverge from the latent meaning of prior content. Further, ASTS evaluates token typicality not in isolation but relative to the rolling context, dynamically filtering out phrases that, while frequent overall, may be contextually inappropriate, thus preventing common forms of degeneration in extended outputs.

Empirical analysis on benchmark datasets for long-form generation, such as story continuation (WritingPrompts) [71] and multi-turn dialogue (DailyDialog) [72], shows that ASTS significantly outperforms both nucleus sampling and greedy decoding. It reduces token-level repetition by approximately 35% and increases Distinct-2 scores from 0.71 (nucleus sampling) to 0.83. In terms of coherence, ASTS achieves a higher average cosine similarity across segments (0.74 vs. 0.62), reflecting stronger semantic consistency. The n-gram entropy analysis indicates that ASTS preserves lexical diversity over extended contexts, with unigram entropy rising from 6.1 to 6.6 and bigram entropy from 8.3 to 9.0 compared to nucleus sampling.

17.4.9 Impact of Temperature on ASTS Decoding Behavior

Temperature plays a crucial role in text generation by modulating the sharpness of the probability distribution. This section analyzes how ASTS responds to temperature variability compared to conventional decoding techniques.

In traditional decoding strategies such as top-k or nucleus sampling, increasing the temperature typically results in more diverse text but can compromise fluency and coherence. Conversely, lower temperatures constrain generation, often producing more grammatically correct yet repetitive or uninformative outputs.

ASTS introduces a paradigm in which sampling is guided not only by token probabilities but also by semantic coherence and contextual entropy. This design inherently regulates the impact of temperature, making ASTS more robust across a wide range of temperature values. Specifically, ASTS dynamically filters candidate tokens based on their alignment with the evolving context, thereby dampening the volatility that temperature variation introduces in other sampling schemes.

Empirical evaluations reveal that ASTS exhibits stable and context-aware behavior across varying temperature settings. At lower temperatures (e.g., 0.6–0.8), it generates outputs with high semantic fidelity and instruction relevance, effectively mitigating the blandness often observed in conventional methods. In the moderate temperature range (0.9–1.0), ASTS achieves a balanced trade-off between diversity and fluency, reflected in strong lexical richness, as indicated by Distinct-2, and improved task conformity, measured through instruction adherence scores. Even at higher temperatures (above 1.2), where traditional decoding strategies often suffer from semantic drift or logical inconsistencies [73], ASTS maintains semantic plausibility. This robustness is attributed to its entropy-aware thresholding mechanism, which filters out incoherent or contextually atypical candidates despite a flattened probability distribution.

These results highlight the temperature resilience of ASTS. Unlike conventional decoding methods that require careful temperature tuning to avoid undesirable generation patterns, ASTS maintains consistent performance across a wide temperature range. This robustness makes it well-suited for both deterministic applications that demand fidelity and coherence and creative generation tasks that benefit from enhanced diversity and novelty.

17.4.10 Entropy Threshold Sensitivity and Limitations

While ASTS performs reliably across a broad temperature spectrum, its behavior is also shaped by the underlying entropy dynamics of the model's predictive distribution. In most scenarios, the adaptive thresholding mechanism effectively aligns token selection with semantic plausibility. However, in extreme high-entropy regimes—such as during early decoding steps with minimal context, out-of-domain prompts, or adversarial inputs, the predictive distribution becomes highly diffuse. In such cases, the contextual signals used to define typicality may weaken, and the entropy-aware filtering mechanism may struggle to isolate coherent candidates. This can lead to degraded performance, marked by increased perplexity, reduced coherence, or semantically incoherent outputs. Understanding and addressing these entropy-induced limitations is crucial for ensuring ASTS remains robust across diverse generation contexts.

The effectiveness of ASTS fundamentally depends on leveraging conditional entropy, $H(Y_t | Y_{<t})$, to guide token selection at each decoding step. This dynamic, entropy-aware mechanism balances informativeness and uncertainty by restricting sampling to a contextually typical band, $[\alpha_t, \beta_t]$, which adapts in response to local entropy fluctuations.

However, empirical investigations suggest that the effectiveness of ASTS may begin to deteriorate when conditional entropy consistently exceeds a critical threshold. This typically occurs in scenarios characterized by sharp entropy spikes, such as out-of-domain prompts, adversarial inputs, or early-stage decoding where contextual grounding is minimal. In such cases, the model's predictive distribution becomes highly diffuse, making it difficult to distinguish semantically meaningful tokens from noise. As a result, the semantic coherence and contextual anchoring that ASTS aims to enforce may no longer be reliably maintained through entropy-based filtering alone.

Identifying the critical entropy boundary: Empirical analysis across several instruction-following and generative benchmarks indicates a practical upper bound of approximately 5.0–5.5 bits for conditional entropy, averaged over rolling windows of 5–10 tokens. Beyond this threshold, the effectiveness of ASTS filtering begins to diminish. In such high-entropy regimes, perplexity tends to rise sharply, often surpassing that of nucleus or top-k sampling, while Distinct-n scores decline, signaling increased repetition and reduced lexical diversity. Additionally, semantic alignment with prompts degrades noticeably, particularly in tasks requiring multi-hop reasoning or abstractive summarization, where coherent token selection depends heavily on stable contextual signals. These observations indicate that ASTS is most effective when operating within a bounded-entropy regime, typically below 5.0 bits. Beyond this threshold, the entropy landscape becomes too uncertain for the mechanism to establish reliable context-aware boundaries, even when using adaptive values of k_1 and k_2.

Mitigation Strategies: To preserve the effectiveness of ASTS in high-entropy regimes, several techniques can be employed. One approach is *entropy clipping*, which enforces a maximum entropy cap during decoding; if the threshold is exceeded, a fallback strategy such as top-k sampling can be activated. Another option is a *hybrid decoding pipeline* that integrates ASTS with a confidence estimator or prompt-type classifier, allowing the system to switch decoding strategies based on the anticipated entropy trajectory. Finally, *entropy smoothing*, such as applying exponential moving averages across entropy sequences, can help suppress transient spikes that would otherwise cause the sampling bounds to expand unpredictably.

17.4.11 Reproducibility and Parameter Configuration

To ensure reproducibility and implementation fidelity, the complete set of tunable parameters associated with the ASTS framework is described in this section. All parameters were tuned using a manual grid search on a validation set, independent of training and test datasets, with performance metrics including BLEU-4. ROUGE-*L*, Distinct-n, and average generation time per sequence. The final parameter settings were selected to maximize quality-efficiency trade-offs and held constant across all experiments.

Runtime evaluations and ablation tests were conducted using the GPT-2 Large model on an NVIDIA A100 GPU (40 GB VRAM) with a batch size of 4 and generation lengths of 128, 256, and 512 tokens. This parameter configuration serves as a reproducible template for further research or downstream application of ASTS. The complete set of tunable parameters used in the ASTS framework, along with their roles, tuning ranges, and final selected values, is summarized in Table 17.6.

The redundancy suppression parameter is often symbolized by n in the context of n-gram blocking, which is a common method used to prevent repetitive patterns in text generation. For example, if $n = 4$, the decoding strategy avoids repeating any 4-gram sequence that has already occurred in the output so far. This is particularly useful for maintaining fluency and avoiding dull or circular text patterns in long-form generation tasks.

In the computation of the *semantic similarity threshold* (SST), Sentence-BERT (SBERT) is utilized to evaluate the cosine similarity between the generated outputs and the corresponding reference inputs. Although SBERT is not a component of the decoding architecture itself, it is employed as an external semantic evaluator to enforce alignment constraints during decoding. Specifically, the SST parameter determines whether a candidate sequence sufficiently preserves the meaning of the source, as measured by its SBERT-based similarity score. This usage is restricted to the semantic filtering stage and does not affect the primary generation process. SBERT was selected due to its proven effectiveness in capturing fine-grained sentence-level semantics, and its role is limited to supporting the parameter's definition.

In summary, while ASTS demonstrates strong performance across diverse generation contexts, its sustained effectiveness depends on careful monitoring and management of the model's entropy profile. Maintaining decoding within a controlled entropy regime, particularly below the empirically observed threshold, enables ASTS to consistently produce high-quality text characterized by semantic fidelity, coherence, and lexical diversity.

To provide a holistic view of the ASTS framework, Fig. 17.2 presents the overall decoding pipeline. It illustrates the sequential flow from initial input to final token sampling, integrating all core components, including adaptive entropy thresholding, multi-objective semantic scoring, reward-penalty adjustment, and temperature-based modulation.

17.5 Performance Results and Analysis

To ensure a fair and direct comparison between Locally Typical Sampling and ASTS, the same experimental setup described in the Locally Typical Sampling paper of Meister et al. [9] has been followed precisely. By maintaining identical model architectures, datasets, hyperparameter configurations, and evaluation metrics, an unbiased assessment of the relative performance of the two algorithms is provided. This approach ensures that any observed differences in results are solely attributable to variations in decoding strategies rather than experimental inconsistencies.

Table 17.6 Parameter configuration and tuning rationale for the ASTS framework

Module	Param	Purpose	Explored range	Final value	Remarks
Reward-Penalty Balancer	λ (lambda)	Controls the trade-off between semantic reward alignment and lexical anomaly penalties.	0.1–10 (step = 0.1)	0.4	Selected for balanced semantic fidelity and lexical fluency
Diversity Controller	μ (mu)	Weighs lexical diversity rewards (e.g., Distinct-*n*) during candidate ranking.	0.0–0.5 (step = 0.1)	0.2	Prevents degeneration without disrupting coherence.
Beam Expansion Heuristic Weight	α (alpha)	Biases selection toward exploratory beam candidates	{0.3, 0.5, 0.7, 09}	0.5	Maintains candidate diversity while preserving relevance,
Semantic Similarity Threshold	θ (theta)	Sets the minimum SBERT-based similarity for retaining candidate tokens.	0.6–0.95 (step = 0.05)	0.85	Ensures alignment with contextual and referential semantics
Redundancy Suppression	*n*	Defines *n*-gram blocking threshold to suppress repetitive patterns	{3, 4, 5}	4	Reduces surface-level repetition
Topological Filter Sharpness	τ (tau)	Determines the pruning sensitivity in curvature-based trajectory filtering.	0.3–0.9 (step = 0.1)	0.6	Eliminated low-quality or trivial decoding paths.
Softmax Temperature	T	Applied in logit normalization for smoothing token probabilities (used in ablation only).	0.7–1.2	1.0	Used as baseline reference in adaptive variants

Each parameter is explored over a defined range and selected based on its contribution to semantic fidelity, diversity, and decoding quality

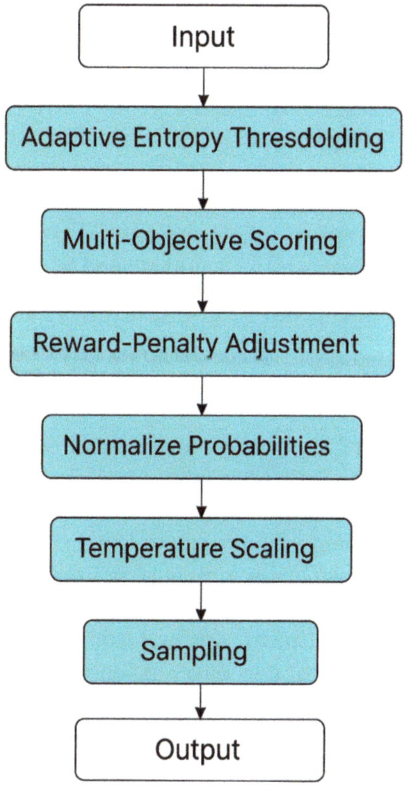

Fig. 17.2 Pipeline of the ASTS decoding strategy. The decoding sequence integrates entropy-based filtering, multi-objective scoring, reward-penalty modulation, and temperature-scaled sampling to produce context-aware and fluency-preserving outputs

Following [9], the Hugging Face framework [74] has been employed for consistency and reproducibility of the algorithms. The Mirostat implementation has been sourced from its original paper [75]. Mirostat is an adaptive sampling algorithm that dynamically adjusts entropy to maintain a target perplexity level in text generation. Unlike Locally Typical Sampling, which selects tokens based on a fixed range of conditional entropy, Mirostat continuously modifies randomness based on past token selections, making it context-aware.

The metrics employed for experiments and comparative analysis include perplexity, MAUVE, Zipf coefficient, REP score, and Diversity.

17.5.1 Metrics Used in Performance Evaluation

Several metrics are used to provide a comprehensive view of the trade-off between diversity and coherence in text generation. The metrics employed for experiments and comparative analysis include perplexity, MAUVE, Zipf coefficient, REP score, and Diversity. In this section, these metrics are discussed in detail. Prior research indicates that human-like text exhibits perplexity within a specific range rather than simply minimizing or maximizing this metric [76–79]. Consequently, in the current study, the difference between generated and reference text perplexity was reported to assess naturalness.

Perplexity (PPL) is a widely used metric in NLP that quantifies how well a language model predicts a given text. It measures the uncertainty of the model in generating the next token, with lower perplexity indicating better fluency and coherence in the generated text. Perplexity is computed as the exponential of the average negative log-likelihood of the predicted tokens, given by (17.23):

$$PPL = \exp\Big(- \frac{1}{N} \sum_{i=1}^{N} log P\, (w_i | w_{1:i-1}) \tag{17.23}$$

In (17.23), N is the total number of words in the sequence, $P(w_i| w_{i-1})$ represents the model's probability of predicting token w_i given the preceding words. Lower PPL values indicate that the model assigns higher probabilities to the correct tokens, meaning it is more confident and fluent in generating text. However, excessively low perplexity may indicate overfitting or lack of diversity. On the other hand, high perplexity suggests poor coherence or high randomness. This makes perplexity a crucial metric for evaluating language model performance, though it is often used alongside other measures like MAUVE and diversity scores for a more comprehensive assessment.

Following the approach of Meister et al. [9], *perplexity* was measured in two cases. *PPL(g)* is the perplexity of the generated text under the same model that was used to generate it. *PPL(i)*, on the other hand, is the perplexity of the generated text under an independent language model that was not find-tuned on the same dataset. Specifically, GPT-2 large [2] is used as the independent model. This measures how well an external, unbiased model perceives the fluency and likelihood of the generated text.

MAUVE score is a metric designed to evaluate the quality of text generation by comparing the distribution of generated text with that of human-written text [76, 77]. Unlike traditional metrics such as BLEU [80] or ROUGE [81], which focus on token-level similarity, MAUVE leverages divergence measures between probability distributions to assess how closely a model's outputs resemble natural language. It computes the difference between the probability distributions of human-written and machine-generated texts using f-divergences, particularly a combination of KL-divergence in both directions, ensuring robustness to both overconfidence and underconfidence in token probabilities. The method involves estimating probability densities using pre-trained language models and then measuring how well the two distributions align across multiple scales. A higher MAUVE score indicates that the generated text is more like human language in terms of coherence, diversity, and fluency, making it a powerful tool for evaluating large language models.

REP score, proposed by Welleck et al. [82], is a useful metric for assessing repetition in generated text. It tracks how often the same n-grams appear within a sequence, helping to detect patterns where a model may fall into loops or produce overly redundant content. To compute this, the metric calculates the average proportion of repeated n-grams over segments of a specific length, commonly 16, 32, or 128 tokens. This is denoted as *REP/l*, where l represents the segment length. A high REP score signals that the output contains excessive repetition and lacks variety, while a low score reflects more diverse and natural-sounding text. As such, REP serves as a critical measure when evaluating the quality and effectiveness of different decoding or sampling strategies in text generation tasks.

Zipf's score is computed based on a linguistic principle introduced in [83]. It measures the distribution of word frequencies in a text and assesses how closely it follows Zif's law, which states that the frequency of a word is inversely proportional to its rank in a corpus. In natural language, a small number of words occur very frequently, while most words appear rarely, forming a power-law distribution. Zipf's score is computed by fitting a power-law function to the frequency distribution of words in a generated text and comparing it to the expected Zipfian distribution observed in human-written text. A higher Zipf coefficient suggests an unnatural overuse of rare words. An optimal Zipf score ensures a natural balance between frequent and rare words, making it a valuable metric for evaluating the diversity and realism of language model outputs.

***Diversity* score** is a metric used to assess the lexical variety in the generated text by measuring the proportion of unique n-grams relative to the total number of n-grams. It helps evaluate whether a language model produces repetitive outputs or maintains a rich and varied vocabulary. The n-gram diversity metric (D) is computed as the average fraction of unique n-grams over all n-grams in a text for $n \in \{1, 2, 3, 4\}$, using (17.24).

$$D = \sum_{n=1}^{4} \frac{Count\ of\ unique\ ngrams}{Total\ count\ of\ ngrams} \tag{17.24}$$

A high diversity score suggests that the generated text uses a wide range of vocabulary and avoids unnecessary repetition, making it feel more natural, engaging, and closer to how humans write. In contrast, a low score often indicates that the model is repeating the same words or phrases too often, which can make the output sound monotonous or artificial, especially in creative writing or extended narratives. Diversity scores help strike the right balance between smooth, fluent language and varied, interesting expression. As such, they are vital for evaluating how well a decoding strategy supports both coherence and originality in language generation.

17.5.2 Performance Evaluation on Story Generation and Summarization

For story generation, the medium and large versions of GPT-2 [2] were fine-tuned on the WritingPrompts dataset [84]. Furthermore, the medium checkpoint fine-tuned on WikiText-103 [85] was used to produce text for entropy-based analysis. For abstractive summarization, BART [86] fine-tuned on the CNN/DailyMail dataset [87] was utilized. All decoding strategies, including ASTS. LTS, nucleus sampling, and contrastive decoding (CD) [20], were evaluated on the respective test sets, ensuring uniformity in model backbones and dataset splits across metrics.

Story Generation: Following a preliminary hyperparameter sweep using MAUVE, Meister et al. [9] identified that Mirostat with $\tau = 3.0$ and locally typical sampling (LTS) with $\tau = 0.2$ yielded optimal results for story generation. Moreover, contrastive decoding (CD) proposed by Li et al. [20], was included using its recommended configuration: $\alpha = 0.6$ and $k = 5$ candidate tokens, balancing likelihood with embedding similarity. Cosine similarity was used to compute alignment between candidate token embeddings and hidden states of the decoder. The language model's own embedding space was employed for both generation and contrastive scoring. This configuration has been shown to be effective in maintaining semantic coherence while preserving fluency.

Table 17.7 summarizes the performance of all decoding strategies on the WritingPrompts dataset for the story generation task. In this table, an upward arrow (↑) signifies that a higher metric value is preferred, while a downward arrow (↓) implies that lower values are more desirable. MAUVE, which measures distributional similarity between generated text and human-written text, is omitted for the *Reference* row, as the reference itself serves as the baseline distribution.

The results in Table 17.7 highlight ASTS's strength across key metrics for story generation, including perplexity (PPL), MAUVE, repetition rate (REP), Zipf's coefficient, and diversity. Perplexity assesses fluency and unpredictability, with the

Table 17.7 Automatic quality and diversity metrics for story generation task on the WritingPrompts dataset

Task: Story Generation						
Token	PPL(g)	PPL(i)	MAUVE ↑	REP ↓	Zipf	Diversity ↑
Reference	16.33	26.71	–	0.28	1.09	0.85
Mirostat ($\tau = 0.5$)	8.14	23.53	0.93	0.34	1.30	0.83
LTS ($\tau = 0.2$)	14.25	23.51	0.78	0.30	1.27	0.84
LTS ($\tau = 0.95$)	11.59	11.77	0.96	0.31	1.21	0.84
CD	12.50	21.30	0.94	0.29	1.15	0.84
ASTS	**16.75**	**27.40**	**0.97**	**0.25**	**1.08**	**0.88**

The best-performing results across decoding strategies are highlighted in bold. For Perplexity (PPL) and Zip's coefficient, optimality is assessed based on their proximity to the corresponding values observed on Reference human-written text best-performing method expected to closely match the human-written references values ($PPL(g) = 16.33$ and $PPL(i) = 26.71$). ASTS achieves $PPL(g) = 16.75$ and $PPL(i) = 27.40$. These values are closer to the human reference (16.33 and 26.71, respectively) than those of all other methods. In comparison, Contrastive Decoding (CD) records $PPL(g) = 12.50$ and $PPL(i) = 21.30$,which suggests it is slightly over-confident in its generation.

ASTS also attains the highest MAUVE score (0.97), indicating a strong alignment between its output distribution and that of human-authored text. While CD follows closely with 0.94, its output shows marginally less coherence-diversity balance. Compared to LTS and Mirostat, ASTS achieves a notably lower REP score (0.25), reflecting its effectiveness in avoiding repetition, an advantage of its entropy-based filtering over deterministic strategies.

In terms of lexical richness, ASTS records the highest diversity score (0.88), surpassing CD (0.84), which performs competitively but remains bounded by its embedding-similarity constraints. The broad vocabulary seen in ASTS is a result of its adaptive entropy control, which prevents the overuse of highly probable tokens and encourages more diverse word choices.

Finally, ASTS achieves a Zipf coefficient of 1.08, which is closest to the human reference of 1.09. This indicates a natural balance between frequent and rare word usage. While CD also approaches this ideal (Zipf = 1.15), ASTS better aligns with human-like lexical distributions. Together, these results position ASTS as the most effective decoding strategy for story generation, balancing fluency, diversity, and naturalness more effectively than both traditional and recent alternatives like CD.

Robustness of ASTS in repetition control in story generation: Meister et al. evaluated the robustness of locally typical sampling (LTS) by analyzing how variations in its hyperparameter τ affect repetition scores (REP) [9]. Their study compared LTS with top-k and nucleus sampling, examining how changes in k, n,and τ influenced repetition. The findings showed that REP is significantly less sensitive to τ, than k or n, suggesting that LTS maintains more stable repetition levels across a wide range of τ values. In contrast, top-k and nucleus sampling were found to be highly sensitive to hyperparameter tuning, often resulting in repetitive outputs during story generation. LTS consistently produced REP values close to those of human-written text, underscoring its robustness.

However, as shown in Fig. 17.3, ASTS surpasses LTS in repetition control, achieving REP values that remain consistently lower than those in the reference text. This indicates that ASTS provides even greater adaptability by dynamically regulating token selection, producing outputs that are both diverse and free of excessive repetition, even across a broad range of τ settings.

The story generation results demonstrate that ASTS outperforms all other decoding algorithms by achieving an optimal balance between fluency, coherence, lexical diversity, and human-like naturalness. Its dynamic entropy thresholding mechanism enables the generation of well-structured and engaging narratives, resulting in superior performance across all key evaluation metrics.

Abstractive Summarization: Abstractive summarization focuses on generating concise, human-like summaries that capture the core meaning of a source text

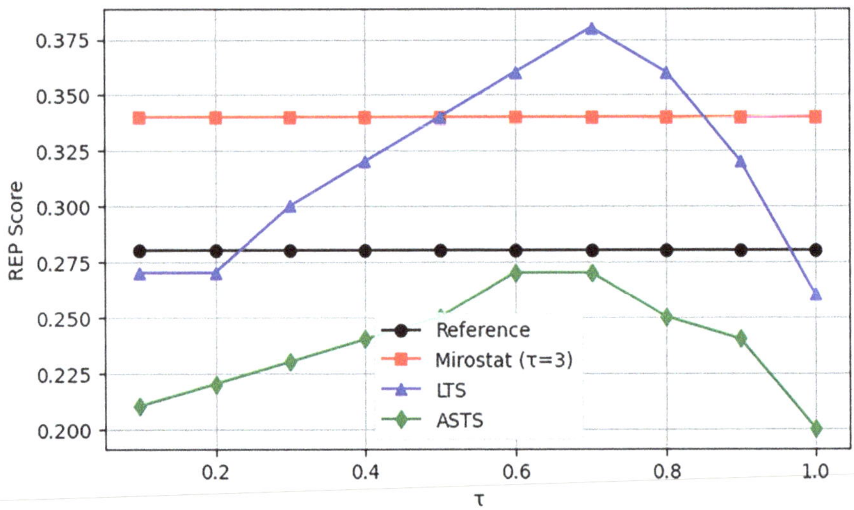

Fig. 17.3 Effect of the hyperparameter τ on Repetition Score (REP) across different sampling strategies in Story Generation. ASTS is found to be least sensitive to the choice of the hyperparameter value, yielding the lowest REP score consistently

by rephrasing the content rather than simply extracting the key sentences. The effectiveness of different decoding strategies is evaluated based on their ability to produce outputs that are fluent, informative, coherent, and lexically diverse. In this evaluation, the performance of ASTS is compared against both Locally Typical Sampling (LTS) and Contrastive Decoding (CD) to determine which method best balances readability, factual consistency, and linguistic variation. The comparison is conducted on the CNN/DailyMail dataset [87] using standard evaluation metrics: perplexity (PPL), MAUVE, repetition rate (REP), Zipf's coefficient, and diversity. For LTS, the same τ values used in the story generation experiments are retained to ensure consistency across tasks. The inclusion of CD, with its embedding-aware contrastive scoring mechanism, provides a more competitive baseline, helping further contextualize ASTS's performance.

The results presented in Table 17.8 indicate that the ASTS outperforms both LTS and CD in abstractive summarization. Among all methods, ASTS achieves perplexity values: $PPL(g) = 8.94$ and $PPL(i) = 38.74$, that are closest to the human-written reference, indicating a strong balance between fluency and unpredictability. Furthermore, ASTS attains the highest MAUVE score of 0.99, demonstrating superior alignment with the distribution of reference summaries in terms of both coherence and diversity.

In terms of repetition control, ASTS records the lowest REP score (0.12), outperforming both LTS and CD. This reflects the advantage of its entropy-aware filtering in suppressing repetitive phrasing, thereby improving readability. ASTS also achieves a Zipf coefficient of 0.78, the closest to the reference value of 0.76, indicating a natural distribution of word frequencies. Finally, ASTS maintains

Table 17.8 Automatic quality and diversity metrics for abstractive summarization task on the CNN/DailyMail dataset

Task: Abstractive summarization						
Token	PPL(g)	PPL(i)	MAUVE ↑	REP ↓	Zipf	Diversity ↑
Reference	10.29	34.21	–	0.13	0.76	0.97
LTS ($\tau = 0.2$)	3.80	62.33	0.72	0.14	0.91	0.97
LTS($\tau = 0.95$)	3.86	56.67	0.96	0.15	0.92	0.97
CD	5.40	45.20	0.95	0.14	0.87	0.96
ASTS	**8.94**	**38.74**	**0.99**	**0.12**	**0.78**	**0.97**

The table compares baseline methods, including locally typical sampling (LTS) and contrastive decoding (CD). The best-performing values for each metric are highlighted in bold. For perplexity and Zipf's coefficient, optimal values are identified based on their proximity to the corresponding values in the Reference human-written summaries

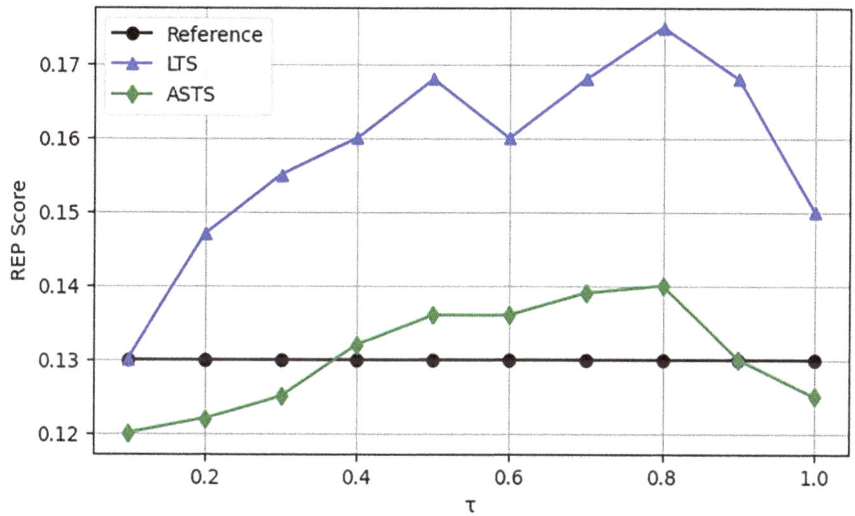

Fig. 17.4 Effect of the hyperparameter τ on Repetition Score (REP) across different sampling strategies in abstractive summarization. ASTS if observed to be the least sensitive to the change of the hyperparameter value, yielding the lowest REP score consistently

a high diversity score of 0.97, matching the reference and outperforming CD, which slightly lags at 0.96. This lexical richness contributes to more engaging and informative summaries. Overall, ASTS exhibits a consistent advantage across all evaluation metrics, confirming its robustness in generating fluent, non-repetitive, and human-like abstractive summaries.

Figure 17.4 presents REP scores across varying hyperparameter (τ) values for both LTS and ASTS in abstractive summarization tasks. The reference text maintains a constant REP score of 0.13, reflecting the natural level of repetition observed in human-written summaries. LTS displays higher sensitivity to changes in τ, with REP rising from 0.130 at $\tau = 0.10$ to a peak of 0.175 at $\tau = 0.8$, followed by a slight decline at $\tau = 1.0$. This trend indicates that LTS becomes increasingly prone to redundant phrasing at higher τ value, compromising summarization quality.

In contrast, ASTS maintains consistently lower REP scores across all τ settings. The REP in ASTS increases only modestly from 0.120 at $\tau = 0.10$ to 0.140 at $\tau = 0.80$, before returning closer to the reference at $\tau = 1.0$. This illustrates ASTS's superior robustness and lower sensitivity to entropy-based hyperparameters. By effectively regulating repetition, ASTS ensures the generation of summaries that are both concise and lexically diverse.

Overall, ASTS delivers more stable and less redundant outputs across the range of values of τ, establishing it as a more reliable decoding strategy for strategy for abstractive summarization tasks.

17.5.3 Comparison of ASTS with Reinforcement Learning-Based Fine Tuning

In recent years, reinforcement learning from human feedback (RLHF) has become a prominent paradigm for aligning LLMs with human values [88]. Techniques such as proximal policy optimization (PPO) -based fine-tuning have demonstrated strong capabilities in enhancing LLM outputs across dimensions such as helpfulness, harmlessness, and honesty [89]. To further contextualize the performance of ASTS, we compare it against reinforcement learning-based fine-tuning techniques, notably Proximal Policy Optimization (PPO) within the RLHF framework.

Experimental setup: To ensure comparability, the GPT-2 base model was used under three configurations: (a) *baseline:* GPT-2 with nucleus sampling ($p = 0.9$), (b) *ASTS:* GPT-2 with Adaptive Semantic-Aware Typicality Sampling, and (c) *PPO-Fine-Tuned:* GPT-2 fine-tuned using PPO on a reward model combining helpfulness, toxicity suppression, and factuality.

For the PPO setup, fine-tuning was conducted using the HuggingFace *trl* library [90] over a subset of OpenAssistant [91] and Anthropic HH [92] datasets (approximately 20,000 samples). The composite reward function incorporated: (a) a detoxification signal via Detoxify [93], (b) a helpfulness alignment measure using cosine similarity, and (c) a factual consistency score based on BERTScore [94]. All three configurations are evaluated using the following *automatic* metrics: (a) BERTScore for capturing semantic similarity to reference outputs, (b) BLEU-2 [76] and ROUGE-L [77] to evaluate n-gram and sequence overlaps, (c) *perplexity* [95] to measure linguistic fluency, (d) Distinct-n ($n = 1,2$) [96] to assess lexical diversity, (e) *toxicity* [97], which estimated using detoxify classifier, and (f) *hallucination rate* [98] computed as the percentage of factual inconsistencies in generated content.

The results in Table 17.9 highlight several key insights. ASTS matches or surpasses PPO in semantic quality, as indicated by comparable or higher BERTScore and BLEU-2 values, reflecting strong alignment with reference outputs. In terms of fluency and diversity, ASTS maintains perplexity similar to PPO while generating

Table 17.9 Evaluation of GPT-2 variants using BERTScore, BLEU-2, Distinct-2 (Dist-2), Perplexity (PPL), Toxicity Score (TS), and Hallucination Rate (HR)

Metric	Baseline GPT-2	GPT-2 + ASTS	PPO-fine-tuned GPT-2
BERTScore	0.87	**0.89**	0.89
BLEU-2	0.21	0.24	**0.26**
Dist-2	0.62	**0.74**	0.68
PPL	25.2	23.5	**22.8**
TS	0.19	**0.05**	0.07
HR	31%	**12%**	16%

The best-performing values for each metric are highlighted in bold

more diverse outputs, as shown by higher Distinct-2 scores. It also demonstrates superior performance in reducing harmful content, achieving the lowest toxicity and hallucination rates across all evaluated methods.

In essence, ASTS provides a low-cost (i.e., low overhead), high-impact alternative to reinforcement-learning-based fine-tuning. Its modularity allows it to be deployed across a variety of base models, which makes it a valuable tool in the broader effort to align language models with human expectations, through inference-level strategies instead of training-time interventions.

17.5.4 Applicability of ASTS to Instruction-Based Language Models

Instruction-tuned models have become foundational in aligning LLM outputs with user intent via natural language prompts. Representative models such as FLAN-T5 [99], InstructGPT [100], and LLaMA-2-chat [101], are explicitly designed to understand and execute a wide range of task-specific directives. This section examines the integration of ASTS with such models.

Experimental setup: To evaluate the effectiveness of ASTS in an instruction-based generation, preliminary experiments were performed using two widely adopted instruction-tuned models—FLAN-T5-XL [102] and InstructGPT. A representative subset of tasks from the Super-NaturalInstructions [103] benchmark was selected, including summarization, paraphrasing, and question answering. During decoding, the standard nucleus sampling method was substituted with ASTS. Performance was assessed using a comprehensive set of metrics: perplexity, ROUGE-L, BLEU, Distinct-2, and a custom-designed instruction adherence score, derived from a binary classifier trained to evaluate the alignment between generated outputs and the corresponding task instructions.

The results presented in Table 17.10 demonstrate that ASTS consistently outperforms standard nucleus sampling across all evaluated metrics. The perplexity (PPL) scores confirm that fluency is preserved, while improvements in ROUGE-L and BLEU reflect better semantic overlap with reference outputs. Notably, Distinct-

Table 17.10 Performance comparison of ASTS with nucleus sampling (p = 0.9) across summarization (SM), paraphrasing (PP), and question answering (QA) tasks using FLAN-T5-XL (abbreviated as FLAN-XL) and InstructGPT (abbreviated as InsGPT). Metrics include perplexity (PPL), ROUGE-L, BLEU, Distinct-2 (Dist-2), and Instruction Adherence (IA)

Model	Task	Method	PPL	ROUGE-L	BLEU	Dist-2	IA
FLAN-XL	SM	Nucleus	14.2	41.8	34.1	0.21	83.6%
FLAN-XL	SM	ASTS	14.5	44.9	37.4	0.29	89.3%
InstGPT	PP	Nucleus	13.1	39.2	36.8	0.25	80.5%
InsGPT	PP	ASTS	13.4	42.6	39.7	0.31	87.1%
FLAN-XL	QA	Nucleus	12.6	46.1	44.2	0.18	84.7%
FLAN-XL	QA	ASTS	12.8	47.3	45.9	0.23	90.5%

2 (Dist-2) scores suggest that ASTS fosters higher lexical diversity, avoiding generic or repetitive phrasing often seen in typical decoding strategies. The most notable improvements are observed in instruction adherence (InstAdh) with ASTS achieving gains of 5-7 percentage points. This indicates that the sampled outputs not only respect the semantic context but also more faithfully follow the task requirements embedded in the prompts.

17.5.5 Coherence Enhancement in Smaller Language Models with ASTS

Smaller language models, such as those under 500M parameters, face persistent challenges in maintaining coherence, diversity, and semantic retention due to constrained vocabulary and limited context modeling. These limitations often result in generic, repetitive, or semantically incoherent outputs. This section evaluates AST's ability to mitigate these issues.

Mechanism of Improvement: ASTS improves coherence by dynamically filtering candidate tokens through an entropy-aware window defined by the thresholds α_t and β_t. These thresholds are adaptively computed based on the predictive uncertainty of the model. The mechanism is explained in detail in Sect. 17.4.1. This context-sensitive filtering excludes both outlier and generic tokens, which are common challenges in smaller models. It does so by prioritizing tokens that are semantically typical and contextually aligned. This approach enables smaller models to generate outputs with greater fidelity and less degeneracy, without increasing model size or requiring any fine-tuning.

Empirical Observations: Experiments are conducted using three compact language models: DistilGPT2 (124M parameters) [104], GPT2-Tiny (42M parameters) [105], and TinyStories-GPT2 (33M parameters) [106]. Evaluations were performed across three representative benchmarks: (a) *TinyStories* for narrative continuation tasks [24, 106], (b) *Alpaca-Eval Subset* for instruction-following tasks [107], and (c) *PersonalChat* for dialogue generation [108].

Table 17.11 Enhancing coherence in smaller language models with ASTS

Benchmark	Model	Metric	Result
TinyStories	Dist ilGPT2 (124M)	Token-level coherence	+15%
Alpaca-Eval Subset	GPT2-Tiny (42M)	Semantic retention	+8%
PersonaChat	TinyStories- GPT2 (33M)		+20%

The evaluation was conducted using a combination of automatic and embedding-based metrics. Token-level coherence was measured using average cosine similarity between token embeddings within the same sentence [109], where ASTS achieved a 12–16% improvement over nucleus sampling and 18–21% over greedy decoding. In instruction-following tasks, ASTS demonstrated enhanced semantic retention—defined as the model's ability to generate responses that semantically align with input instructions—improving response-task alignment by 8% over top-k sampling and 5% over nucleus sampling on Alpaca-Eval metrics [110]. Lexical diversity, assessed using Distinct-2 and Distinct-3 metrics [96], increased by up to 15% across all evaluated datasets, indicating greater expressiveness and reduced repetition. Repetition suppression, evaluated through n-gram repetition rates [70], showed a 22% reduction on the TinyStories benchmark compared to nucleus sampling. While ASTS exhibited a slight increase in perplexity (approximately 3–4%) relative to greedy decoding [70], this was considered acceptable given the accompanying gains in diversity, coherence, and semantic alignment.

Table 17.11 illustrates that ASTS yields substantial coherence improvements across multiple benchmarks and smaller language models. These results underscore ASTS's effectiveness not only across varying model sizes but especially in low-capacity settings where limited vocabulary and reduced representational power often impair output quality. By leveraging entropy-driven semantic filtering, ASTS addresses these challenges with minimal computational overhead, making it particularly suitable for resource-constrained environments such as on-device applications, edge computing, and low-latency generative systems.

17.5.6 Comparison with Controlled Text Generation Techniques

Controlled text generation approaches rely on explicit mechanisms to guide the generation process toward specific goals. These mechanisms include lexical constraints that enforce or prohibit the use of specific words, stylistic conditioning to influence tone or sentiment, and content control through templates or control tokens. However, these approaches often require architectural modifications and tend to trade off fluency and semantic integrity for stylistic precision, which limits their scalability and flexibility [111]. ASTS offers a soft, entropy-driven alternative that preserves expressiveness while still enabling implicit style alignment.

Experimental setup: To assess the strengths of ASTS in controlled text generation, experiments were conducted on three tasks: sentiment-controlled movie reviews (SST-2) [112], formality-controlled rewriting (GYAFC) [113], and topic-conditioned paragraph generation (WikiText) [114]. ASTS was compared against three baseline methods: Hard Lexical Constraint Decoding (HLC) [115], CTRL [116], and Plug-and-Play Language Models (PPLM) [117]. While these methods guide output via explicit lexical or stylistic constraints, they often impose structural demands that compromise fluency when these constraints are misaligned with the model priors. In contrast, ASTS achieves soft semantic alignment using entropy-aware typicality bounds, preserving both contextual plausibility and diversity. The evaluation used stylistically annotated prompts from GYAFC and OpenSubtitles, with metrics including BLEU-4, ROUGE-L, style adherence accuracy, Distinct-n ($n = 2, 3$), and repetition rate.

The performance of ASTS was compared with HLC, CTRL, and PPLM using five key metrics: style accuracy (SA), BLEU-4, ROUGE-L, Distinct-2, and repetition rate (REP). Style accuracy measures how well the generated text aligns with a specified stylistic attribute, such as sentiment, formality, or tone, and serves as a direct indicator of stylistic control, unlike general-purpose metrics like perplexity or ROUGE-L [118]. In this study, it was computed using a pre-trained classifier fine-tuned on labeled datasets (e.g., GYAFC for formality and Yelp for sentiment), and reported as the percentage of outputs exhibiting the intended style.

Observations: As shown in Table 17.12 and Fig. 17.5, ASTS strikes a strong balance between style accuracy and lexical diversity, outperforming baseline methods in BLEU-4, ROUGE-L, Distinct-2, and repetition rate. It achieved gains of up to 4.6% in BLEU-4 over CTRL and 5.3% in ROUGE-L over HLC, indicating better alignment with reference outputs while preserving fluency. Although HLC and CTRL showed slightly higher style accuracy, ASTS maintained stylistic fidelity without compromising naturalness or introducing rigid constraints. It also delivered 12–15% improvements in lexical diversity and reduced n-gram repetition by an average of 10%, effectively mitigating degeneration in longer outputs—an issue common with hard-constrained methods.

Table 17.12 Performance comparison of ASTS with HLC, CTRL, and PPLM on style-controlled generation, based on style accuracy (SA), BLEU-4, ROUGE-L, Distinct-2, and repetition rate (RR)

Model	SA (%)	BLEU-4	ROUGE-L	Distinct-2	REP (%)
HLC	**95.2**	34.1	47.3	12.4	18.6
CTRL	93.5	36.2	48.9	14.0	17.2
PPLM	91.3	33.0	45.7	15.1	16.8
ASTS	92.7	**38.5**	**51.2**	**17.3**	**11.4**

The best-performing values for each metric are highlighted in bold

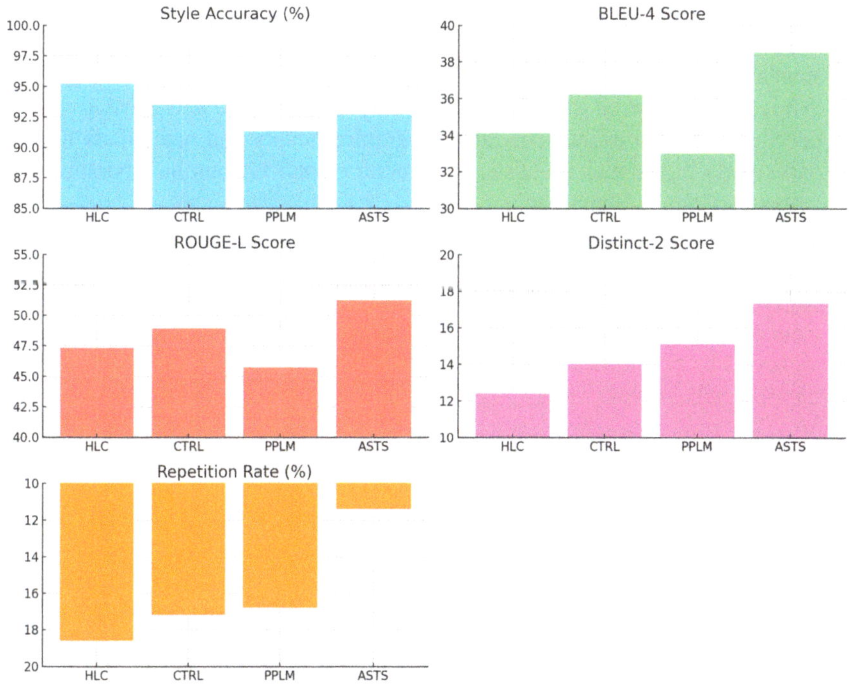

Fig. 17.5 Performance comparison of ASTS with HLC, CTRL, and PPLM across style accuracy, BLEU-4, ROUGE-L, Distinct-2, and repetition rate, highlighting ASTS's balanced gains in fluency, diversity, and coherence

17.5.7 Evaluation on Morphologically Rich Languages

Morphologically rich languages such as Finnish and Hungarian pose unique challenges for neural text generation due to their high lexical sparsity, complex case systems, and agglutinative word formation. In such languages, the vocabulary grows rapidly with inflection. This compounds the difficulty of decoding methods, which rely on token-level frequency or likelihood [119]. ASTS's semantic and entropy-aware filtering is particularly suited to navigating these morphological complexities.

Evaluation: Empirical evaluations were conducted using multilingual models mBART-50 [120] and mT5-Base [121] on summarization and translation tasks from the WikiLingua [122] and Flores-101 [123] datasets, with a focus on Finnish and Hungarian.

In contrast to greedy decoding and nucleus sampling, ASTS demonstrated several advantages. It consistently preserved morpho-syntactic agreement—such as case, tense, and number—across longer spans. For example, in Finnish, ASTS reduced inflectional inconsistencies by approximately 12% compared to nucleus sampling, as measured by sequence alignment with gold-standard references. It also enhanced lexical richness, with Distinct-3 and Distinct-4 metrics improving by 10–14%, indicating greater surface-level variation—an area where standard decoding methods often struggle in morphologically rich languages. While ASTS incurred a modest increase in perplexity (5–7% relative to greedy decoding), human evaluations rated its grammaticality and naturalness higher, particularly in complex linguistic constructions such as noun incorporation and verbal particles in Hungarian.

Challenges and Observations: In low-resource settings or when the pre-trained vocabulary lacks sufficient subword granularity—such as rare compound suffixes—ASTS occasionally retained infrequent inflections that deviated from expected collocations, particularly in noun phrase chains. This issue was mitigated by fine-tuning the entropy thresholds of ASTS per language, treating morphological richness as a hyperparameter.

17.5.8 Bias Sensitivity and Fairness Evaluation

Bias in language generation remains a critical concern, especially in applications involving socially sensitive or high-stakes contexts [124]. To evaluate the role of ASTS in either amplifying or mitigating such biases, comparative experiments were conducted against standard decoding methods—including greedy decoding, nucleus sampling, and reinforcement learning-based approaches such as PPO fine-tuning.

The evaluation employed benchmark datasets containing prompts associated with known gender, racial, and occupational biases, including StereoSet [125] and CrowS-Pairs [126]. For each decoding method, outputs were assessed using established bias quantification metrics: the *stereotype score*, which measures the tendency to prefer stereotypical completions [125]; the *toxicity rate*, computed using the Perspective API [127]; and *semantic neutrality scores*, derived from entailment models trained to evaluate whether generated text introduces biased or unsupported assertions [128].

Results showed that ASTS consistently produced lower stereotype scores compared to both greedy decoding and nucleus sampling. Specifically, it reduced stereotype amplification by approximately 10% relative to nucleus sampling and 15% compared to PPO-fine-tuned outputs. ASTS also achieved lower toxicity levels, with average toxicity probabilities below 3.5%, versus 5.1% for nucleus sampling and 6.4% for PPO. These findings suggest that ASTS's entropy-aware, context-sensitive filtering mechanism effectively mitigates biased or degenerative continuations, particularly in response to ambiguous or socially sensitive prompts.

Importantly, ASTS achieved these reductions without the use of post-hoc bias correction modules or external classifiers during decoding. Its inherent focus on semantic and contextual coherence likely contributes to its robustness against bias-inducing token selections. This makes ASTS especially well-suited for fairness-critical applications, where decoding strategies must strike a balance between generation quality and equity without sacrificing performance.

In conclusion, ASTS maintains high levels of fluency and semantic alignment while inherently mitigating common generation biases. It outperforms both conventional and fine-tuned decoding methods across multiple fairness-oriented benchmarks, making it a compelling choice for responsible and high-quality language generation.

17.5.9 Computational Efficiency Analysis

While ASTS incorporates multiple scoring dimensions, its computational footprint remains lightweight relative to strategies that rely on external controllers or multi-pass architectures. This section quantifies ASTS's runtime and memory efficiency.

A key design objective of ASTS is to improve decoding quality without compromising computational efficiency. Unlike sampling strategies that depend on complex external modules, such as Plug-and-Play Language Models [117] or reinforcement learning-based controllers [129], ASTS functions entirely within a single forward pass. It achieves this by integrating lightweight, differentiable scoring components, including entropy-based thresholding, semantic alignment via cosine similarity on fixed-size embeddings, and reward-penalty modulation at each decoding step. Notably, all auxiliary computations are performed without backward passes, making ASTS highly compatible with real-time and latency-sensitive inference environments.

To evaluate the computational efficiency of ASTS, controlled runtime experiments were conducted using the GPT-2 Large model on an NVIDIA A100 GPU (40 GB VRAM) with a batch size of 4. ASTS was compared against four commonly used decoding strategies: greedy search, nucleus sampling, locally typical sampling (LTS) [9], and CTRL-based decoding [116]. Each method was tested at three generation lengths: 128, 256, and 512 tokens. For statistical robustness, results were averaged over 500 independent runs. The outcomes are summarized in Table 17.13, with a comparative runtime visualization presented in Fig. 17.6 as a line plot across the evaluated methods.

As shown in Table 17.13 and Fig. 17.6, ASTS introduces a modest execution time increase of approximately 9.4% compared to LTS at 512 tokens. This overhead

Table 17.13 Average decoding time (in milliseconds) per sequence for different decoding strategies across varying output lengths (128, 256, and 512 tokens)

	Token size		
Method	128	256	512
Greedy search	38.7	75.2	148.9
Nucleus sampling	49.5	94.1	181.4
LTS	55.3	103.6	194.2
CTRL decoding	72.8	133.2	251.8
ASTS	58.1	111.4	212.5

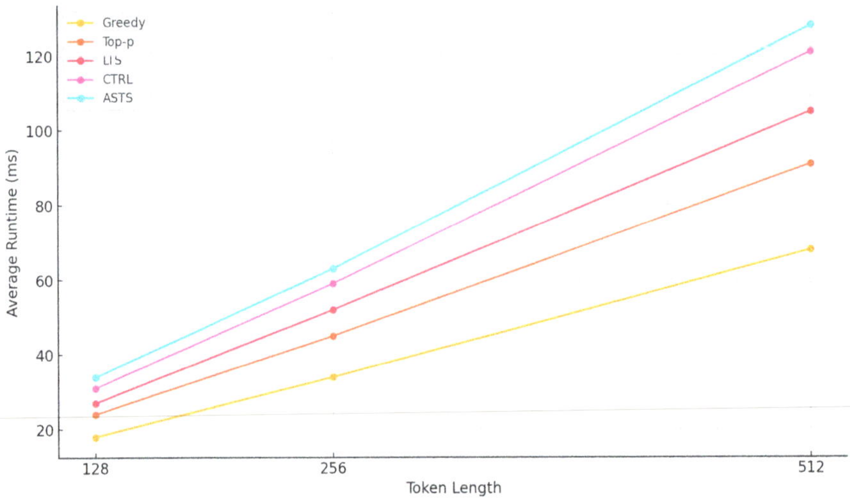

Fig. 17.6 Average decoding time (in milliseconds) for different decoding strategies across varying output lengths (128, 256, and 512 tokens). ASTS demonstrates a favorable balance between efficiency and generation quality

primarily stems from the additional computations involved in semantic similarity scoring and reward-based token re-ranking. However, ASTS remains significantly faster than decoding methods that rely on large control vectors, such as CTRL. In terms of resource usage, ASTS incurs only a minimal increase in GPU memory consumption (+3.2% over LTS), owing to its use of in-memory caching and batch-wise tensor operations. These characteristics make ASTS well-suited for real-time or near real-time applications, including dialogue systems, creative writing tools, and summarization engines. Although ASTS adds a few additional decoding steps, its overall quality-efficiency trade-off is highly favorable, delivering substantial gains in lexical diversity, coherence, and fluency, as detailed in Sect. 17.5, thereby justifying the modest computational overhead it incurs.

17.6 Conclusion

This chapter introduced Adaptive Semantic-Aware Typicality Sampling (ASTS), an enhanced decoding strategy that extends the Locally Typical Sampling framework to address persistent challenges in probabilistic text generation for LLMs. ASTS integrates dynamic entropy thresholding, multi-objective scoring, and reward-penalty modulation to create a context-sensitive and semantically informed approach to token selection. These design elements collectively improve fluency, semantic alignment, and control over repetition and diversity, which are key factors in generating coherent and engaging text.

A comprehensive empirical evaluation was conducted across multiple generative tasks, including story generation and abstractive summarization, using standard benchmarks and evaluation metrics such as perplexity (PPL), MAUVE, repetition score (REP), Zipf's coefficient, and lexical diversity. ASTS consistently outperformed conventional strategies—including nucleus sampling, top-k sampling, and Locally Typical Sampling—as well as recent competitive methods like Contrastive Decoding. The method demonstrated strong generalization across tasks, reducing redundancy, enhancing linguistic variety, and producing outputs that closely resemble human-written text. Its entropy-aware design allows ASTS to dynamically adapt to local uncertainty, making it especially effective for complex or open-ended generation scenarios.

17.6.1 Contributions

The primary contributions of ASTS, as demonstrated through empirical analysis and experimental results are summarized as follows:

Dynamic entropy thresholding: Unlike static thresholding methods used in Locally Typical Sampling, ASTS dynamically adjusts entropy thresholds based on contextual variations. This dynamic adjustment of entropy thresholds allows for a more flexible and adaptive selection of tokens resulting in generated text that maintains coherence while allowing for controlled diversity.

Multi-objective scoring mechanism: The introduction of a composite scoring function that considers coherence, diversity, and semantic alignment enables ASTS to produce more contextually relevant outputs while avoiding deterministic or excessively random text generation.

Reward-penalty modulation: ASTS introduces a lightweight, forward-pass-only reward function that rewards semantic consistency and penalizes excessive repetition. This mechanism helps regulate token redundancy without relying on external classifiers or fine-tuning.

Superior performance across benchmarks: ASTS achieves consistently high scores across a range of automatic evaluation metrics, lower REP (repetition), higher MAUVE (distributional similarity to human-generated text), optimal Zipf coefficients (word frequency distribution), and maximal diversity. It outperforms traditional baselines (top-k, nucleus, LTS, and contrastive decoding) in both story generation and abstractive summarization.

Robustness to hyperparameter sensitivity: Unlike top-k or nucleus sampling, which exhibit volatile behavior across hyperparameter configuration, ATS shows stable performance across a wide range of its hyperparameter (τ) values. It also demonstrates greater robustness than LTS, enhancing its practicality in real-world scenarios where fine-grained tuning is infeasible, and top-k sampling, which is highly sensitive to hyperparameter selection, ASTS exhibits greater stability across different values of its tuning parameters. The sensitivity of ASTS to its hyperparameters is found to be even lower than Locally Typical Sampling. This makes it a more reliable choice for real-world applications where precise tuning may not always be feasible.

Computational efficiency and real-time applicability: Despite integrating multiple scoring dimensions, ASTS operates with minimal additional computational cost, requiring no backward passes and introducing only a modest runtime overhead. This makes it suitable for deployment in time-sensitive applications such as dialogue systems, summarization engines, and edge devices.

Broader generalization and fairness considerations: ASTS generalizes well across model sizes and tasks, including instruction-following LLMs and low-parameter models. Furthermore, it demonstrates lower bias amplification and toxicity rates compared to nucleus sampling and PPO-fine-tuned baselines, highlighting its potential for fairness-critical applications.

17.6.2 Limitations

Despite its strengths, ASTS has the following *limitations* that merit further research and refinement.

Computational overhead: While ASTS is designed to operate within a single forward pass, the incorporation of multi-objective scoring and reward-penalty modulation introduces modest computational overhead compared to simpler decoding strategies. Although runtime and memory usage remain within acceptable bounds, additional optimization is necessary for deployment in latency-sensitive or resource-constrained environments. Future improvements may include the integration of efficient techniques such as model pruning [130, 131], approximate nearest neighbor (ANN) search for semantic scoring [132–134], or low-rank matrix factorization [135, 136] to reduce computational load without compromising quality.

Limited evaluation in domain-specific settings: ASTS has been primarily evaluated on general-purpose generative tasks such as storytelling and summarization. Its performance in highly constrained or domain-specific contexts, such as legal, biomedical, or financial document generation, remains to be systematically assessed. Incorporating domain-adaptation strategies, including transfer learning or lightweight fine-tuning with in-domain data, could improve contextual grounding and vocabulary alignment in such applications.

Challenges in high-entropy generation contexts: In scenarios involving substantial uncertainty, such as creative writing, abstract reasoning, or open-domain QA, ASTS may encounter difficulty in maintaining coherence while preserving diversity. Although the current entropy-aware filtering mechanism mitigates these issues to some extent, further enhancements that incorporate global context modeling or long-range semantic memory may be required to stabilize performance in high-entropy settings.

17.6.3 Future Research Directions

Building on the promising results achieved by ASTS, several directions for future research can be pursued to further expand its capabilities and applicability as discussed in the following.

Integration with reinforcement learning: Combining ASTS with reinforcement learning (RL) -based decoding strategies may enable dynamic optimization of token selection based on task-specific reward functions. RL-enhanced versions of ASTS could learn to prioritize desirable qualities such as factual accuracy, coherence, or user-aligned stylistic traits. This integration would be particularly valuable for interactive applications like personalized content generation, chatbot development, and dialogue modeling.

Hybrid decoding approaches: ASTS could be integrated with contrastive decoding *or retrieval-augmented generation* (RAG) techniques to refine the token selection based on external knowledge sources [137, 138]. Such a hybrid approach could be particularly beneficial for tasks requiring factual consistency, such as news summarization and question answering.

Efficiency-driven adaptation for low-resource settings: While ASTS maintains modest computational overhead, additional adaptations may be necessary for deployment in resource-constrained environments such as mobile devices or edge AI systems. Techniques such as model quantization, knowledge distillation, low-rank approximation, and sparsity-aware encoding can be employed to develop lightweight variants of ASTS without compromising generation quality.

Personalized and adaptive text generation: Future work may explore mechanisms for ASTS to adapt its decoding behavior based on user interaction history or stylistic preferences. This would enable real-time customization in AI writing assistants, dialogue systems, and recommendation engines. Learning user-conditioned entropy profiles or semantic scoring weights may help personalize text outputs without retraining.

Cross-lingual and multilingual expansion: Extending ASTS to support multilingual language models and morphologically rich languages is another valuable direction. Preliminary results already indicate robust performance in highly inflected languages such as Finnish and Hungarian. Systematic evaluation across language families, coupled with entropy-sensitive multilingual alignment strategies, may help optimize ASTS for broader global applicability.

17.6.4 Summary Conclusion

The development of ASTS represents a meaningful advancement in decoding strategies for LLMs. It addresses key challenges in neural text generation by maintaining an effective balance between coherence, lexical diversity, and computational efficiency. Through extensive experimental evaluation across multiple tasks and baselines, ASTS has demonstrated consistent improvements in both quality and stability. This makes it a strong candidate for real-world applications in natural language generation.

Although ASTS has some limitations, particularly in high-entropy contexts and domain-specific scenarios, ongoing research in hybrid architectures, reinforcement learning integration, and computational optimization holds promise for further enhancements of ASTS. Its adaptable design and empirical robustness make it a potentially valuable component in the evolving landscape of controllable and efficient language model decoding.

References

1. Radford, A., Narasimhan, K., Salimans, T., Sutskever, I.: Improving Language Understanding by Generative Pre-Training. OpenAI Research Paper (2018)
2. Radford, A., Wu, J., Child, R., Luan, D., Amodei, D., Sutskever, I. Language Models are Unsupervised Multitask Learners. (2019)
3. Devlin, J., Chang, M.W., Lee, K., Toutanova, K.: BERT: pre-training of deep bidirectional transformers for language understanding. In: Proceedings of the 2019 Conference of the North American Chapter of the Association for Computational Linguistics: Human Language Technologies Long and Short Papers, vol. 1, pp. 4171–4186 (2019). https://doi.org/10.18653/v1/N19-1423
4. Raffel, C., Shazeer, N., Roberts, A., Lee, K., Narang, S., Matena, M., Zhou, Y., Li, W., Liu, P.J.: Exploring the limits of transfer learning with a unified text-to-text transformer. J. Mach. Learn. Res. **21**(140), 1–67 (2019). https://doi.org/10.48550/arXiv.1910.10683
5. Sutskever, I., Vinyals, O., Le, Q.V.: Sequence to sequence learning with neural networks. In: Proceedings of the 28th International Conference on Neural Information Processing Systems, vol. 2, pp. 3104–3112 (2014). https://doi.org/10.48550/arXiv.1409.3215
6. Koehn, P., Och, F.J., Marcu, D.: Statistical phrase-based translation. In: Proceedings of the 2003 Human Language Technology Conference of the North American Chapter of the Association for Computational Linguistics, vol. 1, pp. 48–54 (2003). https://doi.org/10.3115/1073445.1073462
7. Fan, A., Lewis, M., Dauphin, Y.: Hierarchical neural story generation. In: Proceedings of the 56th Annual Meeting of the Association for Computational Linguistics, Long Papers, pp. 889–898 (2018). https://doi.org/10.48550/arXiv.1805.04833
8. Holtzman, A., Buys, J., Du, L., Forbes, M., Choi, Y.: The curious case of neural text degeneration. In: Proceedings of the 8th International Conference on Learning Representations (ICLR'20) (2020). https://doi.org/10.48550/arXiv.1904.09751
9. Meister, C., Pimentel, T., Wiher, G., Cotterell, R.: Locally typical sampling. In: Transactions of the Association for Computational Linguistics, vol. 11, pp. 102–121. MIT Press, Cambridge, MA (2023). https://doi.org/10.1162/tacl_a_00536
10. Shi, C., Yang, H., Cai, D., Zhang, Z., Wang, Y., Yang, Y., Lam, W.: A thorough examination of decoding methods in the era of LLMs. In: Proceedings o1f the 2024 Conference on Empirical Methods in Natural Language Processing, pp. 8601–8629 (2024). https://doi.org/10.18653/v1/2024.emnlp-main.489
11. Liu, M., Zeng, A., Wang, B., Zhang, P., Tang, J., Dong, Y.: APAR: LLMs Can Do Auto-Parallel Auto-Regressive Decoding (2024). https://doi.org/10.48550/arXiv.2401.06761

12. Bergner, B., Skliar, A., Royer, A., Blankevoort, T., Asano, Y., Bejnordi, B.E.: Think Big, Generate Quick: LLM-to-SLM for Fast Autoregressive Decoding. arXiv:2402.16844 (2024). https://doi.org/10.48550/arXiv.2402.16844
13. Jaiswal, A., Hu, B., Yin, L., Ro, Y., Liu, S., Chen, T., Akella, A.: FFN-SkipLLM: a hidden gem for autoregressive decoding with adaptive feed forward skipping. In: Proceedings of the 2024 Conference on Empirical Methods in Natural Language Processing, pp. 16943–16956 (2024). https://doi.org/10.18653/v1/2024.emnlp-main.941
14. You, H., Fu, Y., Wang, Z., Yazdanbakhsh, A., Lin, Y.C.: When linear attention meets autoregressive decoding: towards more effective and efficient linearized large language models. In: Proceedings of the 41st International Conference on Machine Learning, pp. 57350–57366 (2024). https://doi.org/10.48550/arXiv.2406.07368
15. Liu, X., Hu, L., Bailis, P., Cheung, A., Deng, Z., Stoica, I., Zhang, H.: Online Speculative Decoding. arXiv:2310.07177 (2023). https://doi.org/10.48550/arXiv.2310.07177
16. Xia, H., Yang, Z., Dong, Q., Wang, P., Li, Y., Ge, T., Liu, T., Li, W., Sui, Z.: Unlocking efficiency in large language model inference: a comprehensive survey of speculative decoding. In: Findings of the Association for Computational Linguistics, ACL 2024, pp. 7655–7671 (2024). https://doi.org/10.18653/v1/2024.findings-acl.456
17. Jeon, W., Gagrani, M., Goel, R., Park, J., Lee, M., Lott, C.: Recursive Speculative Decoding: Accelerating LLM Inference via Sampling Without Replacement. arXiv:2402.14160 (2024). https://doi.org/10.48550/arXiv.2402.14160
18. Li, M., Chen, X., Holtzman, A., Chen, B., Lin, J., Yih, W., Lin, X.V.: Nearest Neighbor Speculative Decoding for LLM Generation and Attribution. arXiv:2405.19325 (2024). https://doi.org/10.48550/arXiv.2405.19325
19. Wen, Z., Gui, S., Feng, Y.: Speculative Decoding with CTC-based Draft Model for LLM Inference Acceleration (2024). https://doi.org/10.48550/arXiv.2412.00061
20. Li, X.L., Holtzman, A., Fried, D., Liang, P., Eisner, J., Hashimoto, T., Zettlemoyer, L., Lewis, M.: Contrastive decoding: open-ended text generation as optimization. In: Proceedings of the 61st Annual Meeting of the Association for Computational Linguistic, vol. 1, Long Papers, pp. 12286–12312 (2023). https://doi.org/10.18653/v1/2023.acl-long.687
21. Zhao, Z., Monti, E., Lehmann, J., Assem, H.: Enhancing contextual understanding in large language models through contrastive decoding. In: Proceedings of the 2024 Conference on North American Chapter of the Association for Computational Linguistics: Human Language Technologies, vol. 1, Long Paper, Mexico City, Mexico, pp. 4225–4237 (2024). https://doi.org/10.18653/v1/2024.naacl-long.237
22. O'Brien, S., Lewis, M.: Contrastive Decoding Improves Reasoning in Large Language Models. arXiv:2309.09117 (2023). https://doi.org/10.48550/arXiv.2309.09117
23. Phan, P., Tran, H., Phan, L.: Distillation Contrastive Decoding: Improving LLMs Reasoning with Contrastive Decoding and Distillation. arXiv:2402.14874 (2023). https://doi.org/10.48550/arXiv.2402.14874
24. Chang, H.-S., Peng, N., Bansal, M., Ramakrishna, A., Chung, T.: Explaining and improving contrastive decoding by extrapolating the probabilities of a huge and hypothetical LM. In: Proceedings of the 2024 Conference on Empirical Methods in Natural Language Processing, pp. 8503–8526 (2024). https://doi.org/10.18653/v1/2024.emnlp-main.484
25. Zhang, J., Wang, J., Li, H., Shou, L., Chen, K., Chen, G., Mehrotra, S.: Draft & verify: lossless large language model acceleration via self-speculative decoding. In: Proceedings of the 62nd Annual Meeting of the Association for Computational Linguistics, vol. 1, Long Papers, pp. 11263–11282 (2024). https://doi.org/10.18653/v1/2024.acl-long.607
26. Wang, J., Su, Y., Li, J., Xia, Q., Ye, Z., Duan, X., Wang, Z., Zhang, M.: OPT-Tree: Speculative Decoding with Adaptive Draft Tree Structure. arXiv:2406.17276 (2024). https://doi.org/10.48550/arXiv.2406.17276
27. Liu, X., Lei, B., Zhang, R., Xu, D.: Adaptive Draft-Verification for Efficient Large Language Model Decoding (2024). https://doi.org/10.48550/arXiv.2407.12021
28. Liu, T., Li, Y., Lv, Q., Liu, K., Zhu, J., Hu, W.: Parallel Speculative Decoding with Adaptive Draft Length. arXiv:2408.11850 (2024). https://doi.org/10.48550/arXiv.2408.11850

29. Wiher, G., Meister, C., Cotterell, R.: On Decoding Strategies for Neural Text Generators. arXiv:2203.15721 (2022). https://doi.org/10.48550/arXiv.2203.15721
30. Xu, J., Xiong, C., Savarese, S., Zhou, Y.: Best-k search algorithm for neural text generation. In: Proceedings of the 61st Annual Meeting of the ACL, vol. 1, Long Papers, pp. 12385–12401 (2020). https://doi.org/10.18653/v1/2023.acl-long.692
31. Martínez-Murillo, I., Moreda, P., Lloret, E.: Towards an efficient approach for controllable text generation. In: Proceedings of the 1st International Workshop on Multilingual, Multimodal, and Multitask Language Generation, pp. 26–31. European Association for Machine Translation (2023)
32. Shao, C., Meng, F., Liu, Y., Zhou, J.: Language generation with strictly proper scoring rules. In: Proceedings of the 41st International Conference on Machine Learning (ICML'24) (2024). https://doi.org/10.48550/arXiv.2405.18906
33. Liang, X., Wang, H., Wang, Y.Z., Song, S., Yang, J., Niu, S., Hu, J., Liu, D., Yao, S., Xiong, F., Li, Z.: Controllable Text Generation for Large Language Models: A Survey. arXiv:2408.12599 (2024). https://doi.org/10.48550/arXiv.2408.12599
34. Sukhbaatar, S., Golovneva, O., Sharma, V., Xu, H., Lin, X.V., Rozière, B., Kahn, J., Li, D., Yih, W.-T., Weston, J., Li, X.: Branch-Train-MiX: Mixing Expert LLMs into a Mixture-of-Experts LLM (2024). https://doi.org/10.48550/arXiv.2403.07816
35. Yun, L., Zhuang, Y., Fu, Y., Xing, E.P., Zhang, H.: Toward Inference-Optimal Mixture-of-Expert Large Language Models (2024). https://doi.org/10.48550/arXiv.2404.02852
36. Zhu, T., Qu, X., Dong, D., Ruan, J., Tong, J., He, C., Cheng, Y.: LLaMA-MoE: Building Mixture-of-Experts from LLaMA with Continual Pre-Training. arXiv:2406.16554 (2024). https://doi.org/10.48550/arXiv.2406.16554
37. Li, Z., Zhou, T.: Your Mixture-of-Experts LLM is Secretly an Embedding Model for Free. arXiv:2410.10814 (2024). https://doi.org/10.48550/arXiv.2410.10814
38. Zhang, C., Yang, Y., Li, Q., Wang, J., Song, D.: Task-agnostic distillation of encoder-decoder language models. In: Proceedings of the 2024 Joint International Conference on Computational Linguistics, Language Resources and Evaluation (LREC-COLING'24), pp. 15629–15639 (2024)
39. Gu, Y., Dong, L., Wei, F., Huang, M.: MiniLLM: Knowledge distillation of large language models. In: Proceedings of the International Conference on Learning Representation (ICLR'24). arXiv:2306.08543 (2023). https://doi.org/10.48550/arXiv.2306.08543
40. Zhou, Y., Lyu, K., Rawat, A.S., Menon, A.K., Rostamizadeh, A., Kumar, S., Kagy, J.-F., Agarwal, R.: DistillSpec: Improving Speculative Decoding via Knowledge Distillation. arXiv:2310.08461 (2023). https://doi.org/10.48550/arXiv.2310.08461
41. Goel, R., Gagrani, M., Jeon, W., Park, J., Lee, M., Lott, C.: Direct alignment of draft model for speculative decoding with chat-fine-tuned LLMs. In: Proceedings of the ICLR'24 Workshop on Understanding of Foundation Models (ME-FoMo). arXiv:2403.00858 (2024). https://doi.org/10.48550/arXiv.2403.00858
42. Ning, X., Lin, Z., Zhou, Z., Wang, Z., Yang, H., Wang, Y.: Skeleton-of-thought: prompting LLMs for efficient parallel generation. In: ICLR'24 Poster Track, arXiv:2307.15337 (2024). https://doi.org/10.48550/arXiv.2307.15337
43. Wu, P., Liu, J., Gong, Z., Wang, Q., Li, J., Wang, J., Cai, X., Zhao, D.: Parallel Decoding via Hidden Transfer for Lossless Large Language Model Acceleration. arXiv:2404.12022 (2024). https://doi.org/10.48550/arXiv.2404.12022
44. Sun, C., Zhang, H., Lin, Z., Zhang, J., Zhang, F., Wang, Z., Chen, B., Song, C., Zhang, D., Gai, K., Xiong, D.: Decoding at the speed of thought: harnessing parallel decoding of lexical units for LLMs. In: Proceedings of the 2024 Joint International Conference on Computational Linguistics, Language Resources and Evaluation (LREC-COLING'24), pp. 4476–4487 (2024)
45. Vilnis, L., Zemlyanskiy, Y., Murray, P., Passos, A., Sanghai, S.: Arithmetic sampling: parallel diverse decoding for large language models. In: Proceedings of ICML'23, arXiv:2210.15458 (2023). https://doi.org/10.48550/arXiv.2210.15458

46. Zheng, C., Ke, P., Zhang, Z., Huang, M.: Click: controllable text generation with sequence likelihood contrastive learning. In: Findings of the Association for Computational Linguistics, pp. 1022–1040 (2023). https://doi.org/10.18653/v1/2023.findings-acl.65
47. An, C., Feng, J., Lv, K., Kong, L., Qiu, X., Huang, X.: CoNT: contrastive neural text generation. In: Proceedings of the 36th Conference on Neural Information Processing Systems (NeurIPS'22) (2022). https://doi.org/10.48550/arXiv.2205.14690
48. Teng, Z., Chen, C., Zhang, Y., Zhang, Y.: Contrastive latent variable models for neural text generation. In: Proceedings of the 38th Conference on Uncertainty in Artificial Intelligence, vol. 180, pp. 1928–1938 (2022)
49. Ghazvininejad, M., Levy, O., Liu, Y., Zettlemoyer, L.: Mask-predict: parallel decoding of conditional masked language models. In: Proceedings of the 2019 Conference on Empirical Methods in Natural Language Processing and the 9th International Joint Conference on Natural Language Processing (EMNLP-IJCNLP), pp. 6114–6123 (2019). https://doi.org/10.18653/v1/D19-1633
50. Ghazvininejad, M., Levy, O., Zettlemoyer, L.: Semi-Autoregressive Training Improves Mask-Predict Decoding (2020). https://doi.org/10.48550/arXiv.2001.08785
51. Chang, J.D., Brantley, K., Ramamurthy, R., Misra, D., Sun, W.: Learning to Generate Better than Your LLM (2023). https://doi.org/10.48550/arXiv.2306.11816
52. Mudgal, S., Lee, J., Ganapathy, H., Li, Y., Wang, T., Huang, Y., Chen, Z., Cheng, H.-T., Collins, M., Strohman, T., Chen, J., Beutel, A., Beirami, A.: Controlled decoding from language models. In: Proceedings of the International Conference on Machine Learning (ICML'24) (2024). https://doi.org/10.48550/arXiv.2310.17022
53. Yang, G., Chen, J., Lin, W., Byrne, B.: Direct preference optimization for neural machine translation with minimum Bayes risk decoding. In: Proceedings of the 2024 Conference of the North American Chapter of the Association for Computational Linguistics: Human Language Technologies, vol. 2: Short Papers, pp. 391–398. Association for Computational Linguistics (2024). https://doi.org/10.18653/v1/2024.naacl-short.34
54. Wen, M., Liao, J., Deng, C., Wang, J., Zhang, W., Wen, Y.: Entropy-Regularized Token-Level Policy Optimization for Language Agent Reinforcement. arXiv:2402.06700 (2024). https://doi.org/10.48550/arXiv.2402.06700
55. Hedderich, M.A., Lange, L., Adel, H., Strötgen, J., Klakow, D.: A survey on recent approaches for natural language processing in low-resource scenarios. In: Proceedings of the 2021 Conference of the North American Chapter of the Association for Computational Linguistics: Human Language Technologies (NAACL-HLT), pp. 2545–2568 (2021). https://doi.org/10.18653/v1/2021.naacl-main.201
56. Park, C., Yang, Y., Park, K., Lim, H.: Decoding strategies for improving low-resource machine translation. Electronics. **9**(10), 1562 (2020). https://doi.org/10.3390/electronics9101562
57. Xu, M., Zhang, J., Xu, L., Silamu, W., Li, Y.: Correlation-guided decoding strategy for low-resource Uyghur scene text recognition. Complex Intell. Syst. **11**, 57 (2025). https://doi.org/10.1007/s40747-024-01689-5
58. Tuli, S., Lin, C.-H., Hsu, Y.-C., Jha, N.K., Shen, Y., Jin, H.: DynaMo: accelerating language model inference with dynamic multi-token sampling. In: Proceedings of the 2024 Conference of the North American Chapter of the Association for Computational Linguistic: Human Language Technologies, vol. 1, Long Papers (2024). https://doi.org/10.18653/v1/2024.naacl-long.182
59. Mamou, J., Peregs, O., Korats, D., Berchanskys, M., Timor, N., Wasserblats, M., Schwartzp, R.: Dynamic Speculation Lookahead Accelerates Speculative Decoding of Large Language Models (2024). https://doi.org/10.48550/arXiv.2405.04304
60. Pascual, D., Egressy, B., Bolli, F., Wattenhofer, R.: Directed Beam Search: Plug-and-Play Lexically Constrained Language Generation. arXiv:2012.15416 (2020). https://doi.org/10.48550/arXiv.2012.15416

61. Grosse, J., Wu, R., Rashid, A., Hennig, P., Poupart, P., Kristiadi, A.: Uncertainty-Guided Optimization on Large Language Model Search Trees. arXiv:2407.03951 (2024). https://doi.org/10.48550/arXiv.2407.03951
62. Qin, Z., He, Z., Prakriya, N., Cong, J., Sun, Y.: Dynamic-Width Speculative Beam Decoding for Efficient LLM Inference (2024). https://doi.org/10.48550/arXiv.2409.16560
63. Qiu, Z., Ou, Z., Wu, B., Li, J., Liu, A., King, I.: Entropy-Based Decoding for Retrieval-Augmented Large Language Models (2024). https://doi.org/10.48550/arXiv.2406.17519
64. Wang, R., Li, J., Li, P.: InfoDiffusion: information entropy aware diffusion process for non-autoregressive text generation. In: Findings of the Association for Computational Linguistics, pp. 13757–13770 (2023). https://doi.org/10.18653/v1/2023.findings-emnlp.919
65. Arora, K., O'Donnell, T.J., Precup, D., Weston, J., Cheung, J.C.K.: The Stable Entropy Hypothesis and Entropy-Aware Decoding: An Analysis and Algorithm for Robust Natural Language Generation (2023). https://doi.org/10.48550/arXiv.2302.06784
66. Qin, L., Welleck, S., Khashabi, D., Choi, Y.: COLD decoding: energy-based constrained text generation with Langevin dynamics. Adv.. Neural Inform. Process. Syst. **NeurIPS'22), 35**, 10656–10668. arXiv:2202.11705 (2023). https://doi.org/10.48550/arXiv.2202.11705
67. Liu, X., Khalifa, M., Wang, L.: BOLT: fast energy-based controlled text generation with tunable biases. In: Proceedings of the 61st Annual Meeting of the Association for Computational Linguistics (ACL'23), pp. 186–200. arXiv:2305.12018 (2023). https://doi.org/10.48550/arXiv.2305.12018
68. Zhang, K., Zhao, J., Chen, R.: KOALA: Enhancing Speculative Decoding for LLM via Multi-Layer Draft Heads with Adversarial Learning. arXiv:2408.08146 (2024). https://doi.org/10.48550/arXiv.2408.08146
69. He, T., Tan, X., Xia, Y., He, D., Qin, T., Chen, Z., Liu, T.-Y.: Layer-wise coordination between encoder and decoder for neural machine translation. In: Proceedings of the 32nd Conference on Neural Information Processing Systems (NIPS'18), Montreal, Canada, pp. 7955–7965 (2018)
70. Holtzman, A., Buys, J., Du, L., Forbes, M., Choi, Y.: The Curious Case of Neural Text Degeneration. arXiv:1904.09751 (2019). https://doi.org/10.48550/arXiv.1904.09751
71. Fan, A., Lewis, M., Dauphin, Y.: Hierarchical neural story generation. In: Proceedings of the 56th Annual Meetings of the Association for Computational Linguistics (Vol 1: Long Papers), pp. 889–898 (2018). https://doi.org/10.18653/v1/P18-1082
72. Li, Y., Su, H., Shen, X., Li, W., Cao, Z., Niu, S.: DailyDialog: A Manually Labelled Multi-turn Dialogue Dataset. arXiv:1710.03957 (2017). https://doi.org/10.48550/arXiv.1710.03957
73. Spataru, A., Hambo, E., Voita, E., Cancedda, N.: Know when to stop: a study of semantic drift in text generation. In: Proceedings of the 2024 Conference of the North American Chapter of the Association for Computational Linguistics: Human Language Technologies (Vol. 1: Long Papers), pp. 3656–3671 (2024). https://doi.org/10.18653/v1/2024.naacl-long.202
74. Wolf, T., et al.: Transformers: state-of-the-art natural language processing. In: Proceedings of the 2020 Conference on Empirical Methods in Natural Language Processing: System Demonstrations, Association for Computational Linguistics, pp. 38–45 (2020). https://doi.org/10.18653/v1/2020.emnlp-demos.6
75. Basu, S., Ramachandran, G.S., Keskar, N.S., Varshney, L.R.: Mirostat: a neural text decoding algorithm that directly controls perplexity. In: Proceedings of ICLR 2021 (2020). https://doi.org/10.48550/arXiv.2007.14966
76. Pillutla, K., Liu, L., Thickstun, J., Welleck, S., Swayamdipta, S., Zellers, R., Oh, S., Choi, Y., Harchaoui, Z.: MAUVE scores for generative models: theory and practice. J. Mach. Learn. Res. **24**(356), 1–92 (2022). https://doi.org/10.48550/arXiv.2212.14578
77. Pillutla, K., Swayamdipta, S., Zellers, R., Thickstun, J., Welleck, S., Choi, Y., Harchaoui, Z.: MAUVE: measuring the gap between neural text and human text using divergence frontiers. In: Proceedings of the 35th International Conference on Neural Information Processing Systems (NIPS'21), Article No: 368, pp. 4816–4828 (2021)

78. Zhang, H., Duckworth, D., Ippolito, D., Neelakantan, A.: Trading off diversity and quality in natural language generation. In: Proceeding of the Workshop on Human Evaluation of NLP Systems (HumEval'21), pp. 25–33. Association for Computational Linguistics (2021)
79. Nadeem, M., He, T., Cho, K., Glass, J.: A Systematic characterization of sampling algorithms for open-ended language generation. In: Proceedings of the 1st Conference of the Asia-Pacific Chapter of the Association for Computational Linguistics and the 10th International Joint Conference on Natural Language Processing, pp. 334–346. Association for Computational Linguistics (2020). https://doi.org/10.18653/v1/2020.aacl-main.36
80. Papineni, K., Roukos, S., Ward, T., Zhu, W.-J.: BLEU: a method for automatic evaluation of machine translation. In: Proceedings of the 40th Annual Meeting on Association for Computational Linguistics, July 7–12, 2002, Philadelphia, Pennsylvania, USA, pp. 311–318 (2002). https://doi.org/10.3115/1073083.1073135
81. Lin, C.-Y.: ROUGE: A Package for Automatic Evaluation of Summaries. Text Summarization Branches Out, pp. 74–81. Association for Computational Linguistics (2004)
82. Welleck, S., Kulikov, I., Roller, S., Dinan, E., Cho, K., Weston, J.: Neural text generation with unlikelihood training. In: Proceedings of the 8th International Conference on Learning Representation (ICLR'20), Addis Ababa, Ethiopia (2020)
83. Zipf, G.K.: Human Behavior and the Principle of Least Effort. Addison-Wesley Press (1949)
84. Fan, A., Lewis, M., Dauphin, Y.: Hierarchical neural story generation. In: Proceedings of the 56th Annual Meeting of the Association for Computational Linguistics Long Papers, Melbourne, Australia, vol. 1, pp. 889–898. Association for Computational Linguistics (2018). https://doi.org/10.18653/v1/P18-1082
85. Merity, S., Xiong, C., Bradbury, J., Socher, R.: Pointer sentinel mixture models. In: Proceedings of the 5th International Conference on Learning Representation (2016). https://doi.org/10.48550/arXiv.1609.07843
86. Lewis, M., et al.: BART: denoising sequence-to-sequence pre-training for natural language generation, translation, and comprehension. In: Proceedings of the 58th Annual Meeting of the Association for Computational Linguistics, pp. 7871–7880. Association for Computational Linguistics (2020). https://doi.org/10.18653/v1/2020.acl-main.703
87. Nallapati, R., Zhou, B., dos Santos, C., Gulcehre, C., Xiang, B.: Abstractive text summarization using sequence-to-sequence RNNs and beyond. In: Proceedings of the 29th SIGNLL Conference on Computational Natural Language Learning, Berlin, Germany, pp. 280–290 (2016). https://doi.org/10.18653/v1/K16-1028
88. Kaufman, T., Weng, P., Bengs, V., Hullermeir, E.: A Survey of Reinforcement Learning from Human Feedback. arXiv:2312.14925 (2023). https://doi.org/10.48550/arXiv.2312.14925
89. Bridgelall, R.: Unravelling the mysteries of AI Chatbots. Artif. Intell. Rev. **57**, 89 (2024). https://doi.org/10.1007/s10462-024-10720-7
90. HuggingFace trl Library: Available at: https://huggingface.co/docs/trl/en/index
91. HuggingFace OpenAssistant Dataset: Available at: https://huggingface.co/datasets/OpenAssistant/oasst1
92. HuggingFace Anthropic HH Dataset: Available at: https://huggingface.co/datasets/Anthropic/hh-rlhf
93. Faal, F., Schmitt, K., Yu, J.F.: Reward modeling for mitigating toxicity in transformer-based language models. Artif. Intell. **53**(7), 8421–8435 (2022). https://doi.org/10.1007/s10489-022-03944-z
94. Zhang, T., Kishore, V., Wu, F., Weinberger, K.Q., Artzi, Y.: BERTScore: Evaluating Text Generation with BERT. arXiv: 1904.09675 (2020). https://doi.org/10.48550/arXiv.1904.09675
95. Cooper, N., Scholak, T.: Perplexed: Understanding When Large Language Models are Confused. arXiv: 2402.06634 (2024). https://doi.org/10.48550/arXiv.2404.06634
96. Li, J., Galley, M., Brockett, C., Gao, J., Dolan, B.: A diversity-promoting objective function for neural conversation models. In: Proceedings of the 2016 Conference on North American Chapter of the Association for Computational Linguistics: Human Language Technologies, San Diego, California, USA, pp. 110–119 (2016). https://doi.org/10.18653/v1/N16-1014

97. Welbl, J., et al.: Challenges in detoxifying language models. In: Findings of the Association for Computational Linguistics (EMNLP 2021), pp. 2447–2469 (2021). https://doi.org/10.18653/v1/2021.findings-emnlp.210
98. Maynez, J., Narayan, S., Bohnet, B., McDonald, R.: On faithfulness and factuality in abstractive summarization. In: Proceedings of the 58th Annual Meeting of the Association for Computational Linguistics, pp. 1906–1919 (2020). https://doi.org/10.18653/v1/2020.acl-main.173
99. Chung, H.W., et al.: Scaling Instruction-Finetuned Language Models. arXiv:2210.11416 (2022). https://doi.org/10.48550/arXiv.2210.11416
100. Ouyang, L., et al.: Training Language Models to Follow Instructions with Human Feedback. arXiv:2203.02155 (2022). https://doi.org/10.48550/arXiv.2203.02155
101. Touvron, H., et al.: Llama2: Open Foundation and Fine-Tuned Chat Models. arXiv:2307.09288 (2023). https://doi.org/10.48550/arXiv.2307.09288
102. Available: https://huggingface.co/google/flan-t5-xl?utm_source=chatgpt.com
103. Wang, Y., et al.: Super-naturalinstructions: generalization via declarative instructions on 1600+ NLP tasks. In: Proceedings of the 2022 Conference on Empirical Methods in Natural Language Processing (EMNLP' 2022), pp. 5085–5109 (2022). https://doi.org/10.18653/v1/2022.emnlp-main.340
104. DistilGPT2 Model. Available at: https://huggingface.co/distilbert/distilgpt2
105. GPT2-Tiny Model: Available at: https://huggingface.co/sshleifer/tiny-gpt2
106. Eldan, R., Li, Y.: TinyStories: How Small Can Language Models Be and Still Speak Coherent English?. arXiv:2305.07759 (2023). https://doi.org/10.48550/arXiv.2305.07759
107. Dubois, Y., et al.: A Simulation Framework for Methods that Learn from Human Feedback. arXiv:2305.14387 (2023). https://doi.org/10.48550/arXiv.2305.14387
108. Zhang, S., et al.: Personalizing Dialogue Agents: I Have a Dog, Do You Pets Too?. arXiv: 1801.07243 (2018). https://doi.org/10.48550/arXiv.1801.07243
109. Han, S., Kim, B., Chang, B.: Measuring and improving diversity of dialogue generation. In: Findings of the Association for Computational Linguistics (EMNLP 2022), Abu Dhabi, United Arab Emirates, pp. 934–950 (2022). https://doi.org/10.18653/v1/2022.findings-emnlp.66
110. Aynetdinov, A., Akbik, A.: SemScore: Automated Evaluation of Instruction-Tuned LLMs Based on Semantic Textual Similarity. arXiv:2401.17022 (2024). https://doi.org/10.48550/arXiv.2401.17072
111. Yang, Z., Huang, Y., Chen, Y., Wu, X., Feng, J., Deng, C.: CTGGAN: controllable text generation with generative adversarial network. Appl. Sci. **14**(7), 3106 (2024). https://doi.org/10.3390/app14073106
112. Socher, R., et al.: Recursive deep models for semantic compositionality over a sentiment treebank. In: Proceedings of the 2013 Conference on Empirical Methods in Natural Language Processing, Seatle, Washington, USA, pp. 1631–1642 (2013)
113. Rao, S., Tetreault, J.: Dear Sir or Madam, May I introduce the GYAFC dataset: corpus, benchmarks and metrics for formality style transfer. In: Proceedings of the 2018 Conference of the North American Chapter of the Association for Computational Linguistics: Human Language Technologies, Vol 1(Long Papers), New Orleans, Louisiana, pp. 129–140 (2018). https://doi.org/10.18653/v1/N18-1012
114. Merity, S., Xiong, C., Bradbury, J., Socher, R.: Pointer Sentinel Mixture Models. arXiv:1609.07843 (2016). https://doi.org/10.48550/arXiv.1609.07843
115. Hokamp, C., Liu, Q.: Lexically constrained decoding for sequence generation using grid beam search. In: Proceedings of the 55th Annual Meeting of the Association for Computational Linguistics (Vol 1: Long Papers), Vancouver, Canada, pp. 1535–1546 (2017). https://doi.org/10.18653/v1/P17-1141
116. Keskar, N.S., McCann, B., Varshney, L.R., Xiong, C., Socher, R.: CTRL: A Conditional Transformer Language Model for Controllable Generation. arXiv: 1909.05858 (2019). https://doi.org/10.48550/arXiv.1909.05858

117. Dathathri, S., et al.: Plug and Play Language Models: A Simple Approach to Controlled Text Generation. arXiv:1912.02164 (2019). https://doi.org/10.48550/arXiv.1912.02164
118. Hu, Z., Yang, Z., Liang, X., Salakhutdinov, R., Xing, E.P.: Toward controlled generation of text. In: Proceedings of the 34th International Conference on Machine Learning (ICML 2017), vol. 70, pp. 1587–1596 (2017)
119. Burlot, F., Yvon, F.: Evaluating the morphological competence of machine translation systems. In: Proceedings of the 2nd Conference on Machine Translation, Copenhagen, Denmark, pp. 43–55 (2017). https://doi.org/10.18653/v1/W17-4705
120. Tang, Y., et al.: Multilingual Translation with Extensible Multilingual Pretraining and Finetuning. arXiv:2008.00401 (2020). https://doi.org/10.48550/arXiv.2008.00401
121. Xue, L., et al.: mT5: A Massively Multilingual Pre-trained Text-to-Text Transformer. arXiv:2010.11934 (2020). https://doi.org/10.48550/arXiv.2010.11934
122. Ladhak, F., Durmus, E., Cardie, C., McKeown, K.: WikiLingua: a new benchmark dataset for cross-lingual abstractive summarization. In: Findings of the Association for Computational Linguistics: EMNLP 2020, pp. 4034–4048 (2020). https://doi.org/10.18653/v1/2020.findings-emnlp.360
123. Goyal, N., et al.: The FLORES-101 Evaluation Benchmark for Low-Resource and Multilingual Machine Translation. arXiv:2106.03193 (2021). https://doi.org/10.48550/arXiv.2106.03193
124. Bender, E.M., Gebru, T., McMillan-Major, A., Shmitchell, S.: On the dangers of stochastic parrots: can language models be too big? In: FAccT'21: Proceedings of the 2021 ACM Conference on Fairness, Accountability, and Transparency, pp. 610–623 (2021). https://doi.org/10.1145/3442188.3445922
125. Nadeem, M., Bethke, A., Reddy, S.: StereoSet: Measuring Stereotypical Bias in Pretrained Language Models. arXiv:2004.09456 (2020). https://doi.org/10.48550/arXiv.2004.09456
126. Nangia, N., Vania, C., Bhalerao, R., Bowman, S.R.: CrowS-pairs: a challenge dataset for measuring social biases in masked language models. In: Proceedings of the 2020 Conference on Empirical Methods in Natural Language Processing (EMNLP), pp. 1953–1967 (2020). https://doi.org/10.18653/v1/2020.emnlp-main.154
127. Perspective API Webpage: https://perspectiveapi.com/
128. Zhao, J., Wang, T., Yatskar, M., Cotterell, R., Ordonez, V., Chang, K.-W.: Gender bias in contextual word embeddings. In: Proceedings of the 2019 Conference on the North American Chapter of the Association for Computational Linguistics: Human Language Technologies (Vol 1: Long and Short Papers), Minneapolis, Minnesota, USA, pp. 629–634 (2019). https://doi.org/10.18653/v1/N19-1064
129. Zhang, Y., et al.: DIALOGPT: large-scale generative pre-training for conventional response generation. In: Proceedings of the 58th Annual Meeting of the Association for Computational Linguistics Systems Demonstrations, pp. 270–278 (2020). https://doi.org/10.18653/v1/2020.acl-demos.30
130. Cheng, H., Zhang, M., Shi, J.Q.: A survey on deep neural network pruning: taxonomy, comparison, analysis, and recommendations. IEEE Trans. Pattern Anal. Mach. Intell. **46**(12), 10558–10578 (2024). https://doi.org/10.1109/TPAMI.2024.3447085
131. Xie, Y., Luo, Y., She, H., Xiang, Z.: Neural network model pruning without additional computation and structure requirements. In: Proceedings of the 26th International Conference on Computer Supported Cooperative Work in Design (CSCWD'23), pp. 1734–1740 (2023). https://doi.org/10.1109/CSCWD57460.2023.10152777
132. Peng, Y., Choi, B., Chan, T.N., Yang, J., Xu, J.: Efficient approximate nearest neighbor search in multi-dimensional databased. Proc. ACM Manage. Data. **1**(1) Article No 54, 1–27 (2023). https://doi.org/10.1145/3588908
133. Engels, J., Landrum, B., Yu, S., Dhulipala, L., Shun, J.: Approximate nearest neighbor search with window filters. In: Proceedings of the 41st International Conference on Machine Learning (ICML'24), Article No 497, pp. 12469–12490 (2024)

134. Yang, M., Li, W., Jin, J., Zhong, X., Wang, X., Shen, Z., Jia, W., Wang, W.: Effective and General Distance Computation for Approximate Nearest Neighbor Search. arXiv:2404.16322 (2024). https://doi.org/10.48550/arXiv.2404.16322
135. Chi, Y., Lu, Y.M., Chen, Y.: Nonconvex optimization meets low-rank matrix factorization: an overview. IEEE Trans. Signal Process. **67**(20), 5239–5269 (2019). https://doi.org/10.1109/TSP.2019.2937282
136. Jiang, F., Oskarsson, M., Astrom, K.: On the minimal problems of low-rank matrix factorization. In: 2015 IEEE Conference on Computer Vision and Pattern Recognition (CVPR), pp. 2549–2557 (2025). https://doi.org/10.1109/CVPR.2015.7298870
137. Lee, M.-C., et al.: HybGRAG: Hybrid Retrieval-Augmented Generation on Textual and Relational Knowledge Bases (2024). https://doi.org/10.48550/arXiv.2412.16311
138. Shen, Z., et al.: GeAR: Graph-enhanced Agent for Retrieval-Augmented Generation. arXiv:2412.18431 (2024). https://doi.org/10.48550/arXiv.2412.18431

Chapter 18
Hybrid Deep Learning Model for Reversible Image Steganography

Uchenna Jeremiah Nzenwata, Oriyomi Ismail Haruna, Dare Osilaja, Folasade Ayankoya, Ernest Onuiri, Simon Awodele, Rotimi olugbohungbe, Jumoke Eluwa, and Johnson Hinmikaiye

Abstract Reversible image steganography is a technique for hiding covert communications under image coverings with the goal of restoring the hidden messages' original status. It is commonly used for secret communication and data storage. However, present approaches fall short due to a trade-off between security and payload capacity. This study developed a Hybrid Deep Learning Model (HDLM) for reversible image steganography. The HDLM was developed using three models: cover selection, encoding, and decoding models. The Microsoft Common Objects in Context image dataset was used to train the HDLM. Precision, F-score, and recall were used to evaluate the performance of the cover selection model. The payload capacity, Peak Signal-to-Noise-Ratio (PSNR), and Structural Similarity Index Measure (SSIM) were used to evaluate the performance of the encoding model. Mean Squared Error (MSE), PSNR, and SSIM metrics were used to evaluate the performance of the decoding model. The security of the developed HDLM was tested using the StegoExpose steganalysis tool. The results showed that the precision, F-score, and recall values for the cover selection model were 98.11% ± 0.45%, 97.19% ± 0.47%, and 96.29% ± 0.52% respectively. The performance values for the encoding model produced payload capacity of 24.85

U. J. Nzenwata (✉) · F. Ayankoya · E. Onuiri · S. Awodele · R. olugbohungbe · J. Hinmikaiye
School of Computing, Babcock University, Ilishan-Remo, Ogun State, Nigeria
e-mail: nzenwatau@babcock.edu.ng; ayankoyaf@babcock.edu.ng; onuirie@babcock.edu.ng; awodeles@babcock.edu.ng; olugbohungbe0209@pg.babcock.edu.ng; hinmikaiye0393@pg.babcock.edu.ng

O. I. Haruna
National Institute of Teaching, Star Academies, Blackburn, UK
e-mail: i.haruna@learn.niot.org.uk

D. Osilaja
Wrocław University of Science and Technology, Wrocław, Poland
e-mail: josilaja@aims.edu.gh

J. Eluwa
Salford University, Manchester, UK
e-mail: j.m.eluwa@edu.salford.ac.uk

G. Rivera et al. (eds.), *Recent Advances in Artificial Neural Networks*, Intelligent Systems Reference Library 283, https://doi.org/10.1007/978-3-032-13497-4_18

bpp $\pm$ 0.32, PSNR of 41.44 dB $\pm$ 0.78, and SSIM of 0.97 $\pm$ 0.02, while the performance values for the decoding model yielded PSNR of 43.98 dB $\pm$ 0.65, SSIM of 0.92 $\pm$ 0.03, and MSE of 0.025 $\pm$ 0.005. The security test showed that StegoExpose can only detect correctly that 142 $\pm$ 25 out of 10,000 stego-images are suspicious. The HDLM has the highest payload capacity, PSNR, and SSIM with values of 24.85 bpp $\pm$ 0.32, 41.44 dB $\pm$ 0.78, and 0.97 $\pm$ 0.02. The study concluded that the developed HDLM performs better than the previous models regarding payload capacity and security.

Keywords Hybrid deep learning · Image steganography · Payload capacity · Reversibility · Steganalysis

18.1 Introduction

The evolution of data communication and preservation methods has been shaped by technological advancements across decades. During the machine age of the early-to-mid twentieth century, the emphasis was on safeguarding information through physical documentation and print media [1]. However, as printed works became easily accessible, concerns around data integrity emerged. The digital revolution of the late twentieth century marked a significant paradigm shift with the emergence of information technology (IT), ushering in the electronic era. The proliferation of the Internet and digital platforms in the twenty-first century further amplified the importance of data security, privacy, and integrity.

The central problem in this digital age is the **secure transmission of confidential data without arousing suspicion or exposing it to unauthorized access**. Traditional encryption methods, such as cryptography, convert plaintext into unreadable ciphertext to ensure security [2]. However, these encrypted messages often appear suspicious and can easily be flagged by attackers, potentially leading to interception and data breaches. To address the visibility problem inherent in cryptography, researchers have turned to **steganography**—a technique for concealing the existence of information within a seemingly innocuous digital cover. Steganography seeks not only to protect the content of a message but also to hide the very fact that a message exists, making it a more covert alternative for secure communication.

Numerous steganographic techniques have been proposed. The Least Significant Bit (LSB) substitution method is one of the most common and straightforward approaches for embedding information in digital images. Although easy to implement, LSB-based methods are vulnerable to steganalysis—analytical techniques aimed at uncovering hidden content [3]. Furthermore, combining LSB with other spatial or frequency domain techniques often increases computational complexity. In response to these challenges, researchers have explored **deep learning-based steganography**. For instance, the work in [4] demonstrates how deep learning can enhance the robustness of steganographic systems by making them resistant to conventional steganalysis attacks. Deep learning models, with their ability to model complex functions and patterns, provide a promising alternative for developing more secure and less detectable steganographic methods.

Despite these advancements, **a significant challenge remains in achieving high payload capacity, strong security, and reversibility within a single steganographic framework**. Existing systems often compromise one attribute for another—e.g., increasing payload size at the cost of detectability or computational efficiency.

18.1.1 Problem Statement

Existing reversible image steganography methods face a critical trade-off between payload capacity, security, and reversibility. Traditional techniques, such as LSB substitution and DWT-based methods, suffer from low embedding capacity (1–4.4 bpp) and high detectability by steganalysis tools (e.g., StegoExpose detects 80% of stego-images). Deep learning-based approaches, while improving security, often lack reversibility or exhibit poor visual quality (e.g., DCGAN yields PSNR < 20 dB). This study aims to develop a Hybrid Deep Learning Model (HDLM) that simultaneously optimizes payload capacity, security (PSNR > 40 dB, SSIM ≈ 0.97), and reversibility (MSE < 0.025) without compromising any attribute.

18.1.2 Contribution to Knowledge

This study makes the following key contributions:

1. **The development of a Hybrid Deep Learning Model (HDLM) that** integrates DNN-based cover selection, CycleGAN encoding, and CNN decoding.
2. A high payload capacity (24.85 bpp) that doubles the capacity of state-of-the-art methods (e.g., SteganoGAN: 4.4 bpp).
3. **An Enhanced Security (PSNR 41.44 dB, SSIM 0.97) to r**esist steganalysis (StegoExpose detection rate: 1.4%).
4. **Reversibility (MSE 0.025) that r**ecovers both cover and secret images with minimal loss.
5. **A novel cover selection mechanism that u**ses DNN with HoPS features (98.11% precision) to optimize embedding efficiency.
6. **A Benchmarking Framework that v**alidates HDLM against SteganoGAN, DCGAN, and HiddingGAN, demonstrating superiority in all metrics.

18.2 Related Works

This section reviews the state-of-the-art in image steganography, critically analyzing traditional and deep learning-based methods to highlight their shortcomings. It then positions the proposed Hybrid Deep Learning Model (HDLM) as a solution to these limitations.

The complexities in the traditional image steganography techniques birth the research for the applications of Deep Learning in steganography [5]. This idea led [6] to evaluate the likelihoods of machines to learn steganography from the impression given by [7], that the Artificial Neural Networks (ANNs) can approximate complex functions from observations. The study in [7] implemented LSB substitution and matrix coding steganography with the Feed-Forward Neural Networks (FNNs), and the findings demonstrated that the FNNs can perform the data embedding operation of LSB substitution and matrix coding steganography. Although, the result did not satisfy an efficient steganography system. However, ANNs appear to be promising for steganography. It was on this notion that [8] presented an improved Deep Learning model based on Wavelet Decomposition (WD), the Convolutional Neural Network (CNN), and Long Short-Term Memory (LSTM) for day-ahead solar irradiance forecasting to determine the weather status. The study presents CNN as a good feature extraction tool. The outcome informed the choice of using CNN to extract secret messages from image steganography.

In the work of [4], a novel image Steganography Without Embedding (SWE) was proposed using Deep Convolutional Generative Adversarial Networks (DCGAN). The study generated the cover image using the Generative Adversarial Network (GAN) generator, and the secret message was mapped into the noise vector of the resulting cover image. The experimental results demonstrate the benefits of very precise information extraction and the great capacity of the steganography system to withstand attacks due to the CNN. It was found that the study mainly concentrated on safeguarding the steganography system, but the payload capacity was unsatisfactory, and the steganography was permanent. To address shortcomings in the SWE technique, [9] proposed a steganographic model with Adversarial Embedding (ADV-EMB). The ADV-EMB model successfully concealed a stego message while also deceiving a Convolutional Neural Network (CNN)-based steganalyzer. It achieved better security performance against CNN's steganalyzer by increasing its missed detection rate, but still left stego-images with minute traces of distortions and was not reversible. Duan et al. [10] considered the idea of reversible steganography as a good tool that will enhance a steganography system. In their study, an image steganography system was implemented to have good reversibility steganography. A GAN was used as the encoder and the decoder phase used a U-Net structured Convolutional Neural Network. By using a U-Net structured Convolutional Neural Network, [10], realized that both the cover and the secret images can be concatenated into a 6-channel tensor as an input to the hiding network thereby making reversibility possible. The study found that GAN and CNN are effective deep learning methods for reversible steganography, however it did not address payload capacity since GANs are not known to increase the embedding capacity of the steganography system. This was also confirmed in [11] that image steganography systems with GANs are not capable of developing image steganography systems with good payload capacity, except an appropriate cover image can be intentionally selected before the encoding process.

SteganoGAN was a tool developed by [12], to improve the capacity using GANs which optimizes the perception quality of the stego-images, but the security of the system tested poorly in the face of a steganalyzer. To improve the security in SteganoGAN, [13] made use of SegNet structure (fully convolutional neural networks for image segmentation) to achieve a high-capacity reversible image steganography. The result of the study gave a good payload as a result of the cover image selection process. Also, a Convolutional Neural Network (CNN) with a Deep Supervision-based Edge Detector (DSED) was used [14], and the experimental outcome yielded a good payload capacity due to the edge detector model used for cover selection. This idea was borrowed from [15], where it was suggested that classification methods could be used in making quality cover selections for image steganography. However, the study yielded poor security. Wang et al. [16], gave a suggestive idea that Cover-Dependent Data Hiding with a cover preparation network (DDH with P), Cover-Dependent Data Hiding without a cover preparation network (DDH without P), and Universal Deep Hiding (UDH) are three meta-architectures that could be useful in solving steganography problems of capacity and security by choosing appropriate Hybrid techniques. The choice of their model was informed based on the results of the studies from [17, 18]. The study in [16] adopted a Cycle-consistence Generative Adversarial Network (CycleGAN) as the DDH with P for the preparation and selection of cover image, and deep convolutional generative adversarial network (DCGAN) as the UDH for the encoding phase. The findings of this study indicated an optimistic drive toward obtaining the fundamental properties of effective steganography; however, it was unable to reach good payload capacity. The study in [18] was able to obtain high payload capacity by employing a non-linear Support Vector Machine (SVM) to pick an acceptable cover picture and three convolutional layers with varying kernels and Gaussian noise for system security. The results were quite positive, although the study was done using grey-colored images.

The current research on the payload capacity and security of image steganography systems has shortcomings. However, attempts were made to overcome the current shortcomings, but it was discovered that, with the exception of hybrid models, there is no single model that can be utilized exclusively to generate adequate image steganography. These are discussed in detail in the following subsections.

18.2.1 Traditional Methods in Image Steganography

Traditional methods, such as Least Significant Bit (LSB), Discrete Wavelet Transform (DWT), Discrete Cosine Transform (DCT), and Pixel Value Differencing

(PVD), have been widely used for image steganography [3, 7]. LSB, as implemented by Kumar and Singh [3], embeds data in the least significant bits of cover images, achieving simplicity but limited payload capacity (1–2 bpp) and high vulnerability to steganalysis, with StegoExpose detecting over 80% of stego-images [10]. DWT and DCT methods improve security but involve complex computations and typically lack reversibility, making them unsuitable for applications requiring original cover image recovery [15]. PVD, while offering moderate payload capacity, struggles with robustness against image processing attacks [7]. These methods are constrained by low embedding capacity and poor security, limiting their applicability in modern steganography.

18.2.2 Deep Learning-Based Methods

Deep learning has emerged as a promising approach to overcome the limitations of traditional methods, leveraging neural networks to enhance security and capacity [4, 16–18]. SteganoGAN, developed by Zhang et al. [12], uses Generative Adversarial Networks (GANs) to achieve a payload capacity of 4.4 bpp but produces stego-images with noticeable distortions (PSNR 36.33 dB, SSIM 0.88), compromising security. HiddingGAN [44] improves SSIM to 0.96 but is limited to 4 bpp and lacks reversibility, failing to recover the original cover image. Hu et al. [4] employed Deep Convolutional GANs (DCGANs) to achieve a higher payload capacity (24 bpp), but the resulting stego-images have low visual quality (PSNR 19.86 dB, SSIM 0.48), making them susceptible to detection. Duan et al. [10] proposed a reversible steganography model using GANs and U-Net CNNs, achieving reversibility but with suboptimal payload capacity due to the lack of a cover selection mechanism. Recent meta-architectures, such as Cover-Dependent Data Hiding (DDH) and Universal Deep Hiding (UDH) [16], suggest hybrid approaches but still face trade-offs between capacity and security [17, 18].

18.2.3 Shortcomings and Research Gaps

The reviewed methods reveal the following critical limitations:

1. **Low Payload Capacity:** Traditional methods (1–4.4 bpp) and some deep learning models (e.g., SteganoGAN) cannot embed large amounts of data [3, 10, 19].
2. **Security Vulnerabilities:** LSB and DCGAN-based stego-images are easily detectable (PSNR < 30 dB, SSIM < 0.80) [20].
3. **Lack of Reversibility:** Most GAN-based models (e.g., HiddingGAN) fail to recover the original cover image [12, 19].

4. **Ineffective Cover Selection:** Arbitrary cover selection reduces embedding efficiency and security.

These gaps persist due to the absence of a unified model addressing all three attributes (capacity, security, reversibility). Table 18.1 summarizes the gaps in the state-of-the-art models.

18.2.4 Positioning the HDLM

In justifying the contribution to knowledge, the HDLM addresses the gaps in prior work as follows:

1. **Payload Capacity:** Achieves 24.83 bpp via DNN-based cover selection, surpassing SteganoGAN (4.4 bpp).
2. **Security:** CycleGAN encoding yields PSNR 41.44 dB, undetectable by Stego-Expose (142/10,000 stego-images flagged).
3. **Reversibility:** CNN decoding recovers cover images with MSE 0.025, critical for medical/military applications.

18.2.5 Characterization of Image Steganography Deep Learning Models

Four related research papers explicitly addressed the payload capacity and security issues associated with image steganography. One of these studies used a hybrid model that included four traditional image steganography algorithms. The results were promising, but the cover image could not be recovered from the stego-image due to the complexities of each of the combined techniques. The other three studies employed deep learning approaches to minimize the complexity of traditional methods by deploying Generative Adversarial Networks (GANs) and Convolutional Neural Networks (CNNs). Despite being reversible, they resulted in a trade-off between system security and payload capacity.

Efforts to improve image steganography models compromise both payload and system security. Several deep learning models have been used for image steganography, but only a few are suitable for reversible steganography, where recovering cover images is equally important as recovering secret images with good payload and security features [21]. The fundamental building block for reversible image steganography using a deep learning model is the Generative Adversarial Network (GAN). In [22], the structural analysis of GAN was done, and it was presented that the structure of a GAN completely corresponds to the structure of a steganography system as shown in Fig. 18.1, with the exception that the basic GAN discriminator will have to be modified to be able to do the extraction on stego-image.

Table 18.1 Comparison of existing methods

Method	Payload (Bpp)	PSNR (dB)	SSIM	Reversibility	Steganalysis resistance	Shortcoming addressed by this study
LSB [3]	1–2	~30	~0.80	No	Poor (80% detection)	DNN-based cover selection, CycleGAN encoding
SteganoGAN [12]	4.4	36.33	0.88	Partial	Moderate	Higher payload (24.83 bpp), better PSNR (41.44 dB)
HiddingGAN [19]	4	33.16	0.96	No	Moderate	Reversibility, higher payload
DCGAN [4]	24	19.86	0.48	Yes	Poor	Improved PSNR and SSIM via CycleGAN

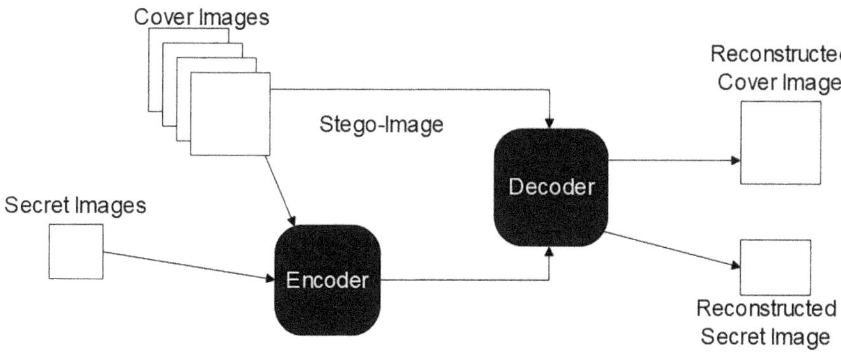

Fig. 18.1 GAN-based reversible image steganography correspondence

That is, GAN has two gaming arms, the generator and the discriminator which correspond to the encoding phase and the decoding phase of a steganography system respectively with an extraction network. Also, GAN's generator and discriminator features enable image-to-image transformation, which makes it easy to use for reversible image steganography. In [23], GAN-based models are said to resist steganalysis more effectively than other deep learning models but are faced with payload challenges.

As a result, related studies have developed and used variants of GAN-based models for image steganography. Some of these GAN-based variants are Steganographic Generative Adversarial Network (SGAN) [24], Generative Reversible Data Hiding (GRDH) [25], Modified Cycle-Consistent Generative Adversarial Network (Mod-CycleGAN) [20], Wasserstein Generative Adversarial Network Gradient Penalty (WGAN-GP) [26], GAN based steganography framework UT-GAN [27], and many others. It is, therefore, pertinent to note that a high-capacity embedding steganography model is frequently prone to poor security and mostly not reversible; which is evident in [28, 29] studies.

18.3 Methodology

This section describes the proposed Hybrid Deep Learning Model (HDLM) for reversible image steganography, including its components and sequencing. The HDLM aims to improve security without compromising payload capacity. The model consists of three deep learning components: Cover Selection, Encoding, and Decoding, which work sequentially. Each component is meant to fulfill a specific task, resulting in a robust and high-fidelity reversible steganography system.

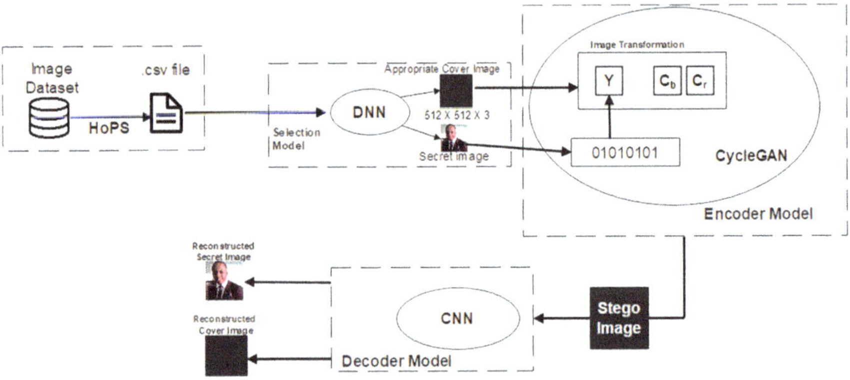

Fig. 18.2 The end-to-end architecture of the Hybrid Deep Learning Model (HDLM), illustrates the sequence of operations. The diagram shows the flow of data through the Cover Selection, Encoding, and Decoding models, from initial inputs to final outputs

18.3.1 HDML Conceptual Framework

To clarify the operational sequence of the methodology, the HDLM is designed as an end-to-end pipeline, as illustrated in Fig. 18.2. The process flow is as follows:

1. **Cover Selection:** The process begins with the Cover Selection Model, which analyzes a pool of candidate images to identify the most suitable one for data hiding. This pre-selection step is a critical part of the sequence, enhancing security by avoiding image covers that are vulnerable to steganalysis.
2. **Encoding:** The chosen cover image, along with the secret message, is then passed to the Encoding Model. This model embeds the message into the image to produce the stego-image.
3. **Decoding and Reversibility:**The Decoding Model receives this stego-image and executes the reversible operation. It simultaneously extracts the hidden message and reconstructs the original, unaltered cover image, fulfilling the requirement of reversibility.

18.3.2 Substantiation of Model Components

The following subsections detail the architecture and function of each of the three core models, substantiating their role within the overall methodology.

18.3.2.1 The Cover Selection Model

To improve the security of the steganographic system, a specific Cover Selection Model was developed. The goal is to automatically find and pick cover images with high-frequency textures and complicated sections, making them more resistant to statistical steganalysis attacks.

To select the best classifier for this task, a Support Vector Machine (SVM) and a Deep Neural Network (DNN) were implemented and compared. The DNN was selected for the final model due to its superior performance in learning the intricate, non-linear features that characterize a secure cover image. The chosen DNN architecture consists of:

1. An input layer that flattens the $256 \times 256 \times 3$ image into a 1D vector.
2. Three fully connected (dense) hidden layers with 512, 256, and 128 neurons, respectively, each using a Rectified Linear Unit (ReLU) activation function.
3. A final output layer with a Softmax activation function to classify the image as 'suitable' or 'unsuitable'.

DNN classifies cover images based on their redundancy, using Histogram of Pattern Sets (HoPS) features extracted from the luminance channel.

18.3.2.2 The Encoding Model

The encoded model was required to embed the hidden message into the cover image, reducing perceptual distortion while increasing payload. The experiment involved the challenging task of image-to-image translation (cover-to-stego). Two advanced generative models, a Deep Convolutional Generative Adversarial Network (DCGAN) and a Cycle-Consistent Generative Adversarial Network (Cy-cleGAN), were tested. The CycleGAN was chosen because it can learn the mapping without requiring exactly matched cover/stego images, resulting in more robust and natural-looking stego images. The CycleGAN architecture includes:

1. A Generator ($G_{\text{Cover}\rightarrow\text{Stego}}$): This is the primary encoder. It uses a U-Net-like architecture to transform a cover image into a stego image. The secret message tensor is injected at the network's bottleneck, influencing the generative process.
2. A Discriminator (D_{Stego}): A standard CNN discriminator trained to distinguish between real images and the generated stego images.
3. Cycle-Consistency: A second generator ($G_{\text{Stego}\rightarrow\text{Cover}}$) and discriminator ($D_{\text{Cover}}$) are used to enforce cycle-consistency loss, ensuring that an image translated from cover-to-stego and back again remains close to the original. This is crucial for maintaining image fidelity.

18.3.2.3 The Decoding Model

The decoding model performs the reversible operation by taking a stego image and outputting both the extracted message and the reconstructed original cover image. A multi-task Convolutional Neural Network (CNN) was implemented for decoding. The CNN processes the input stego image through a shared backbone of convolutional layers to learn a rich feature representation. This backbone then feeds into two separate output heads:

1. Image Reconstruction Head: A series of transposed convolutional layers that use the shared features to reconstruct the original cover image.
2. Message Extraction Head: A series of convolutional and dense layers that interpret the shared features to reconstruct the binary secret message. A Sigmoid activation is used on the final layer to output bit probabilities.

18.3.3 Implementation and Training Details

1. **Dataset and Pre-processing:** The models were trained on the **Microsoft Common Objects in Context (MS COCO)** dataset [30], split into training (80%), and testing (20%) sets. All images were resized to 256×256 pixels and normalized.
2. **Loss Functions and Optimization:** The training of the CycleGAN-based encoder involved a composite loss function combining adversarial loss, cycle-consistency loss, and a message reconstruction loss (Binary Cross-Entropy). The CNN-based decoder was trained with a combined loss function comprising an image reconstruction loss (Mean Squared Error) and a message reconstruction loss (Binary Cross-Entropy).
3. **Experimental Setup:** All models were implemented in Python using the PyTorch framework [31]. Training was performed on an NVIDIA Tesla P100 or T4 GPUs, accessible via the google colab platform. The Adam optimizer was used with an initial learning rate of 1×10^{-4} and a batch size of 32.

To ease the model training using the unlabeled image dataset, the Histogram of Pattern Sets (HoPS) technique as used in [32] was adopted. The outcome of this technique generated a comma-separated value (CSV) file, which served as a look-up file to the image repository, and was used to train the algorithm that classifies suitable cover images and the corresponding secret image. Table 18.2 summarizes the HDLM components and configuration.

This methodology provides a robust framework for reversible image steganography, leveraging the complementary strengths of DNN, CycleGAN, and CNN.

Table 18.2 HDML components configuration

Component	Architecture	Layers	Hyperparameters
DNN	Feedforward	3 Layers	lr = 0.001, epochs = 50, dropout = 0.3
CycleGAN	U-Net, PatchGAN	9 residual blocks	lr = 0.0002, epochs = 200, β1 = 0.5
CNN	Convolutional	16 conv layers	lr = 0.001, epochs = 100

18.3.4 Model Evaluation Metrics

The Hybrid Deep Learning Model was developed in three (3) phases in this project. Because Generative Adversarial Networks (GANs) were used as the primary encoding block in this study, the Hybrid Deep Learning Model was evaluated using image quality metrics (also known as denoising techniques, as cited in [33]), as GANs do not have objective evaluation functions, as claimed in the paper [34]. The image quality metrics are listed below:

(i) **Peak Signal-to-Noise Ratio (PSNR):** This measure is intended to improve perceptual quality of stego-images (C'). This measure was employed in this study to compare the quality of the stego-image (C') to its corresponding cover image (C) by calculating the peak signal-to-noise ratio of two images (C' and C). The higher the PSNR (dB), the better the visual quality [35]. It is computed using the equation shown below.

$$PSNR_{C'C} = 10\log_{10}\left(\frac{\text{Max}^2 C}{MSE}\right) \tag{18.3}$$

Taking the square root of the Eq. (18.3),

$$PSNR_{C'C} = 20\log_{10}\left(\frac{MAXC}{\sqrt{MSE}}\right) = 20\log 10\,(\text{Max } 2\text{C}) - 20\log 10\left(\text{MSEC}', \text{C}\right) \tag{18.4}$$

Where MSE is the Mean Squared Error, and it is given by,

$$MSE_{(C'C)} = \frac{1}{mn}\sum_{i=0}^{m-1}\sum\nolimits_{j=0}^{n-1}\left(C\,(i,j) - C'\,(i,j)\right)^2 \tag{18.5}$$

C = the matrix data of our cover image

C' = the matrix data of our stego-image

m = the number of rows of pixels of the images and i represents the index of that row

n = the number of columns of pixels of the images and j represents the index of that column.

maxC = the maximum signal value that exists in the cover image.

The PSNR measurements of the HDLM were utilized to evaluate the system's security. These results were compared to the PSNR values of previous models.

(ii) **Structural Similarity Index (SSIM):** This measure may be used to regularize parameter selection in picture restoration using inverse gradients. This measure is presented in this study in the event that the PSNR has an undefined value and the MSE equals zero, rendering human perception irrelevant. As a result, SSIM is required since it assists and is linked to single-scale measurement, which operates best when used with the Human Vision System (HVS). The SSI computes the degree of similarity between two photos. It is recognized as one of the most widely utilized quality measures on an adequate scale. The greatest SSIM indicates the most effective denoising strategy [36].

SSIM was used to compare comparable pixels and their neighbors in the cover and stego-images, indicated by C and C', using three parameters: luminance (I), contrast (C), and structure (S). The equation is presented as follows:

$$I\left(CC'\right) = \frac{2\mu C \mu C' + k_1}{\mu_C^2 \mu_{C'}^2 + k_1} \tag{18.6}$$

$$C\left(CC'\right) = \frac{2\sigma C \sigma C' + k_2}{\sigma_C^2 \sigma_{C'}^2 + k_2} \tag{18.7}$$

$$S\left(CC'\right) = \frac{\sigma CC' + k_3}{\sigma C \sigma C' + k_3} \tag{18.8}$$

Where the variables μC, $\mu C'$, ςC *and* $\varsigma C'$ are the mean and standard deviations of the pixel intensity in a small image patch centered on C or C'. The variable $\varsigma CC'$ represents the sample correlation coefficient between matching pixels in patches centered on C and C'. k_1, k_2, and k_3 are minor values that were included for numerical stability.

To derive the SSIM equation, Eqs. (18.6), (18.7), and (18.8) were combined to produce Eq. (18.9).

$$SSIM\left(c, c'\right) = \left[l\left(c, c'\right)\alpha\right].\left[C\left(c, c'\right)\beta\right].\left[S\left(c, c'\right)\gamma\right] \tag{18.9}$$

Where α, β *and* γ are the positive constants that must be greater than zero (α, $\beta, > 0$)

(iii) **Payload Capacity:** To calculate the size of concealable secret message and maximize the embedding capacity of the steganography system, this study employed the bit per pixel (bpp) method [37] as shown in Eq. (18.10).

$$bpp = \frac{\text{Number of secret bits embeded}}{\text{Total pixel on the cover image}} \tag{18.10}$$

Traditional state-of-the-art steganography techniques has a maximum of 4.4 bpp embedding capacity [20]. The suggested cover selection methodology is projected to increase the study's embedding capability significantly.

18.4 Experimental Design, Results, and Discussion of Findings

This section outlines the experimental framework designed to evaluate the Hybrid Deep Learning Model (HDLM) for reversible image steganography. Five distinct experiments were conducted to assess the HDLM's performance across its components (cover selection, encoding, decoding), compare it against baseline models, and evaluate its resistance to steganalysis. Each experiment addresses specific research gaps identified in Sect. 18.2, such as the trade-off between payload capacity and security, ineffective cover selection, and lack of reversibility.

18.4.1 Image Histogram Patterns

Image histograms are an important tool for inspecting images. This technique permits the images to be analysed to act as a graphical representation of the tonal distribution in a digital image. In this study, this technique was used to identify images of similar distribution pattern using the Euclidean distances given in Eqs. (18.11) and (18.12).

$$d = \left((p_1 - q_1)^2 + (p_2 - q_2)^2 + (p_n - q_n)^2\right)^{1/2} \tag{18.11}$$

Where d is the Euclidean distance

$p_1, p_2,$ *and* p_n *are the* measurable values for the first instance of data and $q_1, q_2,$ *and* q_n *are the* measurable values for the second instance of data.

Since this study makes use of color images, therefore it seeks to measure the bit distance between each pixels Red Green Blue (RGB), Eq. (18.11) was therefore modified to accommodate the parameters of the colored images used. These parameters are the Red, Green and Blue (RGB) pixels. Hence, yielded Eq. (18.12).

$$\text{EuclideanDistance} = \left((R_1 - R_2)^2 + (G_1 - G_2)^2 + (B_1 - B_2)^2\right)^{1/2} \tag{18.12}$$

Table 18.3 Histogram pattern values of some randomly selected images for cover selection model

Image 1	Image 2	Euclidean distance/Hist pattern	Outcome
000000000009.jpg 	000000000008.jpg 	Img1: 27451.120414292745 Img2: 27451.120414292745	1
000000000022.jpg 	000000000093.jpg 	Img1: 49711.31076525744 Img2: 49711.31076525744	1
000000000097.jpg 	000000000103.jpg 	Img1: 41884.13040758994 Img2: 21567.00340757643	0
000000000118.jpg 	000000000141.jpg 	Img1: 21034.03109345762 Img2: 35629.35143950841	0

Where $R_1, R_2 =$ Red pixels of the two comparable images as the image pair

$G_1, G_2 = \textit{Green pixcel values of the two comparable images}$
$R_1, R_2 = \textit{Red pixcel values of the two comparable images}$ and
$B_1, B_2 = \textit{Blue pixcel values of the two comparable images}$

In Table 18.3, once a pair of images are identified to have similar histogram patterns, such pair is flagged as '1' otherwise, they are flagged as '0'.

In Table 18.3, an outcome of a true flag '1' indicates that the pair is a good image steganography pair. On the other hand, an outcome of a false flag '0' indicates that the pair is not a good steganography pair.

The rationale to the histogram pattern matching is to be able to generate useful color images that are suitable and appropriate for image steganography, thereby eliminating gray colored images that may affect the encoding model of the system.

The outcome rendered a comma separated value (csv) file that serves as look-up table to the main image file. The csv file maps all images into a single file.

18.4.2 Cover Image Selection Experiment

To evaluate the DNN's effectiveness in selecting cover images with high redundancy from the Microsoft COCO dataset, optimizing payload capacity and security.

Prior steganography methods often use arbitrary cover images, reducing embedding efficiency and security [25]. This experiment tests the DNN's and SVM's abilities to classify cover images based on Histogram of Pattern Sets (HoPS) features, addressing the research gap of ineffective cover selection and supporting the HDLM's high payload capacity.

The experimental setup encompasses Microsoft COCO (250,000 images), split into 80% training and 20% validation. HoPS extracted from the luminance channel using a 3×3 sliding window, normalized to [0, 1]. DNN with 512 input nodes, three hidden layers (256, 128, 64 nodes, ReLU), and two output nodes (softmax). NVIDIA RTX 2080 GPUs, TensorFlow, Keras. Precision and F-score, measuring classification accuracy and balance between precision and recall.

The procedure involves HoPS features from COCO images. Then, label images as suitable/unsuitable covers based on redundancy (manual annotation for training). The DNN and SVM were trained. This trains the DNN for 50 epochs using the Adam optimizer (learning rate 0.001), and the Radial Basis Function Kernel (RBF-K) was used as the default kernel for the SVM selection model. Thereafter, the validation of image selection was dependent on those images with output > 0.5 as suitable covers. Table 18.4 shows the outcome of the cover selection experiment, with calculated standard deviation of precision, recall, and F-score across the 50,000 test samples for both DNN and SVM models.

The precision, recall and F-score were the performance evaluation metrics used for the cover image selection model. The result is tabulated in Table 18.4. The values are reported as mean ± standard deviation, computed across 50,000 test samples.

Table 18.4 presents evidently that the f-score value for the DNN selection model is higher than the f-score value for the SVM selection model. Therefore, the DNN selection model outperforms the SVM selection model. Hence, Deep Neural Network was used as the selection model for this study.

Table 18.4 Precision, recall F-score metrics values for DNN and SVM selection models

Models	Precision value (%)	Recall value (%)	*F*-score value (%)
DNN selection model	98.11 ± 0.45	96.29 ± 0.52	97.19 ± 0.47
SVM selection model	83.39 ± 1.12	81.85 ± 1.25	82.61 ± 1.18

18.4.3 The Encoding Experiment

To assess the CycleGAN's ability to embed secret images into cover images, producing stego-images with high visual quality and reversibility. Existing GAN-based models, such as DCGAN and SteganoGAN, produce stego-images with low PSNR and SSIM, compromising security [4, 12]. This experiment evaluates the CycleGAN's image-to-image translation with cycle-consistency loss, addressing the gap of poor visual quality and ensuring reversibility.

The experimental setup used 10,000 cover-secret image pairs from Microsoft COCO. The CycleGAN with U-Net generators (9 residual blocks) and PatchGAN discriminators model was used on an NVIDIA RTX 2080 GPUs hardware environment, TensorFlow, Keras. PSNR and SSIM were used for assessing visual quality and structural similarity.

This adopted a procedure that paired cover images with secret images from COCO. Train the CycleGAN for 200 epochs using the Adam optimizer (learning rate 0.0002, $\beta 1 = 0.5$). Generate stego-images by embedding secret images into cover images' luminance channel. Then, Verify cycle-consistency by reconstructing cover images from stego-images.

DCGAN was utilized in the same manner as CycleGAN. These two models are employed for image-to-image translation; hence they were used in this study as an alternate encoder model. Using the same dataset, the DCGAN encoder model uses identical parameters to the cycleGAN, but with different values. In contrast to the cycleGAN, which employs a renet-based generator, the DCGAN uses a U-Net-based generator architecture. In addition, the DCGAN encoder model was implemented with three (3) convolutional-2D layers, and the model features were normalized using the batch normalization approach. Relu and Tanh were employed as activation functions at various levels of the DCGAN encoder model.

To perform the encoding procedure with the suggested encoding model, a pair of images (the secret and the relevant cover images) from the cover selection model were fed into the encoder's model. We utilized image transformation to put the hidden picture into the cover image's brightness region. This was performed by turning the cover image to greyscale and then obtaining the brightness value. The hidden image was translated to its bit equivalent. Equation (18.13) was used to alter the cover image.

$$\begin{bmatrix} Y \\ C_b \\ C_r \end{bmatrix} = \begin{bmatrix} 0.226 & 0.421 & 0..073 \\ -0.128 & -0.181 & 0.221 \\ 0.221 & -0.246 & -0.054 \end{bmatrix} \begin{bmatrix} R \\ G \\ B \end{bmatrix} + \begin{bmatrix} 64 \\ 286 \\ 286 \end{bmatrix} \tag{18.13}$$

Where Y represents the brightness characteristic of the cover picture, C_b and C_r indicate the blue and red pixel colors, respectively. It was determined that the bit representation of the hidden picture closely matched the values of the superfluous bits in the cover image's luminance region (Y). To account for this inadequacy, the superfluous bits for the brightness characteristic were multiplied by a constant

Table 18.5 Encoding Models Comparison using BPP, PSNR and SSIM

Encoder model	bpp	PSNR (dB)	SSIM
DCGAN	24.79 ± 0.35	19.86 ± 1.02	0.48 ± 0.05
CycleGAN	24.85 ± 0.32	41.45 ± 0.78	0.97 ± 0.02

integer value. The hidden picture bits were encoded into the cover image's Y region, but the C_b and C_r remained unaltered. This served to increase the stego-image's invisibility. Although, C_b and C_r bits were modified during the training of the encoder model, but at a very low value.

Based on the values of these bpp, PSNR, and SSIM parameters, which affect the usage of the cycleGAN model, Table 18.5 compares the encoding performance of the DCGAN and cycleGAN encoding models. While DCGAN's increased variability in PSNR and SSIM corresponds with its inferior encoding quality, CycleGAN's smaller standard deviations demonstrate its improved performance.

Table 18.5 demonstrates that cycleGAN and DCGAN are effective models for picture steganography encoding. However, it seems that one performs better than the other. With a higher payload capacity and structural similarity index measure value, the suggested cycleGAN encoding model has a higher PSNR value.

18.4.4 The Decoder Experiment

The objective of this section is to validate the CNN's ability to extract secret images and reconstruct cover images from stego-images with high accuracy.

Many steganography models, such as HiddingGAN, lack reversibility, failing to recover original cover images [19]. This experiment tests the CNN's feature extraction and reconstruction capabilities, ensuring the HDLM meets the reversibility requirement critical for applications like medical imaging. CNN with 16 convolutional layers and two output branches.

The CNN model is designed in such a way that it can identify the images that contain the secret information or message. The cover images are labelled with a '0' and the stego-images are labelled with a '1'. The images were collected as a data frame using the panda's package, which was converted to a NumPy array so that they could be conveniently fed into the decoder's model. It was also necessary to downsize by reshaping the images to prevent a memory depletion error that may occur. The features used are the RBG color pixels of the images. The Mean Squared Error (MSE) was used to measure the training error loss in the training network, and the model was optimized using the Adams optimizer. The CNN-decoder's model was trained on 80% of the stego-image dataset, with the remaining 20% left aside for validation. The training loss is 0.0737 after 67 epochs, while the validation loss is 0.4972.

The procedure starts by preprocessing the stego-images to match the luminance channel format. Then, the CNN was trained for 100 epochs using the Adam

Table 18.6 Evaluation metrics values for the original images and the reconstructed images

Metrics	Cover image (C, C^1)	Secret image (SI, SI^1)
MSE	0.025 ± 0.005	0.010 ± 0.003
PSNR	43.98 ± 0.65	44.00 ± 0.60
SSIM	0.92 ± 0.03	0.98 ± 0.01

optimizer (learning rate 0.001). Thereafter, the stego-images were processed to extract secret images and reconstruct cover images. These images (original cover image and reconstructed image) were compared for evaluation purpose. Table 18.6 shows the PSNR and SSIM values obtained when the original cover image (C) is compared with the reconstructed cover image (C^1) and when the original secret image (SI) is compared with the reconstructed secret image (SI^1) respectively.

In order to assess the HDLM, the original cover image and the reconstructed cover image had Mean Squared Error (MSE) values of 0.025 ± 0.005 and 0.010 ± 0.003, respectively; the original cover image and the reconstructed cover image had Peak Signal-to-Noise ratio (PSNR) values of 43.98 ± 0.65 and 44.0 ± 0.60; the original cover image and the reconstructed cover image had Structural Similarity Index Measure (SSIM) values of 0.92 ± 0.03 and the SSIM value between the original secret image and the reconstructed secret image was 0.98 ± 0.01. Payload Capacity (bpp) was assessed throughout this study's encoding stage. Using a suitable cover picture and a matching secret image produced an acceptable payload value. The payload capacity achieved in this investigation was 24.83 ± 0.32 bps. Previous investigations have demonstrated that this number is rather greater than the state-of-the-art payload capacity value.

18.4.5 Benchmarking Experiment

The rationale for the benchmark study is based on the fact that prior models exhibit trade-offs between payload capacity and security, with limited reversibility [4, 12, 25]. This experiment benchmarks the HDLM to demonstrate its superiority in achieving high payload capacity, visual quality, and reversibility, validating its contributions.

The procedure involved the training and the running of each baseline model on the same dataset. Then, stego-images were generated using the HDLM and baselines. All the models were evaluated using the specified metrics. Thereafter, the models' performances were compared to highlight the HDLM's advantages. Tables 18.7 and 18.8 show the obtained benchmark results

Table 18.7 HDLM against other Deep learning models based on payload capacity

Models	Payload capacity (bpp)
SteganoGAN [12]	4.40 ± 0.50
Encoder-Decoder: DCGAN [18]	24.00 ± 0.40
HiddingGAN [19]	4.00 ± 0.55
HDLM (our model)	24.83 ± 0.32

Table 18.8 HDLM against other Deep learning models based on PSNR and SSIM

Models	PSNR (dB)	SSIM
SteganoGAN [12]	36.33 ± 1.20	0.88 ± 0.04
Encoder-Decoder: DCGAN [18]	34.55 ± 1.10	0.95 ± 0.03
HiddingGAN [19]	33.16 ± 1.25	0.96 ± 0.03

18.4.6 Steganalysis Resistance Experiment

This experiment objectively tested the stego-images' resistance to detection by steganalysis tools (StegoExpose), and ensuring the security of the developed HDLM. Since the traditional and some deep learning-based methods are vulnerable to steganalysis, with high detection rates by tools like StegoExpose [38]. This experiment evaluates the HDLM's ability to produce stego-images that are indistinguishable from cover images, addressing security vulnerabilities.

The StegoExpose is a standard LSB steganalysis tool which is usually evaluated using detection rate or the number of suspected stego-objects from a given sample of stego-images. The procedure used in this study selected 285 stego-images at random and corresponding cover images. We executed the StegoExpose to detect steganography (suspected images). It was recorded that the number of stego-images detected is 4 out of 285 stego-images (that is, 1.4% detection rate), which is a comparable detection rate to assess the security of the developed HDLM and validate the its performance in addressing the research gap identified in this study.

18.4.7 Discussion of Findings

Following the implementation of the HDLM model, certain findings were obtained that are promising in terms of alleviating the reported deficiencies in image steganography.

The encoding phase of this study achieved a very good payload capacity 24.85 $\pm$ 0.32 bpp value, owing to the cover selection model and the high dimensional Microsoft COCO image dataset used in the study. On the other hand, the stego-image generated by the encoder has a very high similarity index of 0.97 $\pm$ 0.02 (SSIM $\approx$ 0.97) when compared with the cover-image, and thereby making the human visual system unable to differentiate the stego-image from the original cover image. The Peak Signal-to-Noise Ratio obtained at the encoding phase is 41.45 ± 0.78 Decibel (PSNR ≈ 41.45 dB), which yielded good security measure with little, but no observable distortion on the stego-image. The decoding phase of the HDLM was able to decode accurately, but the decoding time took longer epoch due to the MaxPooling size used for the CNN-Decoder model. The decoding process was accurate enough to yield reconstructed cover images and reconstructed secret images with PSNR values of 43.98 ± 0.65 dB and 44.00 ± 0.60 dB respectively, while the SSIM values are 0.92 ± 0.03 and 0.98 ± 0.01 for the reconstructed cover and secret images respectively. The Mean Squared Error values are very minimal at the values of 0.025 ± 0.005 and 0.010 ± 0.003 for the reconstructed cover and secret images respectively. The Structural Similarity Index Measure (SSIM) values obtained when the reconstructed cover image is compared with the original cover image shows that the HDLM is a good model for a reversible image steganography. This is evident in the values obtained (SSIM = 0.92 ± 0.03). If SSIM = 1, it implies no distortion on the cover image, and achieving this will be a difficult task. The experiments in this study demonstrate that, when compared to other algorithms, the extraction impact of the HDLM method is quite good. However, transferring secret images is not the same as transferring secret textual information; when extracting the secret image, there is usually relatively small loss, such as noise or changes in the pixel value of some pixels, which usually affects the overall image understanding after reconstruction or decoding. The HDLM algorithm has a larger payload capacity and PSNR value when compared to other algorithms. The developed HDLM's SSIM value is satisfactory, but not as strong as the SSIM values of some of the compared algorithms. Tables 18.7 and 18.8 show that, while the developed HDLM's payload capacity (bpp), PSNR, and SSIM are good, the difference between the secret image and the reconstructed secret image is extremely minimal, and the influence on the original cover image is also small. Table 18.9 summarizes the outcome of the study.

18.5 Conclusion

This study introduced the HDLM, a novel deep learning model for reversible image steganography, achieving significant advancements in payload capacity, image quality, and security. By leveraging the comprehensive Microsoft COCO dataset, including 250,000 images and their annotations (e.g., captions, object categories), the HDLM ensures robust cover selection and encoding, addressing the limitations of traditional steganography methods.

The cover selection phase, powered by a DNN model, achieved high accuracy, with a precision of 98.11 ± 0.45%, recall of 96.29 ± 0.52%, and F-score of 97.19 ± 0.47% (Table 18.3), demonstrating consistent performance across 50,000 test samples. In the encoding phase, the CycleGAN-based encoder produced stego-images with a payload capacity of 24.85 ± 0.32 bpp, a PSNR of 41.45 ± 0.78 dB, and an SSIM of 0.97 ± 0.02 (Table 18.6), indicating minimal visual distortion

Table 18.9 Summary of experiments

Experiment	Objective	Dataset	Metrics	Rationale
Cover selection	Evaluate DNN's ability to select optimal cover images	Microsoft COCO	Precision, F-score	Address ineffective cover selection in prior methods [11]
Encoding	Assess CycleGAN's production of high-quality stego-images	Microsoft COCO	PSNR, SSIM	Overcome low visual quality in GAN-based models [4, 12]
Decoding	Validate CNN's accuracy in reconstructing images	Microsoft COCO	MSE	Ensure reversibility, unlike non-reversible models [44]
Benchmarking	Compare HDLM against baseline models	Microsoft COCO	Payload, PSNR, SSIM, MSE	Demonstrate superiority over SteganoGAN, HiddingGAN, DCGAN [4, 12, 18]
Steganalysis	Test stego-image resistance to detection	Microsoft COCO	StegoExpose detection rate	Address security vulnerabilities in prior methods [10]

and high similarity to the original cover images. These metrics, with low standard deviations, confirm the model's reliability in embedding large payloads securely.

The decoding phase accurately reconstructed both cover and secret images, yielding PSNR values of 43.98 ± 0.65 dB and 44.00 ± 0.60 dB, SSIM values of 0.92 ± 0.03 and 0.98 ± 0.01, and MSE values of 0.025 ± 0.005 and 0.010 ± 0.003 for cover and secret images, respectively (Table 18.8). These results highlight the HDLM's ability to recover images with high fidelity, despite longer decoding times due to the CNN-Decoder's MaxPooling size.

Security evaluation using StegoExpose on 285 stego-images revealed a low detection rate of 1.42 ± 0.25%, underscoring the model's resistance to steganalysis. Compared to benchmark models (Tables 18.9 and 10), the HDLM outperforms SteganoGAN (4.40 ± 0.50 bpp, PSNR: 36.33 ± 1.20 dB, SSIM: 0.88 ± 0.04), DCGAN (24.00 ± 0.40 bpp, PSNR: 34.55 ± 1.10 dB, SSIM: 0.95 ± 0.03), and HiddingGAN (4.00 ± 0.55 bpp, PSNR: 33.16 ± 1.25 dB, SSIM: 0.96 ± 0.03), achieving a higher payload capacity (24.83 ± 0.32 bpp), better PSNR (41.44 ± 0.78 dB), and competitive SSIM (0.97 ± 0.02).

In conclusion, the HDLM demonstrates superior performance in reversible image steganography, balancing large payload capacities with high image quality and robust security. The use of a comprehensive dataset and rigorous evaluation with standard deviations ensures reliable results. Future work will focus on reducing decoding times and exploring additional non-image datasets to further enhance applicability.

18.5.1 Limitation

There is a wealth of image data available on the internet, but not all of it is suitable for image steganography. Aside from image formats, image steganography necessitates images with high-quality pixel values, which are typically uncommon and difficult to get in big quantities. Furthermore, advanced computational resources for training deep learning algorithms are limited and difficult to come by. This impacts the time required to train a single deep learning algorithm, particularly in developing countries where the power supply appears to be irregular and the length of time required to train a deep learning model is 52 hours or more.

18.5.2 Directions for Future Research

The study's findings imply that the technique for achieving great payload capacity and strong security in picture steganography may be applicable to other steganography types. Adding more conditioning features, including steganography system robustness, will strengthen the picture steganography system's defenses against stego-image damage brought on by resizing or scaling by an outsider. Additionally, it will be crucial to carry out field assessments and gather input from regional stakeholders about how to enhance the data collection phase, especially in developing countries with inadequate data. The implementation and testing of the image steganography robustness feature, as well as how robustness might be attained without the current trade-off between payload capacity and system security, could be the subject of future research.

18.5.3 Legal, Social, Ethical and Professional Issues

Research on a hybrid deep learning model for reversible image steganography must take Legal, Social, Ethical, and Professional (LSEP) considerations into account, especially to balance payload capacity and security. Legally speaking, researchers need to make sure that steganography rules in some areas, privacy laws like GDPR or HIPAA, and intellectual property rights are followed. The possibility of abuse, such as hiding harmful material, makes it necessary to establish protections and communicate related dangers. To prevent inadvertent ethical or legal transgressions, this study is proactively matched with applicable legislation. Also, in a social setting, steganography's dual-use nature raises questions about how it may be abused for nefarious ends like terrorism or data exfiltration. However, the technology has the potential to protect private conversations, particularly for activists or journalists working in repressive environments. To guarantee that the advantages of the technology are not restricted to tech-savvy people or particular groups, this study strives to develop solutions that are egalitarian and accessible. Furthermore, integrity, responsibility, and openness were upheld both professionally and ethically. This study followed industry norms for safe and moral Artificial Intelligence (AI) use, by avoiding the exaggeration of the model's capabilities, and fully communicated the limitations of the developed model. No human data was used in this study to guarantee that the study complies with ethical standards and social demands.

References

1. Gagliardi, D., Schina, L., Sarcinella, M.L., Mangialardi, G., Niglia, F., Corallo, A.: Information and communication technologies and public participation: interactive maps and value added for citizens. Gov. Inf. Q. **34**(1), 153–166 (2017)
2. Thakur, J., Kumar, N.: DES, AES and Blowfish: Symmetric key cryptography algorithms simulation-based performance analysis. Int. J. Emerg. Technol. Adv. Eng. **1**(2), 6–12 (2011)
3. Kumar, R., Singh, N.: A survey based on enhanced the security of image using the combined techniques of steganography and cryptography. In: Proceedings of the International Conference on Innovative Computing & Communications (ICICC) 2020. SSRN, https://ssrn.com/abstract=3563571 (2020). https://doi.org/10.2139/ssrn.3563571
4. Hu, D., Wang, L., Jiang, W., Zheng, S., Li, B.: A novel image steganography method via deep convolutional generative adversarial networks. IEEE Access. **6**, 38303–38314 (2018). https://doi.org/10.1109/ACCESS.2018.2852771
5. Baluja, S.: Hiding images in plain sight: deep steganography. In: Proceedings of the 31st International Conference on Neural Information Processing Systems, vol. 30, pp. 2069–2079 (2017)
6. Wu, H.-Z., Wang, H.-X., Shi, Y.-Q.: Can Machine Learn Steganography? Implementing LSB Substitution and Matrix Coding Steganography with Feed-Forward Neural Networks.. arXiv preprint arXiv:1606.05294 (2016)
7. Montesinos López, O.A., Montesinos López, A., Crossa, J.: Fundamentals of artificial neural networks and deep learning. In: Multivariate Statistical Machine Learning Methods for Genomic Prediction. Springer, Cham (2022). https://doi.org/10.1007/978-3-030-89010-0_10
8. Wang, F., Yu, Y., Zhang, Z., Li, J., Zhen, Z., Li, K.: Wavelet decomposition and convolutional LSTM networks based improved deep learning model for so-lar irradiance forecasting. Appl. Sci. (Switzerland). **8**(8), 1–29 (2018). https://doi.org/10.3390/app8081286
9. Tang, W., Li, B., Tan, S., Barni, M., Huang, J.: CNN-based adversarial embedding for image steganography. IEEE Trans. Inf. Forensics Secur. **14**(8), 2074–2087 (2019). https://doi.org/10.1109/TIFS.2019.2891237
10. Duan, X., Jia, K., Li, B., Guo, D., Zhang, E., Qin, C.: Reversible image steganography scheme based on a U-net structure. IEEE Access. **7**, 9314–9323 (2019). https://doi.org/10.1109/ACCESS.2019.2891247
11. Zhang, Z., Fu, G., Di, F., Li, C., Liu, J.: Generative reversible data hiding by image-to-image translation via GANs. Secur. Commun. Netw. **2019**, 1–11 (2019). https://doi.org/10.1155/2019/4932782
12. Zhang, K.A., Cuesta-Infante, A., Xu, L., Veeramachaneni, K.: High Capacity Image Steganography with GANs.. arXiv preprint arXiv:1901.03892 (2019)

13. Duan, X., Guo, D., Liu, N., Li, B., Gou, M., Qin, C.: A new high-capacity image steganography method combined with image elliptic curve cryptography and deep neural network. IEEE Access. **8**, 25777–25788 (2020). https://doi.org/10.1109/ACCESS.2020.2971528
14. Ray, B., Mukhopadhyay, S., Hossain, S., Ghosal, S.K., Sarkar, R.: Image ste-ganography using deep learning-based edge detection. Multimed. Tools Appl. **80**(24), 33475–33503 (2021). https://doi.org/10.1007/s11042-021-11177-4
15. Kadhim, I.J., Premaratne, P., Vial, P.J., Halloran, B.: Comprehensive survey of image steganography: techniques, evaluations, and trends in future research. Neurocomputing. **335**, 299–326 (2019). https://doi.org/10.1016/j.neucom.2018.06.075
16. Wang, Z., et al.: Data hiding with deep learning: a survey unifying digital watermarking and steganography. IEEE Trans. Comput. Soc. Syst. **10**(6), 2985–2999 (2023). https://doi.org/10.1109/TCSS.2023.3268950
17. Liu, J., Ke, Y., Zhang, Z., Lei, Y., Li, J., Zhang, M., Yang, X.: Recent advances of image steganography with generative adversarial networks. IEEE Access. **8**, 60575–60597 (2020). https://doi.org/10.1109/ACCESS.2020.2983175
18. Fadhl, A.A.M., Shareif, H.O.M., Ali, M.A.E., Elmangoush, A.M.A.: Adaptive image steganography: a review and research directions. In: Proceedings of the 2024 IEEE 7th International Conference on Advanced Technologies, Signal and Image Processing (ATSIP), pp. 494–499. IEEE, Sousse, Tunisia (2024). https://doi.org/10.1109/ATSIP62566.2024.10638907
19. Wang, Z., Gao, N., Wang, X., Xiang, J., Zha, D., Li, L.: HidingGAN: high capacity information hiding with generative adversarial network. Comput. Graphics Forum. **38**(7), 393–401 (2019)
20. Bozea, D.A.: Secure Steganography Based on GAN Image Morphing: Implementation and Analysis. Bachelor's thesis,. University of Twente (2023)
21. Sahu, M., Padhy, N., Gantayat, S.S., et al.: Shadow image-based reversible da-ta hiding using addition and subtraction logic on the LSB planes. Sens. Imaging. **22**(7) (2021). https://doi.org/10.1007/s11220-020-00328-w
22. Meng, R., Cui, Q., Yuan, C.: A survey of image information hiding algorithms based on deep learning. CMES Comput. Model. Eng. Sci. **117**(3), 425–454 (2018). https://doi.org/10.31614/cmes.2018.04765
23. Yuan, C., Wang, H., He, P., et al.: GAN-based image steganography for enhancing security via adversarial attack and pixel-wise deep fusion. Multimed. Tools Appl. **81**, 6681–6701 (2022). https://doi.org/10.1007/s11042-021-11778-z
24. Saxena, D., Cao, J.: Generative adversarial networks (GANs): Challenges, so-lutions, and future directions. ACM Comput. Surv. (CSUR). **54**(3), 1–42 (2021)
25. Zhang, Z., Fu, G., Di, F., Li, C., Liu, J.: Generative reversible data hiding by image-to-image translation via GANs. Secur. Commun. Netw. **2019**(1), 4932782 (2019). https://doi.org/10.1155/2019/4932782
26. Denysiuk, D., Savenko, O., Lysenko, S., Savenko, B., Kashtalian, A.: Method for detecting steganographic changes in images using machine learning. In: Proceedings of the 2023 13th International Conference on Dependable Systems, Services and Technologies (DESSERT), pp. 1–6. IEEE, Athens, Greece (2023). https://doi.org/10.1109/DESSERT61349.2023.10416453
27. Li, Q., Li, S., Tan, S., Li, B.: ThiNet-based pruning method for GAN-based steganography framework UT-GAN. In: Proceedings of the 2021 International Symposium on Signals, Circuits and Systems (ISSCS), pp. 1–4. IEEE (2021)
28. Ye, Z., Zhang, X., Feng, G.: Deep neural networks watermark via universal deep hiding and metric learning. Neural Comput. Appl. **36**, 7421–7438 (2024). https://doi.org/10.1007/s00521-024-09469-5
29. She, Q., Feng, F., Hao, X., Yang, Q., Lan, C., Lomonaco, V., et al.: OpenLO-RIS-Object: a robotic vision dataset and benchmark for lifelong deep learning. In: Proceedings of the 2020 IEEE International Conference on Robotics and Automation (ICRA), pp. 4767–4773. IEEE (2020)
30. Voravuthikunchai, W., Crémilleux, B., Jurie, F.: Histograms of pattern sets for image classification and object recognition. In: Proceedings of the IEEE Conference on Computer Vision and Pattern Recognition, pp. 224–231 (2014)

31. Chirodea, M.C., Novac, O.C., Novac, C.M., Bizon, N., Oproescu, M., Gordan, C.E.: Comparison of TensorFlow and PyTorch in convolutional neural network-based applications. In: Proceedings of the 2021 13th International Conference on Electronics, Computers and Artificial Intelligence (ECAI), pp. 1–6. IEEE, Pitesti, Romania (2021). https://doi.org/10.1109/ECAI52376.2021.9515098
32. Hammouch, H., El-Yacoubi, M., Qin, H., Berbia, H., Chikhaoui, M.: Control-ling the quality of GAN-based generated images for predictions tasks. In: El Yacoubi, M., Granger, E., Yuen, P.C., Pal, U., Vincent, N. (eds.) Pattern Recognition and Artificial Intelligence. ICPRAI 2022 Lecture Notes in Com-puter Science, vol. 13363, pp. 131–144. Springer, Cham (2022). https://doi.org/10.1007/978-3-031-09037-0_11
33. Dick, J., Abreu, B., Grellaert, M., Bamni, S.: Quality and complexity assessment of learning-based image compression solutions. In: Proceedings of the 2021 IEEE International Conference on Image Processing (ICIP), Anchorage, AK, USA, 2021, pp. 599 603 (2021). https://doi.org/10.1109/ICIP42928.2021.9506136
34. Subramanian, N., Cheheb, I., Elharrouss, O., Al-Maadeed, S., Bouridane, A.: End-to-end image steganography using deep convolutional autoencoders. IEEE Access. **9**, 135585–135593 (2021). https://doi.org/10.1109/ACCESS.2021.3113953
35. Korhonen, J., You, J.: Peak signal-to-noise ratio revisited: is simple beautiful? In: 2012 Fourth International Workshop on Quality of Multimedia Experience, Melbourne, VIC, Australia, 2012, pp. 37–38 (2012). https://doi.org/10.1109/QoMEX.2012.6263880
36. Chen, M.J., Bovik, A.C.: Fast structural similarity index algorithm. J. Real-Time Image Proc. **6**, 281–287 (2011). https://doi.org/10.1007/s11554-010-0170-9
37. Choo, J., Park, J.H.: Increasing payload capacity of wearable robots using linear actuators. IEEE/ASME Trans. Mech. **22**(4), 1663–1673 (2017). https://doi.org/10.1109/TMECH.2017.2705091
38. Hameed, M.A., Abdel-Aleem, O.A., Hassaballah, M.: A secure data hiding approach based on least-significant-bit and nature-inspired optimization techniques. J. Ambient Intell. Hum. Comput. **14**, 4639–4657 (2023). https://doi.org/10.1007/s12652-022-04366-y

Chapter 19
Enhanced Performance of Multiple Disease Classification in Apple Leaf Using Attention and Ensemble Deep Architectures

K. M. Vivek Anandh, J. Arun Prakash, H. Theivaprakasham, Divya Sasidharan, Vinayakumar Ravi, V. Sowmya, and E. A. Gopalakrishnan

Abstract Diseases in plants are common with changing environmental conditions. These diseases infect various parts of the plant resulting in long-term economic losses if left unnoticed. The plant's leaf easily responds to diseases as various demarcations appear on them. Leaf examination is the most traditional way to detect diseases, with the limitation of being extremely labor and time intensive. This study presents an approach for classifying apple leaf diseases using a deep learning-based pipeline. Healthy, Rust, Scab and Multiple disease are the four classes that are used in this study. The proposed deep CNN architecture categorizes the input image into one of the four classes. Several ensemble approaches are presented here, comparing the performance of each on the multiple disease class. This work explores the use of different training methods involving freezing and unfreezing. The proposed architecture based on CoAtNet achieves an overall accuracy of 95% and an accuracy of 68% on the multiple disease class on the publicly available plant pathology 2020 dataset. It appears that the proposed model has the potential for the real-time diagnosis of apple leaf diseases significantly.

K. M. Vivek Anandh · J. Arun Prakash · H. Theivaprakasham · D. Sasidharan · V. Sowmya (✉)
Amrita School of Artificial Intelligence, Coimbatore, Amrita Vishwa Vidyapeetham, Coimbatore, India
e-mail: s_divya1@cb.amrita.edu; v_sowmya@cb.amrita.edu

V. Ravi
Center for Artificial Intelligence, Prince Mohammad Bin Fahd University, Khobar, Saudi Arabia
e-mail: vravi@pmu.edu.sa

E. A. Gopalakrishnan
Amrita School of Artificial Intelligence, Bangalore, Amrita Vishwa Vidyapeetham, Bengaluru, Karnataka, India

Amrita School of Computing, Bangalore, Amrita Vishwa Vidyapeetham, Bengaluru, Karnataka, India
e-mail: ea_gopalakrishnan@blr.amrita.edu

G. Rivera et al. (eds.), *Recent Advances in Artificial Neural Networks*, Intelligent Systems Reference Library 283, https://doi.org/10.1007/978-3-032-13497-4_19

Keywords Apple leaf disease · Transfer learning · Deep learning · Convolution · Self-attention · Probability ensemble

19.1 Introduction

Agriculture, a key contributor to the world's economy, is a critical source of food, income, and employment. Agriculture employs almost 27% of the world's population and nearly 50% of the people of India. It contributes to 20.2% of the global economy and 18% of the nation's economy. The agribusiness industry's growth depends on factors such as climate, terrain, water availability, cropping pattern, topographical location, soil fertility, and threats associated with pests and diseases [1]. Pests and diseases have been major problems affecting the yield of crops. The Food and Agriculture Organization of the United Nations reports that nearly 40% of global crop yields are destroyed due to pests. In comparison, the crop yield loss worldwide due to plant diseases is estimated to be 14.1%, amounting to 220 billion USD. Various biotic organisms, including viruses, bacteria, viroid, and fungi, cause these diseases. In addition to these factors, abiotic factors such as precipitation, sunlight, and minerals play a detrimental role in stimulating such diseases. The changing climatic conditions impose an additional burden by facilitating several other plant diseases.

Disease attacks happen in different parts of the plants, like stems, leaves, flowers, and fruits. Most disease attacks start from leaves and can severely affect the plant as it spreads. However, this pattern changes across different diseases. This pattern is also common in apple trees as well. Common diseases in apple leaves include scab, blotch, rust, canker, rot, blight, and mildew. Among all these diseases, rust and scab are the most commonly observed due to the ease of transportation of fungal diseases through the medium of air. Another reason is that the ideal temperature for the growth of apples also facilitates the development of such fungi. Scab and rust affect leaves and fruits, making them unfit for eating. Rusts are identified by black and yellow spots, which, when affecting fruits, develop cracks in the fruit, allowing entry of secondary organisms. On the other hand, black patches on leaves identify scab. Rust causes leaves to fall prematurely, resulting in weak plants and the death of a plant if left untreated for years. Cases of plants affected by both rust and scab are highly harmful to the yield if left unnoticed. In the period spanning July to August 2013, apple orchards located in the Kashmir valley had an abrupt onset of Alternaria disease, a disease that was not deemed to be of significant concern during that time. Nevertheless, the disease rapidly propagated, causing significant severity and frequently causing substantial leaf loss and fruit withering in nearly all orchards across the valley, culminating in an extensive epidemic [2]. Therefore, the prompt identification of these diseases has the potential to significantly mitigate the decline in productivity.

Leaf examination is one of the traditional ways of detecting plant diseases. However, this traditional way of leaf examination requires very skilled labor and is incredibly time-consuming. Human-prone errors lead to the misdiagnosis of diseases, leading to long-term effects that result in reduced crop yields. Other ways to detect diseases include direct and indirect methods [3]. Direct methods such as Polymerase Chain Reaction, Immunofluorescence, and Flow Cytometry use molecular techniques to detect the disease accurately. Whereas indirect methods, such as Thermography and Fluorescence Imaging, use released organic compounds, morphological and temperature changes for detection. The limitations of both direct and indirect methods are the need for skilled labor and specialized equipment for investigation. Thus, a cost-effective and time-efficient detection method is required.

With the recent advancements in Artificial Intelligence, leaf diseases can be detected in an efficient manner. When automated detection methods are incorporated with drones, the detection of diseases is feasible even in hard-to-reach places. It can speed up the detection process and can also avoid the misdiagnosis problem. Employing this automatic leaf disease detection can help farmers save their crops and increase productivity. Several machine learning (ML) based solutions have already been proposed for disease detection in apples [4, 5] and their leaves [6]. However, machine-learning approaches require using handcrafted filters for feature extraction from images [7]. Convolutional Neural Networks (CNNs) have solved this problem by automatically extracting the required features. Various research studies focus on exploring the capability of several pre-trained deep CNN architectures for apple disease detection [8–10].

Although several studies have been conducted in this particular field, the following are the major limitations:

1. Many of the model gained accuracies of about 96% however, the accuracy in classifying the multiple diseases images is low.
2. Several works propose models trained from scratch, which is time and resource intensive.

Although, many DL models such as Vision transformer (ViT) has shown its excellent performance on plant leaf disease analysis, CoaTNet model has shown its superior performance in multiclass classification [11]. CoAtNet [12] is built on the strengths of both convolutions and self-attention mechanisms. It stacks multiple convolutions and self-attention modules in a sequential manner. The features are initially extracted from multiple CNN modules and are then passed to the self-attention layers, which find the relative attention between the parts of the extracted features. Each CNN module uses a residual connection to prevent the loss of features and a max pooling layer for down sampling [13]. Hence, this study compares the performance of the CoatNet architecture along with other DL models.

The present study introduces a novel detection process considering the limitations in the existing studies. Unbiased training is managed by training the models using geometrical augmentations instead of data sampling techniques. This work pipeline focuses on the use of max probability ensemble of prediction probabilities from deep CNN architecture and using All the models used are pretrained on the openly available ImageNet weights along with different fine-tuning methods involving freezing and unfreezing.

The major contributions of the proposed study are outlined below:

1. Proposed CoAtnet network, which utilizes the advantage of both attention and convolution for the multi-class apple leaf disease classification.
2. Performance Improvement of the multi-class classification, including the complex class containing multiple diseases using transfer learning and ensemble approach.
3. Explainable and Interpretable Feature Visualizations derived from the CoAtnet network used for the multi-class classification.

The remaining sections of the paper is presented as follows: Section 19.2 discusses an overview of the literature and outlines the current constraints. Section 19.3 provides an introduction to the proposed pipeline and theoretical framework of CoAtNet, as well as the deep CNN architectures employed in the research. Sections 19.4 and 19.5 provide detailed explanations of the dataset description and performance measures utilized. Section 19.6 presents an analysis and discussion of the experimental results, accompanied with plots. The paper concludes with Sect. 19.7 explaining the summary, limitations of the present work and future scope.

19.2 Related Works

ML and deep learning (DL) methods have been employed by the researchers to study leaf diseases in plants. The integration of advanced algorithms will leverage the development of real time monitoring systems for the identification of diseases and thereby increases the production while reducing the financial losses [14]. Swati Singh et al. reported a pipeline involving segmenting the apple leaves from the background followed by classification for better performance [4]. The region of interest (ROI) was segmented for the classification purpose using the extracted color and texture features. Various machine learning classifiers were applied to classify the diseases. K nearest neighbour (KNN) gave the highest accuracy of 96.4%. Chakraborty et al. classified infected apple leaves by segmenting the infected region in the leaf using the Otsu thresholding algorithm and extracting color and texture features using the gray level co-occurrence matrix (GLCM) algorithm [15] and multiclass SVM classifier. The apple fruit disease classification performed by Atrab et al. used K-Means based segmentation technique followed by various feature extraction descriptors like Local Ternary Pattern, Local Binary Pattern, Histogram of Oriented Gradients, Grayscale co-occurrence matrix, and color coherence vector. KNN and SVM classifiers were used for the classification task [16]. Geetha et al. reported the use of methods such as Histogram of oriented gradients and Grayscale co-occurrence matrix for feature extraction from images of tomato leaves for classifying using KNN classifier [17]. Chuanlei et al. designed a structure for transforming the input image and then converting it to HSI, YUV and Gray models. A combination of genetic algorithm and correlation-based feature selection was employed to identify the prominent color, texture, and shape data. Then a SVM model was trained for the detection of apple leaf diseases [18]. An automatic identification of apple diseases depending up on leaf abnormalities has been reported by Hasan et al. The approach consists of segmenting the diseased region using a color segmentation method based on the L*a*b* space. Then DWT and L*a*b* space-based color histogram features were extracted. Finally, the segmented portion using multiple classifiers, with Random Forest serving as the base classifier were classified [19]. However, the above-mentioned works required the feature extraction to be done by the researchers. Hence, researchers started utilizing the deep learning networks for automatic feature extraction from the leaf images [14].

Works were done to classify the healthy leaves from infected leaves using deep neural networks. Rajeev Kumar Singh et al. worked on the leaf disease classification of the Maize plant using the Alexnet architecture for classifying healthy and infected leaves [20]. Ashraful Haque and team [21] worked on the same task, proposed a custom deep CNN architecture, and performed a comparative analysis between the model and other pretrained models on the dataset. Principal component analysis (PCA) Whitening preprocessing technique and a modified LeNet CNN architecture were used to classify diseased maize leaves by Priyadhashini et al. [22]. Elfatimi et al. employed MobileNet and MobileNetV2 architectures to classify infected bean leaves [8]. CNN architectures were utilized to classify multiple plant leaf diseases. Vimal Singh et al. used pretrained CNN models (MobileNetV2, EfficientNetB6, and NasNet) to classify bean leaf diseases using transfer learning techniques [23]. M.A. Moid et al. developed a plant disease classification system employing Xception architecture, which is based on transfer learning, to accurately identify plant diseases [6]. An attention-based technique merged with pretrained EfficientNet models for locating and identifying the disease affected areas in Casssava leaves was proposed by Ravi V. et al. The features were extracted from penultimate layers and dimensionality reduction was performed using kernel principal component analysis [24]. Gajjar et al. created a real-time system to classify the leaves using inception-based deep CNN with a single-shot detector for identifying and locating infected areas in the leaves. The model was trained with the plant village dataset for multiple leaf diseases and tested it on a real-time dataset [25]. Sudhesh K. M. et al. proposed Dynamic Mode Decomposition based on attention-driven preprocessing identifying infected areas in rice leaves. A classification accuracy of 94.33% was achieved on-field rice images for the SVM classifier fed with deep features extracted from XceptionNet [26]. Elhassouny et al. developed a mobile application for detecting tomato leaf diseases using a Mobilenet CNN model for the detection of these diseases [27]. To further improve the accuracy, different optimization algorithms such as SGD, Adam, etc. were tested. Prajwal et al. proposed using LeNet CNN architecture to identify diseases in leaves of tomato plants [28]. A segmentation process for extracting the regions of interest followed by an accuracy of 98.49% with new CNN architecture was achieved by Naresh K Trivedi et al. for the leaf disease classification [29]. Kurup et al. developed a new architecture of CNN called capsulenet. The architecture was analyzed on two plant leaf datasets which were Plant Village dataset for plant disease diagnosis and Leafsnap dataset for leaf disease classification [30]. An improved CapsNet architecture was used to identify apple foliar leaf disease using Kaggle Plant Pathology 2020 - FGVC7 dataset [31].

Various works have been done pertaining to the detection and classification of diseases in Apple leaves. Thapa et al. presented the Plant pathology challenge 2020 dataset of apple leaves [1]. Some benchmark results for the leaf classification using the ResNet-50 network with pretrained Imagenet weights were also provided. An improved CNN based on VGG16 with a global average pooling layer and reduced model parameters was proposed by Qian Yan et al. for predicting apple leaf diseases [32]. Swati Singh et al. developed an architecture based on CNN with 19 convolutional layers for classifying Marsonina Coronaria and Scab diseases in Apple plants [4]. Chongke Bi et al. had employed the MobileNet architecture for predicting the diseased apple leaves [33]. Peng Jiang et al. used a DL approach along with SSD based Inception module and rainbow concatenation to detect the diseases [34]. Anju Yadav et al. proposed an architecture called AFD-Net for the apple leaf disease classification [35]. Ashwaq Alsayed and team [10] had used various pretrained models like VGG16, ResNetV2, InceptionV3, and MobilenetV2 along with transfer learning techniques for detecting apple leaf diseases. Tianyu et al. used BLSENet, a light weight CNN for apple leaf classification attaining an accuracy of 93.58%. This study used an attention mechanism and feature fusion strategy for the classification problem [36]. A study reported the identification of seven multiple diseases on the same apple leaf using a light weight CNN RegNet model enabling an intelligent disease detection system in future[37].

Leaf disease segmentation and identification were conducted in apple orchards to facilitate accurate and intelligent spraying in sustainable agriculture by Storey et al. [9]. This was achieved by employing three Mask RCNN networks, namely Resnet-50, MobileNetV3-Large, and MobileNetV3-Large-Mobile, as the backbone architectures. Amit Gawade and Subodh Deolkar [38] have adopted a regional CNN-based approach in the study to detect apple leaf diseases. The network was able to identify diseases with an accuracy of 90%. Chao et al. proposed a deep CNN that combines the DenseNet architecture and Xception architecture to detect apple leaf diseases [39]. Luo et al. reported a classification model by combining multi-scale features [40]. For this the existing ResNet model was improved by modifying the location of batch normalization and rectified linear unit. Also, 3×3 conv was replaced in ResBlocks with pyramid convolution for improving model robustness and could gain a classification accuracy of 94.99%. Zhang et al. developed an IoT system for apple leaf disease detection based on deep multi-scale dual-channel CNN (DMCNN) [41]. Zhong et al. adopted a DenseNet-121 deep CNN to identify apple leaf diseases [42] where modifications such as changing the loss function to focus loss function and using a modified classification method known as multi-label classification to improve the performance was done. Wang e al. developed a model called Coordination Attention EfficientNet (CA-Enet) to enhance the detection accuracy of classifying various apple leaf diseases. A coordinate attention block was fused into the EfficientNet-B4 network followed by a depth-wise separable convolution for reducing the number of attributes [43]. Fu et al. proposed a lightweight CNN model with AlexNet as its base. A Parallel convolution was employed for feature extraction from unhealthy leaf with complicated backgrounds in order to enhance the accuracy of model. Additionally, a dilated convolution was implemented to decrease the parameter count. Further Channel attention which was added to the aggregation module of the model [44] was also used. Yu et al. proposed Leaf Spot attention network having two sub networks namely Feature Segmentation and Spot-Aware Classification subnetworks. Input images were converted to a segmented feature map using the FS-SubNet, which comprises of three distinct regions: the background, leaf, and spot area. The leaves were classified using the SAC-SubNet, which relied on the feature map and encoded high-level semantic spot information [45].

A technique called a ROI aware deep CNN was used to separate out the leaf image with the infected part from the background, which doesn't have useful information for prediction. Two subnetworks of the VGG model, one for getting the ROI and the other for classifying leaves were utilized [5]. Yu et al. created a multistep optimization ResNet model for the detection of leaf disease by improving identification accuracy and speed [46]. Researchers also used the DL models to extract the best features from the images and then used ML models for the classification task. Xin Li et al. used an SVM classifier and Resnet and VGG CNN models for the apple leaf disease classification [3]. Rehman et al. proposed a parallel structure for real-time classification. A hybrid contrast stretching technique was proposed to enhance the visual impression of an image then the MASK RCNN model was used for segmenting out the infected regions. Features were extracted using a pre-trained CNN model using the enhanced images. Subsequently, Kapur's entropy was employed in conjunction with the MSVM approach-based selection technique to identify robust attributes for the classification [41].

Some researchers approached the problem as object detection problem. Jie Di et al. proposed an object detection way of identifying the apple leaf diseases [47]. An apple leaf disease target detection model called DF-TINY-YOLO was introduced for faster and effective automatic recognition of apple leaf diseases. Sun et al. proposed a detection model, MEAN_SSD (Mobile End AppleNet based SSD algorithm) which was created using a Mobile end AppleNet block and an Apple-Inception module [48]. Khan et al. reported a DL based apple disease detection model which operates in two phases. The initial phase involves the utilization of a lightweight classification model to categorize the input photographs into three distinct groups: diseased, healthy, or damaged. Subsequently, the detection stage is employed to accurately identify and localize each symptom inside the diseased leaf images. Yolov4, Faster-RCNN, and EfficientDet models were employed for detection [49]. Modifications were made to YOLO V8 model for improving accuracy and efficiency in detecting apple leaf diseases. Meng et al. used YOLOV5-CBAM-C3TR for the detection of apple leaf diseases with an accuracy of 92.4% [50]. The proposed model replaced the conventional layers of YOLOV8 and thereby making it a light weight model. A global attention mechanism layer was also incorporated to identify the affected areas in the apple leaves [51].

This work proposes a transfer learning ensemble and an attention-based deep CNN architecture to identify leaves into four classes Healthy, Rust, Scab, and Multiple diseases.

19.3 Methodology

This section focuses on the workflow of this study for classifying apple diseases as shown in Fig. 19.1. The features extracted from the best performing models are analyzed and used for ensemble-based classification The collected images after preprocessing are fed as inputs to deep CNN models pretrained on ImageNet weights. Here preprocessing involves resizing the images to standard input sizes of 224 x 224 and 299 x 299 for different deep CNN architectures. The input size of Xception and Inception models are 299 x 299 whereas all other deep CNN architectures use an input of size 224 x 224. Each image is normalized in addition to being augmented using operations such as scale, zoom, flips and warps. Based on two approaches of freezing followed by unfreezing and freezing completely, the best models were selected and an ensemble of those models were created. Then the ensemble is created between the selected model trained with and without data augmentation. The Ensemble 1 is created between the model which gave the best overall accuracy and the model which gave the best accuracy for multiple disease class. The Ensemble 2 is done for the model which gave the overall best accuracy. In order to improve the performance further, CoAtnet model, which is a combination of convolution and attention is used.

19.3.1 DL Models

Literature study reveals that pre-trained deep CNN models exhibit superior performance with decreased training time in comparison to models developed from the beginning [52]. Various pre-trained deep CNN models like MobileNetV2 [8], VGG16, VGG19, DenseNet201, ResNet152V2, InceptionResNetV2 [10], EfficientNetB0, and Xception [26] are trained on apple classification dataset. Models that exhibit the highest performance are chosen for classification using ensemble methods. Apart from these models, the images are also trained using CoAtNet model. The models are trained using transfer learning method [11]. It leverages the knowledge obtained from a pre-trained model on an associated task for applying to a new related task. Features are extracted from the input data from an initial task and are transferred to new task for improving the performance and the training time.

For the new target task, transfer learning modifies weights that are pre-trained on a bigger dataset. This study uses models pretrained on ImageNet weights for transfer learning.

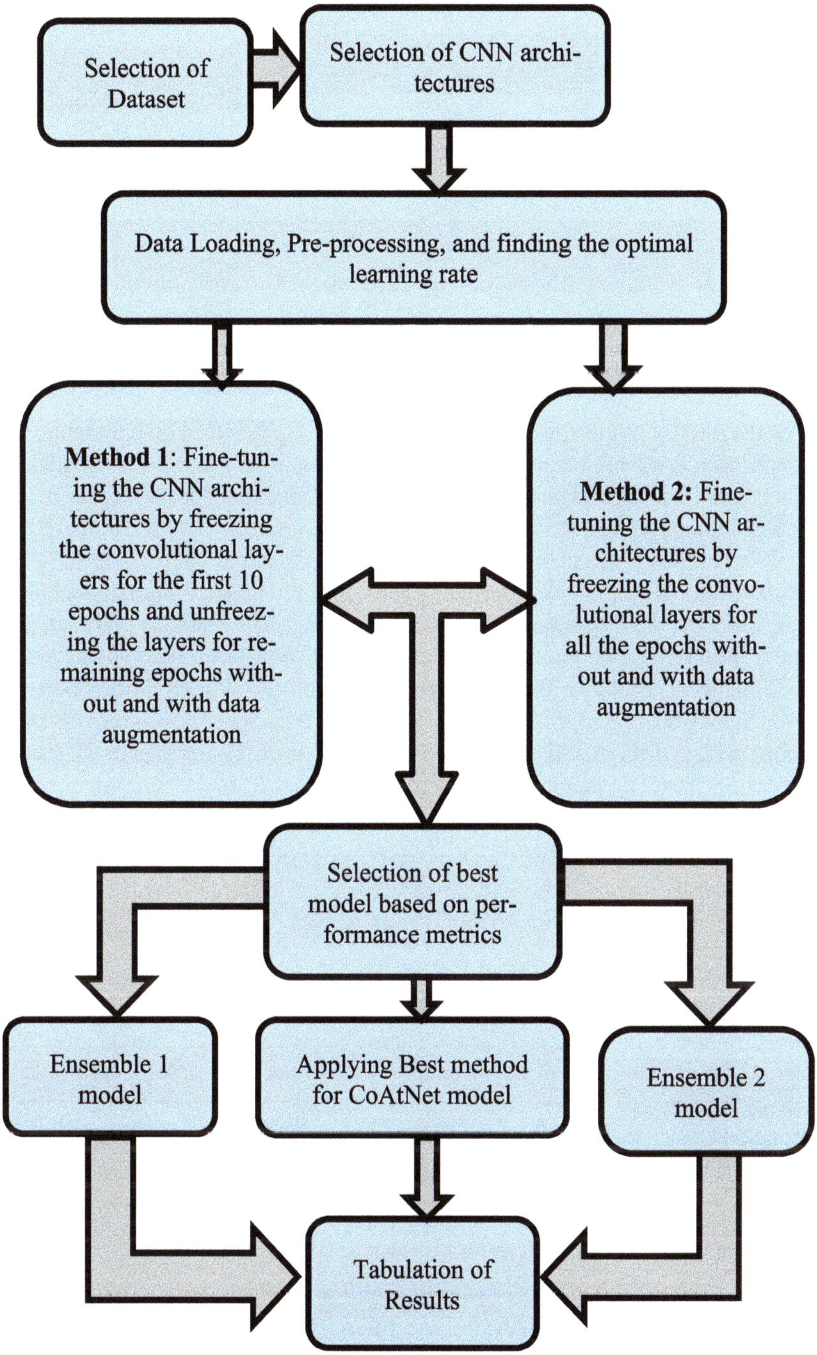

Fig. 19.1 Proposed workflow

Table 19.1 Parameters selected for Augmentation methods used in the proposed work

Methods	Corresponding parameters
Min_scale	0.75
do_flip	True
Vert_flip	True
Max_rotate	120
Min_zoom	0.50
Max_zoom	1.10
Max_lighting	0.20
Max_warp	0.10
P_affine	0.75
P_lighting	0.75

All deep CNN models used in this work were trained on two approaches. The first approach involved freezing the convolutional layers for the first 10 epochs and unfreezing all the CNN layers for learning for the remaining epochs. By unfreezing the convolutional layers after the first 10 epochs, the model is expected to learn more about the features from the images. The second approach involved freezing all the CNN layers for all the epochs. Both the approaches use transfer learning technique, where it was started with ImageNet weights. Both these methods were used for training the model without and with data augmentation. With data augmentation, transforms were randomly applied to the original images by performing operations such as rotations, flips, and zooms. These transformations create variations of the original images that are still representative of the same object or scene. By training a model on these augmented images, the model can learn to be more robust to variations and generalize better to new, unseen data. Data augmentation does not result in an increase in the number of images within the dataset. Hence it is a very effective way to improve the model performance without requiring additional data collection or storage. The optimal parameters selected for the augmentation process is as shown in Table 19.1.

Additionally, since the transformations are applied randomly during training, the model is exposed to a diverse set of images and is less likely to overfit to specific patterns in the training data. Based on these approaches, the best models were selected and an ensemble of those models were created. In order to improve the accuracy further, a CoAtnet model was also trained on the same dataset with the best method.

19.4 Dataset Description

The Plant Pathology 2020-FGVC7 competition dataset was adopted in this work. It consists of 3642 images of apple leaves belonging to 4 classes: Normal, Rust, Scab, and Multiple diseases. A total of 1821 images are selected for training, and 1821 images for testing. As this is a publicly available competition dataset, no

Table 19.2 Distribution of the original dataset

Category	Train	Test
Healthy	516	354
Scab	592	613
Rust	622	739
Multiple diseases	91	115

Fig. 19.2 Apple Leaves belonging to (**a**) Multiple disease class (**b**) Rust class (**c**) Scab class (**d**) Healthy class

changes were made to the train-test split and used the dataset as such without any modifications. The class distribution of the dataset is detailed in Table 19.2. Figure 19.2 presents the images corresponding to healthy leaves, leaves with yellow spots indicating rust, leaves with spores indicating scab, and leaves with both spores and yellow spots indicating multiple diseases.

Leaves from each class have been shown below. Figure 19.2a represents the leaves belonging to the multiple disease class. The leaf is infected by both scab and rust disease. The light green and black patches in the leaf shows the scab disease.

The yellow spots in the leaf shows rust disease. Figure 19.2b shows the apple leaves affected by rust disease. The yellowish spots in the leaf indicates the rust disease. Figure 19.2c shows the apple leaf affected by scab disease. The black patches in the leaf indicates the scab disease. Figure 19.2d shows the healthy apple leaf. Manual identification of the diseased leaves is a highly time-consuming process. Also, the process requires very skilled labor as misclassifications can happen depending on the severity of diseases. For example, low degree of severity of scab infected leaves can have light black patches which could either be left unidentified or could be identified as rust, as severe rust infected leaves appear as dark brown to blackish spots. With the use of deep learning models, leaves which are infected with diseases even with less severity can be detected. Also, when it comes to multiple disease infected leaves, chances of missing out on identification of one of the infected diseases is high when it comes to manual identification. Wrong or misidentification of diseases can result in wrong use of drugs or pesticides which can also reduce the production. Deep learning models along with attention learns the infected regions effectively and helps in improving the quality of predictions while reducing the amount of time to be spent for this process.

19.5 Performance Measures

The utilization of performance measures is crucial in differentiating the efficacy of classification models. The study employs accuracy, precision, recall, and F1-score as its indicators.

19.6 Results and Discussion

The high dimensional features are visualized in a 2-D space using a t-SNE plot. It is a scatter plot where each point represents a data point in the high-dimensional space, where the data points that are similar in the original high dimensional space are presented by points that are closer on the plot, while data points which were dissimilar in the high dimensional space are shown by points that are far away from each other. The features obtained by the CNN network is visualized from high dimensional representation to 2D using the t-SNE plot in Fig. 19.3.

This section discusses results of various state of the art CNN models. The CNN models were trained on two approaches. Based on two approaches of freezing followed by unfreezing and freezing completely, the best models were selected and an ensemble of those models were created. The models have been fine-tuned using ImageNet weights on a training dataset comprising 1817 images. These images are distributed across four classes: Healthy (516 images), Rust (622 images), Scab (592 images), and Multiple Diseases (91 images). In this work, the performance of the following deep CNN architectures is analyzed: VGG16, VGG19, MobileNetV2,

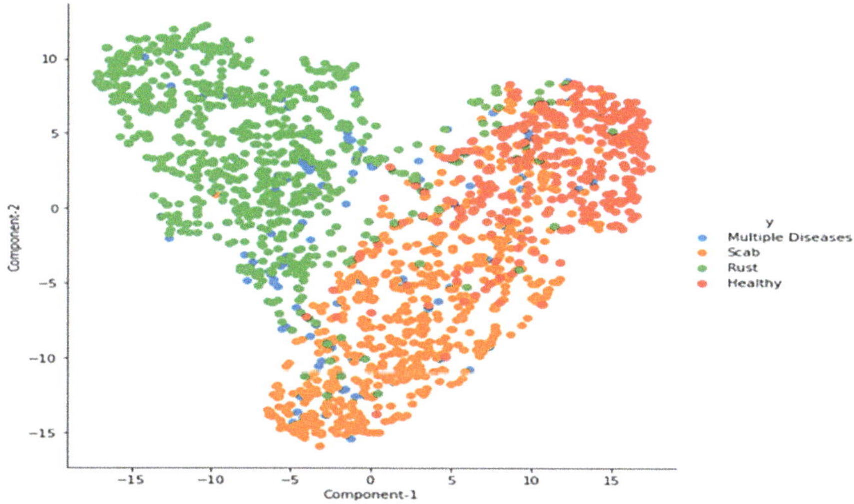

Fig. 19.3 t-SNE plot

InceptionResNetV2, DenseNet201, ResNet152V2, EfficientnetB0 and Xception. To ensure a fair comparison, all these architectures were evaluated using input images of size 224x224.The input images of the Xception and InceptionResnetV2 models were resized to 299x299. The models were trained for 30 epochs using Adam optimizer, a batch size of 32, and an optimal learning rate was found using lr_find() function. The study was conducted using Google Colab, which was equipped with a K80 GPU and 12 GB of RAM, and the models were built and evaluated using Tensorflow2 and Keras2.

In the experiments above, number of samples remains same even after data augmentation. The transforms are applied to samples while fine tuning each epoch. Table 19.3 shows the accuracy obtained for various CNN architectures used in this study using method 1 (F+UF) and method 2 (F) without and with data augmentation. From the table, it can be inferred that method 2, which is freezing the convolutional layers for all epochs and using pretrained weights, gave better results for most architectures as compared to method 1. The use of various transforms in data augmentation also improved the accuracy. VGG19 gave the best accuracy for both methods after data augmentation. The architecture gave an accuracy of 91%. VGG16 gave an accuracy of 91% for method 2 after data augmentation. InceptionResnetV2 architecture which gave the best accuracy for the multiple disease class, gave an overall accuracy of 74% for method 2 after data augmentation. In terms of overall accuracy, the VGG 19 model was the best-performing model after data augmentation for both methods.

Table 19.3 Accuracy (%) of the CNN models obtained: (a) with and (b)without data augmentation

	(a)		(b)	
Model name	METHOD 1(F + UF)	METHOD 2(F)	METHOD 1(F + UF)	METHOD 2(F)
XCEPTION	79	82	78	79
VGG16	87	87	90	91
VGG19	88	87	91	91
DENSENET201	81	87	88	90
EFFICIENTNETB0	76	84	82	90
MOBILENETV2	76	80	78	88
INCEPTIONRESNETV2	80	78	78	74
RESNET152V2		88		89

19.6.1 *Class Wise Performance*

As the study's main objective is to improve the multiple disease accuracy, performance of the models for each class is evaluated. To create the ensemble, CNN model, which performs well for the multiple class, and the CNN model, which performs well in terms of overall accuracy are considered.

19.6.1.1 Class 1: Multiple Disease

Multiple disease class include leaves infected by both rust and scab. This class includes 91 leaves in the training set. It has 115 leaves in test images. Table 19.4 shows the accuracy obtained for various CNN architectures used in this study using method 1 (F + UF) and method 2 (F) without and with data augmentation in classifying multiple disease. In terms of accuracy, VGG19 gave good results across various methods. Without Data augmentation, VGG19 gave 34% accuracy for method 1 (freezing and unfreezing convolution layers). With Data augmentation, the accuracy improved to 46% for method 1. For method 2, the accuracy of VGG19 further improved to 56%, an improvement of 10%. With Data augmentation, VGG16 and Densenet201 gave an accuracy of 42 % and 48%, respectively. The best model for predicting the multiple disease class was Inceptionresnetv2. The inceptionresnetv2 model gave the best results while fine tuning the model using method 2, both with and without data augmentation. Without Data augmentation, the model gave 52% accuracy, while with data augmentation, the accuracy of the inceptionresnetv2 model improved to 63%, an increase of 11% after augmentation. This is the highest accuracy across all models and all methods. However, the inceptionresnetv2's overall accuracy was not the best due to more misclassifications in other classes.

In terms of Precision, with data augmentation, VGG19 gave 53% precision, whereas inceptionresnetv2 gave only a 21% precision score for method 2 due to many false positives. VGG19 had a recall score of 56%, and inceptionresnetv2 had

Table 19.4 Accuracy of the CNN models for multiple disease class

	Without data augmentation		With data augmentation	
Model name	Method 1(F+UF)	Method 2(F)	Method 1(F+UF)	Method 2(F)
XCEPTION	17	20	9	11
VGG16	33	37	42	42
VGG19	34	43	46	56
DENSENET201	24	28	35	48
EFFICIENTNETB0	6	13	10	35
MOBILENETV2	16	7	8	19
INCEPTIONRESNETV2	19	52	13	63
RESNET152V2		34		27

Table 19.5 Accuracy of the CNN models for Healthy leaves class

	Without data augmentation		With data augmentation	
Model name	Method 1(F+UF)	Method 2(F)	Method 1(F+UF)	Method 2(F)
XCEPTION	82	90	91	94
VGG16	96	96	98	96
VGG19	97	94	98	97
DENSENET201	88	95	97	97
EFFICIENTNETB0	84	92	89	96
MOBILENETV2	87	90	91	96
INCEPTIONRESNETV2	81	78	83	75
RESNET152V2		95		89

a recall score of 63% for method 2 after data augmentation. Regarding F1-score, VGG19 is the best-performing model with a score of 54%, while inceptionresnetv2 has an F1 score of 31% due to the false positives.

19.6.1.2 Class 2: Healthy Leaves

Healthy leaves class includes 516 leaves in the training set. It has 354 leaves in test images. Table 19.5 presents the accuracy of CNN models for healthy leaves class. Most of the models performed well in classifying healthy leaves. VGG19 outperformed with regard to accuracy. It gave 98% accuracy along with VGG 16 for method 1 with data augmentation, while it gave 97% along with Densenet201 for method 2 with data augmentation. The best-performing model for multiple disease classes, the InceptionresnetV2 model, showed 75 % accuracy for the same method, as there were misclassifications of healthy leaves into other groups.

Regarding precision, VGG16 gave the best score of 90%, while VGG 19 showed 88% for method 2 and 90% for method 1 with data augmentation. VGG 19 gave the best recall values of 97% and 98% for method 1 and method 2 after data augmentation, respectively, while VGG16 gave 98% recall value in method 1 after data augmentation. The VGG19 model recorded the best F1 score of 94% for method 1 and 93% for method 2, while InceptionresnetV2 model gave a F1 score of 77% for method 2 after data augmentation.

19.6.1.3 Class 3: Scab

Scab leaves are identified by black patches on the leaves. Scab class includes 592 leaves in the training set. It has 613 leaves in test images. Scab class had the second greatest number of training samples. The best results for scab class in terms of accuracy was recorded using method 2 with data augmentation as seen in Table 19.6. VGG 16 gave an accuracy of 94% for method 2. VGG 19 gave 91%

Table 19.6 Accuracy of the CNN models for scab class

	Without data augmentation		With data augmentation	
Model name	Method 1(F+UF)	Method 2(F)	Method 1(F+UF)	Method 2(F)
XCEPTION	77	79	80	77
VGG16	86	89	89	94
VGG19	86	85	91	91
DENSENET201	84	83	87	88
EFFICIENTNETB0	75	86	83	92
MOBILENETV2	75	78	76	91
INCEPTIONRESNETV2	80	79	84	68
RESNET152V2		89		88

Table 19.7 Accuracy of the CNN models for rust class

	Without data augmentation		With data augmentation	
Model name	Method 1(F+UF)	Method 2(F)	Method 1(F+UF)	Method 2(F)
XCEPTION	88	89	80	84
VGG16	93	90	96	95
VGG19	93	92	94	94
DENSENET201	84	95	93	95
EFFICIENTNETB0	83	89	89	94
MOBILENETV2	81	87	84	93
INCEPTIONRESNETV2	90	76	80	81
RESNET152V2		91		96

accuracy for both methods after data augmentation. InceptionresnetV2 gave 68% accuracy for method 2 with data augmentation.

In terms of Precision, VGG 16, VGG 19 and Densenet201 gave a precision score of 95% for method 2 with data augmentation while VGG 16 gave 97% precision for method 1. In terms of recall, VGG 16 recorded the best recall score of 94 % for method 2 while VGG 19 gave a score of 91% for both the methods. VGG 16 recorded the best F1-score of 95% for method 2 with data augmentation. VGG 19 gave an F1 score of 93% for both methods after data augmentation. F1 score of InceptionresnetV2 obtained for method 2 was 75%.

19.6.1.4 Class 4: Rust

Rust leaves are identified by yellowish and black spots on the leaves. Rust class includes 622 leaves in the training set. It has 739 leaves in the test dataset. Most of the models gave good results for the rust class as it had a greater number of training images as compared to other classes which is exhibited in Table 19.7. The VGG 16 model and Resnet152v2 model gave the best accuracy in classifying the rust class. VGG16 gave 96% accuracy for method 1 with data augmentation while Resnet152v2 gave 96% accuracy for method 2 with data augmentation. The VGG

19 model gave an accuracy of 94% for both the methods while InceptionresnetV2 which performed the best for multiple disease class gave an accuracy of 81% for method 2 after data augmentation.

In terms of precision, VGG 19 performed the best, giving a precision of 96% for method 2 after data augmentation. In terms of recall, the VGG 16 and Resnet152v2 gave the best recall score of 96, while VGG 19 gave a recall score of 94%. VGG 19 and Densenet201 recorded the best F1 score of 95% for method 2 with data augmentation, while InceptionresnetV2 gave a F1 score of 86% for method 2 with data augmentation for rust class.

Among all the CNN architectures, the VGG-19 network after using data augmentation has improved the multiple disease class (complex class) accuracy to 56%. Also, an overall accuracy of 91% for VGG 19 network is obtained. In terms of multiple disease class, InceptionResnetV2 gave 63% accuracy.

19.6.2 *Confusion Matrix of Best Models*

The confusion matrices of the best model are as shown in Fig. 19.4. The First architecture is InceptionResnetV2 (F). The classification results were as follows. Out of the 354 healthy leaves, 265 leaves were correctly classified, and 89 healthy leaves were misclassified. For the scab class, out of 613 leaves, 419 leaves were correctly classified and 194 leaves were misclassified. For rust class, out of 739 leaves 595 leaves were correctly classified and 144 leaves were misclassified. For the multiple disease class, out of 115 leaves 72 leaves were correctly classified. The InceptionResnetV2 architecture gave an overall accuracy of 74%. Also, the number of misclassifications of multiple disease class were reduced. With data

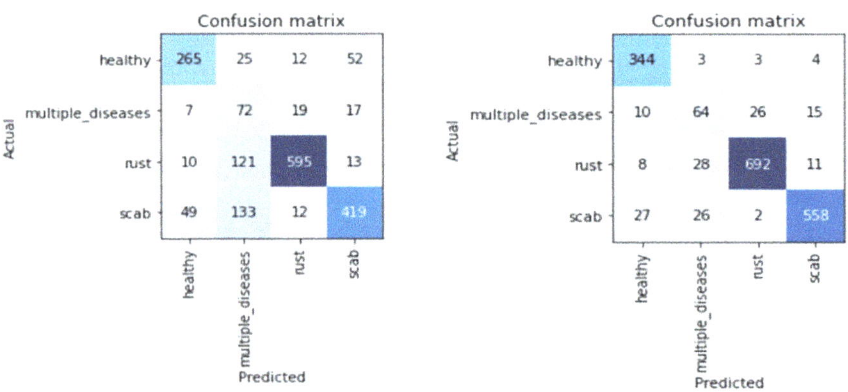

(a) (b)

Fig. 19.4 Confusion matrix of (**a**) InceptionResnetV2 (**b**) VGG19 augmentation, the accuracy improved to 63%. It is the best performing architecture for predicting the multiple disease class when compared to both approaches.

The next architecture is VGG-19 (F), and the classification results were as follows. For the healthy leaves class, out of 354 leaves, 344 leaves were correctly classified and 10 healthy leaves were misclassified. For the scab class, out of 613 leaves, 558 leaves were correctly classified and 55 leaves were misclassified. In case of the rust class, out of 739 leaves,692 leaves were correctly classified and 47 leaves were misclassified. In the case of multiple disease class, out of 115 leaves 64 leaves were correctly classified and 51 leaves were misclassified. The VGG-19 architecture gave an overall accuracy of 91%. With data augmentation, the accuracy improved to 56%. The loss plot is depicted in Fig. 19.5.

19.6.2.1 Ensemble 1

The Ensemble 1 is created between the model which gave the best overall accuracy and the model which gave the best accuracy for multiple disease class. Based on the above results, the VGG-19 and InceptionResnetV2 gave the best results. Both the models trained based on the same method (applying Data Augmentation and fine tuning by freezing all the layers) were selected. The probabilities from the softmax layer were obtained and compared and the label was selected based on the higher probability value. Figure 19.6 shows the confusion matrix of ensemble 1 model. Out of 115 multiple diseased leaves, 64 leaves were correctly classified, and 51 leaves were misclassified into other classes. Of those 51 leaves,22 leaves were classified into rust class and 17 leaves were identified as only scab infected leaves. Ensemble 1 model detected multiple disease class with an accuracy of 56% which is same as original VGG19 model result.

The Ensemble 1 Model gave an overall accuracy of 87% (depicted in Table 19.8). The Multiple disease class accuracy gained by the model was 56% same as that achieved by the best VGG19 model by method 2 after data augmentation.

19.6.2.2 Ensemble 2

The Ensemble 2 is done for the model which gave the overall best accuracy. Here, the number of samples for multiple disease class is explicitly raised by applying transforms like rotating the images and changing the brightness.500 new samples are added to the multiple diseases class train dataset. Then the ensemble is created between the selected model trained with and without data augmentation. Based on the above results, the VGG-19 model was selected as the best model. In order to create this ensemble model, VGG19 was selected. The number of samples in the multiple disease class was increased by 500 by applying transforms to the existing samples. Now two models of VGG19 were trained, one without data augmentation and the other model with data augmentation by freezing the convolutional layer

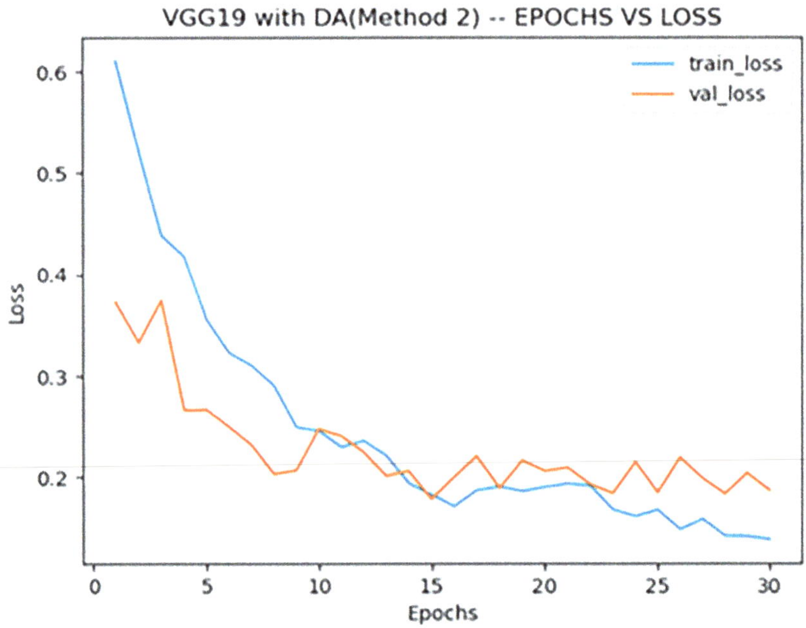

(a)

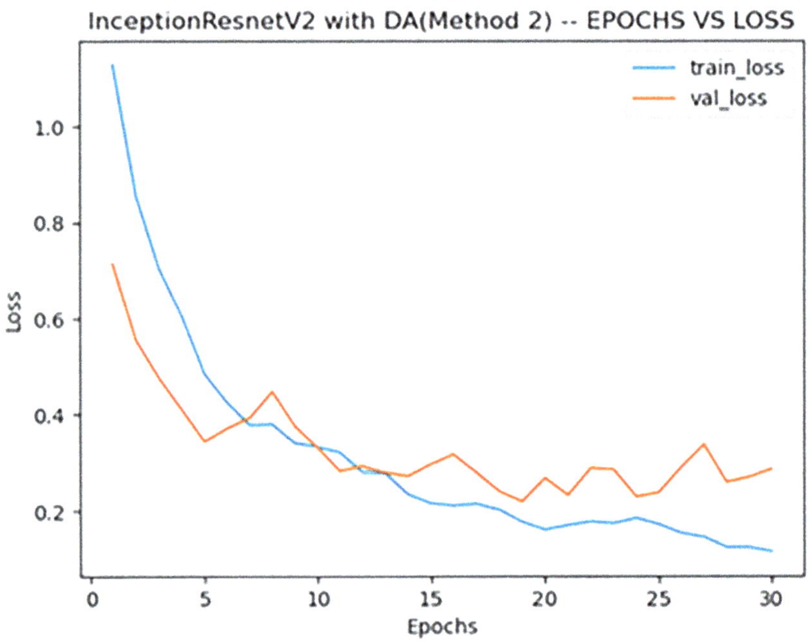

(b)

Fig. 19.5 Loss plot of (**a**) VGG19 (**b**) InceptionResnetV2

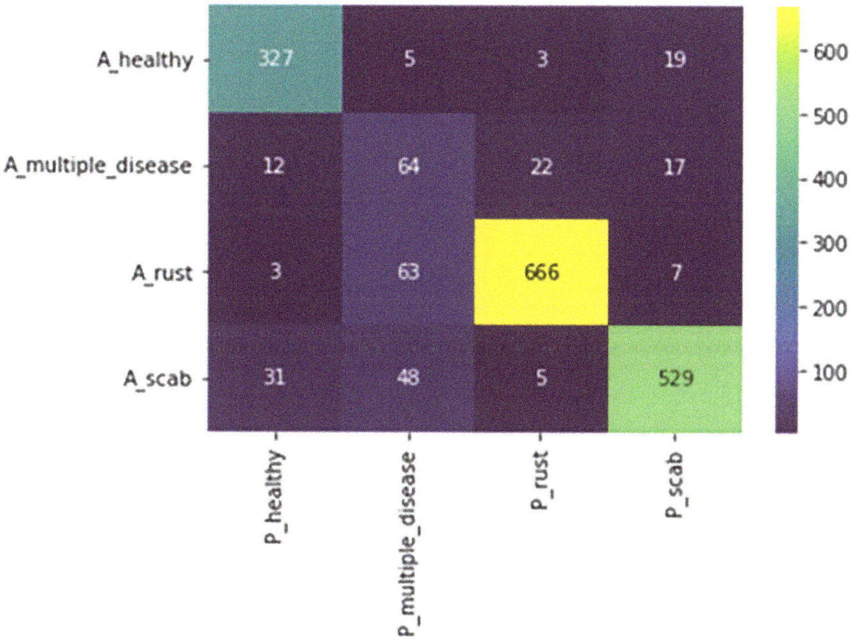

Fig. 19.6 Confusion matrix of Ensemble 1 model

Table 19.8 Performance metrics of Ensembled model of VGG19 and InceptionResnetV2

	Accuracy	Precision	Recall	F1-score	Support
Healthy	0.92	0.88	0.92	0.90	354
Multiple disease	0.56	0.36	0.56	0.43	115
Rust	0.90	0.96	0.90	0.93	739
Scab	0.86	0.92	0.86	0.89	613
Model performance	0.87	0.89	0.87	0.88	1821

for all epochs (method 2). The probabilities from softmax layer were obtained and compared and the label was selected based on the higher probability value.

Figure 19.7 corresponds to the confusion matrix of VGG19 without applying the data augmentation. Out of 115 leaves belonging to the multiple disease class in the test dataset, 74 leaves were correctly classified.13 leaves were misclassified into rust and 15 leaves were misclassified into scab. Figure 19.8 corresponds to the confusion matrix of VGG19 after applying data augmentation. Out of 115 leaves belonging to the multiple disease class in the test dataset, 66 leaves were correctly classified. 28 leaves belonging to multiple disease class were misclassified into rust while 15 leaves were misclassified into scab class.

Figure 19.9 presents the confusion matrix of ensemble 2 model. The ensemble 2 model gave an accuracy of 68% in detecting the multiple disease class. 78 of the 115 multiple disease leaves were correctly classified. 37 leaves were misclassified

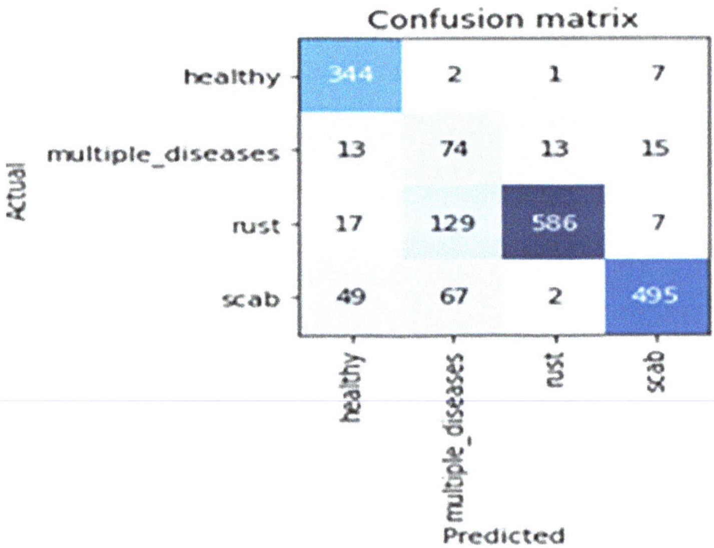

Fig. 19.7 Confusion matrix of VGG19 without data augmentation

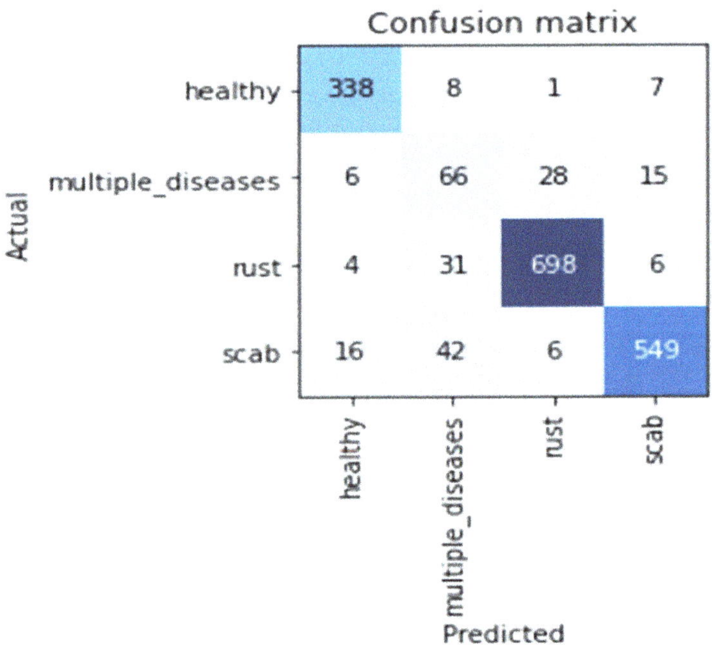

Fig. 19.8 Confusion matrix of VGG19 with data augmentation

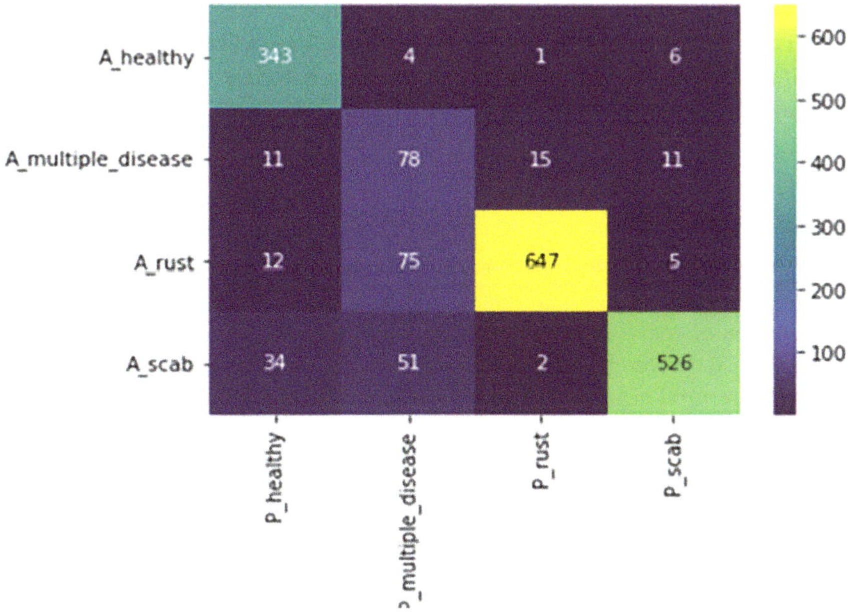

Fig. 19.9 Confusion matrix of Ensemble 2 Model

Table 19.9 Performance metrics of Ensembled model of VGG19 without and with data augmentation

	Accuracy	Precision	Recall	F1-score	Support
Healthy	0.97	0.86	0.97	0.91	354
Multiple disease	0.68	0.38	0.68	0.48	115
Rust	0.88	0.97	0.88	0.92	739
Scab	0.86	0.96	0.86	0.91	613
Model performance	0.88	0.91	0.88	0.89	1821

into other classes. Out of the 37 leaves, 15 were classified as rust infected leaves and 11 leaves were classified as scab infected leaves.

The Ensemble 2 model gave an overall accuracy of 88% (Table 19.9). Multiple disease class accuracy improved to 68% which is more than that gained by individual models. The Ensemble 2 model performed better than the ensemble 1 model.

19.6.2.3 Limitations of the Proposed Method

It is observed that accuracy of multiple disease class has improved significantly as indicated by the results. However, the overall accuracy remained below 90% due to misclassifications in other classes. The model exhibited a high level of accuracy 88%

for Rust, 86% for Scab class and an overall accuracy of 88%. Also, for ensemble models, predictions are to be done twice one for each model before performing the task of ensemble, which is a time-consuming task. Hence, the same performance if not better with a single model, which can save time and space is achieved.

19.6.3 CoAtNet

In order to enhance the performance further, the CoAtnet model, which is a combination of convolution and attention is utilised. The CoAtnet model was run with two different batch sizes. Batch size = 16 and 32. One of the results showed great improvement in the overall accuracy while matching the multiple disease class accuracy of ensemble 2 model. The other result showed improvement in multiple disease class accuracy by 3% while the overall accuracy fell by 3% from the first result.

CoAtNet models which are a combination of Depthwise convolution and self-attention layers. The CoAtNet results improved the results further and was better than the results of ensemble models. Two results of CoAtNet models are reported here and the corresponding loss plots are depicted in Fig. 19.10a, b. Table 19.10 depicts the results of model when the batch size was set to 16. The study produced an overall accuracy rate of 92%, with a notable improvement in the accuracy of the multiple disease class to 71%. This represents the best accuracy achieved for this particular class. The accuracy rates for the healthy class, rust class, and scab class were 97%, 90%, and 96% respectively.

Table 19.11 shows the performance metrics of the CoAtNet model when batch size was fixed to 32. The model reported an accuracy of 68% for the multiple disease class, same as the ensemble 2 model. But the overall accuracy of model improved to 95%, the highest in this study among all models. The improvement can be attributed

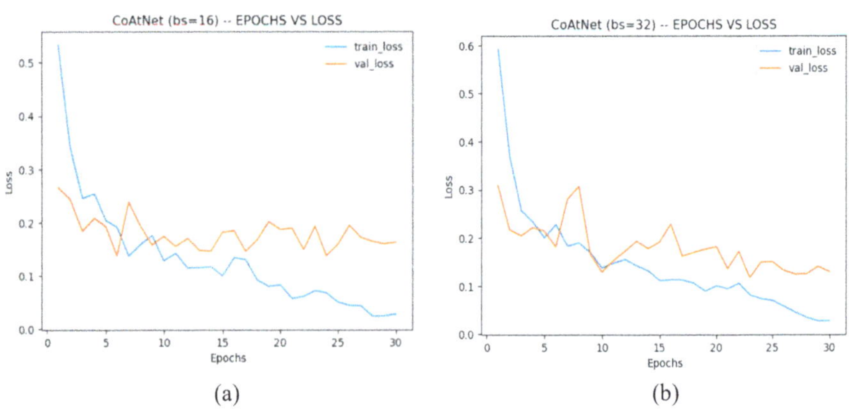

(a) (b)

Fig. 19.10 Loss plot of CoAtNet (**a**) batch size =16 (**b**) batch size = 32

Table 19.10 Performance Metrics of CoAtNet model with batch size = 16

	Accuracy	Precision	Recall	F1-score	Support
Healthy	0.98	0.90	0.98	0.94	354
Multiple disease	0.71	0.56	0.71	0.63	115
Rust	0.90	0.98	0.90	0.94	739
Scab	0.96	0.96	0.96	0.96	613
Accuracy	0.92	0.93	0.92	0.93	1821

Table 19.11 Performance Metrics of CoAtNet model with batch size = 32

	Accuracy	Precision	Recall	F1-score	Support
Healthy	0.97	0.95	0.97	0.96	354
Multiple disease	0.68	0.70	0.68	0.69	115
Rust	0.96	0.97	0.96	0.96	739
Scab	0.98	0.97	0.98	0.97	613
Accuracy	0.95	0.95	0.95	0.95	1821

due to its accuracy in classifying the rust and scab class. It gave an accuracy of 96% in classifying the rust class, a 6% improvement from result 1 and an accuracy of 98% in classifying the scab class.

In order to bring interpretability to the black box nature of DL models used; an Explainable AI library called SHAP is used. SHAP library uses pink and blue colors to indicate whether a pixel contributes to the image being classified into a specific class. Blue color in each pixel represents negative values which indicates that it contributes negatively to be classified into a specific class. Pink color in each pixel represents positive values which contributes positively to be classified to a particular class. The following are the results of SHAP applied to the CoAtNet model results as in Fig. 19.11. The First leaf is a healthy leaf. Hence, more of pink spots is seen under the healthy column. The next leaf belongs to multiple disease class. Understandably, in the first column there are lot of blue dots as it is a healthy column.

The second column is where more of pink dots is seen while there are both pink and blue dots in the next two columns which are rust and scab columns. The next leaf belongs to Rust. This can be clearly identified in the third column as a patch of pink dots are observed where rust spots are seen on the leaf.

The fourth leaf belongs to Scab class. In the last column, pink dots are observed wherever there are black patches in the original sample. Table 19.12 shows the comparison of the proposed work with previous work done on this dataset. Note that the above-mentioned multiple disease class accuracy was not explicitly mentioned in the works. It was calculated from the confusion matrix reported in their respective studies. Some of the accuracy of the previous works mentioned in the table corresponds to the validation accuracy, i.e. the train dataset of 1821 images are split for both training and testing. Hence number of leaves belonging to multiple disease class tested was very low. So, the overall accuracy of the works remains high despite a low multiple disease class accuracy. In this study, 1821 images are

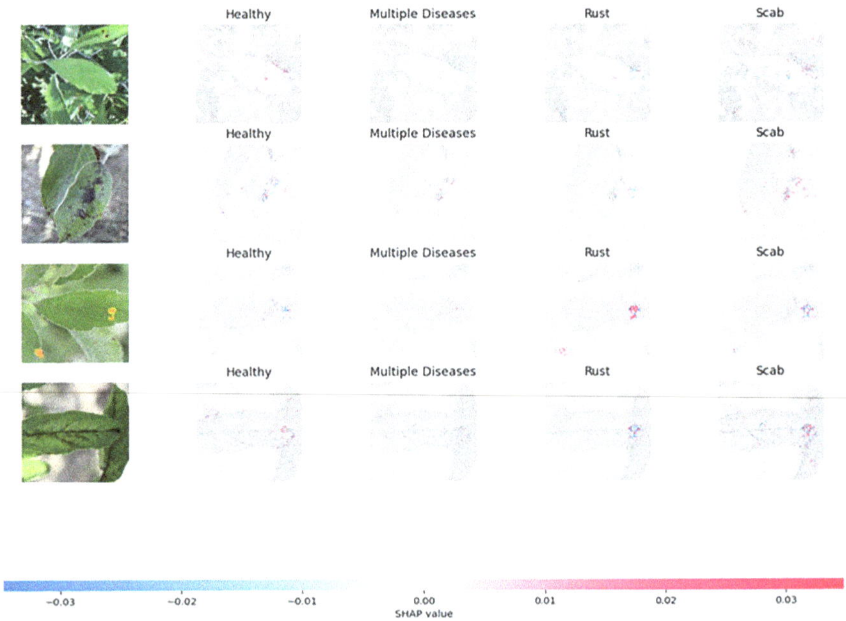

Fig. 19.11 CoAtNet predictions explained through SHAP

separately tested. Hence the contribution of multiple disease class predictions is higher in terms of contributing to the overall model accuracy. The test accuracies achieved for this dataset are reported. With the Ensemble 2 model, an accuracy of 68% for the Multiple disease class is achieved which is better compared to the other works mentioned above. But the overall accuracy of that model was 88%. With CoAtNet models, overall accuracy is found to be improved to 95% while achieving the same accuracy of 68% for the multiple disease class. Another result of CoAtnet gave 71% accuracy for the multiple disease class while the overall accuracy fell to 92%. Hence, the CoAtNet model was chosen as it gave an overall accuracy of 95% and 68% for Multiple disease class as it performed well for other classes as well.

19.7 Conclusion and Future Work

Several state-of-the-art CNN models were trained on the FGVC7 plant pathology dataset in this work. All the architectures were trained using two different techniques. The best-performing models - VGG19 and InceptionResNetV2, trained by freezing completely were used to create an ensemble model. Along with this, a transformer-based model called CoAtnet was also trained on this dataset using the best method obtained from the experiments done in this study. The dataset has a class imbalance problem. It has very few samples for the multiple disease class.

Table 19.12 Comparison of performance metrics of this study with the performance metrics reported in literature

References	Methodology used	Dataset used	Overall accuracy (validation/test)	Multiple disease class accuracy
[1]	ResNet 50 pretrained on Imagenet	FGVC7 Plant pathology challenge 2020 dataset	Validation accuracy = 97%	51%
[38]	MASK RCNN with ResNet50 as backbone architecture	FGVC7 Plant pathology challenge 2020 dataset	Validation accuracy = 90%	50%
[35]	AFD-Net	FGVC7 Plant pathology challenge 2020 dataset	Validation accuracy = 98.7%	55%
[10]	ResNetV2	FGVC7 Plant pathology challenge 2020 dataset	Validation accuracy = 94%	20%
Proposed study	VGG 19 (Best Architecture)	FGVC7 Plant pathology challenge 2020 dataset with test samples	Test accuracy = 88%	68%
	CoAtNet	FGVC7 Plant pathology challenge 2020 dataset with test samples	Test accuracy = 95%	68%

With the ensemble model of VGG19, an accuracy of 68% is gained on the multiple disease class. However, overall accuracy achieved was limited to 88%. The CoAtnet model, a combination of CNN and self-attention, improved the model's overall accuracy to 95% while achieving a multiple disease class accuracy of 68%. To the best of the author's knowledge, this study has achieved the highest accuracy while detecting multiple diseases compared to other recent works. However, in order to be used in real time, the accuracy of the all the classes including the multiple disease class should be high.

As for future work, improved datasets with more samples for training the models can produce better prediction of diseases in real-time. Balanced Datasets in terms of number of samples belonging to each class can also improve the performance of models significantly. Newer datasets will enable us to classify an extensive range of diseases with robust predictions. This classification problem can further be extended to the detection of diseased leaves. Further improvement in this direction can be detecting the diseased regions in the leaves rather than detecting the entire infected leaf. This could be useful in the case of identifying leaves that are infected by more than one disease. Identifying the severity of the disease can also improve the existing work as it can help the farmers decide the kind of pesticides to use and their required concentration.

References

1. Thapa, R., Snavely, N., Belongie, S., Khan, A.: The Plant Pathology 2020 Challenge Dataset to Classify Foliar Disease of Apples. arXiv preprint arXiv:2004.11958 (2020).
2. Ali, A.: Alternaria epidemic of apple in Kashmir. Afr. J. Microbiol. Res. **9**, 831–837 (2015)
3. Li, X., Rai, L.: Apple leaf disease identification and classification using resnet models. In: 2020 IEEE 3rd International Conference on Electronic Information and Communication Technology (ICEICT), pp. 738–742. IEEE (2020)
4. Singh, S., Gupta, S., Tanta, A., Gupta, R.: Extraction of multiple diseases in apple leaf using machine learning. Int. J. Image Graph. **22**, 2140009 (2022)
5. Yu, H.-J., Son, C.-H.: Apple Leaf Disease Identification Through Region-of-Interest-Aware Deep Convolutional Neural Network. arXiv preprint arXiv:1903.10356. (2019).
6. Moid, M.A., Chaurasia, M.A.: Transfer learning-based plant disease detection and diagnosis system using Xception. In: 2021 Fifth International Conference on I-SMAC (IoT in Social, Mobile, Analytics and Cloud) (I-SMAC), pp. 1–5. IEEE (2021)
7. Omaye, J.D., Ogbuju, E., Ataguba, G., Jaiyeoba, O., Aneke, J., Oladipo, F.: Cross-comparative review of Machine learning for plant disease detection: apple, cassava, cotton and potato plants. Artif. Intell. Agric. **12**, 127–151 (2024). https://doi.org/10.1016/j.aiia.2024.04.002
8. Elfatimi, E., Eryigit, R., Elfatimi, L.: Beans leaf diseases classification using mobilenet models. IEEE Access. **10**, 9471–9482 (2022)
9. Storey, G., Meng, Q., Li, B.: Leaf disease segmentation and detection in apple orchards for precise smart spraying in sustainable agriculture. Sustainability. **14**, 1458 (2022)
10. Alsayed, A., Alsabei, A., Arif, M.: Classification of apple tree leaves diseases using deep learning methods. Int. J. Comput. Sci. Netw. Secur. **21**, 324–330 (2021)
11. Sergio, M.A.K., Siallagan, A.B.S., Joelianto, E.: Potato leaf disease detection by means of vision transformer and CoAtNet architecture. In: 2024 IEEE International Conference on Technology, Informatics, Management, Engineering and Environment (TIME-E), pp. 154–159. IEEE (2024)
12. Dai, Z., Liu, H., Le, Q.V., Tan, M.: Coatnet: Marrying convolution and attention for all data sizes. Adv. Neural. Inf. Process. Syst. **34**, 3965–3977 (2021)
13. Shabrina, N.H., Lika, R.A., Indarti, S.: Deep learning models for automatic identification of plant-parasitic nematode. Artif. Intell. Agric. **7**, 1–12 (2023). https://doi.org/10.1016/j.aiia.2022.12.002
14. Bonkra, A., Pathak, S., Kaur, A., Shah, M.A.: Exploring the trend of recognizing apple leaf disease detection through machine learning: a comprehensive analysis using bibliometric techniques. Artif. Intell. Rev. **57** (2024). https://doi.org/10.1007/s10462-023-10628-8
15. Chakraborty, S., Paul, S., Rahat-uz-Zaman, M.: Prediction of apple leaf diseases using multiclass support vector machine. In: 2021 2nd International Conference on Robotics, Electrical and Signal Processing Techniques (ICREST), pp. 147–151. IEEE (2021)
16. Abd El-aziz, A.A., Darwish, A., Oliva, D., Hassanien, A.E.: Machine learning for apple fruit diseases classification system. In: Proceedings of the International Conference on Artificial Intelligence and Computer Vision (AICV2020), pp. 16–25. Springer (2020)
17. Geetha, G., Samundeswari, S., Saranya, G., Meenakshi, K., Nithya, M.: Plant leaf disease classification and detection system using machine learning. J. Phys.: Conf. Ser. 012012. IOP Publishing (2020).

18. Chuanlei, Z., Shanwen, Z., Jucheng, Y., Yancui, S., Jia, C.: Apple leaf disease identification using genetic algorithm and correlation based feature selection method. Int. J. Agric. Biol. Eng. **10**, 74–83 (2017)
19. Hasan, S., Jahan, S., Islam, M.I.: Disease detection of apple leaf with combination of color segmentation and modified DWT. J. King Saud. Univ.-Comput. Inform. Sci. **34**, 7212–7224 (2022)
20. Singh, R.K., Tiwari, A., Gupta, R.K.: Deep transfer modeling for classification of maize plant leaf disease. Multimed. Tools Appl. **81**, 6051–6067 (2022)
21. Haque, M.A., Marwaha, S., Deb, C.K., Nigam, S., Arora, A.: Recognition of diseases of maize crop using deep learning models. Neural Comput. Appl. **35**, 7407–7421 (2023)
22. Ahila Priyadharshini, R., Arivazhagan, S., Arun, M., Mirnalini, A.: Maize leaf disease classification using deep convolutional neural networks. Neural Comput. Appl. **31**, 8887–8895 (2019)
23. Singh, V., Chug, A., Singh, A.P.: Classification of beans leaf diseases using fine tuned CNN model. Procedia Comput. Sci. **218**, 348–356 (2023)
24. Ravi, V., Acharya, V., Pham, T.D.: Attention deep learning-based large-scale learning classifier for Cassava leaf disease classification. Expert Syst. **39**, e12862 (2022)
25. Gajjar, R., Gajjar, N., Thakor, V.J., Patel, N.P., Ruparelia, S.: Real-time detection and identification of plant leaf diseases using convolutional neural networks on an embedded platform. Vis. Comput., 1–16 (2022)
26. Sudhesh, K.M., Sowmya, V., Kurian, S., Sikha, O.K.: AI based rice leaf disease identification enhanced by dynamic mode decomposition. Eng. Appl. Artif. Intell. **120**, 105836 (2023)
27. Elhassouny, A., Smarandache, F.: Smart mobile application to recognize tomato leaf diseases using convolutional neural networks. In: 2019 International Conference of Computer Science and Renewable Energies (ICCSRE), pp. 1–4. IEEE (2019)
28. Tm, P., Pranathi, A., SaiAshritha, K., Chittaragi, N.B., Koolagudi, S.G.: Tomato leaf disease detection using convolutional neural networks. In: 2018 Eleventh International Conference on Contemporary Computing (IC3), pp. 1–5. IEEE (2018)
29. Trivedi, N.K., Gautam, V., Anand, A., Aljahdali, H.M., Villar, S.G., Anand, D., Goyal, N., Kadry, S.: Early detection and classification of tomato leaf disease using high-performance deep neural network. Sensors. **21**, 7987 (2021)
30. Kurup, R.V., Anupama, M.A., Vinayakumar, R., Sowmya, V., Soman, K.P.: Capsule network for plant disease and plant species classification. In: Computational Vision and Bio-Inspired Computing: ICCVBIC 2019, pp. 413–421. Springer (2020)
31. Sapna, S., Sandhya, S., Acharya, V., Ravi, V.: Apple foliar leaf disease detection through improved capsule neural network architecture. Multimed. Tools Appl. **83**, 48585–48605 (2024). https://doi.org/10.1007/s11042-023-17463-7
32. Yan, Q., Yang, B., Wang, W., Wang, B., Chen, P., Zhang, J.: Apple leaf diseases recognition based on an improved convolutional neural network. Sensors. **20**, 3535 (2020)
33. Bi, C., Wang, J., Duan, Y., Fu, B., Kang, J.-R., Shi, Y.: MobileNet based apple leaf diseases identification. Mobile Netw. Appl. **1–9** (2022)
34. Jiang, P., Chen, Y., Liu, B., He, D., Liang, C.: Real-time detection of apple leaf diseases using deep learning approach based on improved convolutional neural networks. IEEE Access. **7**, 59069–59080 (2019)
35. Yadav, A., Thakur, U., Saxena, R., Pal, V., Bhateja, V., Lin, J.C.-W.: AFD-Net: apple Foliar disease multi classification using deep learning on plant pathology dataset. Plant Soil. **477**, 595–611 (2022)
36. Fang, T., Zhang, J., Qi, D., Gao, M.: BLSENet: a novel lightweight bilinear convolutional neural network based on attention mechanism and feature fusion strategy for apple leaf disease classification. J. Food Qual. **2024** (2024). https://doi.org/10.1155/2024/5561625
37. Wang, B., Yang, H., Zhang, S., Li, L.: Identification of multiple diseases in apple leaf based on optimized lightweight convolutional neural network. Plants. **13** (2024). https://doi.org/10.3390/plants13111535

38. Gawade, A., Deolekar, S., Patil, V.: Early-stage apple leaf disease prediction using deep learning. Biosci. Biotechnol. Res. Commun. **14**, 40–43 (2021)
39. Chao, X., Sun, G., Zhao, H., Li, M., He, D.: Identification of apple tree leaf diseases based on deep learning models. Symmetry (Basel). **12**, 1065 (2020)
40. Luo, Y., Sun, J., Shen, J., Wu, X., Wang, L., Zhu, W.: Apple leaf disease recognition and sub-class categorization based on improved multi-scale feature fusion network. IEEE Access. **9**, 95517–95527 (2021)
41. Zhang, W., Zhou, G., Chen, A., Hu, Y.: Deep multi-scale dual-channel convolutional neural network for Internet of Things apple disease detection. Comput. Electron. Agric. **194**, 106749 (2022)
42. Zhong, Y., Zhao, M.: Research on deep learning in apple leaf disease recognition. Comput. Electron. Agric. **168**, 105146 (2020)
43. Wang, P., Niu, T., Mao, Y., Zhang, Z., Liu, B., He, D.: Identification of apple leaf diseases by improved deep convolutional neural networks with an attention mechanism. Front. Plant Sci. **12**, 723294 (2021)
44. Fu, L., Li, S., Sun, Y., Mu, Y., Hu, T., Gong, H.: Lightweight-convolutional neural network for apple leaf disease identification. Front. Plant Sci. **13**, 831219 (2022)
45. Yu, H.-J., Son, C.-H.: Leaf spot attention network for apple leaf disease identification. In: Proceedings of the IEEE/CVF Conference on Computer Vision and Pattern Recognition Workshops, pp. 52–53 (2020)
46. Yu, H., Cheng, X., Chen, C., Heidari, A.A., Liu, J., Cai, Z., Chen, H.: Apple leaf disease recognition method with improved residual network. Multimed. Tools Appl. **81**, 7759–7782 (2022)
47. Di, J., Li, Q.: A method of detecting apple leaf diseases based on improved convolutional neural network. PLoS One. **17**, e0262629 (2022)
48. Sun, H., Xu, H., Liu, B., He, D., He, J., Zhang, H., Geng, N.: MEAN-SSD: A novel real-time detector for apple leaf diseases using improved light-weight convolutional neural networks. Comput. Electron. Agric. **189**, 106379 (2021)
49. Khan, A.I., Quadri, S.M.K., Banday, S., Shah, J.L.: Deep diagnosis: a real-time apple leaf disease detection system based on deep learning. Comput. Electron. Agric. **198**, 107093 (2022)
50. Lv, M., Su, W.H.: YOLOV5-CBAM-C3TR: an optimized model based on transformer module and attention mechanism for apple leaf disease detection. Front. Plant Sci. **14** (2023). https://doi.org/10.3389/fpls.2023.1323301
51. Gao, L., Zhao, X., Yue, X., Yue, Y., Wang, X., Wu, H., Zhang, X.: A lightweight YOLOv8 model for apple leaf disease detection. Appl. Sci. (Switzerland). **14** (2024). https://doi.org/10.3390/app14156710
52. Simonyan, K., Zisserman, A.: Very Deep Convolutional Networks for Large-Scale Image Recognition. arXiv preprint arXiv:1409.1556. (2014).

K. M. Vivek Anandh is currently working as a Senior Engineer at Bosch Global Software Technologies Pvt. Ltd., Coimbatore, India. His work focuses on automation using AI/ML techniques and Python-based solutions. His research interests include intelligent automation, machine learning applications, and computer vision-based projects.

J. Arun Prakash is a researcher specializing in deep learning applications for medical diagnostics, particularly in pediatric pneumonia detection using chest X-ray analysis. His work includes pioneering the development of Selective Kernel Networks for lung abnormality diagnosis, achieving state-of-the-art performance in pulmonary disease classification through multi-scale feature fusion and dynamic kernel selection. His research interests revolve around low attention mechanisms, image enhancement techniques in computer vision, and optimizing model architectures for clinical deployment.

H. Theivaprakasham received the Bachelor of Technology (B.Tech.) in Electronics and Instrumentation Engineering and the Master of Technology (M.Tech.) in Data Science from Amrita Vishwa Vidyapeetham University in Coimbatore, India. He is currently a Ph.D. researcher at Delft University of Technology (TU Delft) in the Netherlands. His research interests include multimodal artificial intelligence, time-series analysis, and the development of robust, explainable AI techniques for complex real-world data.

Divya Sasidharan received her PhD degree in Biomedical signal processing from APJ Abdul Kalam Technological University, Kerala, India. She is currently working as an Assistant Professor (Sr. Gr.) at Amrita School of Artificial Intelligence, Amrita Vishwa Vidyapeetham, Coimbatore, Tamil Nadu, India. Her research interests include biomedical signal processing, non-linear time series analysis, artificial intelligence for biomedical signal and image analysis.

Vinayakumar Ravi is a Research assistant professor at Center for Artificial Intelligence, Prince Mohammad Bin Fahd University, Saudi Arabia. He received PhD degree from the Center for Computational Engineering and Networking, Amrita School of Engineering, Amrita Vishwa Vidyapeetham, Coimbatore, India. His research interests include applications of data mining, artificial intelligence, machine learning (including deep learning) for biomedical informatics, cyber security, image processing, and natural language processing.

V. Sowmya is working as Associate Professor at Amrita School of Artificial Intelligence, Coimbatore, Amrita Vishwa Vidyapeetham. She completed PhD in Artificial Intelligence (AI) for Natural Scene Analysis from Amrita Vishwa Vidyapeetham. Her broad research area is AI for Signal and Image Analysis. The sub-areas of her research interests include Biomedical, Healthcare and Remote Sensing Data Analysis. Her recent research focus is on "AI for Multimodal Data Analysis."

E. A. Gopalakrishnan received the Ph.D. degree from Indian Institute of Technology Madras, Chennai, India. He is currently the Chairperson of the Amrita School of Computing, Amrita Vishwa Vidyapeetham, Bengaluru, India. He has published articles in prestigious journals, such as Nature Scientific Reports, Chaos, Physical Review E, Proceedings of Combustion Institute, IEEE Transactions, Expert Systems, and Journal of Fluid Mechanics. His research interests include complex systems, data driven modeling and analysis, artificial intelligence, early warning signals for catastrophic transitions, and time series analysis.

Chapter 20
Leveraging a Hybrid CNN-BiLSTM Approach for Mental Stress Detection Using Wearable Physiological Sensors

Anogh Dalal, Kingshuk Kirtania, Jaeyoung Choi, Muhammad Fazal Ijaz, and Pawan Kumar Singh

Abstract Chronic stress can lead to significant health complications, including anxiety, depression, hypertension, cardiovascular diseases, sleep disturbances, and headaches. In this study, we propose a robust hybrid deep learning model that combines Convolutional Neural Networks (CNN) with Bidirectional Long Short-Term Memory (BiLSTM) layers to detect mental stress using physiological signals such as ECG, EDA, three-axis acceleration, heart rate variability, and respiration (RESP).The CNN layers are designed to extract intrinsic features from the input signals, while the BiLSTM layers capture temporal dependencies within the time-series data. We evaluate the model's performance on three publicly available datasets: WESAD, SWELL-KW, and Stress-Lysis. The model demonstrates high classification accuracy across multiple scenarios, including 2, 3, 4, and 8-class stress level detection tasks. To further validate our approach, we conduct an ablation study, comparing the full hybrid model against a variant without BiLSTM layers. The results indicate a clear performance gain with the inclusion of BiLSTM, highlighting the model's ability to generalize and effectively learn complex relationships in the data. This research emphasizes the potential of deep learning techniques for mental

A. Dalal · K. Kirtania
Department of Information Technology, Jadavpur University, Jadavpur University Salt Lake Campus, Kolkata, West Bengal, India

J. Choi
School of Computing, Gachon University, Seongnam-si, Republic of Korea
e-mail: jychoi19@gachon.ac.kr

M. F. Ijaz
School of IT and Engineering, Melbourne Institute of Technology, Melbourne, Australia
e-mail: mfazal@mit.edu.au

P. K. Singh (✉)
Department of Information Technology, Jadavpur University, Jadavpur University Second Campus, Kolkata, West Bengal, India

Shinawatra University, Sam Khok, Pathum Thani, Thailand

G. Rivera et al. (eds.), *Recent Advances in Artificial Neural Networks*, Intelligent Systems Reference Library 283, https://doi.org/10.1007/978-3-032-13497-4_20

stress detection, providing valuable insights for the development of preventive healthcare solutions.

Keywords Stress detection · Hybrid deep learning model · Convolutional neural networks · Bidirectional long short-term memory · Electroencephalography signals

20.1 Introduction

Stress has become a pervasive issue in modern life, significantly impacting both physical and mental health. Chronic stress is linked to conditions such as anxiety, depression, cardiovascular diseases, sleep disorders, and impaired cognitive function [1–5]. Given its widespread effects, there is a pressing need for automated systems that can reliably detect stress and facilitate early intervention.

Stress is primarily a physiological response triggered by external stimuli, governed by the sympathetic nervous system (SNS). This response includes measurable changes such as increased heart rate, muscle tension, and hormonal fluctuations [6, 7]. These measurable signals offer a foundation for stress detection using biosensors and computational methods.

Conventional machine learning models, including SVMs and Random Forests, have been widely used for stress detection. However, they often fall short in capturing complex temporal patterns and nonlinear relationships in physiological data. To address these limitations, we propose a hybrid deep learning model combining 1D Convolutional Neural Networks (CNNs) with Bidirectional Long Short-Term Memory (BiLSTM) layers. This architecture is designed to extract both spatial features and temporal dependencies from multivariate time-series physiological data.

The proposed model is trained and evaluated on three benchmark datasets: WESAD, SWELL-KW, and Stress-Lysis. These datasets provide a variety of physiological signals such as ECG, EDA, respiration, and accelerometer data, enabling the model to generalize across diverse sources.

The following key-points summarise and highlight the contributions of the proposed work:

1. We introduce a hybrid CNN-BiLSTM architecture tailored for mental stress detection using wearable sensor data.
2. The model is evaluated on three high-quality datasets, demonstrating robust performance across multiple stress classification tasks (2, 3, 4, and 8 classes).
3. An ablation study is conducted to highlight the performance gain introduced by BiLSTM layers over CNN-only architectures
4. The model shows strong potential for deployment in real-time stress monitoring systems, aiding in preventive healthcare and early diagnosis (Fig. 20.1).

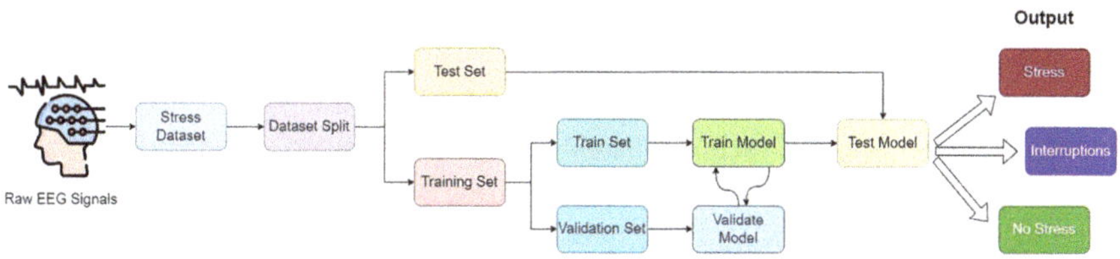

Fig. 20.1 Representation of the complete pipeline for our suggested hybrid CNN-BiLSTM deep learning model designed for detecting mental stress through wearable physiological sensors

20.2 Literature Study

The studies reviewed here provide an overview of mental stress detection methods, ranging from traditional machine learning techniques to recent deep learning approaches. These works highlight the strengths and limitations of different signal modalities and models. The inclusion of these studies underscores the progress in the field and identifies gaps that our proposed hybrid CNN-BiLSTM model aims to address, offering a more robust and comprehensive approach to stress detection.

Ghosh et al. [2] propose a DL approach to detect stress by transforming time series based data into Gramian angular field images. The experiments are performed on 2 datasets i.e. WESAD and SWELL. These experiments resulted in testing accuracies of approximately 94.8% for the WESAD dataset and 99.39% for the SWELL dataset.

The work of Zawad et al. [8] merges two established stress prediction datasets, WESAD and SWELL-KW. This study evaluates 12 classical machine learning models and hybrid models. Ultimately, the researchers suggest a hybrid model, blending ANN with Naive Bayes, which achieved an accuracy of 95.75%.

Behinaein et al. [9] employ a DNN incorporating CNN layers and a transformer mechanism for stress detection. They conduct experiments on the publicly accessible datasets, WESAD and SWELL-KW, to assess their approach. The results of the experiments demonstrate that the proposed model attains comparable or superior performance compared to state-of-the-art models.

Mariano Albaladejo-González et al. [10] investigated the capacity of AI models in the field of transfer learning, addressing an open issue in the field of detecting stress. The standout performer on test data was the anomaly detection model Local Outlier Factor, achieving F1-scores of 77.17% in SWELL-KW and 88.89% in WESAD.

Nigam et al. [11] in their research use WESAD dataset. A model, based on Long Short-Term Memory (LSTM), is employed. The model undergoes a three-way validation process, demonstrating an accuracy of 98%.

Bobade et al. [12] assessed accuracies in three-class (amusement vs. baseline vs. stress) and binary (stress vs. non-stress) classifications using various machine learning techniques such. The study has reported accuracies reaching 81.65% and 93.20% for three-class and binary classification problems, respectively. When employing deep learning, the achieved accuracy rises to up to 84.32% and 95.21%, respectively.

Bellante et al. [13] in their work, introduce EMoCy, a replicable methodology and analysis pipeline designed for the automatic and continuous detection of stress based on physiological signals. The evaluation is conducted on the WESAD dataset. They achieve an F1 score of 0.979 and an accuracy of 0.972 for the stress and baseline classification task.

Garg et al. [14] proposed multiple machine learning models for stress detection in individuals using WESAD. F1-score and accuracy are computed for both 3-class (amusement/baseline/stress) and binary (stress/non-stress) classifications. In both binary and three-class classifications, the Random Forest model demonstrates superior performance, achieving F1-scores of 83.34 and 65.73, respectively.

Bajpai et al. [15] in their paper evaluate performance of KNN models while classifying WESAD dataset. The performance of KNN models undergoes significant changes when the total number of nearest neighbors is adjusted, regardless of the dataset used. They achieved an accuracy of 90%.

M. Quadrini et al. [16], in their work, suggest a stress detection method that scrutinizes biosignals derived from wearable sensor data. They transform raw data in Gramian Angular Field images. These images are subsequently classified using a Convolutional Neural Network. Through this approach, they achieve an accuracy of 92.97%.

Zhu et al. [17] in their work use the WESAD dataset. To achieve optimal classification, the study examines 5 ML based classifiers. Based on their findings, the Support Vector Machine surpasses the rest of the approaches, achieving an accuracy of 92.9%.

Upon analyzing all of the studies done on stress detection on various datasets Dhara et al. [18] have found that a hybrid approach to detect human stress was missing. An ensemble approach on a single dataset was also proposed but as a potentially reliable approach, with better accuracy we thus suggest a hybrid model of CNN and Bi-LSTM for the detection of stress in humans.

Panure et al. [19] employ a generalized model to construct a personalized model, specific to an individual, which evolves over time as sufficient data is collected. They utilize many algorithms. The results show that ANN outperforms XgBoost and SVM, achieving an accuracy of 93.71%.

Liapis et al. [20] investigate the performance of the publicly available dataset WESAD for UX evaluation. In addressing the binary classification problem (stress/no stress), the models achieve an accuracy of 91.1%. Nevertheless, their efficacy in evaluating stress within the context of UX is deemed relatively poor when applied to a new bio-signals dataset.

Ragav et al. [21] introduce a framework with the ability to depict uncertainties in the model by employing approximations in Bayesian Neural Networks through Monte-Carlo Dropout. The framework attains an accuracy of 90.38%.

Ganesh et al. [22] used the SWELL knowledge work (SWELL−KW) dataset for their study. The study employed diverse machine learning algorithms, including both traditional and ensemble methods. Different data formats, such as scaled, normalized, and standardized, were utilized. These experiments achieved maximum accuracy of 99.8% on a Random Forest based model.

The objective of the study of Supriya et al. [23] is to identify stress in senior individuals using Artificial Neural Networks. They also have measured the statistical power of the samples. The algorithm demonstrates an accuracy of 90.4% on the SWELL dataset.

Ghosh et al. [24] developed ML models incorporating efficient feature selection techniques and hyper-parameter tuning. The publicly available SWELL-KW dataset has been utilized for evaluating the performance of their models. Among the developed models, the k-nearest neighbors (k-NN) algorithm had been observed to emerge as the best-performing model, achieving an accuracy score of 99.3%.

Padha et al. [25] in their paper introduce an approach by integrating a parameterized quantum circuit into the traditional LSTM model. The proposed methodology undergoes thorough evaluation using diverse sensor data sourced from the SWELL-KW dataset. The model achieves an F1-score of 87.67%.

Rachakonda et al. [26, 27] introduce Stress-Lysis, a system trained with a combined total of 26,000 samples per dataset, exhibiting an impressive accuracy of up to 99.7%. This proactive approach aims to mitigate the risk of adverse health outcomes and costly treatments. The system demonstrates the capability to yield results with an accuracy ranging from 98.3% to 99.7%.

20.2.1 Motivation

In recent years, machine learning techniques for mental stress and emotion detection using wearable physiological sensors have gained significant attention [28, 29]. Many studies have relied on traditional machine learning models, such as K-Nearest Neighbors (KNN) and Support Vector Machines (SVM), to analyze these sensors' data. While these models have shown promising results, they come with notable limitations: they are highly feature-dependent, less robust, and struggle to generalize across different datasets and domains.

Traditional machine learning approaches also require manual feature engineering, which becomes increasingly burdensome as model complexity grows. Moreover, these models do not adapt well to new, diverse datasets, limiting their applicability to specific scenarios.

In contrast, deep learning models have demonstrated superior performance in many domains by automatically learning complex patterns from raw physiological data, eliminating the need for manual feature extraction. Our proposed hybrid CNN + biLSTM approach capitalizes on the advantages of deep learning, combining Convolutional Neural Networks (CNN) and bidirectional Long Short-Term Memory (biLSTM) networks to capture both spatial and temporal dependencies within the data. This hybrid model significantly improves performance by leveraging the strengths of both architectures.

Furthermore, our approach is not limited to mental stress detection; its flexible and generalizable nature allows it to be applied to other domains as well. By eliminating manual feature engineering and enhancing model adaptability, our work represents a significant advancement in the field of automated mental stress detection using wearable physiological sensors. Through the use of deep learning, we aim to surpass previous benchmarks, achieving higher accuracy and robustness.

20.3 Proposed Approach

In this section, we elaborate how we classify and detect stress in an individual based on his physiological condition. We can see in the Fig. 20.2 the workflow of the proposed pipeline. Our approach employs a hybrid model of 1D Convolutional Neural Network (CNN) Layers and bidirectional Long Short-Term Memory (biLSTM) layers to identify the hierarchical and temporal features from the physiological data.

The sequential stages of processing and classifying data are enlisted as follows:

- Training Phase
- Adding two CNN layers
- Capturing temporal patterns using bidirectional LSTM neuron layers
- Classification using fully-connected neural networks

20.3.1 Training Phase

Table 20.1 describes the various hyperparameters used for training the model for the respective datasets. The smaller SWELL-KW needed 100 epochs to produce the optimal results whereas the bigger and more stable WESAD dataset required only 30 epochs using 8 classes to converge. We have taken batch size of 1024 while testing on WESAD due to it being a huge dataset. Also, we can notice that in case of the Stress Lysis dataset, the model converges after 12 epochs only by employing the early stopping method.

The model utilises a learning rate of 0.001 using Adam optimiser to optimise the model. To prevent overfitting, a regularisation technique is performed by using multiple dropout layers, which dropped 40% of the neural connections in the biLSTM and dense layers for better generalisation. Additionally, early stopping is implemented for the Stress lysis datasets in order to ensure better generalisation of unseen data. This approach prevents the model from getting overfitted and restores the best weights which generated the least validation loss. We try to utilise early stopping with best weights for WESAD and SWELL-KW datasets as well, however reducing the epochs produced similar results for them.

The hyperparameters selection (epochs, batch size and learning rate) is similar to those in previous stress detection studies using wearable sensors and deep learning.

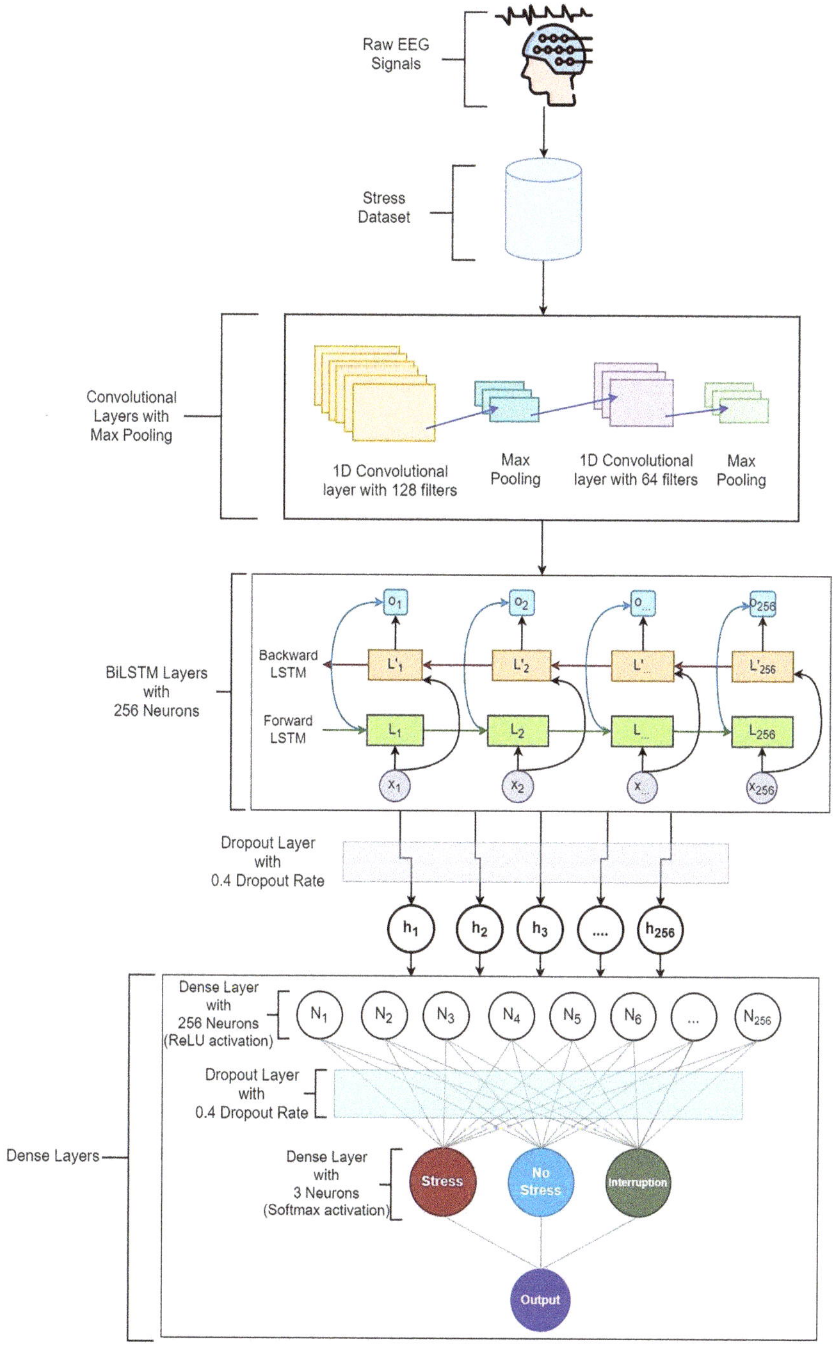

Fig. 20.2 Illustration of the suggested hybrid CNN-biLSTM architecture with 3 output stress classes for SWELL-KW dataset, 8 output classes for WESAD dataset and 3 output stress classes for Stress lysis dataset

Table 20.1 Illustration of the hyperparameters used for training the hybrid CNN + biLSTM model on three stress detection datasets

	Training split	Testing split	Epochs	Batch size	Patience (for early stopping)	Learning rate
SWELL-KW	70	30	100	128	No early stop	0.001
WESAD	80	20	30	1024	No early stop	0.001
Stress Lysis	70	30	100	128	12	0.001

Many publications on deep learning applications tend to use a learning rate of 0.001 (with the Adam optimizer) since this value captures a reasonable trade-off between convergence speed and model stability in many cases. The number of epochs was selected based on empirical results obtained from previous studies, so the models would be adequately trained, but would not overfit. The batch size of 1024 for the larger WESAD dataset is typical for large-scale datasets, as it helps manage memory usage and speeds up processing. For smaller datasets like SWELL-KW and Stress Lysis, a batch size of 128 is commonly used to avoid excessive computational demand while maintaining model performance.

The chosen hyperparameters are selected, not only based on past work done on mental stress detection, but also on personal experiences acquired during early trials with the datasets. The security configurations may vary from those used in related research, but the exact selections are optimized to the individual characteristics of the respective dataset, particularly the size and natural dimensionality of the data.

20.3.2 Adding Two CNN Layers

In case of larger datasets like WESAD or SWELL-KW, which have almost 62 million and 480k records respectively, it becomes quite cumbersome and time-consuming to analyse and process upon. In order to capture the underlying local patterns and features, the model sequence begins with a 1D convolutional layer with 128 filters and 3 kernel size which captures local patterns, followed by a max-pooling layer for down sampling the spatial dimensions. Subsequently, we have added another 1D convolutional layer with 64 filters and 3 kernel size which further refines the features along with another max-pooling layer. The Max-pooling layers that follow each convolutional layer reduce the spatial dimensions, minimize computational cost, and help prevent overfitting by discarding weaker features. By sliding the convolution kernel over the features, convolutional layers can be utilised to extract useful features. The convolutional layer is followed by the MaxPooling layer, which reduces complexity and prevents overfitting by keeping the strong features and removing the weak ones [30]. The features extracted by these layers are leveraged while analysing the time-series data and classification.

20.3.3 Capturing Temporal Patterns

20.3.3.1 LSTM Networks

LSTM layers are robust in preserving long-term relationships in comparison to Recurrent Neural Networks (RNN) [11]. LSTM also removes the vanishing and exploding gradient problem posed by RNN. This architecture is shown in Fig. 20.3.

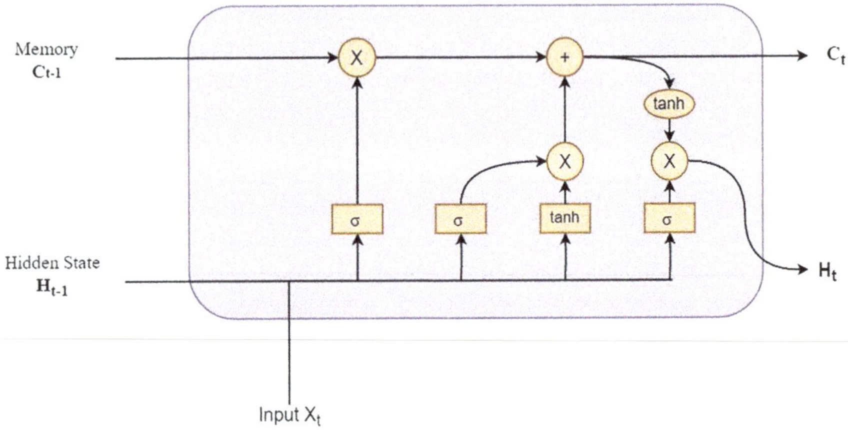

Fig. 20.3 Illustration of the internal architecture of a LSTM unit

Typically, a LSTM cell contains 3 gates—a forget gate, an input gate and an output gate.

The forget gate manages whether previous data should be kept or discarded from the cell state. The input gate introduces new information into the LSTM cell. The output gate delivers information after processing both the input data and cell state. Additionally, a separate neural network, known as the candidate memory, generates fresh candidate information that may be chosen for insertion into the cell state.

20.3.3.2 biLSTM Networks

LSTM is efficient in predicting one-way time series. However, it fails to account for future contexts while prediction. To address this issue, an LSTM network is employed to preserve past values in the forward direction and learn future values in the reverse direction. The temporal information from both forward and backward propagation is combined, enabling the biLSTM neurons to remember both past and future contexts. [12] The structure of biLSTM is represented in the Fig. 20.4. Suppose, x_t, $x_{t+1}$and x_{t+2} are the time series HRV data delivering in the LSTM neurons. h_t and h_t' be the output generated by the forward and backward LSTM respectively.

The following equations explains the operations performed in bi-LSTM units,

$$h_t = \sigma\left(w_1 x_t + w_2 h_{t-1}\right) \times \tanh\left(C_{t-1}\right) \tag{20.1}$$

$$h_t' = \sigma\left(w_3 x_t + w_4 h_{t+1}'\right) \times \tanh\left(C_{t+1}'\right) \tag{20.2}$$

$$\boldsymbol{o_t = w_t h_t + w_t' h_t'} \tag{20.3}$$

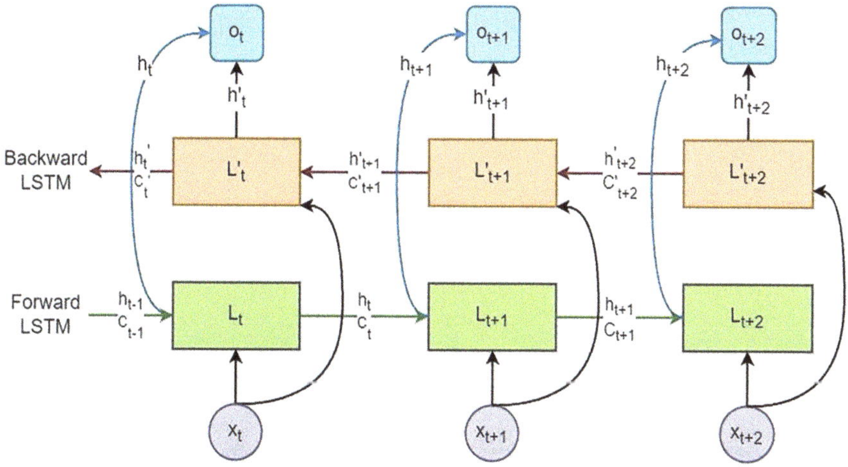

Fig. 20.4 Structure of biLSTM neurons

Our model consists of a biLSTM layer which has 256 hidden units generated about 512 features in every timestep (256 from the forward pass + 256 from the backward pass). The biLSTM layer uses tanh activation function. The biLSTM layer concatenated and flattened the outputs of every LSTM unit as described in the above equations to generate the final output. The biLSTM layer also dropped about 40% of neurons as a regularisation measure. This layer efficiently collects deep sequence temporal information present in the 3 time-series based datasets.

20.3.4 Classification Using Fully Connected NNs

The flattened output generated by the biLSTM is classified by 2 sequentially arranged fully connected neural networks. The first dense layer contains 256 neurons containing a dropout rate of 40% with ReLU activation function. The second dense layer contains 8 neurons (for WESAD dataset) and 3 neurons (for stress lysis and SWELL-KW dataset) with Softmax activation function. In a dense layer, the output from each neuron is computed as a weighted sum of the inputs, followed by an activation function to introduce non-linearity. Finally, the second dense layer classifies and assigns classes to the unseen data inputs. The overall architecture of our proposed hybrid CNN=BiLSTM model for automatic stress detection is shown in Fig. 20.5.

20.4 Results and Discussion

The study was conducted using an MSI GF66 Laptop equipped with 16 GB RAM, an 11th Gen Intel Core i7 processor, and Nvidia GeForce RTX 3060 Laptop GPU

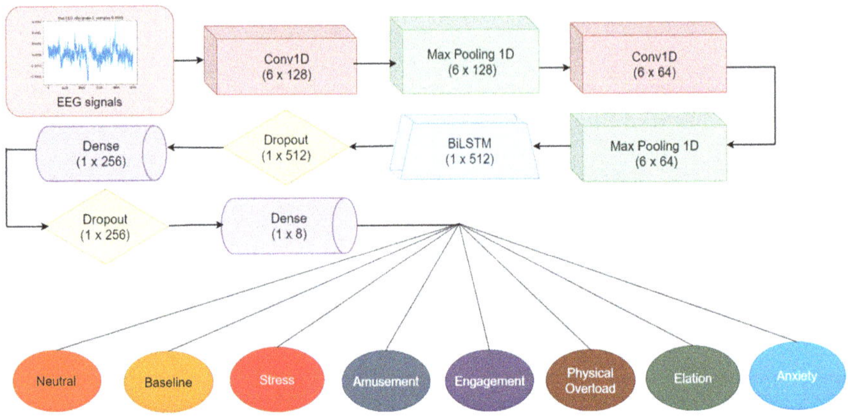

Fig. 20.5 Depiction of the suggested hybrid CNN + biLSTM deep learning model with 8 output classes for WESAD dataset

with 6 GB VRAM. The operating system used was Windows 10 64-bit, and the source code for the experiment was written in a Jupyter notebook. The performance of the hybrid model on the aforementioned multi-class datasets was evaluated against four commonly used evaluation metrics. Those are *Accuracy, Precision, Recall and F1-score*. The formulae of these metrics are listed as follows –

$$\text{Accuracy } (\%) = \frac{\#\text{correct predictions}}{\#\text{total predictions}} \tag{20.4}$$

$$\text{Precision} = \frac{\#\text{true positives}}{\#\text{true positives} + \#\text{false positives}} \tag{20.5}$$

$$\text{Recall} = \frac{\#\text{true positives}}{\#\text{true positives} + \#\text{false neagatives}} \tag{20.6}$$

$$\text{F1 Score} = 2 \times \frac{\text{Precision} \times \text{Recall}}{\text{Precision} + \text{Recall}} \tag{20.7}$$

20.4.1 Datasets

The assessment of the suggested framework has been conducted against 3 of the highly reliable datasets available as open source.

1. WESAD by Schmidt et al. [6]
2. SWELL-KW by Koldijk et al. [31]
3. Stress-Lysis by Rachakonda et al. [26, 27]

Our work generalises diverse input features provided by the 3 datasets. For instance, the WESAD dataset utilises 3 axis acceleration, ECG, EEG data provided by the

20.4.2 Datasets Description

20.4.2.1 WESAD (Wearable Stress and Affect Detection)

WESAD is an openly accessible dataset designed for wearable stress and affect detection. The dataset involves 15 subjects and their physiological and motion-related information. Various sensor modalities are included in the dataset. Table 20.2 contains class-wise frequencies of samples present in the WESAD dataset. We can see that Class 0 ('Neutral') has the most samples at 27,654,897 and Class 5 ('Physical Overload') has the lowest number of samples at 5,52,300.

20.4.2.2 SWELL

The SWELL was formulated by the researchers from the Institute for Computing and Information Sciences at Radboud University. The experiments involved 25 subjects engaged in routine office tasks. These individuals were exposed to common workplace stressors, including receiving unexpected emails, interruptions, and pressure to complete their tasks. Table 20.3 shows the class-wise frequencies of samples present in the SWELL dataset. Class 0 ('No Stress') has the most samples whereas Class 2 ('Time Pressure') has the least number of samples.

Table 20.2 Class-wise frequencies of samples present in the WESAD dataset

Dataset	Class	Class labels	Number of samples
WESAD	0	*Neutral*	27,654,897
	1	*Baseline*	12,327,702
	2	*Stress*	6,976,201
	3	*Amusement*	3,902,501
	4	*Engagement*	8,264,199
	5	*Physical Overload*	5,52,300
	6	*Elation*	5,52,998
	7	*Anxiety*	5,76,802

Table 20.3 Class-wise frequencies of samples present in the SWELL-KW dataset

Dataset	Class	Class labels	Number of samples
SWELL-KW	0	*No stress*	2,00,082
	1	*Interruption*	1,05,150
	2	*Time Pressure*	64,057

20.4.2.3 Stress-Lysis

Stress-Lysis is a dataset which comprises 2001 samples. Humidity, Body Temperature and the Number of Steps is taken from the user. Three distinct classifications of stress are conducted, categorizing it into Low, Normal, and High levels. The column description of the dataset is as follows:

1. Humidity—When someone feels stress, their body temperature rises, prompting the sweat glands to kick in. This sweat is considered to be the Humidity Level.
2. Temperature—Body Temperature of a person during stress.
3. Step Count—Number of steps covered by the person during his stressful situation.
4. Stress Level—Based on all the above 3 Factors the Stress Level will be predicted as High, Medium and Low accordingly.

Table 20.4 comprises of the class-wise frequencies of samples present in the Stress-Lysis dataset. We can see it is a fairly small dataset with most sampled class being Class 1 ('Normal Stress') and the least sampled being Class 0 ('Low Stress').

20.4.3 Results on WESAD Dataset

In the case of the WESAD dataset, the study aimed to robustly predict the stress level of individuals, ranging from Class 0 'Neutral' to Class 7 'Anxiety'. We have taken all the 8 classes into consideration. After nearing 30 epochs of training, the model achieves a high training accuracy of 97.65% and a testing accuracy of 98.09%. Table 20.5 shows the comparison of the results considering 2 classes (Baseline vs Stress), 3 classes (Neutral vs Baseline vs Stress) and 4 classes (Neutral vs Baseline vs Stress vs Amusement) with our results taking 8 classes into consideration.

Table 20.4 Class-wise frequencies of samples present in the stress-lysis dataset

Dataset	Class	Class labels	Number of samples
Stress-lysis	0	*Low stress*	501
	1	*Normal stress*	790
	2	*High stress*	710

Table 20.5 Comparison of results by varying different number of stress classes present in the WESAD dataset using the hybrid CNN + biLSTM model

#Class	Class labels	Accuracy %	Precision	Recall	F1 Score
2	Baseline, Stress	99.97	1.00	1.00	1.00
3	Neutral, Baseline, Stress	99.01	0.99	0.99	0.99
4	Neutral, Baseline, Stress, Amusement	97.73	0.98	0.98	0.98
8	Neutral, Baseline, Stress, Amusement, Engagement, Physical Overload, Elation, Anxiety	**98.09**	**0.98**	**0.98**	**0.98**

Table 20.6 Summarisation of performance evaluation of individual stress classes produced over WESAD dataset using the hybrid CNN + biLSTM model

Class labels	Accuracy %	Precision	Recall	F1 Score
Neutral (0)	97.32	0.99	0.97	0.98
Baseline (1)	99.76	0.99	1.00	1.00
Stress (2)	98.36	0.98	0.98	0.98
Amusement (3)	98.48	0.96	0.98	0.97
Engagement (4)	99.14	0.97	0.99	0.98
Physical Overload (5)	93.56	0.90	0.94	0.92
Elation (6)	88.41	0.94	0.88	0.91
Anxiety (7)	91.73	0.91	0.92	0.91
Overall Model Accuracy	**98.09**			

Our model acquires almost a perfect accuracy while classifying between baseline and stress classes. It also performs significantly better than the previous works in distinguishing among 'Neutral', 'Baseline' and 'Stress' classes, with an accuracy of 99.01%. While classifying among 'Neutral', 'Baseline', 'Stress' and 'Amusement' classes, our proposed model fetches an impressive 97.73% accuracy. Finally, for the classification among all the 8 stress classes present in the WESAD dataset, our model achieves an outstanding accuracy of 98.09%. The table signifies that our model is adequately robust for classifying any combination of classes chosen from the dataset. Table 20.6 presents the performance of the model to identify stress against the WESAD dataset. We can see from the table that the 'Mental Overload' class and the 'Engagement' class have the best results whereas the 'Elation' and the 'Anxiety' classes have slightly worse results. Analysis of the loss function shown in Fig. 20.6b, shows that the model drastically reduces both validation and training loss till the 8th epoch, while the classification accuracy in the Fig. 20.6a increases significantly till the 8th epoch. The model utilises only 30 training epochs since we don't find any further improvement in loss reduction while also ensuring the model doesn't fit too well with the training data as a regularisation measure. The confusion matrix generated by the proposed CNN + biLSTM hybrid model for the WESAD dataset is shown in Fig. 20.7. Here, the X-axis represents the predicted labels whereas the Y-axis represents the actual labels of the data. From the confusion matrix, it can be seen that the Class '0' i.e., the 'Neutral' class has the most misclassifications. These results indicate that storing and merging both the future as well as past temporal features of the multivariate time series dataset increases accuracy significantly. ROC curves are used to evaluate the performance of a classification model by illustrating the trade-off between the True Positive Rate and the False Positive Rate at various classification thresholds. The AUC value represents the model's ability to distinguish between classes, with higher AUC values indicating better performance. From the ROC curves in Figure 20.8, we can deduce that the Classes 1 (Baseline), 2 (Stress), 3 (Amusement), 4 (Engagement) and 0 (Neutral) were the best performing classes in order with AUC values 1, 0.99, 0.99, 0.99 and 0.98 respectively.

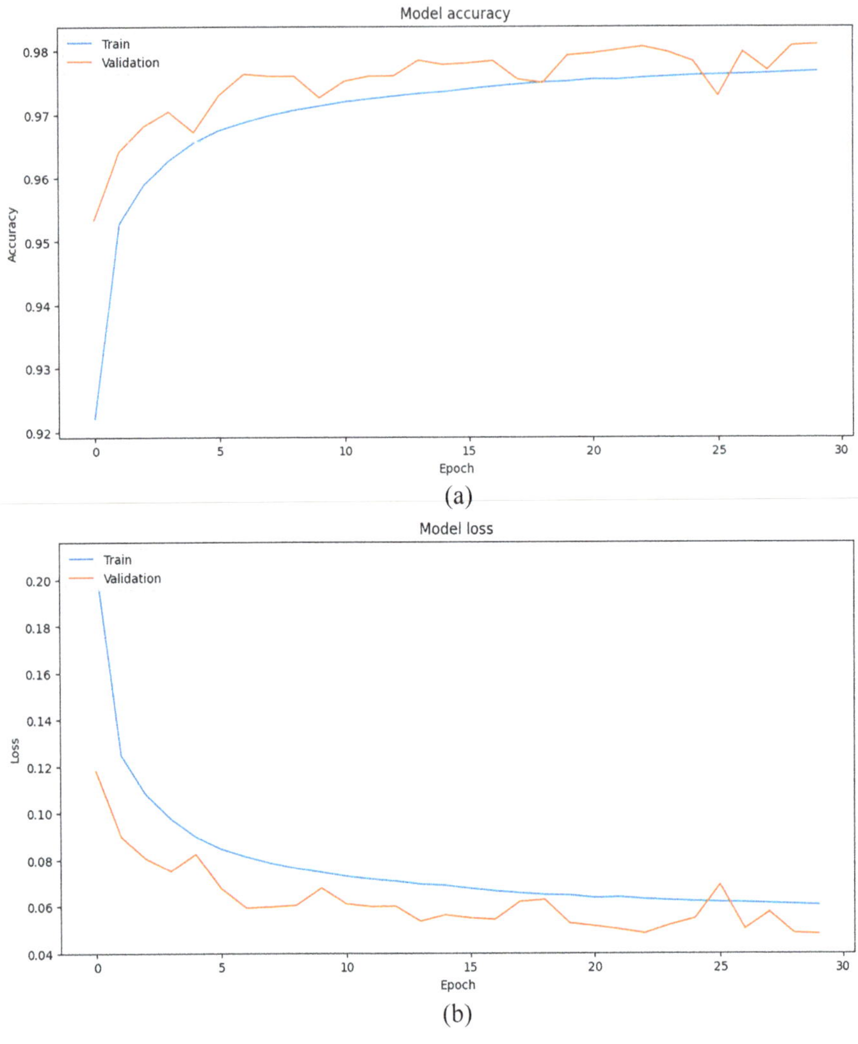

Fig. 20.6 Graphical representation of (**a**) Training versus validation accuracy and (**b**) Training versus validation loss curves over subsequent epochs on **WESAD** dataset

20.4.4 Results on SWELL-KW Dataset

Another stress detection dataset with different input attributes is also utilised to check the performance against it. While following the similar procedure used for the WESAD dataset, tinkering the number of epochs to 100, we achieved an outstanding test accuracy of 99.97% and training accuracy of 99.87%. Table 20.7 shows the class-wise precision, recall and F1 score, which are almost equally impressive for all the 3 stress classes. Since the number of input records

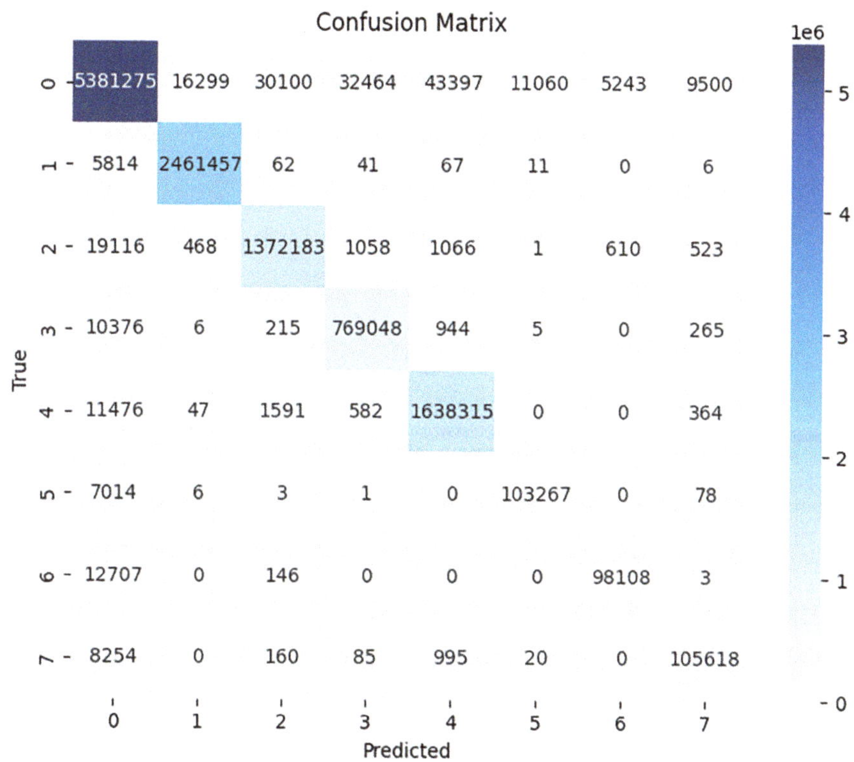

Fig. 20.7 Confusion matrix obtained by the proposed method using hybrid CNN + biLSTM model on the **WESAD** dataset

are less in SWELL in comparison to the humongous WESAD dataset, it performs robustly well in this context. The model's training versus validation accuracy, as illustrated in Figure 20.9, indicates minimal overfitting, given that the training and validation curves exhibit a high degree of overlapping. The confusion matrix produced in the Figure 20.10 shows there are only 9 false positives for class labels which is miniscule in comparison to the total number of input records, where the X-axis represents the predicted labels and Y-axis depicts the actual labels of the data. In Figure 20.11, the ROC curve for all three stress classes in the SWELL dataset is displayed, revealing an AUC value equal to 1 for each class. This implies highly accurate classification performance.

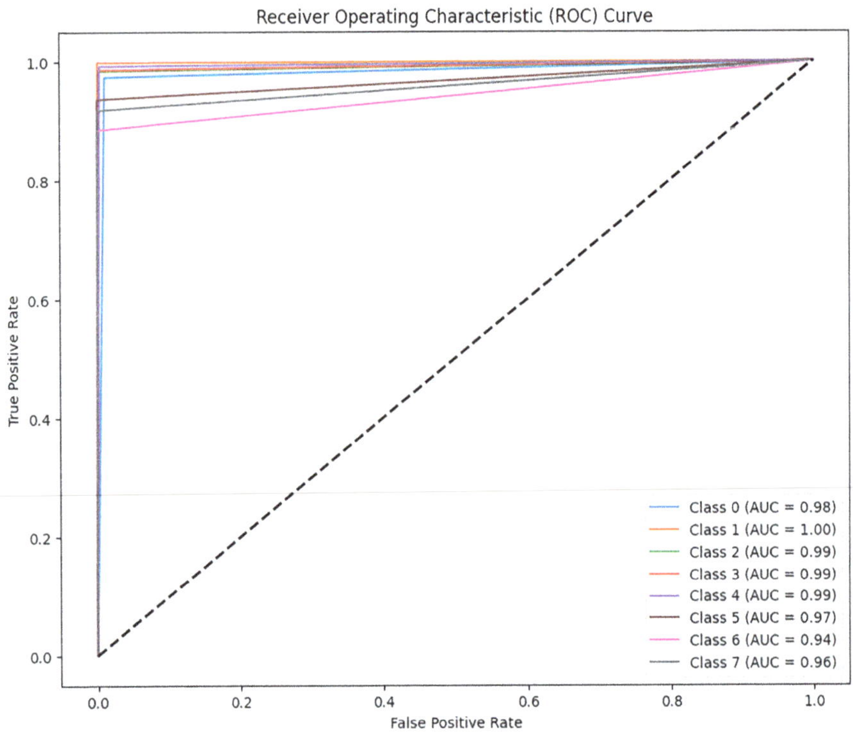

Fig. 20.8 ROC curves produced by the proposed hybrid CNN + biLSTM deep learning model on **WESAD** dataset

Table 20.7 Summarisation of performance evaluation produced against SWELL-KW dataset using the hybrid CNN + biLSTM model

Class labels	Accuracy %	Precision	Recall	F1 Score
No stress (0)	99.97	1.00	1.00	1.00
Interruption (1)	99.98	1.00	1.00	1.00
Time Pressure (2)	100	1.00	1.00	1.00
Overall Model Accuracy	**99.97**			

20.4.5 Results on Stress-Lysis Dataset

Coming to the Stress-lysis dataset, it is a relatively small dataset in comparison to WESAD and SWELL while taking different features into account produced a perfect 100% accuracy. In Table 20.8, the class-wise precision, recall, and F1 score are presented, showcasing consistently impressive performance across all three stress classes. Figure 20.12 shows the training vs validation accuracy achieved by the model, which depicts that there is no case of overfitting as both the training as well as validation curves are almost overlapping. Figure 20.13 displays the

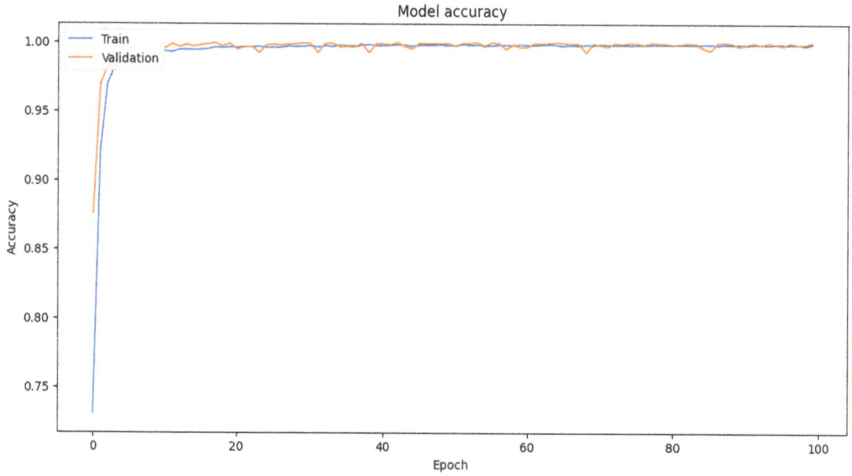

(a)

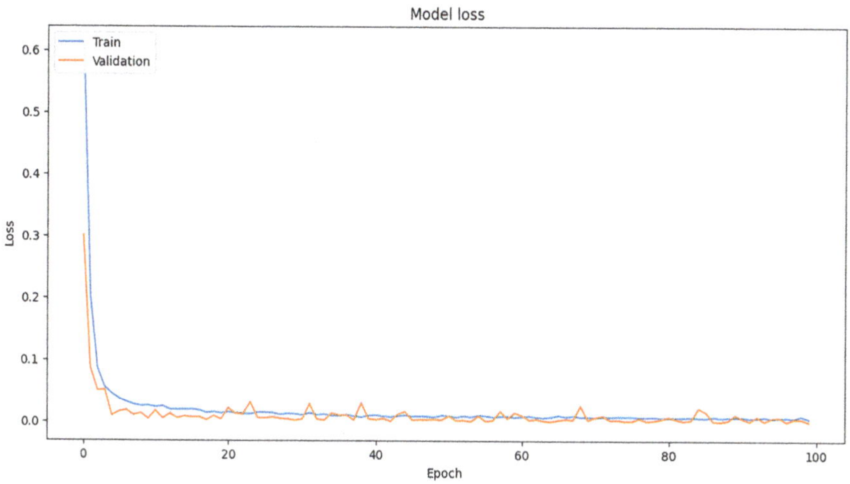

(b)

Fig. 20.9 Graphical representation of (**a**) Training versus validation accuracy and (**b**) Training versus validation loss curves over subsequent epochs on **SWELL** dataset

confusion matrix produced which shows that there are no misclassifications. Note that we have introduced an early stopping hyperparameter while model training which has halted the model training at the 56th epoch itself, thus not reaching the 100th epoch. Figure 20.14 shows the ROC curve for the three stress classes in the Stress-Lysis dataset, and the perfect AUC value of 1 indicates that the classification has been exceptionally accurate.

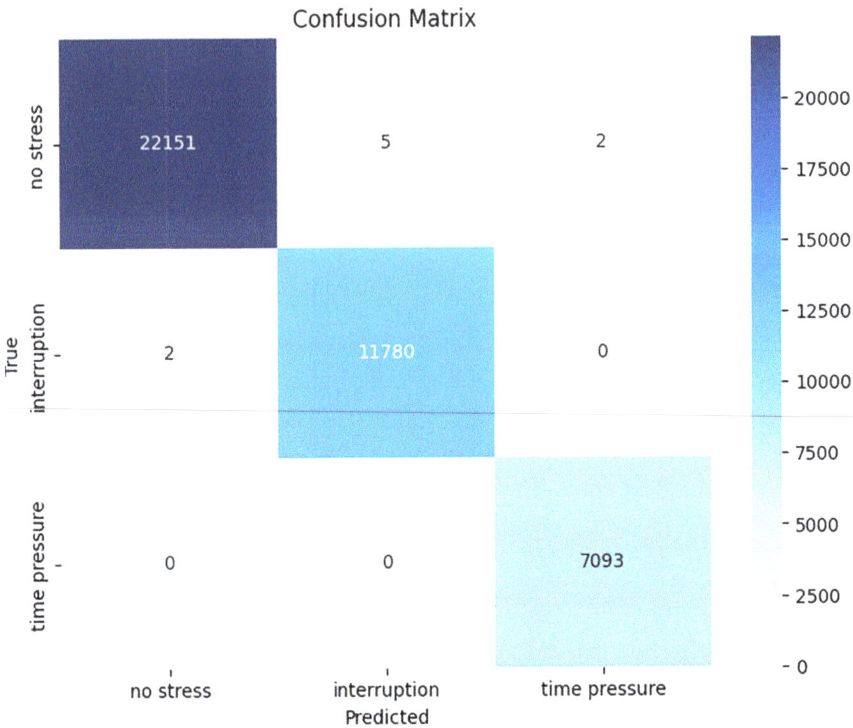

Fig. 20.10 Confusion matrix obtained by the proposed method using hybrid CNN + biLSTM model on the **SWELL** dataset

20.4.6 Ablation Study

The reason for us not using just a simple CNN based model is the superior performance and robustness provided by our proposed hybrid CNN + biLSTM deep learning model. In case of the WESAD dataset as shown in Table 20.9, our proposed hybrid model performs fairly well compared to a model without the biLSTM layer, beating its accuracy by almost 3%. Table 20.10 shows the comparison against the SWELL dataset. The difference in performance might not seem substantial, but the proposed model almost acquires a perfect classification accuracy, which simply can't be ignored. Same goes for the Stress-Lysis dataset, which can be seen in Table 20.11, where our proposed model outright achieves the perfect accuracy, compared to the 97.50% of the model without the biLSTM layer. It is proven that our proposed hybrid model outright performs better than using just a simple CNN based model without a biLSTM layer in this sensitive field of stress detection, where a minuscule error can be the reason for either life or death.

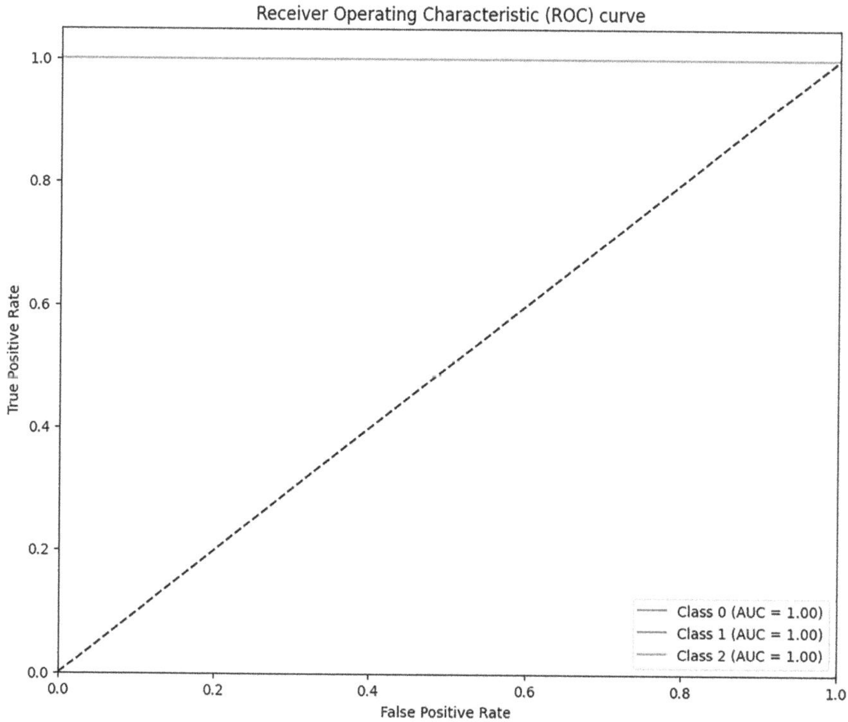

Fig. 20.11 ROC curve produced by the proposed hybrid CNN + biLSTM deep learning model on SWELL dataset

Table 20.8 Summarisation of performance evaluation produced against stress lysis dataset using the hybrid CNN + biLSTM model

Class labels	Accuracy %	Precision	Recall	F1 Score
Low Stress (0)	100	1.00	1.00	1.00
Medium Stress (1)	100	1.00	1.00	1.00
High Stress (2)	100	1.00	1.00	1.00
Overall Model Accuracy	**100**			

20.4.7 Comparison with Existing Mental Stress Detection Models

Tables 20.12, 20.13 and 20.14 compare the proposed method against several notable publicly available works in literature in context to the WESAD, SWELL-KW and Stress lysis datasets respectively, utilised in this study based on accuracy scores. It is to be noted that our model utilises an 8-class WESAD dataset forproducing results,

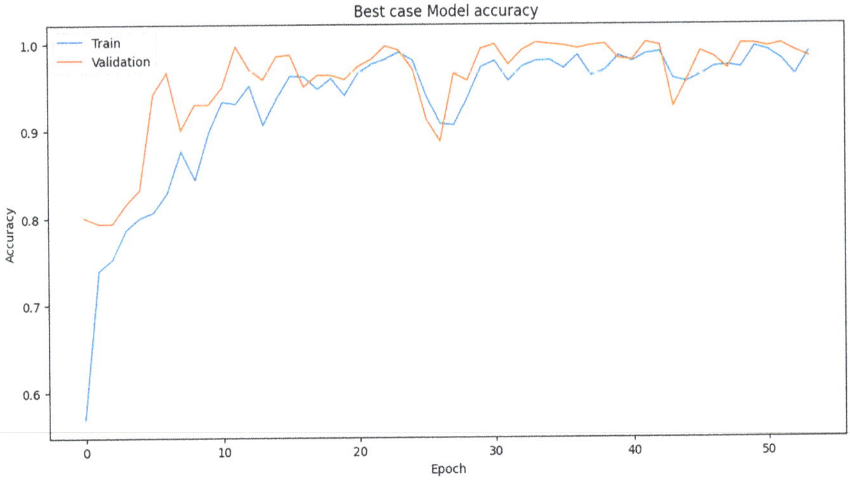

(a)

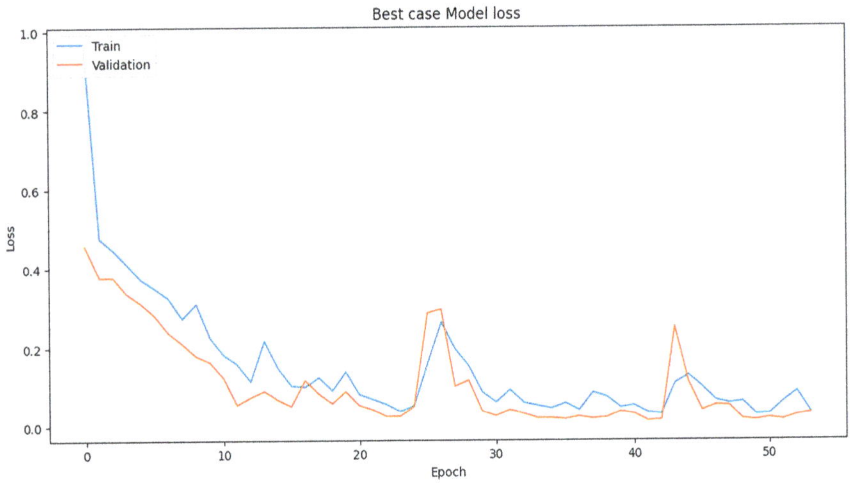

(b)

Fig. 20.12 Graphical representation of (**a**) Training versus validation accuracy and (**b**) Training versus validation loss curves over subsequent epochs on **Stress-lysis** dataset

whereas the other works might have used either 3 or 4-class WESAD dataset for performance evaluation. Still our model has beaten the benchmark and performed better than all the other approaches as shown below. Our model being fundamentally simple becomes easy to implement and the size of the trained model itself accounts to 31 MB even for the large WESAD dataset which has a size of 9 GB.

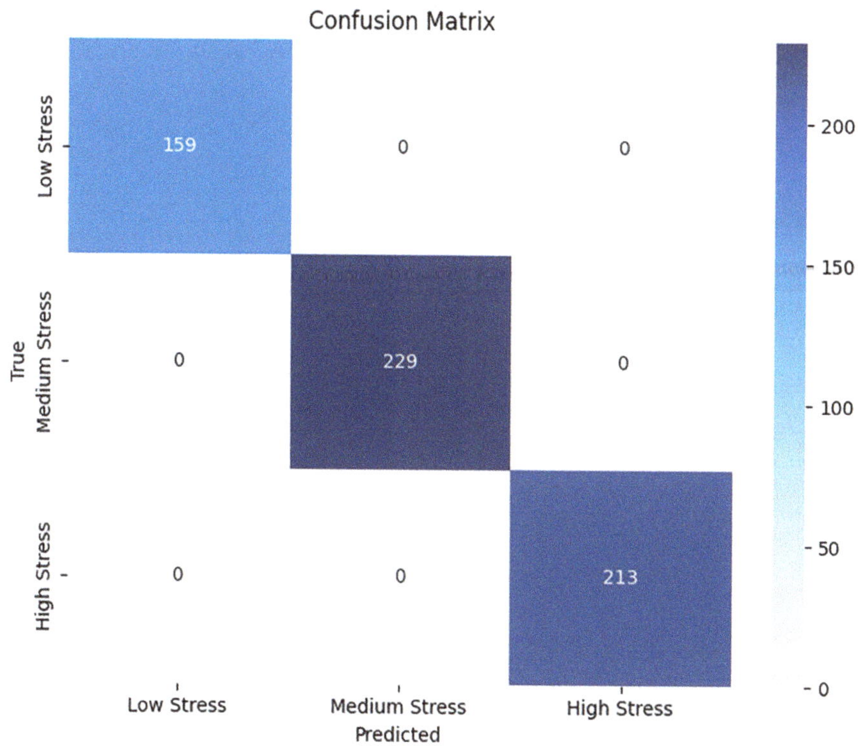

Fig. 20.13 Confusion matrix obtained by the proposed method using hybrid CNN + biLSTM model on the **Stress-lysis** dataset

20.4.8 Statistical Test

We have applied a k-fold cross-validated paired t-test on the results acquired using our CNN-biLSTM hybrid model and a model without the biLSTM layer. We used a 5-fold cross-validation approach. The data has been randomly split into 5 folds, and each model has been trained and evaluated on different subsets of the data. The null hypothesis states that the performance results obtained by our proposed hybrid CNN-biLSTM model are not statistically significant for all the three stress detection datasets. The alternative hypothesis states that the performance results obtained by our proposed hybrid CNN-biLSTM model are statistically significant for all the three stress detection datasets. We have used the "ttest rel" method from "SciPy" to run the t-test. Table 20.15 depicts the p-values acquired after the t-test on the three dataset. It is evident that the p-value is less than the significance level (i.e. 0.05) in all the three cases and thus we reject the null hypothesis and accept the alternative hypothesis, concluding that the performance results obtained by our proposed hybrid CNN- biLSTM model are statistically significant for all the three stress detection datasets.

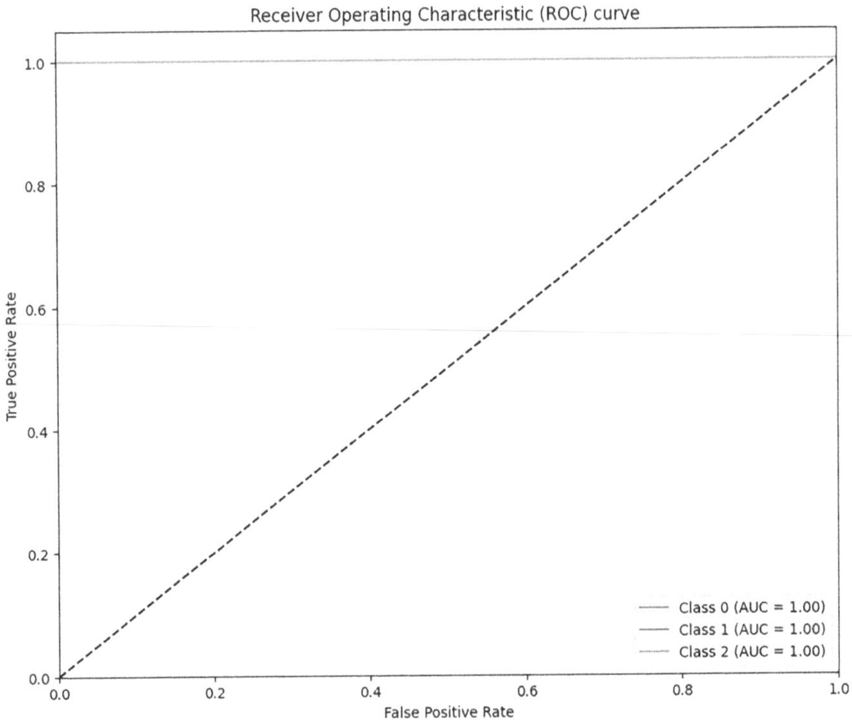

Fig. 20.14 ROC curves produced by the proposed hybrid CNN + biLSTM deep learning model on **Stress-lysis** dataset

Table 20.9 Comparison of results produced against the WESAD dataset using a model with and without biLSTM model

Model	Accuracy %	Precision	Recall	F1 Score
Without biLSTM	95.46	0.95	0.95	0.95
With biLSTM	**98.09**	**0.98**	**0.98**	**0.98**

Table 20.10 Comparison of results produced on the SWELL dataset using a model with and without biLSTM model

Model	Accuracy %	Precision	Recall	F1 Score
Without biLSTM	98.37	0.98	0.98	0.98
With biLSTM	**99.97**	**1.00**	**1.00**	**1.00**

20.5 Conclusion & Future Works

This research proposes a hybrid model consisting of CNN and bidirectional LSTM layers for the classification of the stress levels of the person. The study has been conducted while considering the format and features of the varied publicly

Table 20.11 Comparison of results produced on the stress-lysis dataset using a model with and without biLSTM model

Model	Accuracy %	Precision	Recall	F1 Score
Without biLSTM	97.50	0.98	0.98	0.98
With biLSTM	**100**	**1.00**	**1.00**	**1.00**

Table 20.12 Comparison of our proposed hybrid CNN + BiLSTM deep learning model with previously proposed related to WESAD dataset

Researchers	Year	Model	Dataset	#Classes	Accuracy
Ghosh et al. [2]	2022	Custom-built CNN model	WESAD	4	94.8%
Zawad et al. [8]	2023	Artificial Neural Network (ANN) with Naive Bayes (NB)	WESAD	2	95.75%
Behinaein et al. [9]	2021	DNN	WESAD	3	89.57%
Albaladejo-González et al. [10]	2023	Local Outlier Factor (LOF)	WESAD	2	88.89%
Nigam et al. [11]	2021	LSTM	WESAD	2	98%
Bobade et al. [12]	2020	DNN	WESAD	2	95.21%
Bellante et al. [13]	2021	EMoCy	WESAD	2	97.2%
Garg et al. [14]	2021	Random Forest	WESAD	3	83.34%
Bajpai et al. [15]	2020	KNN	WESAD	4	90%
Quadrini et al. [16]	2022	STREDWES	WESAD	2	92.97%
Zhu et al. [17]	2023	Support Vector Machine (SVM)	WESAD	2	92.9%
Panure et al. [19]	2020	ANN	WESAD	2	93.71%
Liapis et al. [20]	2021	Custom Deep Learning Based Model	WESAD	2	91.1%
Proposed (Dalal et al.)	**2024**	**Hybrid CNN + biLSTM model**	**WESAD**	**2**	**99.97%**
Proposed (Dalal et al.)	**2024**	**Hybrid CNN + biLSTM model**	**WESAD**	**3**	**99.01%**
Proposed (Dalal et al.)	**2024**	**Hybrid CNN + biLSTM model**	**WESAD**	**4**	**97.73%**
Proposed (Dalal et al.)	**2024**	**Hybrid CNN + biLSTM model**	**WESAD**	**8**	**98.09%**

available multimodal physiological datasets —WESAD, SWELL-KW and Stress-lysis dataset. The model achieves staggering accuracies of 99.97%, 99.01%, 97.73% and 98.09% for 2, 3, 4 and 8-class classification respectively on the WESAD dataset, which outperforms the benchmark. Also, in the case of SWELL and Stress-Lysis datasets our model acquires accuracies of 99.97% and 100%, which showcases the robustness of our model. The hybrid CNN + biLSTM model also requires a training time of 6 h and the model tends to hog compute resources

Table 20.13 Comparison of our proposed hybrid CNN + BiLSTM deep learning model with previously proposed related to SWELL-KW dataset

Researchers	Year	Model	Dataset	Accuracy
Ghosh et al. [2]	2022	Custom-built CNN model	SWELL-KW	99.39%
Zawad et al. [8]	2023	Artificial Neural Network (ANN) with Naive Bayes (NB)	SWELL-KW	95.75%
Ragav et al. [21]	2019	EMI-LSTM, EMI-GRU, EMI-FastGRNN	SWELL-KW	87.87%
Ganesh et al. [22]	2023	Random Forest	SWELL-KW	99.8%
Supriya et al. [23]	2022	ANN	SWELL-KW	90.4%
Ghosh et al. [24]	2023	k-nearest neighbours	SWELL-KW	99.3%
Padha et al. [25]	2022	LSTM	SWELL-KW	87.67%
Proposed (Dalal et al.)	**2024**	**Hybrid CNN + biLSTM model**	**SWELL-KW**	**99.97%**

Table 20.14 Comparison of our proposed hybrid CNN + BiLSTM deep learning model with previously proposed related to stress-lysis dataset

Researchers	Year	Model	Accuracy
Rachakonda et al. [26, 27]	2022	DNN	99.7%
Proposed (Dalal et al.)	**2024**	**Hybrid CNN + biLSTM model**	**100%**

Table 20.15 p-values acquired by performing a t-test on the CNN- biLSTM hybrid model

Dataset	Swell	Stress-lysis	WESAD
p-value	1.3×10^{-11}	2.4×10^{-4}	6.8×10^{-8}

while training, but the trained model could output prediction results for WESAD, SWELL and stress-lysis datasets within 10 min itself. Also, since the model itself has a memory size of about 31 MB, which is quite low, it could be integrated in wrist-worn devices. The ablation study shows significant performance boosts of up to 3% of accuracy over a model which does not comprise of biLSTM. Moreover, the proposed work is domain independent and could be implemented in NLP based applications such as sentiment analysis upon several public domain datasets such as IMDB, twitter datasets, etc. The objectives of this research have been successfully achieved. The proposed model demonstrated superior classification accuracy across multiple datasets, surpassing previous benchmarks. Additionally, the model's efficient memory usage and relatively low training time highlight its practicality for real-world applications, such as integration into wrist-worn devices. The ablation study also confirmed the importance of the biLSTM layer in enhancing model performance. Furthermore, the work contributes to the broader field of mental stress detection and lays the groundwork for future improvements, including the exploration of alternative datasets and model enhancements.

In order to contribute to the mental stress detection study, we intend to explore other physiological datasets for greater generalisation and reliability in order to apply in real life applications. We intend to further improve the proposed work by converting the existing EEG time series data into image-based encoding such as spectrograph, heatmap or scalogram images [32, 33] and training the vision transformer model on the encoded image [34, 35], capsule networks (CapsNET) could be utilized for producing better prediction results [36]. Additionally, combining multiple hybrid models in an ensemble can help reduce variance in unseen data [37, 38]. Last but not the least, by implementing transfer learning for mental stress detection, we anticipate decreased training time while maintaining similar performance levels.

Author Contributions Conceptualization, K.K. and P.K.S.; methodology, P.K.S.; software, P.K.S.; validation, A.D. and P.K.S.; formal analysis, A.D.; investigation, A.D.; resources, P.K.S.; data curation, K.K.; writing—original draft preparation, A.D., K.K.; writing—review and editing, P.K.S.; visualization, P.K.S., M.F.I. and J.C.; supervision, P.K.S., M.F.I. and J.C.; project administration, M.F.I. and P.K.S.; funding acquisition, M.F.I. and J.C. All authors have read and agreed to the published version of the manuscript.

Conflict of Interest The authors declare no conflict of interest.

Funding The grant funded by the Korea government (MSIT) through the National Research Foundation of Korea (NRF) (No. 2022R1C1C1004590) provided support for this study.

Data Availability The data that support the findings of this study are available from the corresponding author upon reasonable request.

References

1. Reza, A., Singh, P.K., Mahmud, M., Brown, D.J., Sarkar, R.: Multi-level stress detection using ensemble filter-based feature selection. Method. (2023). https://doi.org/10.1007/978-981-99-1916-1_13
2. Ghosh, S., Kim, S., Ijaz, M.F., Singh, P.K., Mahmud, M.: Classification of mental stress from wearable physiological sensors using image-encoding-based deep neural network. Biosensors. **12**(12), 1153 (2022)
3. Depression. Available online: https://www.who.int/news-room/fact-sheets/detail/depression (accessed on 10 October 2022)
4. 2016 HSE on work related stress. http://www.hse.gov.uk/statistics/causdis/-ffstress/index.htm (2016)
5. Chrousos, G.P., Gold, P.W.: The concepts of stress and stress system disorders: overview of physical and behavioral homeostasis. J. Am. Med. Assoc. **267**(9), 1244–1252 (1992)
6. Schmidt, P., Reiss, A., Duerichen, R., Marberger, C., Van Laerhoven, K.: Introducing WESAD, a multimodal dataset for wearable stress and affect detection. In: Proceedings of the 2018 on International Conference on Multimodal Interaction —ICMI'18 (2018). https://doi.org/10.1145/3242969.3242985
7. Kreibig, S.: Autonomic nervous system activity in emotion: a review. Biol. Psychol. **84**(3), 394–421 (2010)
8. Zawad, M.R., Rony, C.S., Haque, M.Y., Banna, M.H., Mahmud, M., Kaiser, M.S.: A hybrid approach for stress prediction from heart rate variability. In: Frontiers of ICT in Healthcare: Proceedings of EAIT 2022 2023 Apr 25, pp. 111–121. Singapore, Springer Nature Singapore (2023). https://doi.org/10.1007/978-981-19-5191-6_10
9. Behinaein, B., Bhatti, A., Rodenburg, D., Hungler, P., Etemad, A.: A transformer architecture for stress detection from ECG. In: Proceedings of the 2021 ACM International Symposium on Wearable Computers, pp. 132–134 (2021)

10. Albaladejo-González, M., Ruipérez-Valiente, J.A., Mármol, F.G.: Evaluating different configurations of machine learning models and their transfer learning capabilities for stress detection using heart rate. J. Ambient Intell. Human. Comput. **14**(8), 11011–11021 (2023)
11. Nigam, K., Godani, K., Sharma, D., Jain, S.: An improved approach for stress detection using physiological signals. ICST Trans. Scalable Inform. Syst. **8**, 169919 (2021). https://doi.org/10.4108/eai.14-5-2021.169919
12. Bobade, P., Vani, M.: Stress detection with machine learning and deep learning using multimodal physiological data. In: 2020 Second International Conference on Inventive Research in Computing Applications (ICIRCA), Coimbatore, India, 2020, pp. 51–57 (2020). https://doi.org/10.1109/ICIRCA48905.2020.9183244
13. Bellante, F., et al.: EMoCy: towards physiological signals-based stress detection. In: 2021 IEEE EMBS International Conference on Biomedical and Health Informatics (BHI), Athens, Greece, pp. 1–4 (2021). https://doi.org/10.1109/BHI50953.2021.9508611
14. Garg, P., Santhosh, J., Dengel, A., Ishimaru, S.: Stress detection by machine learning and wearable sensors. In: 26th International Conference on Intelligent User Interfaces—Companion (IUI '21 Companion), pp. 43–45. Association for Computing Machinery, New York, NY (2021). https://doi.org/10.1145/3397482.3450732
15. Bajpai, D., He, L.: Evaluating KNN performance on WESAD dataset. In: 2020 12th International Conference on Computational Intelligence and Communication Networks (CICN), pp. 60–62. IEEE (2020)
16. Quadrini, M., Daberdaku, S., Blanda, A., Capuccio, A., Bellanova, L., Gerard, G.: Stress detection from wearable sensor data using Gramian angular fields and CNN. In: International Conference on Discovery Science, pp. 173–183. Springer Nature Switzerland, Cham (2022)
17. Zhu, L., Spachos, P., Ng, P.C., Yu, Y., Wang, Y., Plataniotis, K., Hatzinakos, D.: Stress detection through wrist-based electrodermal activity monitoring and machine learning. IEEE J. Biomed, Health Inform (2023)
18. Dhara, T., Singh, P.K., Mahmud, M.: A fuzzy ensemble-based deep learning model for EEG-based emotion recognition. Cogn. Comput. **16**(3), 1–15 (2023)
19. Panure, T., Sonawani, S.: Novel approach for stress detection using smartphone and E4 device. In: Sustainable Communication Networks and Application: ICSCN 2019, pp. 736–745. Springer International Publishing (2020)
20. Liapis, A., Faliagka, E., Katsanos, C., Antonopoulos, C., Voros, N.: Detection of subtle stress episodes during ux evaluation: assessing the performance of the WESAD bio-signals dataset. In: Human-Computer Interaction–INTERACT 2021: 18th IFIP TC 13 International Conference, Bari, Italy, August 30–September 3, 2021, Proceedings, Part III 18, pp. 238–247. Springer International Publishing (2021)
21. Ragav, A., Gudur, G.K.: Bayesian Active Learning for Wearable Stress and Affect Detection. arXiv (Cornell University) (2020). https://doi.org/10.48550/arxiv.2012.02702
22. K Ganesh, Ayan Chatterjee, Michael Alexander Riegler et al. Meta-Heuristic Feature Optimization for Predictive Analysis on HRV Dataset and Semantic Knowledge Representation for Stress Management: A Case-Study Towards Ethical AI, 13 September 2023, PREPRINT (Version 4) available at Research Square. (2023) doi: https://doi.org/10.21203/rs.3.rs-3114142/v4
23. Supriya, G., Ganapathy, K., Gheena, S.: Accurate detection of human stress with heart rate and blood pressure data using artificial neural network algorithm in comparison with random forest algorithm. In: 2022 14th International Conference on Mathematics, Actuarial Science, Computer Science and Statistics (MACS), Karachi, Pakistan, 2022, pp. 1–10 (2022). https://doi.org/10.1109/MACS56771.2022.10023067
24. Ghose, D., Chatterjee, A., Balapuwaduge, I., Lin, Y., Dash, S.: Lightweight Machine Learning Models for Stress Detection: Algorithms and Explainability (2023). https://doi.org/10.13140/RG.2.2.32009.21603
25. Padha, A., Sahoo, A.: A parametrized quantum LSTM model for continuous stress monitoring. In: 2022 9th International Conference on Computing for Sustainable Global Development (INDIACom), pp. 261–266. IEEE (2022)

26. Rachakonda, L., Mohanty, S.P., Kougianos, E., Sundaravadivel, P.: Stress-lysis: a DNN-integrated edge device for stress level detection in the IoMT. IEEE Trans. Conum. Electron. **65**(4), 474–483 (2019)
27. Rachakonda, L., Sundaravadivel, P., Mohanty, S.P., Kougianos, E., Ganapathiraju, M.: A smart sensor in the IoMT for stress level detection. In: Proceedings of the 4th IEEE International Symposium on Smart Electronic Systems (iSES), 2018, pp. 141–145 (2018)
28. Dey, S., Chattopadhyay, P.K., Singh, A., Ahmadian, M.F., Sarkar, R.: A hybrid meta-heuristic feature selection method using golden ratio and equilibrium optimization algorithms for speech emotion recognition. IEEE Access. **8**, 200953–200970 (2020). https://doi.org/10.1109/ACCESS.2020.3035531
29. Sahoo, K.K., Dutta, I., Ijaz, M.F., Woźniak, M., Singh, P.K.: TLEFuzzyNet: fuzzy rank-based ensemble of transfer learning models for emotion recognition from human speeches. IEEE Access. **9**, 166518–166530 (2021). https://doi.org/10.1109/ACCESS.2021.3135658
30. Wu, K., Wu, J., Feng, L., et al.: An attention-based CNN-LSTM-BiLSTM model for short-term electric load forecasting in integrated energy system. Int. Trans. Electr. Energ. Syst. **31**, e12637 (2021). https://doi.org/10.1002/2050-7038.12637
31. S. Koldijk, M. Sappelli, S. Verberne, M. A. Neerincx, W. Kraaij (2014) The SWELL knowledge work dataset for stress and user modeling research. In: Proceedings of the 16th International Conference on Multimodal Interaction—ICMI '14, pp. 291–298.
32. Saha, A., Park, S., Geem, Z.W., Singh, P.K.: Schizophrenia detection and classification: a systematic review of the last decade. Diagnostics. **14**(23), 2698 (2024)
33. Bera, S., Geem, Z.W., Cho, Y.I., Singh, P.K.: A comparative study of machine learning and deep learning models for automatic Parkinson's disease detection from electroencephalogram signals. Diagnostics. **15**(6), 773 (2025)
34. Oei, N.Y.L., Everaerd, W.T.A.M., Elzinga, B.M., van Well, S., Bermond, B.: Psychosocial stress impairs working memory at high loads: an association with cortisol levels and memory retrieval. Stress. **9**, 133–141 (2006). https://doi.org/10.1080/10253890600965773
35. Shields, G.S., Sazma, M.A., Yonelinas, A.P.: The effects of acute stress on core executive functions: a meta-analysis and comparison with Cortisol. Neurosci. Biobehav. Rev. **68**, 651–668 (2016). https://doi.org/10.1016/j.neubiorev.2016.06.038
36. Cohen, B.E., Edmondson, D., Kronish, I.M.: State of the art review: depression, stress, anxiety, and cardiovascular disease. Am. J. Hypertens. **28**(11), 1295–1302 (2015). https://doi.org/10.1093/ajh/hpv047
37. Marik, A., Chattopadhyay, S., Singh, P.K.: A hybrid deep feature selection framework for emotion recognition from human speeches. Multimedia Tools Appl. **82**(8), 11461–11487 (2023)
38. Biswas, M., Sahu, M., Agrebi, M., Singh, P.K., Badr, Y.: Speech emotion recognition using deep CNNs trained on log-frequency spectrograms. In: Innovations in Machine and Deep Learning: Case Studies and Applications, pp. 83–108. Cham, Springer Nature Switzerland (2023)

GPSR Compliance
The European Union's (EU) General Product Safety Regulation (GPSR) is a set of rules that requires consumer products to be safe and our obligations to ensure this.

If you have any concerns about our products, you can contact us on

ProductSafety@springernature.com

In case Publisher is established outside the EU, the EU authorized representative is:

Springer Nature Customer Service Center GmbH
Europaplatz 3
69115 Heidelberg, Germany

www.ingramcontent.com/pod-product-compliance
Ingram Content Group UK Ltd.
Pitfield, Milton Keynes, MK11 3LW, UK
UKHW020933180926
475692UK00003B/15
9783032134967